The study of flowers provides a unique insight into the biology and evolution of flowering plants as a whole. The comparative study of flowers began in temperate regions and as a consequence most books on flowers are concerned with temperate plants. In contrast, this volume concentrates on tropical flowers which, while sharing many of the features seen in flowers of temperate plants, exhibit phenomena which are purely tropical.

The first part of the volume deals with general structural and biological features of flowers and shows facets of their diversity. The second part focuses on the flowers of selected tropical plant groups and emphasizes their structural and biological idiosyncrasies and evolutionary features. New trends in the study of floral evolution and the role of flowers in the study of flowering plant phylogeny are also outlined.

Diversity and evolutionary biology of tropical flowers

CAMBRIDGE TROPICAL BIOLOGY SERIES

Also in the series

Tomlinson, P. B. *The botany of mangroves*
Lowe-McConnell, R. H. *Ecological studies in tropical fish communities*
Roubik, D. W. *Ecology and natural history of tropical bees*
Benzing, D. H. *Vascular epiphytes*

Diversity and evolutionary biology of tropical flowers

PETER K. ENDRESS
Institute of Systematic Botany, University of Zurich

Drawings by
BRIGITTA STEINER-GAFNER and PETER K. ENDRESS

PUBLISHED BY THE PRESS SYNDICATE OF THE UNIVERSITY OF CAMBRIDGE
The Pitt Building, Trumpington Street, Cambridge CB2 1RP, United Kingdom

CAMBRIDGE UNIVERSITY PRESS
The Edinburgh Building, Cambridge CB2 2RU, UK http://www.cup.cam.ac.uk
40 West 20th Street, New York, NY 10011–4211, USA http://www.cup.org
10 Stamford Road, Oakleigh, Melbourne 3166, Australia

First published 1994
First paperback edition (with corrections) 1996
Reprinted 1998

Typeset in Times

A catalogue record for this book is available from the British Library

Library of Congress Cataloguing in Publication data

Endress, Peter K.
Diversity and evolutionary biology of tropical flowers/by Peter
K. Endress; drawings by Brigitta Steiner-Gafner and by the author.
p. cm. – (Cambridge tropical biology series)
Includes bibliographical references (p.) and index.
ISBN 0 521 42088 1 (hardback)
1. Flowers – Tropics – Evolution. 2. Tropical plants – Evolution.
3. Botany – Tropics – Variation. I. Title. II. Series.
QK653.E57 1994
582. 13′0913–dc20 93–26508 CIP

ISBN 0 521 42088 1 hardback
ISBN 0 521 56510 3 paperback

Transferred to digital printing 2004

WV

CONTENTS

ACKNOWLEDGEMENTS

My work in tropical regions in the past 25 years has been mainly devoted to the study of primitive flowering plants, but I was always fascinated by the entire diversity of the angiosperms, and from the onset I included many of them in my studies. Thus the book had a long gestation, and the manuscript began its life under the heavy rains in the Botanical Garden of Bogor, Java, in January 1991. I thank P.B. Tomlinson for his encouragement to write this book, for his advice and for reading the entire manuscript. It is not possible to mention all the many biologists who inspired my studies but they should become apparent through the text. I thank all the colleagues and friends who helped me in so many ways during my travels. Above all I have to thank B.P.M. Hyland, Atherton, for his hospitality and for his patience with an antipodal bloke who became excited by the living fossils. I also thank B. Gray, G. Stocker, A. Dockrill, J.G. & R. Tracey, and R. & S. Manzanell for field trips in tropical Australia. In Central America I was supported by A. Molina, L.O. Williams and E. Matuda, in India by B.M. Johri, H.Y. Mohan Ram and C.A. Ninan, in Madagascar by E. Rakotobe, in Java by A.J.G.H. Kostermans, in Papua New Guinea by J.L. Gressitt and in New Caledonia by P. Morat, H.S. MacKee and G. MacPherson. Many of the travels were supported by the Georges-und-Antoine-Claraz-Schenkung, and the study of flowers was also supported by the Swiss National Science Foundation, which is gratefully acknowledged. I also profited from several Botanical Gardens in the tropics, especially Bogor, Singapore, Atherton, and Brisbane. Other Botanical Gardens, mainly of the Universities of Zurich and Basel, and the Städtische Sukkulentensammlung Zurich also provided valuable material. I thank the directors of these institutions for allowing me to study flowers. On the home front I thank the greenhouse gardeners of the Botanical Garden of the University of Zurich for their expertise in the cultivation of many species, L. Braun, J.C. Gauteur, R. Gerber, S. Giger, R. Hartmann and W. Philipp. For suggestions or for reading parts of the manuscript I thank C.D.K. Cook, D. Frame, A.M. Juncosa, D.R. Kaplan, S.S. Renner and S. Vogel.

Most of the illustrations are original. A great number of the drawings were done by B. Steiner-Gafner, whose talent brought the flowers to life. I am especially indebted to R. Siegrist for skilful laboratory work, to U. Jauch for ever patient support with the SEM and to A. Zuppiger for careful photographic work. My scientific collaborators at Zurich during the past few years, U. Bamert, M. Buzgó, P. R. Crane, L. Hufford,G. Kappeler, A. Karrer, A. Nandi, E. Ramp, R. Rutishauser, H. Schneider, S. Stumpf, C. Wagner and M. Wolf, helped in many ways. I thank L. Davy for careful editing of the text. My greatest thanks go to Mary Endress for constant encouragement, help and patience.

Note on the second printing

The second printing provides the opportunity to include a few minor corrections and amendments. A number of colleagues have kindly commented on the book. For the communication of errors and suggestions I am especially grateful to A. Bernhard, B. Fischer, M. Luckow, S. S. Renner and D. W. Stevenson.

The manuscript of the book had been concluded in late November 1992. The breakthrough in molecular macrosystematics, which came shortly thereafter, has led to a vigorous new development of higher level phylogenetic studies in flowering plants. This provides a new background for the discussion of evolutionary changes in flowers of larger groups, such as, e.g., Asteridae, which will have to be considered in the future.

1

Introduction

1.1 The study of flowers: the evolutionary context

Since the pioneering classical works by J. W. von Goethe (1790) on comparative morphology and by C. K. Sprengel (1793) on comparative flower biology two hundred years ago the study of flowers has diversified into many branches of botany and its interface with zoology. Floral studies now embrace morphology, anatomy, development, pollination biology, breeding systems, genetics, molecular biology, palaeobotany, and diversity and evolution. The past two decades, especially, have been a time of fascinating discoveries with the elaboration of scanning electron microscopy, the advent of new molecular genetic techniques, the detection of well-preserved early fossils, and broader diversity studies. Mainstreams of research have diverged in the course of time into more profound and at the same time more narrow approaches. Thus it becomes difficult to synthesize a picture of our present knowledge about flowers, not only because of the vast literature, but also because the study of complexity and of diversity necessarily have very different approaches, so that a complete picture will never be possible. However, some new views may emerge.

Flowers are devices to allow sexual processes and the subsequent production of propagules. Sexual processes are pervasive in the living world and each organism group has evolved with this phenomenon elaborating successively more complexity around it. The evolution of flowers with their particular functions has to be envisioned starting from the still flowerless early seed plants. This particular historical situation of the angiosperms (or more broadly: the anthophytes) is expressed in that sexual processes take place within the ovules and in that propagules are seeds, which develop from the ovules. The relatively high complexity of flowers involves a number of special functions that finally lead to the sexually derived propagules.

In the course of plant evolution in the higher evolved groups the more complicated sporophytes dominated more and more over the gametophytes. In the angiosperms this resulted in the highly reduced embryo sacs and pollen tubes as the last remnants of the female and male gametophytes, respectively. The embryo sacs develop within ovules, which are included in carpels. The male gametophytes develop from the microspores, which are produced in the pollen sacs of the stamens. The microspores have to be transported to the carpel surface in order to be able to produce the pollen tubes, which then grow through the carpels to the ovules. Pollen germination takes place on a special site of the carpel surface, the stigma.

In the pathway of the male gametophyte from the pollen sac to the stigma and from the stigma to the ovule there are manifold intricate devices to facilitate or to hinder self or foreign pollen (or gametophytes) in attaining the goal. The very different nature of the two parts of this pathway require (or enable) different means to work. For the first part action of animals or abiotic agents (or sometimes the plant itself) is necessary, resulting in pollination; for the second part physiological and cytological interactions of gametophyte (pollen tube) and sporophyte (carpel and ovule) are required, resulting in fertilization (syngamy) within the embryo sac.

Pollination, especially if effected by animals, requires pollinator attraction by different means. The advent of petals and nectaries early in angiosperm evolution, in addition to many other devices, reflects these needs. And finally, or better, initially, organs for efficient protection of all these parts during the period of their earlier development are required, the more so, the more elaborated (complicated) the flowers become. Sepals have evolved to fulfil these functions. In many cases extrafloral organs, especially bracts, are also involved in flower protection and pollinator attraction.

Owing to their early coevolution with pollinators, flowers were an important factor in the diversification of certain animal groups: bees, butterflies, moths, flies, hummingbirds, bats and others. These animals, in turn, promoted the explosive radiation and diversification of the flowering plants, the flowers being at the centre of this diversity.

Flowers are highly plastic structural systems, so they were able to acquire their overwhelming diversity. On the other hand, there are all kinds of limitations, constraints that prevented the evolution of certain forms.

Therefore, all the flowers of our wildest dreams do not occur in nature. Not only must the forms function in all their respects, but the complicated forms must also arise from an apical meristem by an ordered development. Both these requirements set constraints on the array of conceivable forms.

Table 1.1. *Aspects of flower structure*

	Emphasis		Example: *Thunbergia grandiflora*
Organization (Bauplan)	Historical	Macroevolution	Acanthaceous flower
Construction (Gestalt)	Architectural	↕	Lip-flower
Mode	Ecological	Microevolution	*Xylocopa* bee-flower

Ecology and morphogenesis set outer (ultimate) and inner (proximate) constraints on the forms. By the intervention of the historical dimension, these levels of nature interact in different ways.

Flowers are never completely and ideally adapted to their environment. They are constrained by history: they have a phylogenetic burden. Many levels of their evolutionary history have imprinted their marks on them. They cannot escape them. One can often see traces of earlier phylogenetic (sub)strata in the structure of flowers. One should not forget that each flower, however harmoniously functioning at any time, is a mixture of features that are of different evolutionary ages. Different historical levels are incorporated and work together. The notion of 'evolutionary tinkering' (Jacob 1977) is especially apt for flowers.

A practical approach is to recognize three levels: (1) organization (bauplan, groundplan); (2) construction (architecture, gestalt); and (3) mode (style) (Delpino 1868–74, Vogel 1954) (Table 1.1). These three levels are successively more superficial in terms of their phylogenetic roots. 'Organization' rests on (phylo)genetically deeply rooted patterns that are the least easily changed. 'Construction' relates to proportions that have to do with mechanical properties and architectural features. 'Mode' means relatively superficial adaptations, especially in terms of pollination biology, such as size, colour, scent, indument etc., that are often highly variable at low systematic levels. These three levels do not form closed, mutually exclusive systems, since they are evolutionarily interconnected, and they should not be used in too rigid a manner. They correspond to some extent to the 'historical', 'fabricational' and 'functional' aspects of morphology as used in zoology (Seilacher 1974; see also Reif 1975, Riedl 1978, Gould & Lewontin 1979, Kaplan & Hagemann 1991).

In this book aspects of organization are mainly treated in chapter 2, aspects of construction in chapter 3, and aspects of mode in chapters 4 and 5.

1.2. Tropical flowers

The comparative study of flowers began in temperate regions. Traits of temperate flowers also occur, of course, in tropical ones. In many respects, however, the emphasis is different, and some phenomena are purely tropical. A pioneer of tropical research was F. Müller, whose Brazilian observations brought a wealth of novelties in his time. Although his publications were scattered pieces of information (collected by Möller 1915–21), they were discussed and brought to the attention by his brother H. Müller and by Ch. Darwin. The floral biologist F. Delpino unfortunately had to return early from a travel to South America owing to illness, but he included a number of tropical plants in his classical work (1868–74). Of Darwin's three books on flowers, only the one on orchids (1862) contains information on a fair number of tropical examples, including a balanced treatment from bauplan to pollination biology.

At the end of the last century and in the first half of this century the main activity of tropical botany was in Eastern Asia, centred around the Botanical Gardens in Bogor and Singapore. Most botanists did not devote their work particularly to the study of flowers. They were overwhelmed by the tropical diversity and gave more general accounts including notes on flowers, especially on phenology (e.g. Haberlandt 1893, Massart 1895; for the Neotropics see, for example, Belt 1874, Spruce 1908). Burck's (1890) studies on flowers of Annonaceae are an example of interesting observations with interpretations that did not hold, since they were derived from temperate plants. One of the first authors to give a comparative survey exclusively on some tropical flower phenomena was Beccari (1904) from his observations in Borneo. In Bogor important studies on the anatomy and embryology of tropical flowers were carried out: Treub (1889) and Koorders (1897) investigated the comparative anatomy of water calyces and Treub (1891) detected the chalazogamy in *Casuarina*; Ernst studied tropical parasites and mycotrophic plants, including *Rafflesia* (Ernst & Schmid 1913). Knuth (1898/99) and Knuth *et al.* (1904/05) compiled many observations on tropical pollination biology.

Tropical flower biology also has important roots in Java in the first half of our century with the work of van der Pijl (e.g. 1930), Cammerloher (e.g. 1931), Porsch (e.g. 1936), and Docters van Leeuwen (e.g. 1938). Corner (e.g. 1940), based on the Malayan Peninsula, also included observations on flowers in his general work on tropical plant biology. Other observations on phenology were published, for example, by Coster (1926), Holttum (1940), and van Steenis (1942).

After the second world war, new activities in the study of flowers arose in all parts of the tropics. Corner (e.g. 1946, 1949) and van der Pijl (e.g.

1954) continued their Eastern Asiatic work. Baker (e.g. Baker & Harris 1957) and Vogel (e.g. 1954) studied pollination biology in Africa, Vogel also in South America (e.g. 1957). The general books on tropical plant biology by Richards (1952), Holttum (1954), and Corner (1964) included accounts on flowers.

In the 1960s, studies on floral biology expanded rapidly in all regions of the tropics but with a prominent concentration in the Neotropics. These studies yielded many exciting new insights, which will be discussed in this book. A number of surveys, textbooks, and research programmes of tropical biology, including aspects of flowers, have also been published since that time (e.g. Baker 1964, Baker *et al.* 1982, van der Pijl 1969, Janzen 1975, 1977, 1983a, Ashton 1977, Tomlinson 1977, 1980, 1986, 1990, Croat 1978, Hallé *et al.* 1978, Raven 1980a, Longman & Jenik 1987, Bawa *et al.* 1990, Benzing 1990, Whitmore 1990, Mabberley 1992). Other general works, even if centred in temperate regions, contributed to the advance in our understanding of flowers (e.g. Knoll 1956, Werth 1956a, Jaeger 1959, Meeuse 1961, Percival 1965, Free 1970, Kugler 1970, Proctor & Yeo 1973, Sattler 1973, Frankel & Galun 1977, Faegri & van der Pijl 1979, Guédès 1979, Bernier *et al.* 1981/85, Barth 1982, Rohweder & Endress 1983, Vogel 1983, Willson 1983, Willson & Burley 1983, Meeuse & Morris 1984, Richards 1986, Moncur 1988, Sedgley & Griffin 1989, Weberling 1989, Fahn 1990, Hess 1990, Dafni 1992). In addition, a large number of edited books containing articles on flowers have appeared in this period, which cannot all be mentioned here.

Today, facing the threats to biodiversity, especially in the tropics, better knowledge of the biology of flowers in the widest sense is urgent. The understanding of flowers is a central theme for the phylogenetic reconstruction of the angiosperms at all levels. Better knowledge of phylogenetic history and of interactions between animals and plants is vital for evaluation of conservation actions (e.g. Bobisud & Neuhaus 1975, Janzen 1977, 1986a,b, Tomlinson 1977, Prance 1990, 1991, Soulé 1990, Bawa & Ashton 1991).

1.3 *Delonix regia*, the flamboyant (Caesalpiniaceae): an introductory example

'This tree is a joy of creation, beyond the invention of man' (Corner 1988: *Wayside trees of Malaya* 1, p. 435).

The flamboyant (*Delonix regia*), with its exceedingly attractive flowers, is one of the most familiar ornamental trees in the tropics (Fig. 1.1). The trees flower best in regions of the wet tropics with a longer dry period.

Figure 1.1. *Delonix regia* (Caesalpiniaceae). 1. Inflorescence (× 0.3). 2. First-day flower with expanded flag (× 0.7). 3. Second-day flower with folded flag (× 0.7). 4. Two differently folded flags of older flowers (× 0.7).

Flowering is at the end of the dry season, when the trees may be in flower for a month or so (Ghouse & Hashmi 1981).

But where is the origin of this widely cultivated tree? What is its reproductive biology, and especially the biology of its fulminant flower display? In 1828 the Austrian botanist Wenzel Bojer discovered the species at Foul Point on the east coast of Madagascar (Blatter & Millard 1977). In 1829 the original description (as *Poinciana regia*) accompanied by a coloured plate was published in Curtis's Botanical Magazine (tab. 2884). In 1837 Rafinesque used the name *Delonix regia* for the first time. Bojer introduced the tree in Mauritius, and from here it was distributed to several other tropical regions. In Singapore it is recorded from 1840 (Corner 1988). Later the species seemed to be extinct in the wild and its origin was uncertain. Only after a century, in October 1932, was the plant refound, apparently indigenous, in the forest reserve of Antsingy (West Madagascar) by Léandri (1933, 1936).

Although the tree regularly produces fruits and seeds in cultivation, almost nothing is known about its floral biology. The fruits are conspicuous, large, woody pods and contain numerous seeds. There are only vague indications about pollinators. They have never been observed in the putative natural environment of the species in Madagascar. Only scattered observations from cultivated specimens outside Madagascar mention birds (Winkler 1906, Werth 1915, Arroyo 1981) and butterflies, mainly Papilionidae (Vogel 1954, Owen 1971, DeVries 1983a,b) as flower visitors. Arroyo (1981) rates *Delonix regia* as a 'classic ornithophilous species of the Old World' that acquired ornithophily through butterfly-pollinated intermediaries. Winkler (1906) interprets the adaptational situation of the species as transitional from butterfly to bird pollination but more on the ornithophilous side.

Not even the floral behaviour during anthesis has been studied in detail so far, except for a short note by Corner (1988): 'Though the flowers are not fully open until 9 a.m. or later, the petals begin to emerge from the calyx-bud shortly after midnight. The flowers last only two days, and the standard curls up and fades on the evening of the first day'. One also knows of the peculiar asymmetric nectar gate at the flower base (Lindman 1902, Troll 1951), the amount of nectar production per day of the flowers (Fahn 1949), some features of floral anatomy (Rao & Sirdesmukh 1956), details on pollen structure with its peculiar viscin threads (Cruden & Jensen 1979, Hesse 1984a,b), and the occurrence of a crateriform wet stigma (Owens 1990).

Thus, there are a number of very scattered bits of information, which do not give a coherent understanding of the flowering in this species. It may stand as an example for the paradoxical situation of how poorly known even 'familiar', conspicuous and widely used plants in the tropics are and how they are threatened by extinction in their natural habitat. Some new observations on *Delonix* will be presented in section 8.6.1.

2

Floral organization (bauplan, groundplan)

2.1. Structural units and floral symmetry

Flowers are complicated plant parts. They are differentiated into regions with different structures and functions. Each region contains a number of structural units. Flowers have a closed organization, since the floral apex is no longer active after initiation of the innermost floral organs. Different kinds of organs have different functions in the flower. Repetitive patterns abound in that each kind of organ mostly occurs in sets of variable or fixed numbers (structural units). A flower or floral axis is usually situated in the angle between a subtending leaf (a bract or a foliage leaf) and an axis. This subtending leaf is the pherophyll of the flower. Often the floral axis contains one (adaxial) or two (lateral) bracteoles (prophylls) below the flower.

Many flowers are so attractive to us because of their often high symmetry, be it the bizarre mirror-symmetrical form of an orchid or the 5-parted windwheel of a periwinkle. The two most common symmetry patterns are polysymmetric (actinomorphic, radially symmetric, regular) flowers having several symmetry planes, and monosymmetric (zygomorphic, dorsiventral) flowers having only one symmetry plane. Disymmetric flowers (with two symmetry planes) and asymmetric flowers are more rare.

Symmetry is primarily determined by the number and arrangement pattern of the structural units. Monosymmetry is often superimposed on a primary polysymmetric configuration of the structural units (and asymmetry often on a monosymmetric configuration) by unequal differentiation of floral sectors (see also sections 3.5 and 9.2.1).

The floral organs arise on the floral apex in ordered patterns, basically in a centripetal sequence, as do the organs of the vegetative shoots. At transition from the vegetative into the floral state the apical meristem becomes more massive (deep) and may develop into a 'meristem plug' (Rauh & Reznik 1951). At its upper edges the outermost floral organs are initiated, followed rapidly by the inner ones. The formation of a massive apical meristem is a precondition for the rapid initiation of the floral organs. The floral organs appear as fully meristematic hemispherical or somewhat laterally (tangentially) extended mounds. These first stages are similar for all organ types. Organ initiation often manifests itself as a locally more intensely staining field in the floral meristem. The first periclinal cell divisions that follow organ initiation and are associated with the beginning of bulging up of the organ primordium occur in the second and/or deeper layers of the floral meristem, apparently merely depending on the later thickness of the organ (Rohweder 1963, Guédès 1979). The formation of an organ primordium induces a procambial strand that later becomes the main (middle or dorsal) vascular bundle of the organ. The procambial strand connects with one or more older vascular bundles deeper in the floral base. As the organ grows, more procambial strands may be formed. Procambial strands are initiated where they are needed, i.e. in the direction of main photosynthate attraction by a morphogenetic centre, whereby certain taxon-specific patterns may be superimposed (e.g. Carlquist 1970, Kaplan 1971, Rohweder 1972, Schmid 1972).

During floral development, centres of activity of growth and differentiation arise and are replaced by other ones at other sites. Each centre influences its neighbourhood in that the adjacent regions have to adjust to the newly arisen changes. During the entire development process, regions in the floral base are influenced at different times from different directions, involving differential thickening, tissue differentiation and vascular bundle formation. The structural interpretation of mature organs has to take this into consideration.

With the process of floral elaboration in the course of evolution, floral organs of one kind may fuse postgenitally (in that free parts fuse during ontogeny) or congenitally (in that a common base of several organs develops as a ring wall, which may eventually form a tube). In the most elaborate flowers, organs of different kinds may also fuse (postgenitally or congenitally) to form complicated structures (organ complexes). Other terms for congenital fusion are phylogenetic fusion (Cusick 1966), meristem fusion (Hagemann 1970) and interprimordial growth (Sattler 1978). The growth process of the congenitally united base of organs is also referred to as zonal growth. Mechanisms of postgenital fusion have been reviewed by Verbeke (1992). The terms 'fusion' and 'union' are often used in the same sense, but one tends to use 'fusion' if organs of different kinds are involved, and 'union' if organs of the same kind are involved.

Plants, in contrast to animals, have rigid cell walls, and there is no cell migration during development. Therefore, the surfaces of mature floral

Table 2.1. *Key events in the life history of a flower*

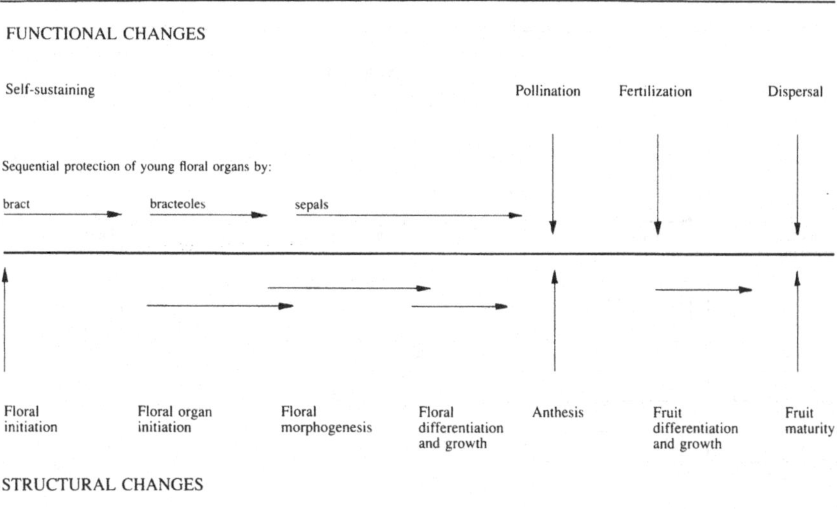

organs are the primary morphological surfaces, in that they correspond topologically to the surface of the initial undifferentiated floral apex. However, at certain sites secondary morphological surfaces occur: (1) where organs have fallen (abscission regions, e.g. the scar of a fallen sepal); (2) where parts of organs have split by special tissue differentiation (dehiscence regions, e.g. the stomium and the inner surface of an open anther; or intercellular spaces, e.g. in the septa of the ovaries of certain taxa). Further, parts of the primary morphological surface may be blurred by postgenital fusions (e.g. in the ventral slit of carpels) or hidden by internalization (e.g. the inner surfaces of ovary walls).

The multitude of functions of a flower are not always performed by the same organs; this is an expression of the overwhelming diversity found in flowers. From the viewpoint of floral evolution at the level of the angiosperms we may distinguish between primary and secondary functions. Primary functions are the development of ovules and seeds, the reception and guiding of male gametophytes by the carpels, and the development and release of pollen by the stamens. Secondary functions are the protection of stamens and carpels in many ways, the enhancement of pollination, and all the elaborations leading to a flexible balance between in- and outbreeding. In these secondary functions all kinds of floral organs outside the flowers can be involved in various ways (Table 2.1). The histology of floral organs is diverse but has not been comparatively studied to a large

Table 2.2. *Floral organs, presence and major functions in different angiosperm groups*

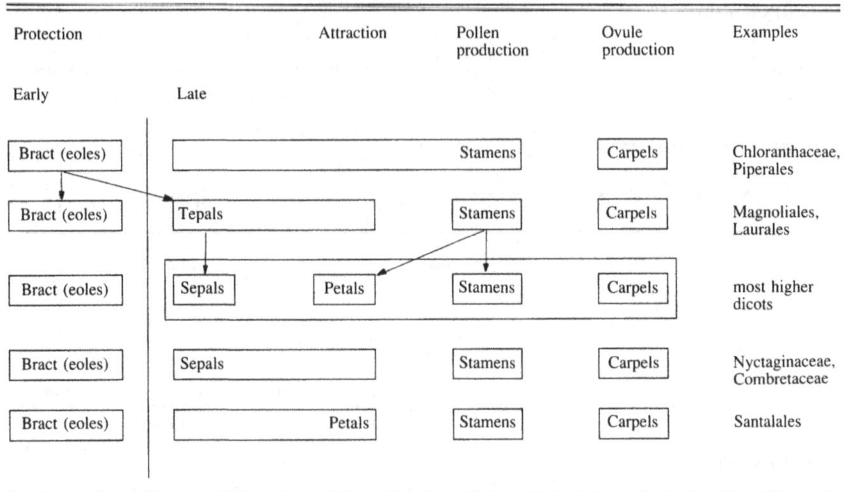

extent. It seems to be variously dependent on development, function, and longevity of the organs, but also on the systematic position of the plant (e.g. Pass 1940).

As a consequence of the primary functions, each angiosperm species with sexual reproduction (the vast majority of the angiosperms) has stamens and carpels in its flowers, sometimes together (bisexual or perfect or hermaphrodite flowers), sometimes separated (unisexual flowers). As a consequence of the secondary functions, additional organs are common in flowers, especially perianth organs, which occur very often as two different kinds: sepals and petals (Table 2.2).

The following presentation of the different kinds of floral organs is not in the order of primary and secondary functions but in the order of the ontogenetic development, i.e. from the periphery to the centre of the flower.

2.2 Perianth

2.2.1 Structure and main functions

Most flowers have peripheral flat, sterile organs which are in their totality referred to as the perianth. Usually they enclose and protect the fertile floral organs in bud, and at anthesis they are often an important part of the floral display in attracting pollinators.

In a number of angiosperms there are two series of perianth parts. They are often quite different in that the outer series has a mainly protective function in bud and the inner one mainly pollinator attraction functions at anthesis. The outer series is the calyx, the structural units of which are the sepals. The inner series is the corolla, the structural units of which are the petals. Accordingly, the sepals are often green and robust; their margins overlap well in bud or they touch each other and may then be postgenitally fused, while the petals are often coloured and sometimes delicate in texture and less regularly overlap or touch each other in bud. The respective position of the sepals or petals in bud is called aestivation. Supporting tissues, which are located mainly over the vascular bundles, are often differentiated, particularly in persistent calyces.

In some groups there is only one series of perianth parts or two series of very similar parts. In this case the perianth is sometimes called a perigon and the structural units tepals. They may have both protective and attractive functions. However, the common use of the term perigon is somewhat unsatisfactory, since it is subordinated and not coordinated to the term perianth (special case of a perianth).

The protective function of the young floral organs may be exerted in a sequence by different organs: at first by the subtending leaf or bract (pherophyll) of the floral primordium and by the bracteoles (prophylls), then by the sepals, and in certain groups later also by the petals or even by parts of the stamens (Tables 2.1 and 2.2).

Attraction of pollinators is usually performed by the petals by means of colours, the pigments being located in the epidermis and/or mesophyll (see section 5.7). The presence of intercellular spaces in the mesophyll and especially of epidermal papillae enables an efficient colour display by differential light reflection. Epidermal papillae may also be involved in scent effusion (see section 5.5). More than half of the 330 species out of 18 families (all extratropical, however) studied by Schubert (1925) had papillate petal surfaces.

2.2.2 Development

According to their different structures and functions, sepals and petals may show strikingly different developmental patterns.

Sepals often appear in a spiral sequence (Figs 8.25, 8.38). They become large in early development and protect the inner organs from an early stage by overlapping of their margins (imbricate aestivation: e.g. Dilleniaceae, Actinidiaceae, Ochnaceae, Theaceae, most Clusiaceae, most Dipterocarpaceae, most Flacourtiaceae, Passifloraceae) or more rarely by mutual contiguity of the margins of neighbouring organs (valvate aestivation: e.g.

some Cunoniaceae, Anisophylleaceae, Combretaceae, Rhizophoraceae, Mimosaceae, Asclepiadaceae–Stapelieae). In some specialized monosymmetric flowers the first-formed sepal may become especially large and may be the main protective organ of the bud (e.g. *Caesalpinia*, Caesalpiniaceae, Tucker *et al.* 1985; *Monophyllaea*, Gesneriaceae, Weber 1976a).

Petals, in contrast, are often initiated almost simultaneously in a whorl (Figs 8.25, 8.38), sometimes also in a very rapid spiral. They are often conspicuously delayed in later ontogeny of the bud; only shortly before anthesis do they resume rapid growth so that they are fully grown at anthesis. Owing to their retardation they often do not overlap (or not completely) in bud. If they overlap, their aestivation is often irregular; it tends to be more regular in groups without petal retardation (see section 2.2.3).

Concomitant with this differential developmental (and functional) behaviour is an often broad base with three vascular traces and acuminate tip in sepals but a small base with only one vascular trace but a broad and sometimes even bifid (emarginate) tip in petals (Hiepko 1965, Rohweder & Endress 1983) (Fig. 2.1). Later growth of the broad sepal primordium is often mainly basal and longitudinal (basiplastic) the tip often remaining relatively undifferentiated, while in the narrower petal primordia the later growth is often more regular in all directions (periplastic) (Hagemann 1970). If the narrow petal base is elongated, it is called a claw and the broad upper part a blade (plate). Histologically the green sepals often resemble foliage leaves in showing a parenchyma with one or two palisade layers, however on their dorsal (and not ventral) side, which is usually exposed (Tschech 1939), while the petals are thinner with a simple parenchyma that often shows large air spaces; this is a consequence of the petals' rapid growth and should also be seen in an ecolo-

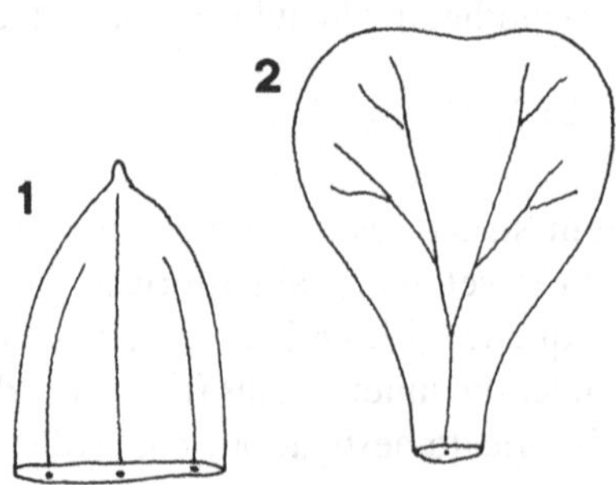

Figure 2.1. Characteristic form and vascular supply of a sepal (1) and a petal (2).

gical context, as the intercellular spaces may enhance the brightness of colours of the petals (as an optical 'tapetum') (see section 5.7).

After anthesis sepals in some groups show further growth and may take part in fruit development (e.g. Dilleniaceae, Dipterocarpaceae, Ochnaceae, some Malvaceae, Lamiaceae, Asteraceae), while petals usually fall off or decay. In some groups the petals are shed by an abscission tissue before wilting; in others they wilt on the flowers. In the abscising type in particular, ethylene seems to play an important part in the process (Mayak & Halevy 1980, Woltering & van Doorn 1988). In many cases pollination or fertilization induce corolla senescence (Halevy 1986). An extreme condition is the rapid decay by autolysis, which occurs in the petals of some short-lived flowers (e.g. Commelinaceae, Suttle & Kende 1978; Convolvulaceae, Matile 1978).

2.2.3 Diversity

The different behaviour of the perianth parts as mentioned before is usually described as characteristic for angiosperms. This is especially true for the middle evolutionary level of the dicotyledons (Rosidae, Dilleniidae, Caryophyllidae) and the lower evolutionary level of the monocotyledons (Alismatidae, Commelinidae). However, it should be pointed out that there is a great diversity in the perianth as seen through the entire angiosperms. It is important to consider this diversity for a better evolutionary understanding of the perianth. The perianth is a good example of the frequent occurrence of 'transference of function', which Corner (1958) explained with the example of *Saraca* (Caesalpiniaceae) (see section 8.6.4).

The largest flowers (*Rafflesia arnoldii* and *Aristolochia grandiflora*), almost 1 m in diameter or more than 1 m in length, owe their dimension largely to perianth size, while the androecium and gynoecium are much smaller. At the other extreme, a perianth may be completely lacking (some Magnoliidae, some Hamamelididae, some Arecidae).

Protective structures

The firm texture of protective perianth members may come about at the morphological and (mainly) at the anatomical–histological level. In contrast to the attractive parts, these organs are often relatively thick; they have thicker cell walls and smaller intercellular spaces. In many cases collenchymatic or sclerenchymatic layers occur (e.g. sepals of *Stachytarpheta cayennensis*, Verbenaceae (Fig. 2.2); *Jacaranda mimosifolia*, Bignoniaceae) (Pass 1940). Tissues containing tannins (sepals of *Delonix regia* and *Bauhinia variegata*, Caesalpiniaceae) (Fig. 2.2), oxalate crystals or stone cells are often also present. The periphery of the protective organs may be covered by an indumentum of sclerified or glandular hairs (*Notothixos*, Viscaceae; *Matudaea*, Hamamelidaceae; *Solanum*) (Fig. 2.3) or by a waxy layer (*Eucalyptus*, Myrtaceae), which may all have reflective properties for irradiation; viscid parts are present, for example, in *Pisonia* (Nyctaginaceae) and *Plumbago*. A broad variety of such devices occurs in palm flowers (Uhl & Moore 1973, 1977). Outer perianth members may have secondary tips ('unifacial' tips) at the apex or at the dorsal side near the apex (Baum 1950). At the bud stage they are situated at the topographical end of the bud. Their tissues may be differentiated early, with a relatively extensive lignified part, and may add to the protective function (e.g. sepals of *Passiflora* and *Costus*; Baum 1950, Weber 1980). Protective parts are often green and in some plants contain palisade or spongy parenchyma (Fig. 2.2); they may be photosynthetic up to fruit maturity.

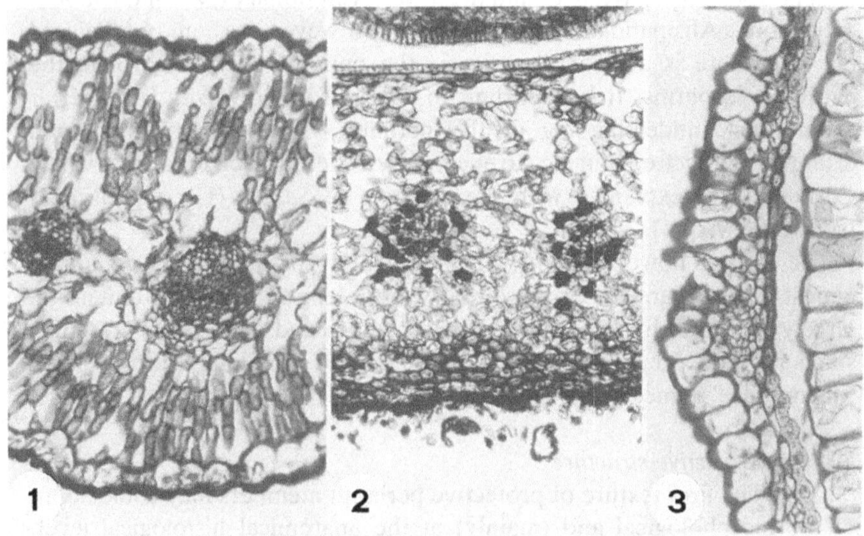

Figure 2.2. Transverse sections of sepals with different histological differentiation. 1. *Peganum harmala* (Zygophyllaceae) with palisade parenchyma on upper and lower side (× 130). 2. *Bauhinia variegata* (Caesalpiniaceae), with spongy parenchyma on upper side and tanniferous tissue at periphery and around vascular bundles (× 100). 3. *Stachytarpheta cayennensis* (Verbenaceae) with sclerenchymatous upper epidermis (on the right) (× 200).

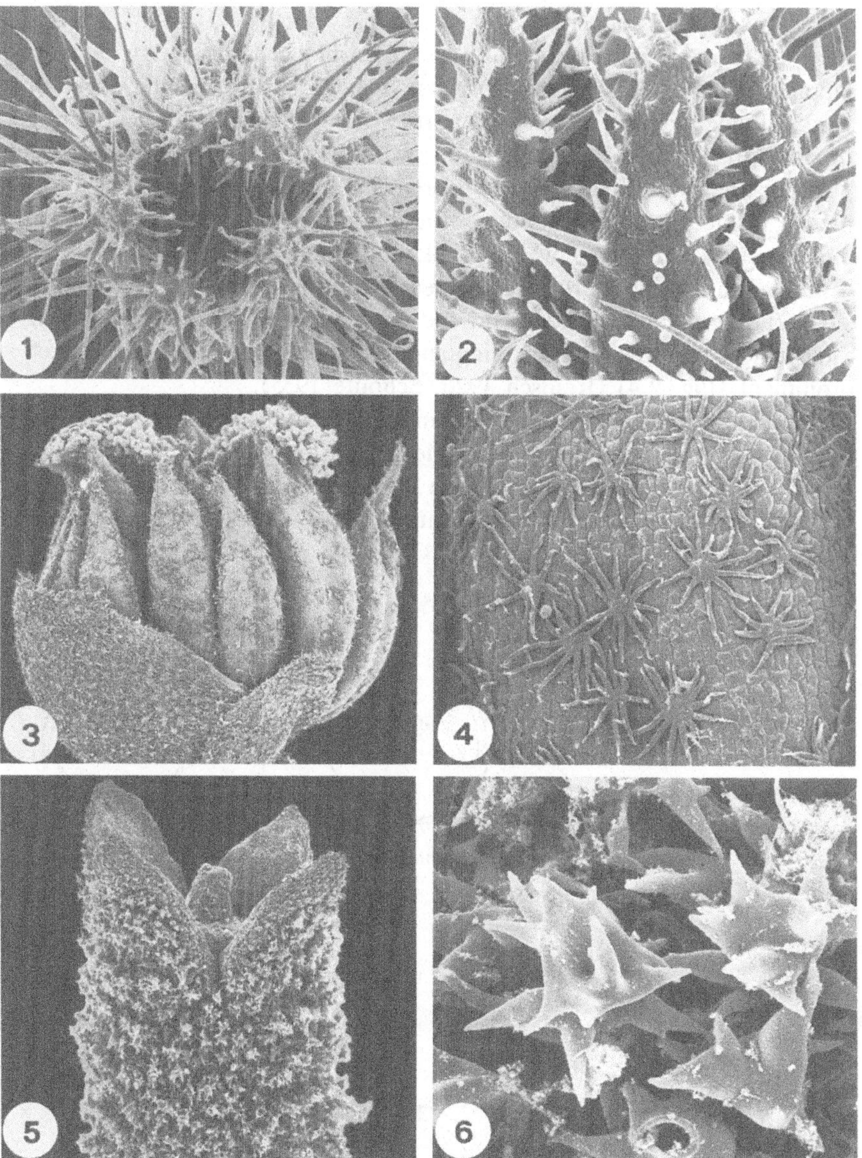

Figure 2.3. Indument of sepals and other peripheral floral organs in bud. 1–2. *Solanum sisymbrifolium*, sepals. 1. From above (× 35). 2. From the side, simple and glandular hairs (× 60). 3–4. *Matudaea trinervia* (Hamamelidaceae), bracts and stamens. 3. Flower from the side (× 13). 4. Stamen with stellate hairs (× 80). 5–6. *Notothixos subaureus* (Viscaceae). 5. Flower from the side (× 30). 6. Floral surface with dendritic hairs (× 300).

Aestivation patterns

Aestivation of the perianth members may be imbricate (with overlapping flanks of neighbouring organs), valvate (with touching edges or flanks of neighbouring organs) or apert (without touching of neighbouring organs). In pentamerous flowers the two extremes among the imbricate patterns are quincuncial (with two organs completely outside, two completely inside, and one in between) and contort (with all organs having one covering and one covered flank and none completely outside or inside). The other patterns are cochlear (with one organ completely outside, one completely inside, and the other three in between). Among the cochlear patterns the ascending and descending cochlear are the most prominent (Fig. 2.4) (Reinsch 1926, Schoute 1935).

An important aspect is that the mutual relationship between neighbouring organs is asymmetric in imbricate but symmetrical in valvate patterns. This relationship has an impact on the potential of synorganization of the organs: in asymmetric configurations it is difficult for two contiguous

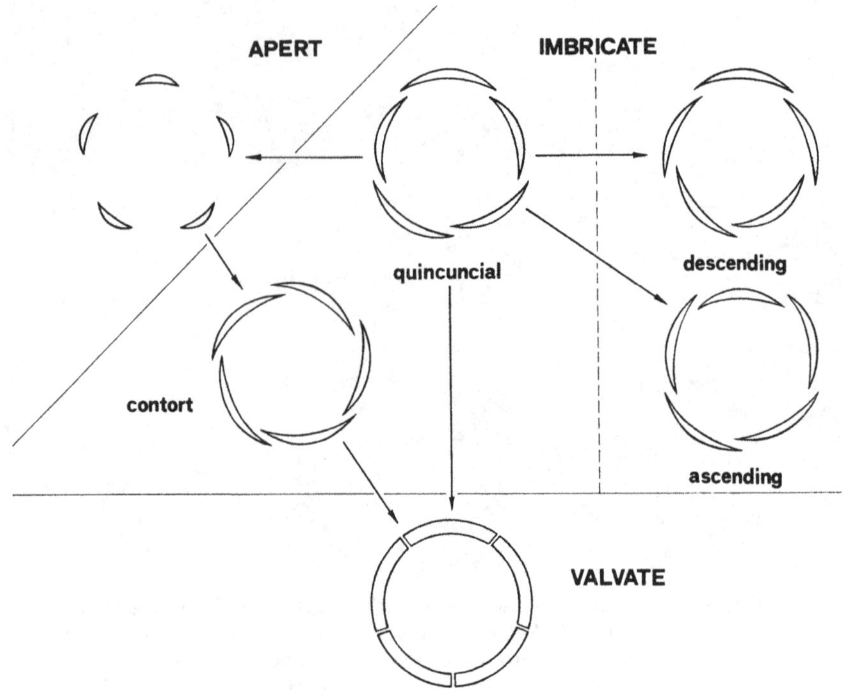

Figure 2.4. Major aestivation patterns of perianth organs and their main evolutionary trends.

organs to fuse because their edges grow in different directions, whereas in symmetrical configurations the potential for closer surface interaction with fusion is present.

Relationships between protective structures and aestivation patterns

The outer cycle, the calyx, fulfils its protective function in different ways. The sepals may be free and are then often imbricate with quincuncial aestivation; more rarely they are valvate and the contiguous margins may then be more or less postgenitally coherent. This coherence may arise by interdentation of epidermal cells or by interdentation of their cuticular surfaces only, or by interlocking hairs, or by firm postgenital fusion (Sigmond 1929). They may also form a tube by zonal growth below the individual sepals (congenital union), which may remain very short and scarcely be apparent at anthesis. If the entire tube expands during floral development, the calyx appears cup-like (retuse) (e.g. some Fabaceae, Bignoniaceae, Acanthaceae, Verbenaceae). If the tip of the tube does not expand, the calyx totally encloses the other organs, and at anthesis it splits either by a circular abscission line (calyptrate calyx; e.g. some Monimiaceae; Myrtaceae; *Barringtonia*, Lecythidaceae) or it splits lengthwise into two (or more) lips (e.g. *Bauhinia*, Caesalpiniaceae; some Bignoniaceae; Acanthaceae; *Faradaya*, Verbenaceae; *Brugmansia*, Solanaceae) (Fig. 3.2). In perianths with imbricate aestivation, secretions at the margins of the organs, sometimes by colleter-like structures, may also have protective functions, but this behaviour is poorly known (e.g. some Monimiaceae, Apocynaceae).

In many groups the second cycle of perianth parts (the petals) also takes part in protection of the inner floral organs. The petals are then robust and not delayed in ontogeny. As in sepals they may have an imbricate or – more frequently than in sepals – a valvate aestivation. In taxa with such petal behaviour, the petal aestivation is more regular than in those with the 'typical' retarded petals and may, then, be constant in larger groups. In these cases the sepals are often shorter; they are the protecting organs only in the earliest stages of the flower, while later in floral ontogeny the function is transferred largely to the petals. Examples of imbricate aestivation of the inner perianth series are: some Myrtaceae, Lecythidaceae (Fig. 8.12). Examples of valvate aestivation of the inner perianth series are: Escalloniaceae, Pittosporaceae, Leeaceae, Vitaceae, Icacinaceae, Olacaceae, Opiliaceae, Loranthaceae, Mimosaceae, many Araliaceae, Stapelieae of Asclepiadaceae) (Fig. 8.42). The corolla aestivation pattern may be stable within a family (e.g. contort in Gentianales, Malvaceae, Oxalidaceae, Linaceae, Plumbaginaceae; ascending vs. descending in Caesalpiniaceae vs. Fabaceae) or diverse and then characteristic of subfamilies, tribes or genera (e.g. Acanthaceae, Scotland *et al.* 1993; Scrophulariaceae, Hartl 1955; Rubiaceae, Bremer 1987). Of the two contort morphs, contort to the right (dextrorse) and contort to the left (sinistrorse), both may occur in equal proportions within a species or individual (e.g. some Oxalidaceae, Linaceae, Plumbaginaceae, Bahadur *et al.* 1984) or one of them may be exclusive for a larger taxon (e.g. Apocynaceae–Plumerioideae: sinistrorse; Apocynaceae–Apocynoideae and Asclepiadaceae–Asclepiadeae: dextrorse, Fig. 8.38).

An extreme example of the protective function of both perianth cycles is *Eucalyptus* (Myrtaceae), where in some subgroups the common base of both cycles forms a calyptra, while the free apices of the sepals and petals cease growing and retain their primordial size (Drinnan & Ladiges 1989). In some Monimiaceae the several tepal cycles show a similar behaviour (Endress 1980a).

In a few groups a further cycle of protective organs, an 'epicalyx', occurs outside the calyx and corolla (e.g. Malvaceae, some Rosoideae). Rarely the bracteoles (or bracts that are not especially differentiated into an epicalyx) protect the floral organs up to anthesis and the calyx may be reduced or lacking. The bracteoles may be congenitally fused into a tubular structure (e.g. Eupomatiaceae, Himantandraceae, Endress 1977b; some Hamamelidaceae, Endress 1978) or two large bracteoles may be postgenitally united (*Thunbergia*, Acanthaceae, Figs. 8.48, 8.50). In *Amherstia* (Caesalpiniaceae) the two large bracteoles are also postgenitally united but, in addition, the sepals act as protective organs in bud.

Relationships between floral symmetry and aestivation patterns

Among the variants of imbricate aestivation the contort pattern is the most symmetrical one, as seen at the level of the entire flower (rotational symmetry). It is often found in the corolla of polysymmetric flowers, where the petals are already relatively large in bud (e.g. Gentianales, many Rubiaceae, Malvaceae, Oxalidaceae). It is even present in some monosymmetric flowers, where monosymmetry is relatively weakly expressed in early development (some Acanthaceae, e.g. *Thunbergia*) (Fig. 8.50). Advantages of contort aestivation may be seen in (1) structural stability for long, slender flowers in bud (e.g. *Oxyanthus*, Rubiaceae), because the symmetry with all petals having the same direction of the flanks allows maximally compact packing; and (2) relatively easy opening (and reclosing in flowers that open several times). In addition, valvate petal aestivation usually goes hand in hand with polysymmetric flowers.

In contrast, in monosymmetric flowers ascending or descending patterns prevail (e.g. Fabales, Scrophulariales, Lamiales). Many families with monosymmetric flowers oscillate between ascending and descending petal aestivation; the pattern may be constant in subfamilies, tribes or genera. More rarely is it uniform in entire families (e.g. only ascending in Caesalpiniaceae, only descending in Fabaceae and Lamiaceae).

Attractive structures

The diversity of petal surface patterns and histological structures with respect to different optical and other properties is surveyed by Kugler (1970), Kay *et al.* (1981) and Kay (1988) and discussed in section 5.7.

More rarely both cycles of perianth organs take part in optical attraction of pollinators. Extreme examples are some genera of the Rubiaceae and Verbenaceae, where the calyx is large and coloured. The most familiar examples are *Mussaenda* and *Warscewiczia* (Rubiaceae) where some peripheral flowers of the inflorescences have an enlarged, usually red sepal (in addition to the four small inconspicuous ones). Histologically, these sepals behave like 'typical' petals with their papillate epidermis and large intercellular spaces in the simple mesophyll (Weber 1955, Hallé 1961, Leppik 1977) (see also section 5.7). Other examples in the Rubiaceae are *Alberta*, *Carphalea*, *Mussaendopsis* and *Nematostylis*, where, in contrast all flowers have one or more enlarged, brightly coloured sepals (Puff *et al.* 1984). In addition, some Verbenaceae show conspicuous calyces that contrast in colour with the corolla (*Petrea*, *Holmskjoldia*, *Clerodendrum* species).

Union of perianth organs

In some large groups the petals are congenitally united (Asteridae, some Rosidae) to form tubular, bell-shaped or salverform flowers. The calyx is often also congenitally united (e.g. many Asteridae, Fabaceae). A congenitally united calyx is synsepalous; a congenitally united corolla is sympetalous. More rarely the petals are postgenitally united at anthesis (e.g. *Correa*, Rutaceae; *Oxalis* species, see Hartl 1957; *Polyosma*, Escalloniaceae; some Rubiaceae, see Robbrecht 1988). Tubular or salverform flowers can also be constructed with free petals, which are held together by a tubular calyx. In these cases the petals have long claws. Sometimes a scale ('ligule') develops at the base of the blade (e.g. Sapindaceae, Erythroxylaceae, Bromeliaceae). If such scales are more elaborated they may form a 'paracorolla'. Claws and ligule are often involved in synorganizations (see section 3.4).

In some families the congenital union of the petals comes about by mediation of the stamens (e.g. Commelinaceae, Rohweder 1969; Bromeliaceae, Müller 1896; Bruniaceae, Leinfellner 1964; *Cuscuta*, Convolvulaceae, Erbar 1991). In other cases the stamens are fused with the corolla tube on its inner side (e.g. Scrophulariales, Lamiales). The fusion product of stamen and petal is called 'stapet' by Ritterbusch (1991). Erbar (1991) distinguishes early and late sympetaly. In early sympetaly the corolla tube originates either before or together with the corolla lobes; in late sympetaly the initially free corolla lobes interconnect by meristem fusion. Erbar (1991) also points to correlations of the two modes of development with larger taxonomic groups: late sympetaly is common among Scrophulariales, Solanales, and Lamiales, while early sympetaly is more concentrated in Campanulales, Asterales, Rubiales and Dipsacales.

The extremely complicated flowers of *Ceropegia*, (Asclepiadaceae), show a combination of lower congenital and upper postgenital union, whereby a middle free region forming lateral windows is intercalated (see section 8.7.1). Many of these groups with united petals have monosymmetric flowers in the form of lip blossoms or flag blossoms (Scrophulariales, Lamiales, some Solanales, some Campanulales, some Fabales). In these monosymmetric flowers the adaxial part of the corolla, the lower lip, usually serves as a landing place for pollinators (mainly hymenopters) and a holding place where pollinators have to use force to gather pollen in some extreme cases (see Brantjes 1981 a,b). In these elaborated and sometimes highly synorganized flower forms the petals are firm in texture and show relatively early expansion so that they are the protective organs of the inner floral parts for a longer time of floral development. Because they are united they also lack the narrow basal portion of many free petals and they may be served by more than one vascular bundle each. This is the case particularly in the Scrophulariales, Lamiales and Solanales but also in Gentianales, Rubiales and Asterales. The situation in sympetalous flowers is further complicated anyway by the congenital fusion of the stamens with the corolla. Fabaceae behave similarly but without congenital petal union.

Flowers with spiral or uniseriate perianth

Another kind of behaviour of the perianth is when it does not occur in two series, but rather all perianth units are spirally arranged and often occur in very variable numbers so that a particular number of series cannot be determined (see also section 2.5.2). In this case there is usually no sharp demarcation between an outer and an inner kind of organ. Although the outer and inner organs may have different qualities or features, the outer merge gradually into the inner.

In such cases there is often also a gradual transition from bracts outside the flower to the outermost floral organs so that it is difficult to 'delimit' the flower. This will be followed up further in the chapter on evolutionary aspects.

This type of perianth occurs mainly in Magnoliidae, where in some groups the spiral phyllotaxis does not cease with the last sepal formed but continues into the centre of the flower so that the flowers are completely spiral. This has been found (partly unpublished results) for Austrobaileyaceae, Illiciaceae, Schisandraceae, Eupomatiaceae, Monimiaceae (*Hortonia*, some Atherospermatoideae), some Ranunculaceae, some Menispermaceae and Calycanthaceae. In spiral flowers aestivation of the protective perianth members is always imbricate, never valvate.

In some of these groups floral phyllotaxis is quite labile (flexible) so that within a family, genus or even species both spiral and whorled floral phyllotaxis may occur. Examples are Monimiaceae (spiral: *Hortonia*, whorled: *Kibara*), Menispermaceae (spiral: *Hypserpa*, whorled: *Carronia*), Lauraceae (spiral and whorled in one and the same species *Endiandra montana*) (Kubitzki 1987, Hyland 1989, personal observation).

In these flowers with spiral phyllotaxis the innermost perianth units often behave similarly to the outer ones in ontogeny in that they are not significantly delayed. This is also true for some other Magnoliidae with a more or less whorled perianth (e.g. Magnoliaceae, Annonaceae, Lauraceae, Winteraceae) and also for many monocotyledons (Liliidae, Zingiberidae). An exception among Magnoliidae with spiral flowers is *Calycanthus*, which shows a slight retardation (Hiepko 1965).

Another kind of flowers are those with a simple perianth and the organs in a single series. This may be an evolutionarily primary situation or it may have arisen from flowers with a double perianth by loss of either the corolla (e.g. some Hamamelidaceae) or the calyx (e.g. some Rubiaceae, perhaps Santalaceae and Viscaceae).

2.2.4 Additional functions

Protective secretions before anthesis

A striking specialization in some tropical plants are so-called water calyces. The young floral organs are covered and protected by an aqueous or mucilaginous liquid that is secreted by a carpet of glandular hairs on the large inner surface of the closed flower buds, mainly the surface of the tepals, which are much larger compared with the other floral organs at this stage (Fig. 8.53). The tepals are firmly closed; they are often congenitally fused almost up to the top where they leave only a tiny orifice, which is, in addition, somehow occluded by postgenital fusion. After the first report on *Spathodea campanulata* (Treub 1889) water calyces were found in many other Bignoniaceae, some related families (such as Verbenaceae,

Solanaceae, Gesneriaceae) and other groups (Melastomataceae, Leeaceae) (de Lagerheim 1891, Raciborski 1895, Kraus 1895, Hallier 1897, Pascher 1960, Weber 1976a). Koorders (1897) gave a comparative description of such flowers. There is no recent work on the significance of water calyces.

In the floral buds of some Apocynaceae and Rubiaceae colleter-like glands ('squamulae') on the inner surface of the sepals secrete a slimy substance, which may have a protective function and be comparable to some extent with the secretion of the water calyces.

Secretions in floral buds are not uncommon but they have rarely been studied (e.g. McConchie 1983 for *Maidenia*, Hydrocharitaceae, and other water plants).

Secretions at anthesis

An additional function to the optical and often olfactorial attraction of petals is attraction by rewards, mainly nectar. Nectariferous petals occur in many groups, mainly Magnoliidae and Liliidae (see also section 5.2). Floral nectaries on the inner side of the sepals are characteristic of Malvales.

Nectaries are frequent on the outside of the sepals of tropical flowers and play an important role in the protection of flowers (and whole plants) by ants against predators (see section 5.2).

Nutritious tissues and food bodies

Food bodies on tepals that are eaten by pollinating beetles have been observed in *Exospermum* and other Winteraceae (Thien *et al.* 1990). This is also the case in some Annonaceae (where the entire tepal tissue may also be partly eaten) (Gottsberger 1970, Schatz 1987).

Pollen presentation and transportation

Secondary pollen presentation occurs on the petals of *Acrotriche*, Epacridaceae (McConchie *et al.* 1986), but in a generalized manner also in *Magnolia* and similar flower types (Vogel 1978a).

In flowers with valvate and postgenitally fused sepals or tepals a tension may be built up by differential growth so that the flower opens explosively if a certain force is exerted. This mechanism is used by some mainly bird-pollinated groups for pollen transmission to the pollinator's body and from the body to the stigma, such as certain Rhizophoraceae (Tomlinson *et al.* 1979, Juncosa & Tomlinson 1989), Proteaceae (Ramsey 1988), and Loranthaceae (Feehan 1985). In some Marantaceae forcible opening of floral buds by bees occurs (see section 8.9.6).

Hyperstigma

An extraordinary function is exerted by the inner tepals of the female flowers of some Monimiaceae. These flowers have the carpels enclosed in a floral cup. The entrance is a narrow pore. The inner tepals are secretory and constitute a 'hyperstigma' where pollen germinates and pollen tubes grow down towards the carpels (Endress 1980a) (see section 8.2.2). These secretory structures may be evolutionarily derived from glands at the tepal margins or tips that function in floral bud protection in less specialized genera (*Hortonia*, *Siparuna*).

Functions after anthesis

In many groups the sepals are additionally involved in fruit differentiation in various ways. A good example is the Dilleniales: in Dipterocarpaceae the sepals become large wings on the nuts; in some *Dillenia* species (Dilleniaceae) they become thick and fleshy and so take part in the formation of a berry-like fruit; in other *Dillenia* species and in *Ochna* (Ochnaceae) they provide a vivid colour contrast with the diaspores.

Even petals, however rarely, may attain new functions after anthesis. In *Coriaria* the petals become fleshy and form the outer layer of the functionally drupe-like fruit.

2.2.5 Evolutionary aspects

A hypothesis elaborated by Čelakovský (1896/1900) that has been most widely accepted is that the sepals of angiosperms are evolutionarily derived from bracts and the petals from stamens. Arguments put forward for this view are the frequent transitions between stamens and petals as well as between bracts and sepals, respectively, the often corresponding number of vascular traces, and the fact that petals and stamens are often retarded in ontogeny (stamens, however, less than petals). In plants with unisexual flowers the petals are often larger in male flowers than in female flowers; this fact at least points to the intimate physiological relationship between corolla and androecium (Baker 1948). The suggestion by Stanton *et al.* (1986) that the evolution of floral signals may be driven by selection on male function focuses the same problem from another perspective.

The petals are therefore seen as highly elaborated staminodes with special optical attraction function. This view is presented in detail by Hiepko (1965) who also discusses other hypotheses on the origin of the perianth parts in angiosperms. In this view the perianth parts in Magnoliales and Laurales – the most primitive angiosperms! – would only represent sepals (owing to their anatomy and ontogeny) even in those frequent cases where they are differentiated into a green, protecting 'sepaloid' and a coloured, attractive 'petaloid' region. However, in Ranunculales true 'petals' would occur (especially in Ranunculaceae, Berberidaceae and Menispermaceae).

This view is, however, difficult to maintain if the angiosperms as a whole are considered, since – mainly in tropical groups – diversity is too great. Some aspects of this problem have already been discussed by Burtt (1961), Hiepko (1965), Rohweder & Huber (1974) and Weber (1980). By comparative research over a wide range of taxa it becomes obvious that organs that most probably correspond to the typical petals of a certain group may attain all structural and developmental attributes of sepals concomitant with changing functions. If treated in isolation, there is no character combination which could stringently prove an organ's nature as a petal or sepal.

What one can do is to trace corresponding organs within a smaller or larger taxonomic group by a broad comparison. As a rule of thumb one may equate the inner cycle (series) with petals if two cycles (series) are present, as is so often the case; if only one series is present, it is usually homologous with a calyx. However, for both cases there may be exceptions. Examples of exception for the first case are *Mirabilis* (Nyctaginaceae), *Alchemilla* (Rosaceae), or *Choriceras* (Euphorbiaceae). Here the inner series corresponds to a calyx, the outer to an epicalyx or a group of bracts, although they are integrated in the flower. Similarly, in *Passiflora* species the attractive organs are petals and sepals and the protective organs are bracts (Fig. 8.17.1). Examples for the second case are the Santalales in the Rosidae. The extremely specialized family Viscaceae and the closely related Santalaceae have a simple perianth (a perigon) of a single cycle of thick valvate organs. Are they derived from petals or sepals? If we compare them with less reduced related groups, the Loranthaceae show outside such valvate organs a whorl of tiny organs, the so-called 'calyculus', the name implying an organ category that does not fit in the usual pattern. Still less reduced related groups are the Opiliaceae, Olacaceae and Icacinaceae. In Icacinaceae (and similarly in Vitaceae) these organs are caducous at anthesis and, in this sense, they behave as expected of petals. They all have a double perianth with thick valvate petals and relatively small (short) sepals. This pattern also appears in some other more primitive Rosidae, such as Pittosporaceae and Escalloniaceae. We may, therefore, tentatively see the evolution as follows: in those groups where the elaboration of the second whorl, the corolla, has gone towards a protecting organ with tight valvate bud closure, the outer series, the original calyx, is reduced, and it may even disappear. This also occurs within the family Rubiaceae, where the usually double perianth is reduced in some cases so that the valvate petals alone are left (e.g. *Galium*,

Pötter & Klopfer 1987). In this light the 'calyculus' of the Loranthaceae may be seen as a reduced calyx (and not as an accessory organ), and the simple perianth of the Viscaceae, Santalaceae and Balanophoraceae may be seen as a corolla that has taken over calyx functions (see section 9.4.3).

In some Balanophoraceae reduction went still further in that the perianth is more reduced, especially in female flowers, and the protective function is often completely taken over by inflorescence bracts (or by transformed flowers?) that may attain a shield-like shape, thus another transference of function (Kujit 1969, Endress 1975, Hansen & Engell 1978).

Another possibility is the transference of ontogenetically early protective function to bracteoles and concomitant reduction or complete loss of the sepals, as occurs in *Thunbergia* (Acanthaceae). However, here nectar is still secreted at the place of the disappeared sepals (van der Pijl 1954).

Aestivation is more often valvate than imbricate in highly synorganized flowers, since the neighbouring organs have symmetrical relationships. A shift from imbricate (contort) to valvate corollas is obvious in Asclepiadaceae (Asclepiadoideae). The corolla of *Asclepias* has low synorganization, while in *Ceropegia* with its trap-flowers and lateral windows it is highly synorganized. Accordingly there is a shift from imbricate (contort) to valvate aestivation from Asclepiadeae to Stapelieae. Such an evolutionary shift is especially striking in Caesalpinieae (Caesalpiniaceae). The valvate calyx of *Delonix* begins ontogeny in an imbricate (quincuncial) pattern – the prevailing pattern in the tribe – and then becomes valvate by excessive dorsal thickening of the sepals (Fig. 8.24). In *Parkinsonia* the overlapping flanks of the imbricate (quincuncial) calyx are coherent by a secretion and slight further organ extension occurs by an incipient thickening (Fig. 8.24). This can be seen as an evolutionary precursor pattern to that in *Delonix*.

Thus there is an evolutionary trend from quincuncial (mainly sepals) or contort (mainly petals) to valvate, and, from apert to contort (mainly petals). In monosymmetric flowers there is a trend from quincuncial to ascending or descending cochlear (Fig. 2.4).

In the families with the most elaborate flowers (Asclepiadaceae, Orchidaceae) the corolla is especially diverse at the morphological and histological level. In both families congenital fusion with the androecium occurs. The corona in the Asclepiadaceae and complicated effigurations of the lip in Orchidaceae seem to be facilitated by this synorganization between corolla and androecium (Brown 1833, Kunze 1982a, Endress 1990a): in both families these complex organs have a firm texture and the surface sculpturing is very diverse (Ehler 1975, 1976; see also chapter 8). Since the effigurations appear relatively late in ontogeny, they have the potential to develop three-dimensionally and to become complicated structures. In a few groups of both families, incredibly bizarre ornamentations have evolved on these organs, one of the most extreme genera being *Bulbophyllum*, Orchidaceae (Fig. 8.71).

2.3 Androecium

2.3.1 Structure and function

The androecium consists of the male functional organs, the stamens, and sometimes additional sterile organs derived from them, the staminodes. Stamens are relatively uniform, so their delimitation usually poses no problems. Most stamens have four pollen sacs (microsporangia) arranged pairwise in two lateral thecae. At maturity each theca usually opens by a longitudinal slit along a preformed dehiscence line, the stomium, through which the contents of both its pollen sacs are liberated (Figs. 2.5, 2.6.3, 2.7.3). Before the stomium dehisces, the septum between the two pollen sacs dissolves so that the two pollen containing chambers derived from the pollen sacs merge into one. In most angiosperms except for some Magnoliidae the two thecae are contracted into a well-defined anther that sits on an often elongated basal part, the filament (Fig. 2.6). Thus, filament and anther are generally the two obvious parts of a stamen. The part between the two thecae, the connective, is often thin. The normally single vascular bundle of the stamen usually extends into the connective.

Not only is this basic construction quite stable but also the size of stamens varies within restricted limits. Anthers rarely become longer than about 2 cm (e.g. in some Liliaceae and allied families, and in Magnoliaceae, Bombacaceae). An extreme case is *Strelitzia* with anthers of 5 cm length. This restriction is probably due to the complicated developmental events during meiosis, requiring a high degree of cellular synchrony, and during pollen development (summary in J. Heslop-Harrison 1972), which considerably constrain size variation. In addition, the filaments are never extensively long because of static architectural constraints (long free filaments or free parts of basally fused filaments do occur in some Bombacaceae, e.g. *Pachira*, or Caesalpiniaceae, e.g. *Baikiaea*, about 10 or more cm long) (see also section 3.3).

2.3.2 Development

During ontogeny the proportions of a stamen change considerably. The more or less hemispherical primordium becomes four-angled and four-lobed. The main part of the young, undifferentiated stamen will

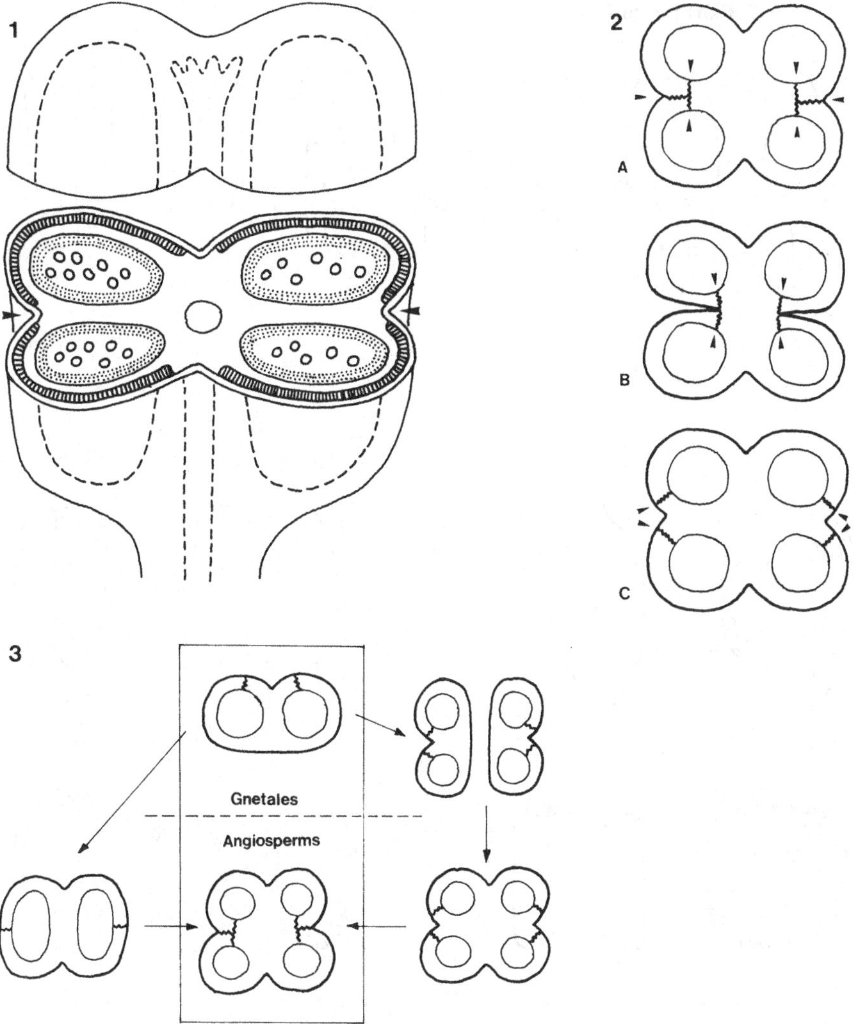

Figure 2.5. 1. Basic anther structure. Surface view and transverse section. Interrupted lines, outline of sporangia and of vascular bundle, in surface view; stippling, tapetum surrounding pollen-producing tissue; hatching, endothecium; arrows, stomia, in transverse section. 2. Forms of anther dehiscence by longitudinal slits (dehiscence regions marked with arrows and wavy lines). A. Common form (most angiosperms). B. Thecal septum reduced (e.g. *Altingia*, Hamamelidaceae). C. Two stomia per theca (e.g. *Strelitzia*). 3. Potential evolutionary pathways of anther structure from early anthophytes (as represented by Gnetales) to angiosperms (dehiscence regions marked with wavy lines).

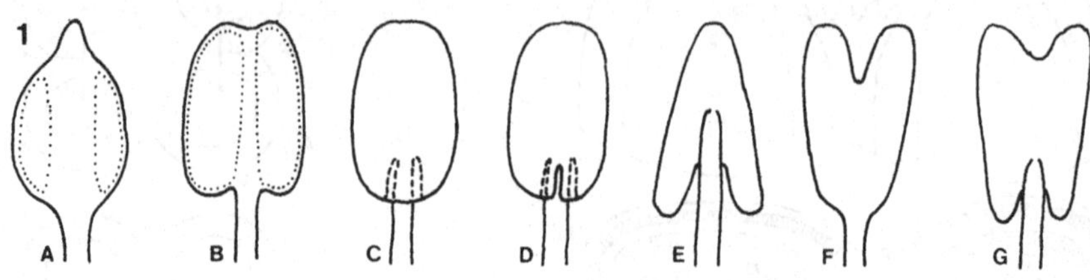

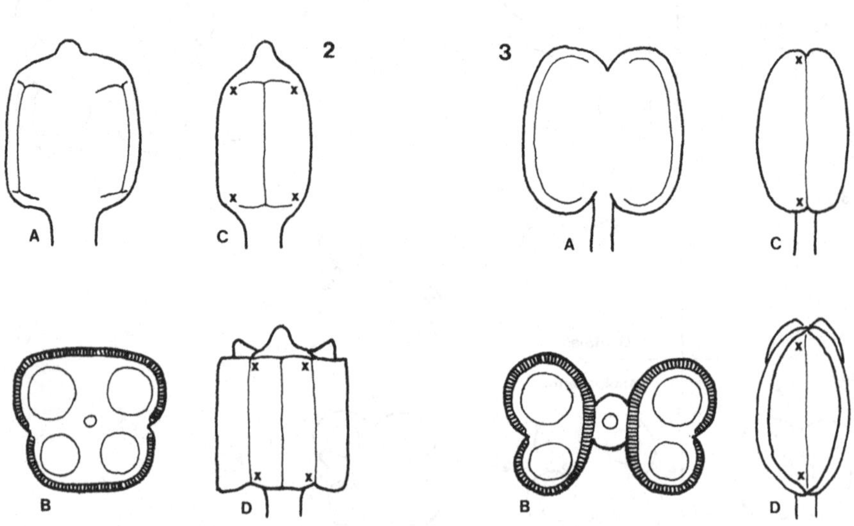

Figure 2.6. 1. Anther shapes (anthers seen from dorsal side). A. With connective protrusion (e.g. *Embolanthera*, Hamamelidaceae), B. With slightly rounded outline (common form). C. With basal pit (e.g. *Aphanopetalum*, Cunoniaceae). D. With basal pseudopit (e.g. *Mangifera*, Anacardiaceae). E. Sagittate (e.g. *Allamanda*, Apocynaceae). F. V-shaped (e.g. *Claoxylon*, Euphorbiaceae). G. X-shaped (e.g. *Nitraria*, Zygophyllaceae). 2–3. Anther dehiscence by valves and by simple longitudinal slits (modified after Hufford & Endress 1989). A. Frontal view of anther. B. Transverse section (hatched: endothecium and endothecium-like tissue). C–D. Lateral view (x: end points of dehiscence region). C. Closed. D. Open. 2. Anther with valvate dehiscence. 3. Anther with simple longitudinal dehiscence.

Figure 2.7. Anthers with thecal organization and simple longitudinal dehiscence. 1. *Ascarina rubricaulis* (Chloranthaceae), two unistaminate male flowers each with subtending bract, slit-like opening (× 13). 2. *Pseudowintera axillaris* (Winteraceae), closed and open stamen with thickened filament (× 17). 3. *Maytenus emarginata* (Celastraceae), anther with widely gaping thecae (× 80). 4. *Casuarina stricta*, slit-like opening (× 30). 5. *Pavonia hastata* (Malvaceae), monothecal anther (× 60). 6. *Rhizophora* × *lamarckii*, polysporangiate anther, from ventral side (× 16).

give rise to the anther, while the filament often remains extremely short for a long time and only immediately before anthesis rapidly elongates.

Anther

In *Lilium*, which has very large anthers, Gould & Lord (1988) found by longitudinal marking that the anthers elongate in basipetally moving waves. It is not known whether this is caused by pulses of a growth substance, by a mechanical trigger correlated with turgor changes, or by electrical impulses.

The early differentiation of the anther is due to the complicated histological development of the pollen sacs, where the sporogenous cells transform into meiocytes that undergo meiosis. The resulting microspores develop into the pollen grains, which have to differentiate the male gametophyte and the complicated sporoderm.

This pollen-producing tissue is completely surrounded by a tapetum, a layer with which it interacts intensively (e.g. Heslop-Harrison 1972, Pacini 1990). Genetic experiments in tobacco have shown that an inactive tapetum leads to male sterility but has no effect on the surrounding tissues (Koltunow *et al.* 1990). At anther maturity the endothecium and the stomium, the histological structures that allow anther dehiscence, are differentiated. The endothecium is usually a subepidermal cell layer with partially thickened (and lignified) cell walls providing mechanical tension in the direction away from the stomium (dehiscence line). Conversely, the subepidermal cells lining the stomium are often not thick-walled, so that the weakness of the dehiscence line is established. Dehiscence is further enhanced by the dissolution of the underlying septum, which takes place shortly before dehiscence. The tissue at the dehiscence region is often full of oxalate crystals, which may be involved with the dissolution process (Bonner & Dickinson 1989, Horner & Wagner 1980, Endress & Stumpf 1990).

Keijzer (1987a,b) distinguishes five steps in the opening process (studied in *Gasteria*, Aloeaceae, and *Lilium*): (1) differentiation of the endothecium; (2) dissociation of the tissue in the septum by enzymatic lysis of the middle lamellae of the cell walls; (3) mechanical rupture of the tapetum by expansion of volume of pollen mass so that both locules of the theca become confluent; (4) mechanical rupture of the small epidermis cells in the stomium by tangential expansion of the epidermis and endothecium and inward bending of the two thecal wall parts (perhaps with the help of enlarged epidermis cells bordering them, which may be pressed together by the expansion and inward bending and rupture the small ones); and (5) outward bending of the two thecal wall parts by dehydration, probably mainly by evaporation (see also Schmid 1976).

The two wall parts of the two pollen sacs open, by the longitudinal dehiscence line through the middle of the theca, in opposite directions and so mutually reinforce the gaping. The detailed mode of anther deformation by this process varies according to anther shape as well as structure and extension of the endothecium.

After dehiscence, the two lips of each theca (corresponding to the former peripheral parts of the two pollen sacs) are more or less widely gaping; these lips are often called 'valves', a term which is here used only in a restricted sense, see section 2.3.3). The endothecium layer is often continuous around the connective side of each theca (whereas it is discontinuous along the stomium). Both lips act thus as mutual counterforces, which may enhance the opening of the theca (Hufford & Endress 1989) (Fig. 2.6.3).

In short: pollen sacs may be seen as important functional units in development, whereas thecae are important functional units at anthesis (at and after dehiscence) (Endress & Stumpf 1990).

Filament

In contrast to the complicated anther structure and development, the filament is simply parenchymatous with a vascular bundle. It can therefore elongate within a shorter time. This is extreme in grasses, where elongation of several millimetres occurs within a few minutes and is the most rapid growth known in flowering plant organs (Askenasy 1879, Percival 1921, Schoch-Bodmer 1939). In grasses new vascular tissue is not formed during this rapid elongation, in contrast to groups with more slowly growing filaments (J. Heslop-Harrison *et al.* 1987).

2.3.3 Diversity

Stamen number and arrangement

However relatively uniform stamen structure may be, no other organ category at the same comparative level has the same diversity in number. Among the Magnoliidae between one (e.g. *Ascarina*, Chloranthaceae) and 2000 (e.g. *Tambourissa*, Monimiaceae) stamens may occur in one flower (Lorence 1985, Endress 1990a). In the Dilleniidae there are also examples with up to 2000 stamens (*Adansonia*, Bombacaceae; Faegri & van der Pijl 1979).

Arrangement patterns of stamens are diverse: spiral, whorled, more complicated or unordered. Differences are also due to centripetal or centrifugal androecium initiation (see section 2.5.3).

Stamen and anther proportions

The two thecal slits of an anther may look towards the centre of the flower (introrse anther, the most frequent behaviour in angiosperms, e.g. most Rosidae) or towards the periphery (extrorse anther, e.g. *Degeneria*; *Galbulimima*, Himantandraceae; Annonaceae; Aristolochiaceae; *Simmondsia*; Iridaceae) or exactly to the sides (latrorse or aequifacial anther). Anther proportions may be fairly constant in larger groups. That introrse anthers are the most frequent ones may be seen in connection with pollination biology: it is advantageous for many flower forms to release the pollen in the direction of the stigmas (Webb & Lloyd 1986). On the other hand, there is a correlation between extrorse anthers and short filaments: if anthers are almost sessile, it is easier to present pollen towards the floral periphery. An extreme case are united anthers, which are extremely extrorse (e.g. Canellaceae, Myristicaceae) (Fig. 2.10).

In some groups of the Magnoliidae, anther and filament are not well delimited. In contrast, the thecae in these cases are well recognizable functional units. In some taxa the thecae are situated at the margin of a scale- or band-like organ (e.g. in some Winteraceae and Chloranthaceae). In others the thecae may be more in the median region of a broad scale (e.g. adaxial in *Austrobaileya* or abaxial in *Degeneria* or *Galbulimima*), in *Austrobaileya* protruding, in *Degeneria* and *Galbulimima* more embedded in the scale (Figs. 2.8, 8.3). In some extreme examples the part above the thecae is longer than the part below it, so the stamens look almost upside down. This occurs in *Galbulimima*, *Doryphora* (Monimiaceae) (Fig. 8.3), some *Magnolia* species (Howard 1948, Nooteboom 1988), among Hamamelididae in the Loropetalinae (Endress 1989), among Dilleniidae in some Dipterocarpaceae (Woon & Keng 1979), and among Liliidae in some Liliales (Dahlgren *et al.* 1985).

Also in many other groups the connective more or less protrudes as a tip on the anther apex (various Magnoliidae; 'lower' Hamamelididae; among Rosidae mostly not constant in larger groups: e.g. Crassulaceae, Cunoniaceae; Asteridae: Asteraceae; Arecidae: Pandanaceae, Arecaceae) (Fig. 2.6.1).

At the other extreme are forms where the anther is very compact with minimal sterile tissue and is extremely delimited from the filament, sometimes by a very thin and narrow joint (Fig. 2.6.1). The joint may be on the dorsal side (dorsifixed anther), ventral (ventrifixed) or at the base (basifixed). Some groups use this joint as an abscission region, where the anther is abscised at anthesis. This is mainly the case in some protandrous flowers that may thus radically remove their own pollen at a certain stage (e.g. some Liliaceae). Dorsifixed and ventrifixed anthers with a thin joint are versatile. They can be moved like a seesaw.

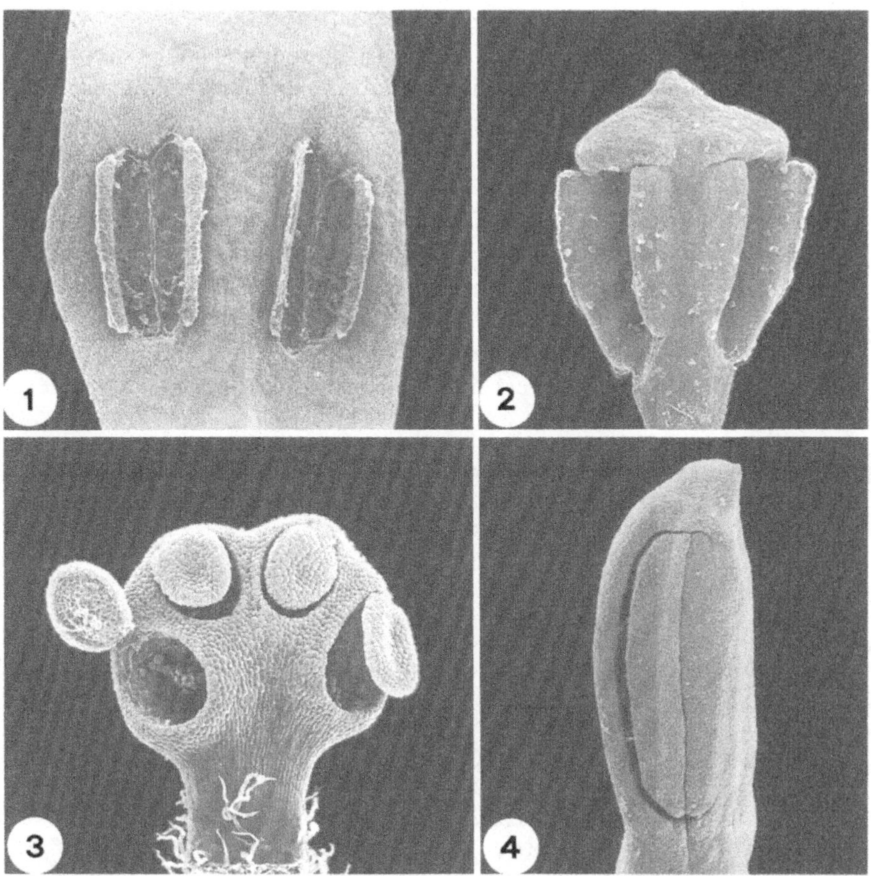

Figure 2.8. Anthers with valvate dehiscence. 1. *Galbulimima belgraveana* (Himantandraceae), stamen from dorsal side (× 30). 2. *Artabotrys hexapetalus* (Annonaceae), stamen from dorsal side (× 25). 3. *Cinnamomum camphora* (Lauraceae), stamen from ventral side (× 50). 4. *Dicoryphe viticoides* (Hamamelidaceae), stamen from the side, ventral side at right (× 30).

The combinations introrse–dorsifixed and extrorse–ventrifixed prevail but are not exclusive. An exception is, for example, *Montinia* (Escalloniaceae) with extrorse but dorsifixed anthers (Endress & Stumpf 1991).

Other variants in shape include sagittate, V-shaped, and X-shaped anthers, and anthers with a basal pit or pseudopit (Endress & Stumpf 1991) (Fig. 2.6.1).

Union of stamens (synandry)

Stamens are mostly free. Congenitally united stamens occur in several Magnoliidae and are common in Malvales and Fabales. In general the filaments are united, in extreme cases the anthers are also (Myristicaceae, Canellaceae, *Stephania* of Menispermaceae, Nepenthaceae, *Phyllanthus cyclanthera* of Euphorbiaceae) (Figs 2.9, 2.10). Union may lead to a tube or, in unisexual flowers, to a compact cylinder. As already pointed out in section 2.2.3, the stamens may be fused with the perianth (as commonly in Asteridae). Fusion with the gynoecium also occurs (as in Aristolochiales or Orchidales).

Postgenitally united anthers occur in some buzz-pollinated flowers (see section 5.1) and in some flowers with secondary pollen presentation (see section 2.4.5).

Figure 2.9. Androecia with united stamens. 1–2. *Cola acuminata* (Sterculiaceae). 1. From the side (× 20). 2. From above (in the centre: reduced gynoecium) (× 18). 3–4. *Nepenthes* sp. 3. From the side (× 25). 4. From above (× 25).

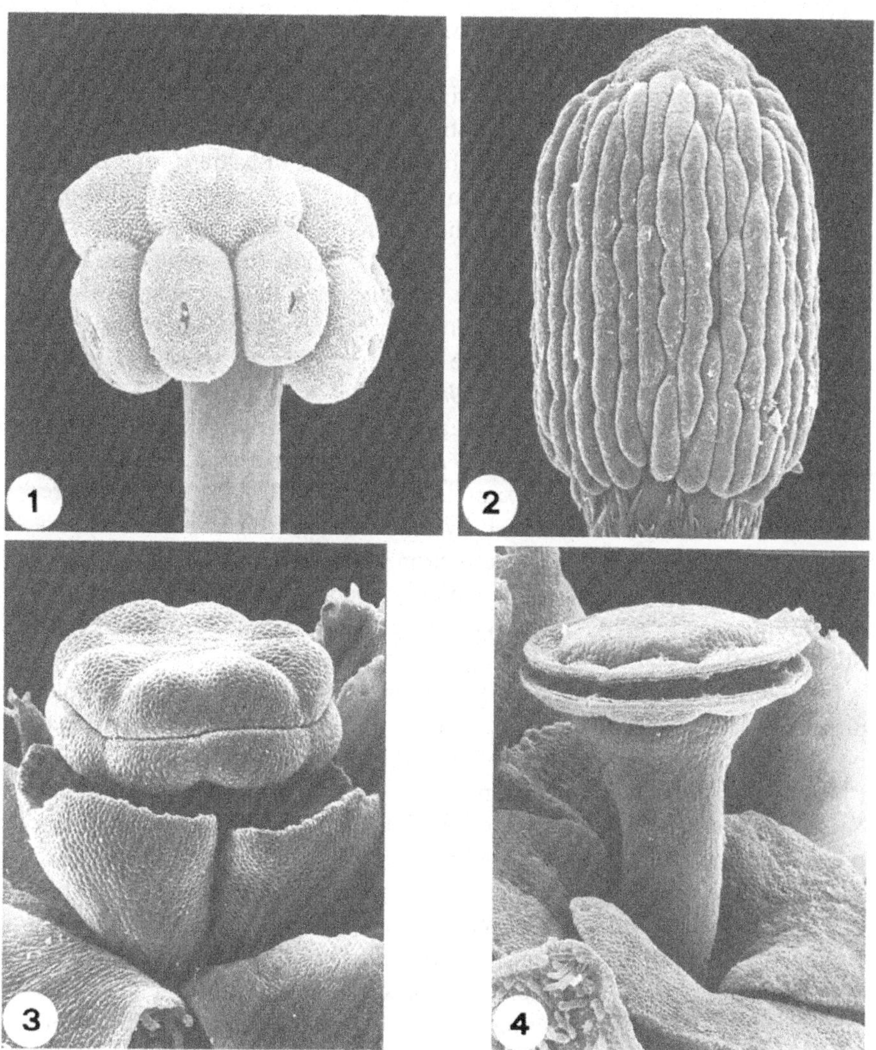

Figure 2.10. Androecia with united stamens, all from the side.
1. *Spathicarpa sagittifolia* (Araceae), thecae with pore-like dehiscence (× 30). 2. *Myristica insipida*, pollen sacs polysporangiate (× 35). 3–4. *Stephania japonica* (Menispermaceae). 3. Closed synandrium (× 50). 4. Dehisced synandrium (× 50).

Correlation between stamen structure and pollination mode

There are many correlations between stamen structure and a particular pollination mode. The following are examples.

(1) Flowers adapted to pollination by larger animals (bats, birds, butterflies, moths) sometimes have large, broad anthers with the stomia facing in the same direction, extremely introrse, slit-like and not widely opening. The narrow opening arises in that the two parts of the thecal wall curve rather inwards than outwards (Figs 8.20, 8.27, 8.46). In combination with versatility (seesaw mechanism) such anthers work like a stamp on bird feathers or butterfly wings. The anther openings are passively directed towards the pollinator's body at the slightest touch and pollen is transferred to it, e.g. in *Delonix*, *Amherstia* (Caesalpiniaceae), *Erythrina* (Fabaceae), *Melianthus* (Melianthaceae), *Cleome* (Capparaceae), *Aeschynanthus* (Gesneriaceae), *Lilium*, *Gloriosa*, *Alstroemeria*, *Pancratium* (Liliaceae *s.l.*). An extreme example is *Globba* (Zingiberaceae) where the anthers are further broadened by four lateral extensions and have an elaborate joint (Müller 1931) (see section 8.9.4) (Fig. 8.62).

(2) Fly-flowers sometimes have anthers opening with pores and pollen in slimy masses (e.g. Araceae, Viscaceae, Rafflesiaceae, see Endress & Stumpf 1990) (see section 8.3.2).

(3) Bee-flowers have many anther forms. A particular melittophilous differentiation is the possesssion of long, narrow thecae with a terminal pore. The bees harvest the pollen by vibrating it out of the pores ('buzz-pollination') as, for example, in *Cassia* (Caesalpiniaceae) and *Solanum* (Solanaceae) (see section 5.1). Some groups with terminal anther pores have reverted from buzz-pollination to other pollination mechanisms. In these groups pollen is released by the stroking and compressing of anthers, which work like bellows (some nectariferous Melastomataceae, Renner 1990; perfume-flowers in *Cyphomandra*, Solanaceae, S. Vogel, personal communication). There is a concomitant architectural change of anther walls from the vibrating- to the bellows-mechanism (S. Vogel, personal communication).

(4) Anemophilous flowers often have relatively large, long anthers compared with their nearest relatives that are zoophilous. The anthers often open with only narrow slits and are not widely gaping (Fig. 2.7.1,4). The filaments may be lax and pendulous as, for example, in Poaceae, Cyperaceae, or in *Acer negundo* (see section 4.2.1).

Dehiscence patterns

Other stamen variations include deviations in the dehiscence pattern and the number of sporangia, which are partly correlated.

By far the most common pattern is dehiscence of each theca by a simple longitudinal slit. This pattern is predominant in all subclasses of the angiosperms (Fig. 2.7). Rarely, each of the two pollen sacs of a theca opens independently by a longitudinal slit. In these cases each pollen sac protrudes individually from the connective and the septum between the two pollen sacs may be short or very broad. This situation occurs, for example, in *Strelitzia* and *Heliconia* (see section 8.9.2) (Fig. 8.59) and in some Rutaceae and Meliaceae (Endress & Stumpf 1991). Conversely, the two stomia of an anther may be confluent over the apex to form a single dehiscence line. In the extreme, the two thecae are also confluent (synthecal anther) or one theca may be wanting (monothecal anther) (Fig. 2.7.5).

In some Magnoliidae and Hamamelididae with relatively massive anthers with thick connectives, the thecae do not open by a single longitudinal slit but rather by two valves. Here the stomium bifurcates at both ends (e.g. *Degeneria*, Annonaceae, many Hamamelidaceae) or, in some groups, each pollen sac even has its own stomium and forms a separate valve (many Laurales) (Fig. 2.8). In several of these groups the pollen sac number per theca has been reduced to one (some Laurales, some Hamamelidaceae).

Both changes may be seen as a consequence of the thick connective. Here, the two mirror image parts of a theca do not act as mutual counterforces. Counterforces are lacking or, if present, they are represented by the neighbouring pollen sacs of *different* thecae. This is also shown by the fact that the endothecial layer in these stamens is often continuous not *within* a theca but *between* the two thecae (Trochodendrales, Hamamelidales) (Fig. 2.6.2). Endothecial continuity may also be lacking in massive, broad stamens (Endress & Hufford 1989, Hufford & Endress 1989, Endress & Stumpf 1990).

The parallel reduction of the pollen sac number per theca from two to one in different groups of the Magnoliidae and Hamamelididae may thus be the result of the lack of functional constraint on the pollen sac pairs.

Polysporangiate and athecal anthers

In a number of scattered groups out of 28 dicot families and perhaps only 3 monocot families, polysporangiate anthers have developed (Endress & Stumpf 1990). In most cases they arise by transverse partitioning of the four pollen sacs. There may be additional longitudinal partitions (Figs 2.11, 2.12). Extreme forms are those of *Kerianthera* (Rubiaceae) (Robbrecht 1988) and *Rhizophora* (Rhizophoraceae) (Juncosa & Tomlinson 1987) with several hundred sporangia (Fig. 2.7.6). These anthers have retained their thecal organization and open like normal ones by two longitudinal slits (Endress & Stumpf 1990).

In a few groups thecal organization of these polysporangiate anthers has been lost (Endress & Stumpf 1990) (Figs 2.11, 2.12). There, each of the many sporangia opens individually by a pore. These stamens have also lost their differentiation into anther and filament. They occur in Viscaceae, *Mitrastemon* (Rafflesiaceae) and *Polyporandra* (Icacinaceae). In *Rafflesia*, all sporangia converge to a single pore. In all these athecal anthers pollen seems to be liberated through pores as slimy masses, and pollination by flies has been recorded (Endress & Stumpf 1990).

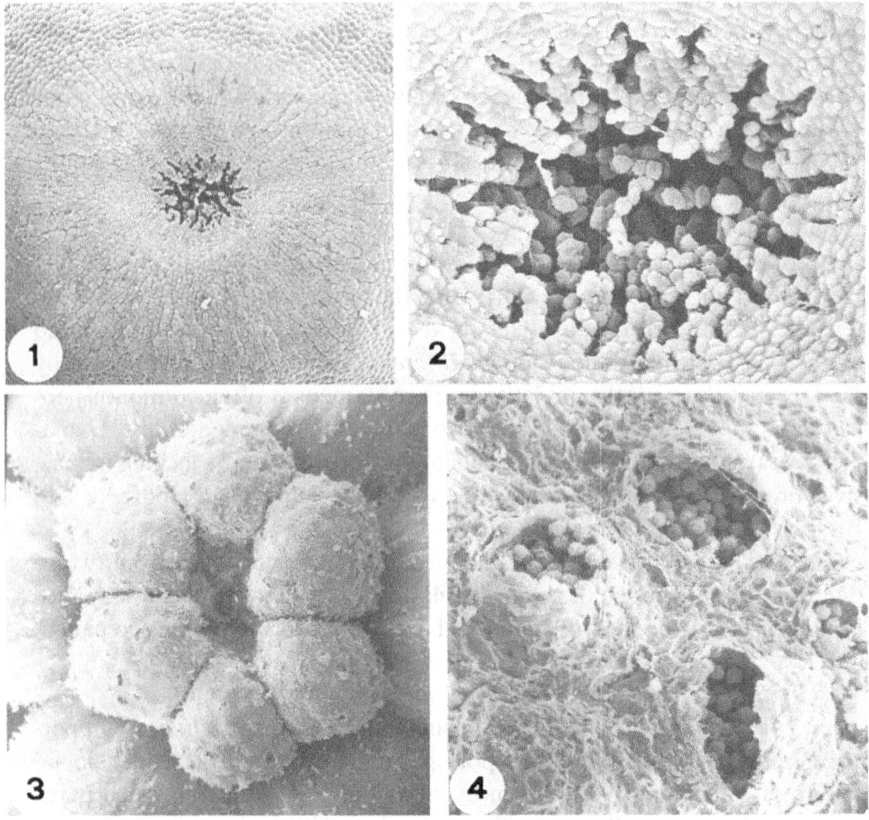

Figure 2.11. Polysporangiate stamens with athecal organization. 1–2. *Rafflesia arnoldii*. 1. Stamen from above, with apical pore (× 18). 2. Pore, magnified (× 80). 3–4. *Polyporandra scandens* (Icacinaceae). 3. Androecium with six dehisced stamens, from above (× 19). 4. Three dehisced sporangia, magnified (× 190).

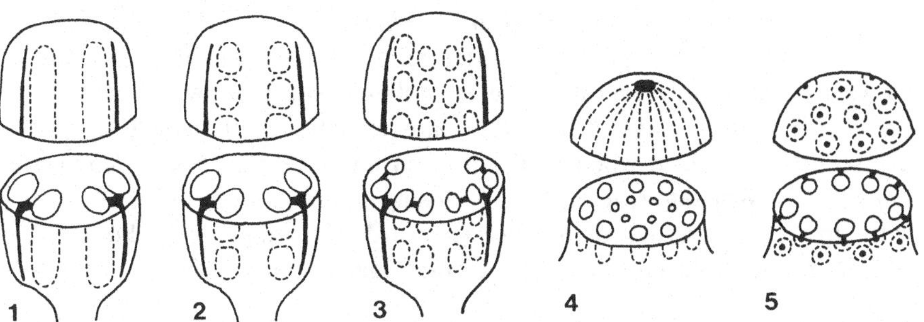

Figure 2.12. Forms of polysporangiate stamens. Surface view and transverse section. Interrupted lines, outline of sporangia in surface view; black, dehiscence sites (modified after Endress & Stumpf 1990). 1–3. Stamens with thecal organization. 1. Normal stamen with four sporangia. 2. Polysporangiate stamen with transverse septa (e.g. *Caloncoba*, Flacourtiaceae). 3. Polysporangiate stamen with transverse and longitudinal septa (e.g. *Rhizophora*). 4–5. Polysporangiate stamens with athecal organization. 4. Stamen with a single pore for all sporangia in the centre (e.g. *Rafflesia*). 5. Stamen with numerous pores, one for each sporangium (e.g. *Polyporandra*, Icacinaceae).

Heteranthery

In some groups two different morphs of stamens occur in the same flower (heteranthery): one set is often cryptic and contains pollen that is mainly used for pollination, while the other set is optically attractive and contains 'fodder' pollen mainly collected by bees, or it may be staminodial (see section 5.1).

In some taxa heteranthery seems to be primarily due to bud architecture. The stamens or anthers of two whorls are then different in bud (many Fabales) (Fig. 8.26.3) and may keep the difference until anthesis (e.g. *Nitraria*, Zygophyllaceae, Endress & Stumpf 1991).

Endothecium

The diversity in the partially thickened endothecium cells has frequently been mentioned (e.g. Noel 1983, Thiele 1988, French 1985) and a certain systematic constancy seems to be present. In Magnoliidae the inner tangential cell walls tend to be completely thickened ('base plate'), while they tend to be thickened only in the centre in other subclasses (Noel 1983). However, the suspected correlation between thickening pattern and anther construction has not been established.

Protection of pollen in dehisced anthers

Many anthers open widely after dehiscence and do not close again. However, there are several means of pollen protection, which may be significant in long-lived flowers and in rainy climates. Anthers with valvate dehiscence may close when wet and open again (e.g. Kerner 1905, for Lauraceae). Repeated opening and closing is also known from anthers with simple longitudinal dehiscence (Edwards & Jordan 1992). Poricidal anthers provide efficient pollen protection against rain.

Different ways of pollen aggregation

The description of single pollen development and structure is beyond the scope of this book. However, means of pollen aggregation to improve pollen transport should be briefly mentioned. Aggregation can be accomplished in four ways: (1) by a coating of pollenkitt, a sticky secretion of the tapetum; (2) by a glue originating from floral parts other than the tapetum; (3) by 'viscin' threads, elastic extensions providing a loose pollen aggregation; or (4) by coherence of the pollen into a more or less firm mass (pollinia) or groups of masses (massulae or polyads) or at least tetrads.

Abundant sticky pollenkitt is present in many animal-pollinated flowers (Dobson 1989).

A glue from other floral parts is more rare (e.g. *Cyclanthera*, Cucurbitaceae, Vogel 1981c; several Marantaceae; *Hedychium*, Zingiberaceae, Vogel 1984; *Eupomatia*, Endress 1984a,b; *Drymonia*, Gesneriaceae; Steiner 1985).

Most flowers with viscin threads are pollinated by large, imprecise pollinators (Lepidoptera, birds, bats). With the help of the viscin threads, pollen grains hang together in loose aggregates, which enhances the chances of pollination. Viscin threads occur in some Annonaceae (*Porcelia*), Caesalpiniaceae (*Delonix*, *Bauhinia*), Passifloraceae (*Tetrastylis*), Onagraceae, Ericaceae (*Rhododendron*), Balsaminaceae and Strelitziaceae (*Strelitzia*) (Cruden & Jensen 1979, Graham *et al.* 1980, Kronestedt & Bystedt 1981, Hesse 1981, 1984a, 1986, Waha 1984, Vogel & Cocucci 1988, Morawetz & Waha 1991, Buzato & Franco 1992) (Fig. 8.26.6). Viscin threads are differentiations of the pollen sacs that arise in various ways. In most groups they originate inside the pollen sacs and they contain sporopollenin; in orchids they are proteinaceous. In *Impatiens* (Balsaminaceae) and *Strelitzia* they are formed by epidermal cells of the thecae and are cellulosic (see section 8.9.2). In plants with viscin threads the pollen:ovule ratio is comparatively low (Cruden & Jensen 1979) (see section 6.8). The clumps of pollen, loosely held together by viscin threads, may be compared to some extent with the 'search vehicles' of water-pollinated plants (see section 4.2).

Pollinia are known from Asclepiadaceae and Orchidaceae (Figs 8.39.6, 8.70), massulae or polyads from some Orchidaceae, Mimosaceae (Fig. 8.31.4) and Annonaceae, and tetrads from many families (e.g. Annonaceae, Winteraceae, Ericaceae, Nepenthaceae, Asclepiadaceae) (Fig. 2.28.3).

It has also been argued that electrostatic forces play a role in the adherence of pollen to the pollinator's body and in the transfer from the pollinator to the stigma, but few details are known (see section 5.1).

2.3.4 Additional functions

Stamens may be involved in optical or olfactory attraction of pollinators. They often considerably enlarge the floral contours, especially by very long filaments (brush-flowers), and so add to the showiness of flowers (e.g. many Myrtaceae, Lecythidaceae–Planchonioideae, Sonneratiaceae, Caryocaraceae, Mimosaceae, Bombacaceae). Many of these flowers are pollinated by relatively large animals (birds, bats, butterflies or moths). Stamens often provide colour contrasts with other floral parts, particularly by the often yellow colour of the anthers.

Stamens (and staminodes) may also be almost the sole optical attractant if the corolla or the entire perianth is very small (e.g. *Calliandra*: Mimosaceae, Lecythidaceae–Planchonioideae, many Myrtaceae) or entirely lacking (e.g. some Chloranthaceae, Eupomatiaceae, Himantandraceae) (or share this function with the gynoecium: *Trochodendron*).

In some groups the stamens (not pollen!) are the sole or main scent-producing parts of the flowers (e.g. Chloranthaceae, Austrobaileyaceae, some Solanaceae: *Cyphomandra*, *Solanum*) (Endress 1980c, 1987b, M. Sazima & Vogel 1989, D'Arcy *et al.* 1990).

Stamens may also be part of the floral construction that canalizes the pollinator's mouth parts towards the nectar holding region. This is often the case in flowers with tubular corollas (e.g. *Brugmansia*, Solanaceae, and many other Asteridae) or flowers where nectar is hidden by the stamens (Caesalpiniaceae, Fabaceae, Sapindales) (Fig. 3.2). In these latter cases the filaments are often hairy, thus providing obstacles to the entrance of small insects as potential nectar thieves (Fig. 8.23). In flowers with a tubular perianth such hairs are often inside the tube.

Parts of stamens may also produce nectar (e.g. Laurales with distinct paired nectaries at the filaments; Lardizabalaceae: *Decaisnea*) (Fig. 8.3.8).

Staminodes may look like stamens without (functional) anthers or they may be elaborated in some way. Staminodes may have no obvious functions in just being parts of the 'bauplan' of the flower that cannot be thrown away easily (e.g. the median stamen in some Scrophulariales or rudiments in *Bauhinia*). In other cases they play a role in the floral display. They are osmophores (*Austrobaileya*) or optically attractive (*Jacaranda*, Bignoniaceae) or, together with the ability to move during anthesis, they may seclude the gynoecium from the outside in some relic Magnoliidae (see section 8.1.2) (Fig. 8.2). This is most spectacular in Eupomatiaceae, where inner staminodes are large and petaloid and also replace the missing perianth. They are also prominent in Austrobaileyaceae, Himantandraceae, Degeneriaceae, and less so in Calycanthaceae, a few Monimiaceae and Annonaceae. Staminodes may also secrete nectar (some Lecythidaceae, also Lauraceae, where the innermost whorl of the androecium is often staminodial). They may also take part in the canalizing apparatus for pollinators' tongues, e.g. in *Heliconia*, where the upper median stamen is still present as a staminode with this function, while in the other genera of the Musaceae sensu lato this stamen is totally lacking (see section 8.9.3).

2.3.5 Evolutionary aspects

Anther structure

Stamens are very stable throughout the angiosperms: there is an overwhelming occurrence of two lateral thecae, each with two pollen sacs. A major evolutionary step was probably the concentration of these thecae in an anther without much sterile tissue. This also restricted the potential dehiscence pattern to the common lateral slits. Only in the Magnoliidae and 'lower' Hamamelididae, where relatively massive anthers occur, could valvate patterns develop in addition to longitudinal ones. The potential to develop broad or thick connectives or connective protrusions was, however, not lost. A certain plasticity allowed these in cases where they would be useful. As examples may be noted the connective protrusions in Asteraceae, which are involved in the specialized dispenser mechanism of secondary pollen presentation, or the broad connectives in Zingiberaceae, which provide a holding device for the lax style that has to be held in median position; or in Asclepiadaceae where the centre of the connective is glued to the style head. Multiple convergent evolution led to the long poricidal anthers that are characteristic for 'buzz-pollinated' bee-flowers with bizarre extremes in the Melastomataceae.

Stamen number

In the number of stamens per flower several trends occur. The most obvious is fixation of a relatively low number of stamens in regular (whorled) flowers and even more in highly synorganized monosymmetric flowers (e.g. constantly one stamen in the Orchidaceae sensu stricto or Zingiberaceae). However, the potential to produce polyandrous androecia has not been lost at the middle and higher evolutionary level of the dicots (and monocots). Here, a low number of primary primordia often secondarily subdivide to give rise to a higher number of secondary stamen primordia (among dicots many Rosidae and Dilleniidae, among monocots Velloziaceae and palms (see section 2.5.3).

Pollen presentation and behaviour of the filaments

In the woody Magnoliidae (Magnoliales, Laurales), i.e. the 'primitive' angiosperms, the stamens behave developmentally in a similar way to the tepals, in that development proceeds equally. If the stamens are crowded in polyandrous androecia and the filaments remain very short, they have attained other means to loosen the anthers for pollen presentation, e.g. by thickening of the bases (Winteraceae, Illiciaceae, Menispermaceae; Carlquist 1981; personal observation) or, very unusual, by abscising from the floral base and being retained only by the extended reinforcing spirals of the tracheids of the vascular bundle (Annonaceae; Endress 1985) (Figs 2.7.2, 8.1.8).

In many higher advanced angiosperms (also including some herbaceous Magnoliidae, such as Ranunculaceae and Papaveraceae) the stamens are developmentally more differentiated in that the long and slender filaments elongate only very shortly before anthesis (e.g. Caryophyllidae, Hamamelidaceae, many Rosidae, Commelinidae). In some groups this late and quick expansion comes almost explosively. However, in some Rosidae the filaments are already elongated in bud and are sometimes incurved before they spread at anthesis (many Myrtaceae, Mimosaceae, Rosaceae, Chrysobalanaceae) (Fig. 3.3). This is also especially the case where the stamens form a close functional unit together with other floral parts that has to undergo a harmonious development (e.g. Fabaceae, Polygalaceae, Asteridae) (Fig. 8.34).

2.4 Gynoecium

2.4.1 Structure and function

Outer and inner surface, main functional parts

The gynoecium, the female organ, is the central and most complicated structure of the angiosperm flower, mainly because parts of it are not freely exposed but enclosed and form a system of canals and chambers (Fig. 2.13). The apical surface, the stigma, receives the pollen grains, and from there the pollen tubes grow through the style into the ovary, which contains the ovules. In textbooks the gynoecium is often oversimplified to show a solid style with the stigma atop it, and a hollow ovary containing ovules. In reality it is much more complicated and many structural aspects are better understood in conjunction with ontogeny and functions.

In its development the gynoecium usually arises as a cup-shaped structure like the vase that emerges on a potter's wheel. Therefore, its primary (morphological) surface continues from the outside over the upper rim down to the inside.

During ontogeny the inner space is partly secluded from the outside. However, the primary inner morphological surface remains completely distinguishable in most plant groups up to anthesis and later. It can easily be detected in stained microtome (transverse) sections.

The primary inner surface is very important because the ovules are differentiated on it, mostly when the inner space is already secluded from the outside. Furthermore, the pollen tube transmitting tract is formed on this inner surface or at least partly there. Its secretory nature and, therefore, special histological differentiation, is the reason why the primary

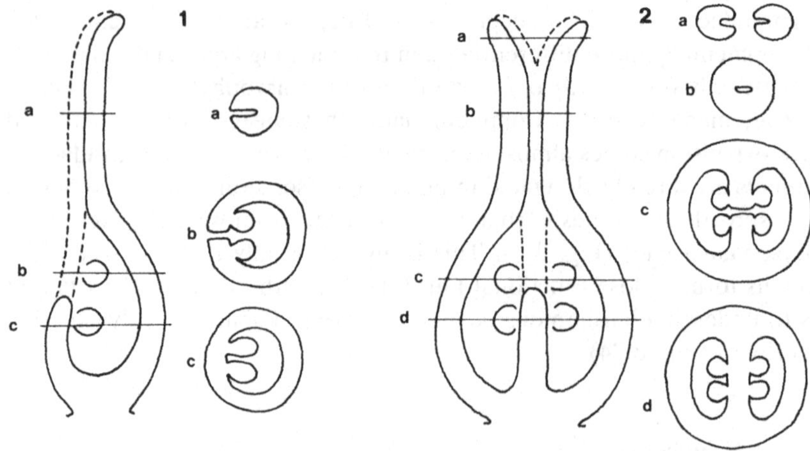

Figure 2.13. Schematic sections of gynoecia. Left: median longitudinal section; right: transverse sections of corresponding levels (marked with a-d). Interrupted line: outline of parts that are not exactly in the median plane. 1. Unicarpellate gynoecium (or single carpel): a–b. Plicate part of style and ovary. c. Ascidiate part of ovary. 2. Syncarpous gynoecium: a. Plicate part. b–c. Symplicate part of style and ovary. d. Synascidiate part of ovary.

inner surface can often be distinguished for such a long time, in spite of its postgenital fusion.

The inclusion of the ovules in an ovary and seclusion from the environment is highly characteristic for the angiosperms (hence the name!). This is effected by postgenital fusion of inner surfaces or at least by secretions that plug open spaces. Hollow styles usually also have at least a short region of complete postgenital fusion (e.g. Liliaceae, see Willemse & Franssen-Verheijen 1986). The presence of 'open carpels' in some primitive angiosperms (*Degeneria*, Winteraceae) as indicated in many textbooks is a myth (Figs 2.14, 2.15, 8.4).

The enclosed pollen tube transmitting tract is a delicate and highly efficient filter for the selection of suitable male gametes before fertilization (e.g. Knox 1984) and for the exclusion of potential parasites (e.g. Jung 1956).

Syncarpy and apocarpy

The gynoecium consists mostly of several structural units, the carpels. In the majority of taxa the carpels are more or less united from the beginning of development (congenitally fused or continuous) to form a unified construction (syncarpous gynoecium; probably more than 80% of

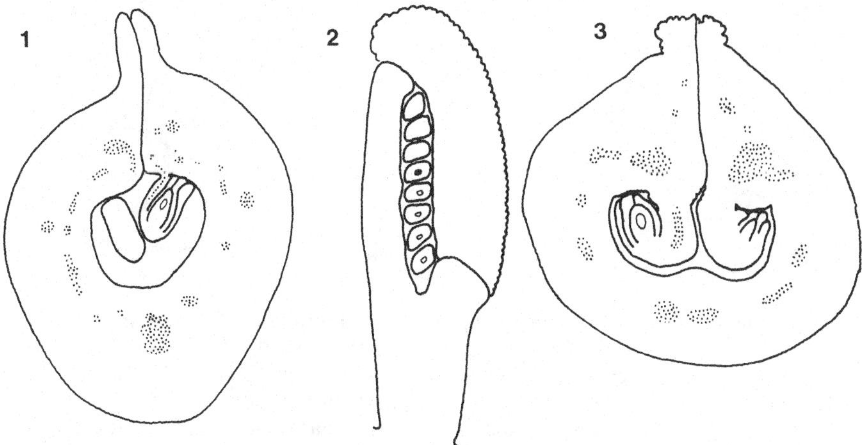

Figure 2.14. 1. *Degeneria vitiensis*. Transverse section of carpel before anthesis, vascular bundles dotted (× 16).
2–3. *Tasmannia piperita* (Winteraceae). 2. Longitudinal section of carpel at anthesis, ventral side at left, stigmatic crest marked with wavy outline (× 16).
3. Transverse section of carpel at anthesis, vascular bundles dotted (× 45).

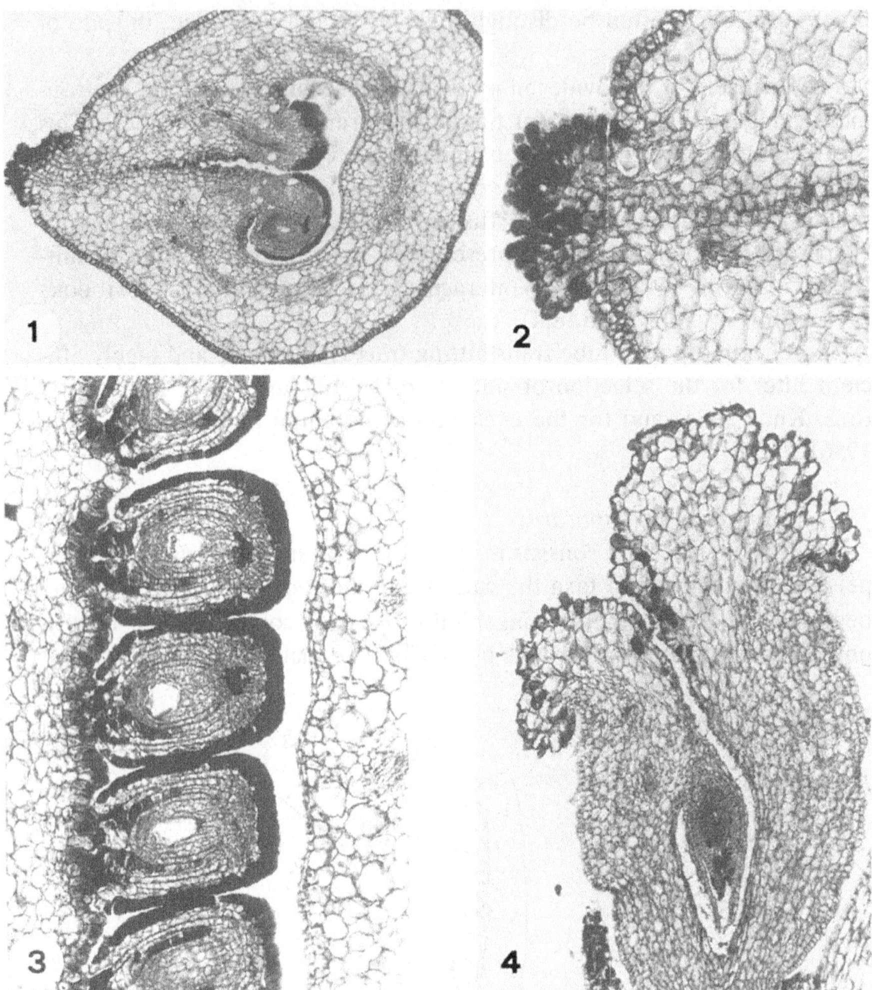

Figure 2.15. 1–3. *Tasmannia piperita* (Winteraceae). Carpels at anthesis. 1. Transverse section (× 50). 2. Transverse section, higher magnification of stigmatic crest and postgenitally united ventral slit (× 100).
3. Longitudinal section, the anatropous ovules with the micropyle appressed to the secretory (dark) surface of the placenta (× 120). 4. *Ascarina rubricaulis* (Chloranthaceae). Median longitudinal section of carpel at anthesis, inner surface not postgenitally fused but cavity filled with secretion; the single ovule pendent, orthotropous (× 60).

the angiosperm species). In some groups the carpels are free (apocarpous or choricarpous gynoecium; *ca.* 10% of the angiosperms), and in a few the gynoecium consists of a single carpel (unicarpellate gynoecium; *ca.* 10% of the angiosperms). The majority of unicarpellate species are in the large group of the legumes (Fabales; about 15 000 species). There are all transitions from the completely free to the completely united state of the carpels. Many morphologists use the term apocarpous only for the completely free state, whereas syncarpous is used for all transitional stages until complete fusion. It seems that in many monocotyledons in the seemingly syncarpous gynoecia the carpels are united only during development (postgenitally), a state of affairs that would affect the proportion of syncarpous and apocarpous angiosperm species. This has been investigated only in very few representatives (e.g. Hartl & Severin 1981, van Heel 1988).

Free carpels

It is appropriate to explain the characteristics of free carpels first (Fig. 2.13.1). Like an entire syncarpous gynoecium, a single carpel has a cup- or bottle-like shape. It has an outer and inner surface, and contains one to numerous ovules. The bottle-like shape can ontogenetically arise in different ways. Either it can grow equally on all sides or unequally in that the ventral side, i.e. the side towards the floral apex (the centre of the flower), is more or less inhibited. In the first case, the transition between inner and outer primary surface is restricted to the top of the carpel: the carpel is totally ascidiate (ascidiform, utriculate, tubular) and its wall is congenitally closed throughout (Fig. 2.16.4). In the second case, the transition between inner and outer primary surface has the form of a slit from the base to the top of the carpel on its ventral side, and the slit is postgenitally closed: the carpel is totally plicate (or conduplicate) (Fig. 2.16.2). Again, there are also all kinds of intermediate forms, where the lower part of the wall is congenitally but the upper part postgenitally closed. And, as in the case of the pair 'apocarpous/syncarpous', most morphologists use the term plicate (or epeltate or impeltate) only for carpels without any trace of congenital fusion on the ventral side, whereas all transitional forms up to completely ascidiate are termed peltate.

In fact, one or other extreme is rarely realized; categorization is therefore somewhat arbitrary. Some debates in the literature seem to be based simply on confusion due to different definitions but not on the facts as such. Some European authors since Čelakovský (1876) have been particularly impressed by the frequent occurrence of congenital closure of the carpel base. If a general homology between carpels and foliage leaves is accepted and both are seen as phyllomes, such a congenital basal closure in carpels is indeed striking, since it occurs much less frequently in foliage leaves. Therefore, there is a tendency in the European literature to call every carpel peltate that has only the slightest manifestation of peltation

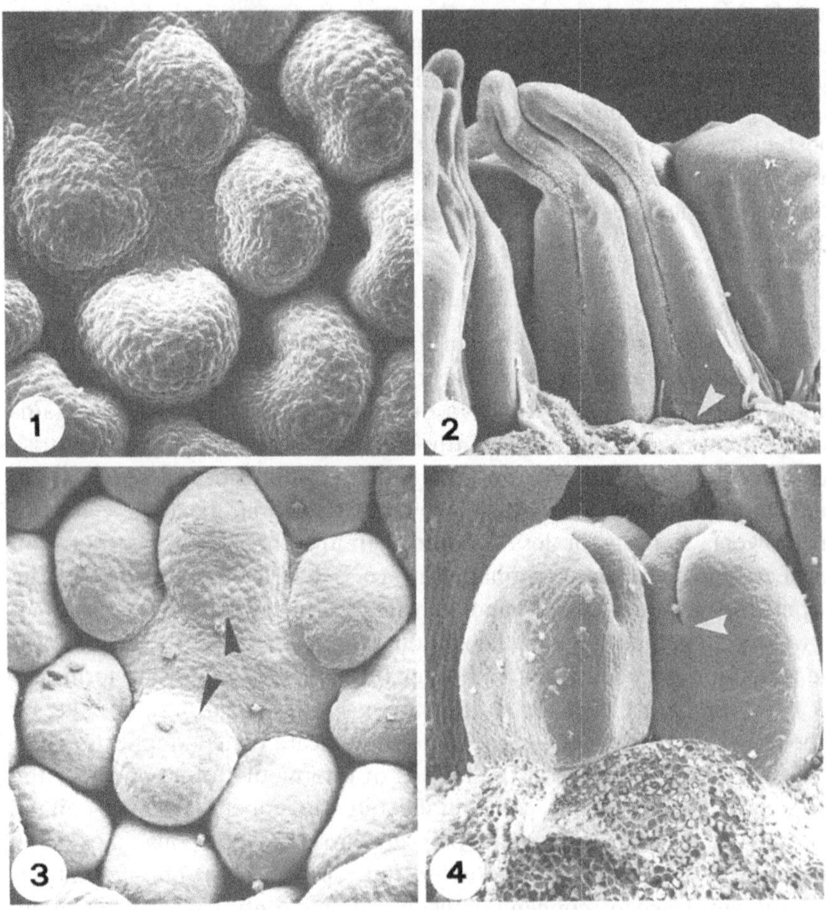

Figure 2.16. Development of epeltate and ascidiate carpels; white arrows in 2 and 4 point to lower end of ventral slit. 1–2. *Artabotrys hexapetalus* (Annonaceae). Epeltate carpels. 1. Very young carpels from above (× 220). 2. Older carpels from ventral side (× 45). 3–4. *Austrobaileya scandens*. Ascidiate carpels. 3. Very young carpels from above; black arrow points to cross zone (× 130). 4. Older carpels from ventral side (× 80).

(e.g. Baum 1952, 1953, Leinfellner 1950, Guédès 1966, 1979, Rohweder 1967, Endress 1972a,b, 1977a, van Heel 1983, 1984). In contrast, other authors, especially in America, concentrated more on the plicate part of the carpel and called carpels conduplicate when they also showed a peltate base (Bailey & Swamy 1951, Eames 1961, Swamy 1949, Swamy & Periasamy 1964, Takhtajan 1959, 1991, Cronquist 1988).

These different focuses also have evolutionary implications. Bailey & Swamy (1951) view the conduplicate carpel form as evolutionarily basal, while for Leinfellner (1969a) the peltate form is basal. It cannot be decided which of these two views is correct, as long as the evolutionary origin of the carpel is unclear (see section 9.3).

Syncarpous gynoecia

A syncarpous gynoecium is, accordingly, more complicated than a single carpel (or a unicarpellate gynoecium) (Figs 2.13.2, 2.20), although the characteristics of the structural units, the carpels, are similar. On the other hand, such a syncarpous gynoecium in some respect may also be superficially similar to a single free carpel. The primordia of the carpels, as the morphogenetic centres of the syncarpous gynoecium, may build carpels that are soon unified in a ring by a ring meristem. The carpellary wall may still enclose a chamber of its own (eusyncarpous gynoecium, with septate ovary) or the single carpels may, so to speak, remain open, so that a common single chamber for the whole gynoecium arises (paracarpous gynoecium, with non-septate ovary) (Troll 1928b). There are all kinds of intermediate forms.

Again, in the ovary base each carpellary chamber may be congenitally closed, as in a single carpel. Accordingly, this region is called synascidiate (Leinfellner 1950). Higher up, there is continuation (communication) between the chambers by the continuous primary morphological surfaces. Here, the carpellary flanks may be postgenitally closed or they may be reduced and not reach the centre so that only one chamber is formed (symplicate region). The free carpel tips are the asymplicate, or simply the plicate region. Superficially, a symplicate region with postgenital fusion and a synascidiate region may be similar. However, the presence of the aforementioned communication is biologically very important, as we will see later.

More rarely, free carpels are postgenitally united to form a gynoecium that is functionally syncarpous (some Gentianales, e.g. Padmanabhan *et al.* 1978, Fallen 1985; some Liliales, e.g. Hartl & Severin 1981, van Heel 1988).

Ovary and ovules

In the lower part of the gynoecium the internal surface usually lines a cavity system and contains the ovules that develop into seeds. This part of the gynoecium is the ovary, which develops (alone or together with other floral parts) into the fruit. The region where the ovules are inserted is the placenta. The most common form of placenta is a longitudinal rim near the margins of each carpel where the ovules sit in a row.

Each ovule consists of a nucellus and one or two integuments arising from the chalaza and covering the nucellus and forming a micropyle. The nucellus usually contains an embryo sac (female gametophyte) with an egg cell that will eventually be fertilized by a sperm cell of a male gametophyte. The male gametophyte is contained within a pollen grain (microspore). If deposited on the stigma, the pollen grain germinates and a pollen tube grows down through the style to the ovary into an ovule, entering it through the micropyle and the nucellus apex, where it delivers two sperm cells into the embryo sac, one of which fuses with the egg cell to give rise to the embryo, the other with the central cell (or polar nuclei) to give rise to the endosperm. This 'double fertilization' is highly characteristic of the angiosperms. However, it also occurs in their putative nearest relatives, the Gnetales (see Friedman 1990a,b, 1992).

There is increasing evidence that the two sperm cells in a pollen tube are not completely identical; there may be differences in organelle numbers. Only one is predetermined to fertilize the egg cell and is therefore the true male gamete. The other, the 'associated cell', is destined to fuse with the central cell to form the primary endosperm cell (Knox & Singh 1990).

The micropylar part of the integuments and the nucellar apex are often secretory and so provide growth medium and direction for the incoming pollen tube. The innermost integument layer adjoining the nucellus is often differentiated as a secretory 'integumentary epithelium' (mainly in ovules with a thin nucellus, where the epidermis is dissolved during embryo sac development). Between the end of the vascular bundle serving the ovule and the base of the embryo sac, a zone with lignified cell walls often develops: the hypostase. It is probably related to the translocation of nutrients into the embryo sac and endosperm (Tilton 1980b).

Stigma

The stigma receives the pollen grains and here it is decided whether or not they will germinate. It contains the receptor sites for pollen recognition (e.g. Dumas *et al.* 1984). In dry stigmas in particular (see section 2.4.4) this pollen–stigma interaction is expressed by a sequence of changes: pollen adhesion and pollen hydration. At the biochemical level, several enzyme activities and lectin-binding sites can be seen (Dumas *et al.* 1988).

Pollen tube transmitting tract and compitum

After pollen germination on the stigma the pollen tubes grow through the pollen tube transmitting tract, a specially differentiated tract between stigmatic surface and micropyle (Arber 1937). This tract develops on the inner morphological surface of the gynoecium, as shown for *Datura* (Solanaceae) by Satina (1944). No case is so far known of its having another position. It encompasses either only the surface region or – in addition – deeper cell layers below the epidermis. The pollen tubes grow along or near the epidermal surface in a secretion of the epidermis, or in the swollen cell walls of the epidermis, or also in deeper cell layers (Sassen 1974). This depends on the size of the lumen of the style and the number of ovules and pollen tubes required, respectively. Therefore, gynoecia with very numerous ovules, such as those of Orchidaceae, Rafflesiaceae and some Ericaceae, have a very extensive pollen tube transmitting tract.

In apocarpous carpels or in syncarpous gynoecia with extremely ascidiate carpels each carpel has its separate, individual pollen tube transmitting tract. In the majority of syncarpous gynoecia, however, there is a zone comprising a common transmitting tract, namely in the symplicate region, where the inner surfaces of all carpels communicate (Figs 2.13.2, 2.20). In this common zone, the compitum, pollen tubes can cross between carpels and a regular distribution of pollen tubes is possible (Carr & Carr 1961). Another functional aspect is that here a centralized pollen tube selection can occur among the entire pollen load deposited on the stigma at one time (Endress 1982); this is superior to pollen tube selection in each individual carpel. The significance of pollen tube selection in angiosperms in general has been extensively demonstrated by D. Mulcahy and collaborators (e.g. Mulcahy 1979) (see also sections 6.8 and 9.3).

Downwards, in the ovary, the compitum ends at least where the synascidiate region begins. The pollen tube transmitting tracts separate again for each carpel or placenta. On a pluriovulate placenta the upper ovules are served by the first pollen tubes, the lower ones later (Marubashi & Nakajima 1981, for *Nicotiana*, Solanaceae; Rocha & Stephenson 1991, for *Phaseolus*, Fabaceae). In the transition to the ovules the epidermis is often papillate. The micropyle usually faces this papillate surface, which is enhanced by the anatropous (or campylotropous) structure of most ovules (see section 2.4.4). The functional counterpart of the placental transmitting tissue on the ovule is the epidermis of the micropyle, which is apparently also secretory, mainly at the top of the integuments and the nucellar apex. This has, however, been investigated in only a limited number of species in some detail (Tilton 1980a, Knox 1984, Yan *et al.* 1991). The details of the further pathway of the pollen tube tip and the events in the release of the pollen tube contents into the embryo sac through one of the synergid cells are at the embryological–cytological level (Knox & Singh 1987) and are therefore beyond the scope of this book.

Direction of the pollen tubes in the pollen tube transmitting tract

After capture on the stigma, hydration and germination of the pollen, the emerging pollen tube grows in mucilaginous secretions from stigma to nucellus to bring the sperm cells into the embryo sac. After penetration into the nucellar apex the pollen tube grows intercellularly and finally enters the embryo sac (Jensen 1969, Wilms 1981). The pollen tube grows by tip extension, whereby cytoplasm is restricted to the tip and is separated from the older parts of the tube by several callose plugs. Pollen tubes are probably at first 'autotrophic' in that they use their own reserve substances (Vasil 1974), and later 'heterotrophic' in that they use nutrients of the mucilage of the transmitting tract (Kroh & Helsper 1974, G. & D. Mulcahy 1982). The part where pollen tube growth is autotrophic is called 'stigma depth' by Cruden & Lyon (1985). They found a positive correlation between stigma depth and pollen grain size (implying the amount of reserve substances). Secretion in the transmitting tract starts long before anthesis (14 days in *Trimezia*, Iridaceae, Bystedt & Venniger-holz 1991a). What defines the route and direction of the pollen tubes is a largely unresolved question, owing to the limited accessibility of the transmitting tract enclosed within the gynoecium. There are different structural levels that should be considered.

(1) Anatomical/histological level: The transmitting tract shows files of elongated cells. Toward the stigmatic surface they often diverge in the manner of a fountain (Fig. 2.20.4). The direction of the walls of these elongated cells seems to define the pathway of pollen tubes in stigma and style (e.g. Capus 1879, J. & Y. Heslop-Harrison 1986). A particularly illustrative example is *Dalechampia* (Euphorbiaceae). While other Euphorbiaceae typically have three simple or divided stigmatic lobes (Figs 2.17.3, 2.26), they are absent in *Dalechampia*. Instead, long strips of stigmatic tissue extend from the gynoecial apex downwards along the stylar surface (Webster & Armbruster 1991) (Fig. 2.17.1–2). The cell files below this stigmatic surface are directed obliquely upwards, so that apparently those pollen tubes coming from the (topographically) lower parts of the stigmatic strips first have to grow upwards before they reach the central

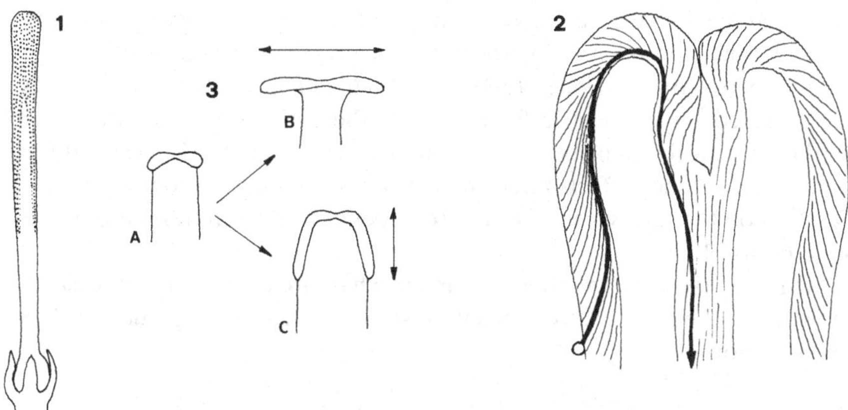

Figure 2.17. *Dalechampia* (Euphorbiaceae). Stigma organization. 1. *D. spathulata*. Female flower at anthesis, from the side. Style with three widely descending stigmatic strips (stippled) (× 6). 2. *D. triphylla*. Median longitudinal section of gynoecium apex with peripheral ascending cell rows of pollen tube transmitting tissue that converge and descend in the centre; black line with arrow indicating presumed pathway of a pollen tube (× 45). 3. Derivation of the stigma form of *Dalechampia* (C) from that of other Euphorbiaceae (B) by different direction of growth (double arrows) of a similar early ontogenetic stage (A).

stylar transmitting tract. The diameter of the pollen tube transmitting tract may decrease towards the ovary concomitant with a decrease in pollen tube number (Cruzan 1986) (Fig. 2.20).

(2) Thigmotropism: Iwanami (1953) has suggested that pollen tube guidance may be thigmotropic by characteristics of cell surfaces in the pollen tube transmitting tract. This was endorsed by Hirouchi & Suda (1975) by using minute network filaments as growth substrates for pollen tubes.

(3) Chemotropism: Chemotropic factors have been proposed that direct the pathway of the pollen tubes (reviewed by J. & Y. Heslop-Harrison 1986, Dumas & Russell 1992). This kind of mechanism can mainly be expected in the ovary where the tubes grow towards the ovules without anatomical guides (Hepher & Boulter 1987); the tubes often make bends before entering a micropyle (Webb & Williams 1988). The micropylar region of the integuments and the ovular apex are secretory regions, often with enlarged cells and nuclei. Tilton (1980a) and Tilton & Lersten (1981) speculate that the thin sheet of exudate covering the micropyle at the time when the synergids differentiate in certain Liliaceae may function to

increase the surface area on which the chemotropic agents synthesized by the synergids and released by the ovular apex are exposed.

(4) Electrotropism: Pollen tubes of *Camellia* (Theaceae) and other taxa are directed by an electric field towards the cathode (Nakamura *et al.* 1991). Electrical signals between stigma and ovules might also account for rapid long-distance effects after pollination by the intervention of plasmodesmata on the transverse walls of the pollen tube transmitting cells in the style (Knox 1984).

Which of these factors operate under what circumstances is not clear. However, there are two recent results, which may guide further investigations.

(1) Pollination-induced polarity.

Pollen inserted in the mid-style of *Nicotiana* grows equally well towards the stigma and towards the ovary. However, if insertion is preceded by a stigmatic pollination, the pollen tubes mainly grow toward the ovary (G. & D. Mulcahy 1987, 1988). To what kind of tropism this apparently long-distance effect has to be related is unclear. Another long-distance effect immediately following pollination is induction of a growth stimulus for pollen tubes by the ovary (G. & D. Mulcahy 1986) and the degeneration of a synergid in the ovules (as reported by Jensen & Fisher 1968 for *Gossypium*, Malvaceae).

(2) Mechanical action of the intercellular matrix in the pollen tube transmitting tissue.

The intercellular matrix may actively transport the tip of the pollen tubes towards the ovary. Sanders & Lord (1989, 1992) showed that, in species of three families studied, latex beads introduced onto the transmitting tracts of the styles were actively translocated to the ovary at a speed similar to that for pollen tubes. The migration of the coloured beads could be directly observed through the semitransparent styles.

2.4.2 Development

Gynoecium initiation mostly encompasses the entire floral apex. Often the subepidermal cell layer disappears by frequent periclinal cell divisions as the gynoecium primordium appears (Rohweder 1963, for Commelinaceae) and the meristem becomes deeper. In cases where a larger number of carpels is initiated, the most central region of the apex may not be incorporated in the gynoecium primordium (e.g. *Kitaibelia*, Malvaceae). In addition, in some unicarpellate gynoecia a small unused remainder of the apex may still be present on the ventral side of the carpel (e.g. *Trimenia*, Trimeniaceae, Endress & Sampson 1983). However, the entire apex is often used up by unicarpellate gynoecia (e.g. *Pseudowintera*, Winteraceae, Sampson & Kaplan 1970; *Cinnamomum*, Lauraceae, Endress 1972a; *Acacia*, Mimosaceae, Derstine & Tucker 1991).

A special situation is also present when in an apocarpous (or syncarpus) gynoecium each carpel is surrounded by a region that has not been incorporated into the formation of the neighbouring carpels (*Tambourissa*, Monimiaceae; *Nelumbo*).

Free carpels

It is characteristic of many free carpels that the more or less hemispherical primordium develops into a chair-like stage (periplastic development) (Endress 1972a,b, Erbar 1983, van Heel 1981, 1983, 1984) (Figs. 2.16.3, 2.21). The ventral edge may be viewed as corresponding to the cross-zone in a peltate leaf, i.e. the region where the margin crosses over the median plane of the organ. In the most simple case there is a single ovule in a median position (most Laurales). It is initiated at the cross-zone and soon becomes overtopped by a secondary cross-zone, which prepares the subsequent closure of the carpel (Fig. 2.18). By intercalary growth of the carpel wall the ovule sinks deeper and deeper into the now developing ovarial cavity.

The carpels of the apocarpous *Siparuna* (Monimiaceae) go through several cycles of meristem expansion and subsequent morphological differentiation of carpellary parts (Endress 1972a,b). After each step of morphological differentiation the part that will undergo another step of morphological differentiation remains meristematic and again builds up a deeper meristem, while the part that does not further differentiate loses its meristematic character (Fig. 2.18.2). This has not been studied in other groups but may also be expected there.

Syncarpous gynoecia

Gynoecium initiation and early ontogeny is often similar in larger groups. However, it may also be quite different despite its later similarity within the same family. An example is Solanaceae; in *Schizanthus pinnatus* the two carpels appear totally ascidiate (synascidiate) at inception in that the common septum primordium arises at the same time as the dorsal regions, while in *Atropa belladonna* septum initiation is much later so that the carpels appear plicate (symplicate) at the beginning (Huber 1980). In later stages both *Schizanthus* and *Atropa* have an ovary with a synascidiate lower and a symplicate upper zone. In many other taxa the early development is intermediate between these two extremes (Fig. 2.19). In syncarpous Rhizophoraceae the carpels may appear separately or as a common annular primordium (Juncosa 1988).

Each carpel primordium, be it free or united with other carpels, initiates a primary longitudinal procambium strand in its centre, which eventually differentiates into the dorsal bundle. Lateral bundles are usually initiated somewhat later as the carpel flanks develop, and the ovular bundles appear following the ovular primordia. Further longitudinal bundles and oblique

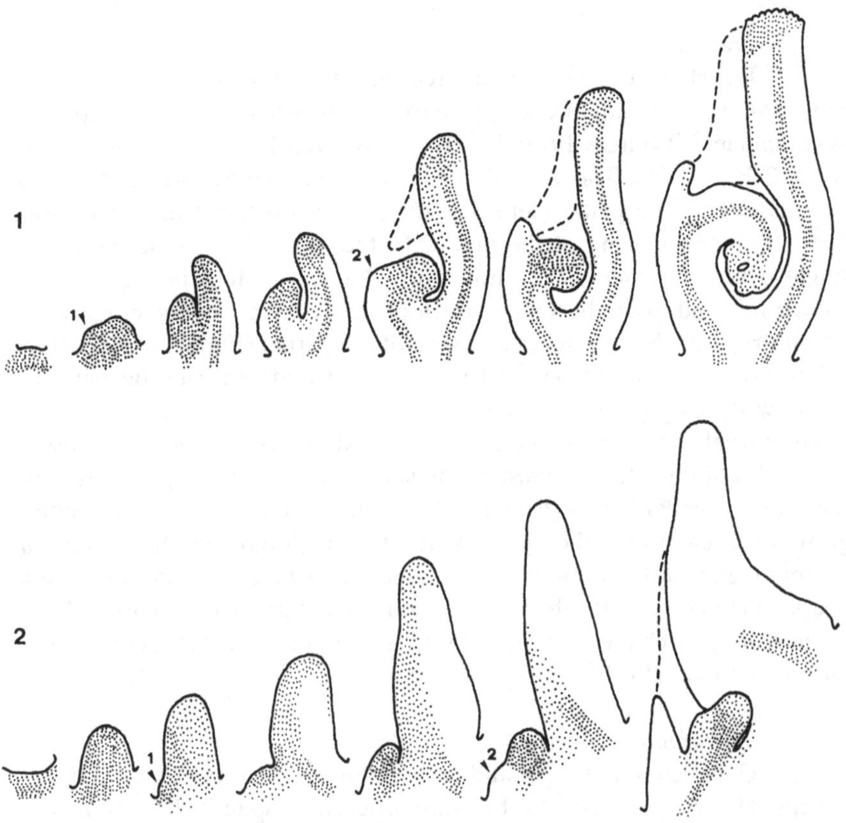

Figure 2.18. Carpel development as seen in median longitudinal sections (ventral side at left). Morphogenetic and histogenetic differentiation with rhythmical increase and decrease of meristematic areas (dotted). The primary cross-zone (arrow 1) gives rise to the ovule. The secondary cross-zone (arrow 2) contributes to carpel closure behind the ovule (× 80) (modified after Endress 1972b). 1. *Cinnamomum camphora* (Lauraceae). 2. *Siparuna andina* (Monimiaceae).

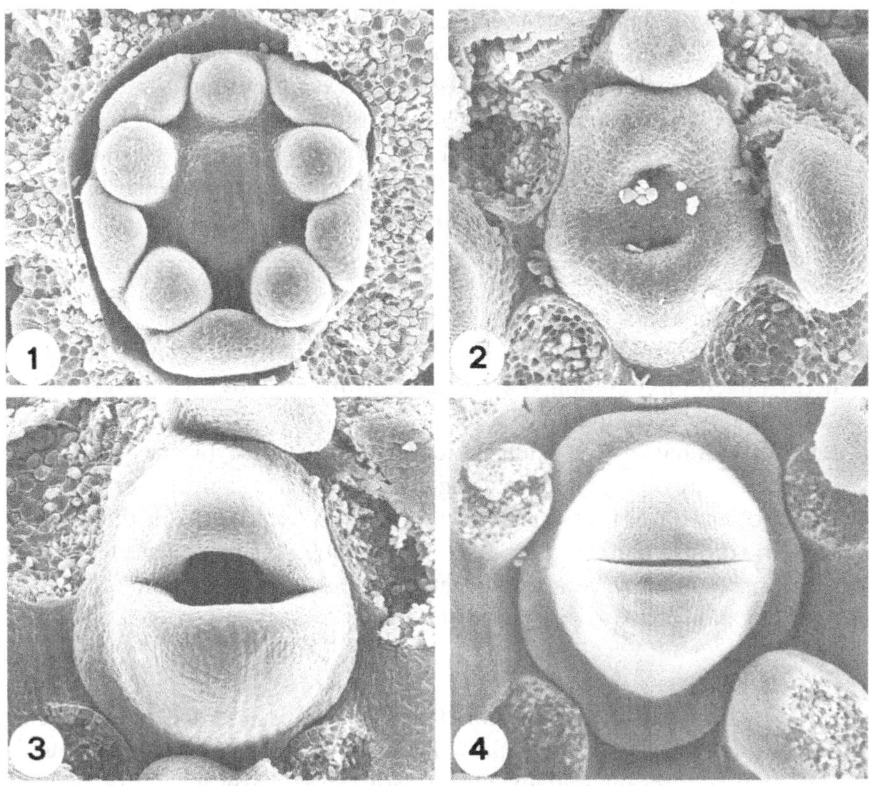

Figure 2.19. *Podranea ricasoliana* (Bignoniaceae). Development of bicarpellate, syncarpous gynoecium. 1. Dome-shaped gynoecium primordium in the centre of the young flower (× 150). 2. Young gynoecium, still open, with the common cross-zone of the two carpels in the centre, which gives rise to the synascidiate basal part (× 200). 3. Older stage, still open (× 180). 4. Gynoecium closed (× 100).

or transverse bundles connecting the main bundles may develop during later development.

In some syncarpous gynoecia where the carpels appear united in a ring-like primordium from the beginning (without protruding median parts) the dorsal bundles are not well developed or are even absent, in contrast to the massive lateral (placental) bundles (e.g. Papaveraceae, Capparaceae, Karrer 1991).

Integrative studies on the development of gynoecia from the primordium up to anthesis are rare. Crone & Lord (1991) found in *Lilium longi-*

florum that gynoecium growth after early development showed three distinct phases: (1) a phase of diffuse growth (between 0.2 and 1 cm length); (2) a phase of predominantly basal stylar growth (between 1 and 10 cm length); and (3) a phase of predominantly terminal stylar growth (between 10 and 13.5 cm length). In the second half of phase (2) cell divisions cease and growth is only by cell expansion. This kind of growth is similar to that in tepals and stamens (Gould & Lord 1988, 1989).

Placentae

The sequence of inception of ovules along a placenta may be acropetal, basipetal or bidirectional. This seems to be correlated with parietal, axile or mixed placentation (Payer 1857, Okamoto 1984). However, this deserves a critical comparative study.

In both laminar-diffuse and protruding-diffuse placentae (see section 2.4.4) there may be a gradient during development so that the peripheral ovules are less developed than the central ones (Fig. 2.25). In the long, protruding-diffuse parietal placentae of *Downingia* (Campanulaceae) new ovular primordia may be inserted between ones already initiated during intercalary elongation of the ovary (Kaplan 1968).

Ovules

In bitegmic ovules the inner integument is initiated before the outer one (Fig. 2.25.3). In ovules with a single integument two integuments may be present initially but growth ceases early in one of them so that it may be obliterated in later stages (e.g. *Peperomia* vs. *Piper*). Alternatively, their common base may extend by intercalary growth, while the free parts become obliterated (e.g. Anacardiaceae, Balsaminaceae) (Bouman & Calis 1977, Bouman 1984, Boesewinkel & Bouman 1991). However, in ovules with a single integument, development usually starts with a single primordium. In ovules with two integuments, the inner integument is often thinner than the outer and develops from the epidermis (dermatogen) alone (Bouman 1984).

2.4.3 *Lycopersicon* (Solanaceae): an example

Structure and function of the gynoecium has been studied in detail in only a few taxa. Among them are various Solanaceae, such as *Nicotiana*, *Petunia* and especially the tomatoes (*Lycopersicon peruvianum* and *L. esculentum*).

As in most Solanaceae the gynoecium contains two carpels that are fused up to the top. The ovary is septate and each cavity contains a broad, protruding placenta with numerous ovules. The gynoecium is synascidiate up to the top of the placental region and symplicate in the uppermost part of the ovary and the style (Fig. 2.20).

The vascular system consists of a median dorsal bundle for each carpel, which extends up to the stigma, and two main placental bundles in transverse position. During development a dense net of secondary bundles develops between the two dorsals and each ovule is connected with the main placental bundles.

Differentiation of the pollen tube transmitting tract was studied by Webb & Williams (1988) for *L. peruvianum*. The entire transmitting tissue is densely cytoplasmic. The capitate stigma consists of unicellular papillae. The stigma is continuous with vertical files of transmitting tissue in the style by the intervention of shorter cells with large intercellular spaces (Dumas *et al.* 1978). These shorter cells, together with the epidermal papillate cells, appear to constitute the main stigmatic secretory tissue, which secretes a hydrophobic exudate. The inner morphological surface of the gynoecium is obliterated in the style during ontogeny. The former inner epidermis and several subepidermal cell layers differentiate as transmitting tissue, which occupies about half the diameter of the style. This region is distinguished by swelling of the cell walls (middle lamellae), whereby vertical files of long, narrow cells are dissociated, which are linked by their end walls but not attached laterally. They are then surrounded by a mucilaginous substance. In the transition region to the ovary the transmitting tissue retracts to the centre of the septum (Fig. 2.20). As soon as the synascidiate region begins it subdivides into two tracts, one for each carpel. Each tract spreads over the entire placental surface, occupying one to three cell layers, and ceases at the lower end of the placenta. The epidermal cells are roundish or slightly papillate. The anatropous, unitegmic, tenuinucellar ovules direct their micropyle towards the placenta. A mucilage is present on the placental surface and between the ovules, which is continuous with the intercellular mucilage in the stylar region and the stigmatic exudate. All these secretions stain positively with ß-Yariv reagent indicating arabinogalactans and arabinogalactan-proteins (Webb & Williams 1988).

In *Lycopersicon peruvianum* the stigma is receptive for 4 to 5 days. Pollen grains on a receptive stigma hydrate within 15 min and germinate after 3 h 30 min in the stigmatic exudate (Pacini & Sarfatti 1978), and pollen tubes grow between the papillae and in the swollen cell walls of the style. In the top of the ovary, which they reach in less than 24 hours, the pollen tubes are distributed on the two placentae. There they grow between the ovules. They may leave the placental surface but remain within the mucilage that lines placenta and ovules. The micropyle of each

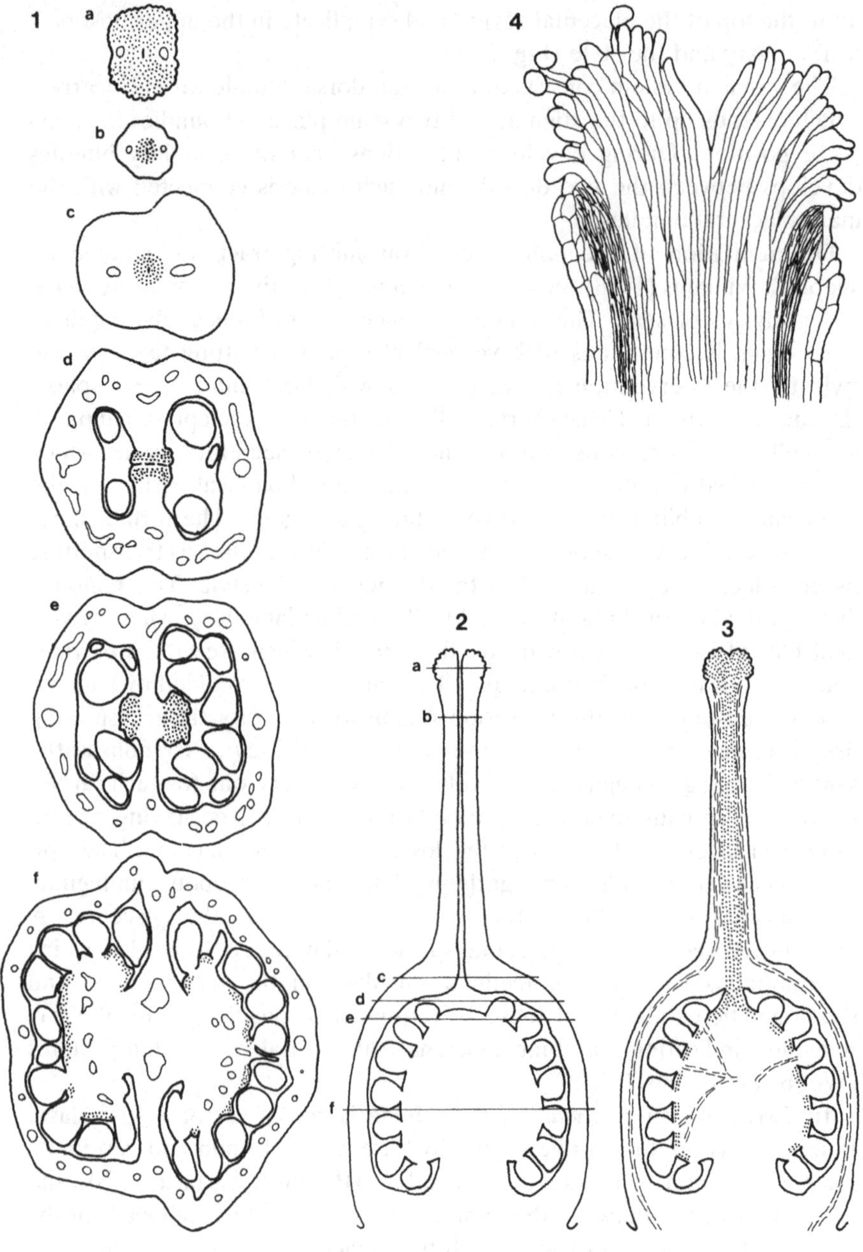

Figure 2.20. *Lycopersicon esculentum* (Solanaceae). Bicarpellate, syncarpous gynoecium at anthesis. 1. Series of transverse sections. Stippling, pollen tube transmitting tissue;

ovule is entered by a single pollen tube. Pollen tubes may make some turns near the micropyle before entry. Thus, in the entire pollen tube transmitting tract, mucilage is the growing medium of the pollen tubes (Webb & Williams 1988).

However, the stigmatic secretion is rich in lipids and is hydrophobic (Pacini & Sarfatti 1978, Webb & Williams 1988), while the stylar mucilage, derived from the intercellular matrix and containing high levels of arabinogalactans and arabinogalactan-proteins, is hydrophilic (Cresti *et al.* 1976). The stigmatic secretion is presumed to play a role in pollen germination and protection of the stigmatic surface from desiccation or wetting (Shivanna 1982), while the intercellular matrix in the style may provide physical and/or nutritional support for pollen tubes (Hoggart & Clarke 1984).

While *Lycopersicon esculentum* is self-compatible, *L. peruvianum* has a gametophytic self-incompatibility system (see section 6.5). Incompatible pollen tubes are arrested in the style. This is controlled by multiple alleles of a single S-gene locus (Tanksley & Loaiza-Figueroa 1985). Mau *et al.* (1986) studied molecular biological aspects of self-incompatibility in this species. Evolutionary aspects were studied by Schemske & Lande (1987).

2.4.4 Diversity

Carpel number and arrangement

The range of diversity is especially large in the gynoecium, owing to its complicated structure. Apocarpous gynoecia have the largest range in carpel number, from one (many Magnoliidae, Proteaceae, Fabales) to almost 2000 (*Tambourissa ficus*, Monimiaceae) (Lorence 1985). In syncarpous gynoecia the carpel number is more restricted because there are architectural constraints on the construction of a functional compitum. Most syncarpous gynoecia have between two and five carpels, arranged in a whorl. If they are numerous, the whorl becomes deformed, flattened (as seen in a vertical projection) in such a way that part of the carpels come to lie in parallel rows (*Actinidia*, Actinidiaceae; *Tupidanthus*, Araliaceae;

Fig. 2.20 (*cont.*)
interrupted lines, postgenitally fused surfaces; thin lines, vascular bundles. a–d. Symplicate part of stigma, style and ovary. e–f. Synascidiate part of ovary (× 45). 2–3. Median longitudinal section (the style is longer in reality) (× 30). 2. Schematic diagram of the inner morphological surface, levels a–f corresponding to sections in 1. 3. Same but with pollen tube transmitting tract and vascular bundles. 4. Median longitudinal section of stigma with germinating pollen tubes and converging cell rows of pollen tube transmitting tract (× 160).

Kitaibelia, Malvaceae). A still greater variation is in ovule number per flower. It varies between one (e.g. Lauraceae, Poaceae, Asteraceae) and several million (!) (some Orchidaceae, Walter 1983).

Carpel structure

Carpels of apocarpous gynoecia show the entire range from epeltate to completely ascidiate. Sometimes the form is relatively constant in a larger group (e.g. epeltate in most Annonaceae, about halfway ascidiate in Lauraceae, extremely ascidiate in Potamogetonaceae, Nymphaeaceae). In other families a wide range can be found from epeltate or weakly ascidiate to strongly ascidiate (e.g. Ranunculaceae, Winteraceae) (Fig. 2.21). This makes one hesitant to assign a fundamental evolutionary difference to these two extremes. Since the proportions of the different regions may also change drastically during the development of a carpel, new attempts

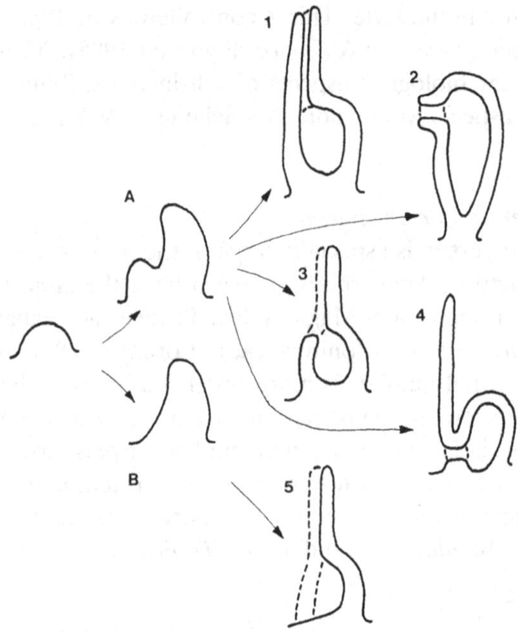

Figure 2.21. Diversity of carpel development as seen in median longitudinal sections. A. Peltate (ascidiate). B. Epeltate. 1. Completely ascidiate. 2. Strongly ascidiate with dorsal bulge. 3. Half ascidiate. 4. Weakly peltate with dorsal bulge. 5. Epeltate. Examples: Austrobaileyaceae (1), Winteraceae (2,3), Monimiaceae (3,4), Annonaceae (3,5).

towards a classification of carpel forms for phylogenetic purposes are desirable (Taylor 1991, Cresens & Smets 1992). Such a reconsideration will be useful, if it is based on broad comparative research on a wide range of angiosperms.

Carpels of syncarpous gynoecia, in contrast, are only exceptionally completely ascidiate. Mostly they have a long symplicate region, while the synascidiate zone may be considerable to lacking. The border between the synascidiate and symplicate region is almost never in the style but usually somewhere in the ovary. This can again be seen as a constraint exerted by the presence of a compitum. Owing to its ontogeny, a compitum is always correlated with a symplicate region and would not be easily possible in a totally synascidiate construction.

In completely syncarpous gynoecia, i.e. those with a single style, the symplicate region (and the compitum) take up the entire style and at least the uppermost part of the ovary (e.g. Asteridae, Pittosporaceae, Escalloniaceae, Meliaceae, Ochnaceae, Primulales, Ericales, Ebenales). In gynoecia with free stylar parts the symplicate region comprises at least the uppermost part of the ovary (e.g. Hamamelidaceae, Saxifragaceae, Geraniales, Caryophyllidae, Dilleniaceae). In paracarpous gynoecia the compitum is situated only above the ovary, although the entire gynoecium is symplicate, since the placentae lie far apart from each other (e.g. Canellaceae, many Violales, Pittosporaceae, some Saxifragaceae, Gesneriaceae, Orchidaceae, Rafflesiaceae). Intermediate forms between paracarpous and eusyncarpous gynoecia occur, where only the lower part of the ovary is septate, or – still more – where the placental ridges are so near each other that the gynoecium looks almost septate at first sight (e.g. Cucurbitaceae).

Each carpel may be bulged out or bulged up dorsally in the ovary region at anthesis. In the first case a syncarpous ovary may superficially look apocarpous. A nice example is *Ochna* (Baum 1951), where the carpels are completely fused up to the stigmatic region and a compitum is present in the style (carpels structurally and functionally syncarpous). However, in the fruiting stage the ovary of each carpel forms a separate black diaspore containing a single seed (functionally apocarpous). In the second case apical septa (Hartl 1962) are formed, which may, in some respects, be seen as the apical counterparts of the basal septa in the synascidiate part of the ovary. However, although this system of apical septa of a gynoecium is congenitally uniform, the centre is occupied by the basally extended stylar canal, which opens into the ovary at the lower end of the septal system. Apical septa occur in many angiosperm groups.

Ovary position

The ovary is superior or inferior. In a flower with a superior ovary the outer floral organs, stamens, petals and sepals are at the base of the ovary; in a flower with an inferior ovary the outer floral organs are on top of the ovary (Fig. 2.22). It looks as if the ovary had congenitally sunken into the flower base. Between the two extreme forms of superior and inferior ovaries all kinds of intermediates also exist, which may be called, for example, semi-inferior or 3/4-inferior. In some larger groups the ovary is constantly superior (e.g. Fabales, Apocynaceae, Acanthaceae) or inferior (e.g. Rubiaceae, Araliaceae, Zingiberaceae, Orchidaceae), while other ones exhibit variability (e.g. Hamamelidaceae, Saxifragaceae, Asclepiadaceae).

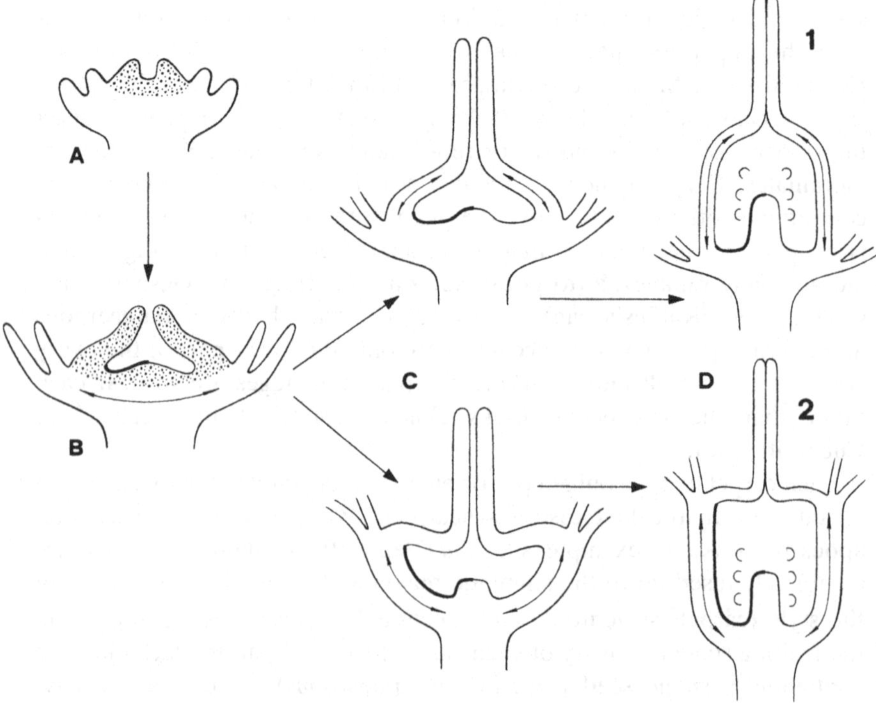

Figure 2.22. Development of superior and inferior ovary by differential expansion (double arrows) as seen in median longitudinal sections. Meristematic tissue is stippled in A and B. The thick black line on the left side of the ovarial cavity in B–D marks corresponding parts. 1. Superior. 2. Inferior. A. Carpel primordia. B. Later stage, common origin of 1 and 2. C. Beginning differentiation. D. Mature stage.

There has been a long and somewhat fruitless debate over whether inferior ovaries were fused with the floral axis or with the bases of the outer floral organs. It is convenient to say that the ovary is congenitally fused with the floral base. However, one has to bear in mind that the source of the confusion is the phased development of the system: (1) the floral organs are initiated (primary morphogenetic phase); (2) the organs attain their shapes by differential growth (secondary morphogenetic phase); (3) the organs differentiate their specific tissues (histogenetic phase). Depending on whether one focuses only on phase (2) or only on phase (3), the interpretation of the inferior ovary may be quite different. Phase (2) shows that the inferior position arises by pronounced dilatation of the ovary base but little dilatation of the ovary top (Leins 1972a). Thereby the outer base comes to lie near the top. Hence, the later outer top of the ovary corresponds to its former outer base (Fig. 2.22). In phase (3) the tissues differentiate. They may become similar irrespective of the morphological nature of the apparent ovary wall now present, be it the ovary wall alone, as in a superior ovary, or a combination of ovary wall and floral base, as in an inferior ovary. The later histological differentiation of certain parts of the floral base may be similar to parts of sepals, petals, stamens or the floral pedicel, irrespective of the earlier position of these parts during morphogenesis. This is not surprising for a region where all these organs meet. Since these two phases of development are superimposed but the result of each of them as seen in isolation is so different, different interpretations of the result are possible.

Further, in phase (1) the organs do not originate just at the surface of the floral apex but deeper layers are also incorporated. Therefore, the organs may be seen as having a common base from the beginning, which becomes involved in the intercalary elongation. Therefore, the notion that the inferior ovary is surrounded by the congenitally fused outer organs is reasonable (Kaplan 1967).

The same applies to the interpretation of a floral cup, which is the common base of all outer floral organs. An ovary may be superior and be surrounded by a floral cup, or it may be inferior and, in addition, be surrounded by a floral cup.

Placentation

In syncarpous gynoecia there are three main forms of placentation (Fig. 2.23). (1) Eusyncarpous ovaries have axile placentation: the ovules are in the angle where the flanks of a carpel meet in the centre of the ovary. This is the most common form of syncarpous gynoecia (e.g. many Dilleniales, Malvales, Euphorbiales, Liliales). (2) Paracarpous ovaries have parietal placentation: the ovules are on the ovary wall where two carpels meet (e.g. many Violales). (3) Another kind of construction is the syncarpous gynoecium with a free central placenta (e.g. Primulales, Lentibulariaceae). Here, there is no septation and the ovules are situated on a massive, often almost spherical, placental column that originates at the base of the ovary.

Apical placentation (as in Combretaceae) may be seen as an extreme kind of parietal placentation, whereby the ovules are restricted to the top of the ovary. Basal placentation (as in Amaranthaceae) may be linked with all three main forms, depending on the specific evolutionary context.

In free carpels with several ovules and in eusyncarpous gynoecia there is often one longitudinal series of ovules (linear placenta) on each carpel flank (on each side of the ventral slit) near the carpel margin (marginal or submarginal placenta) (Fig. 2.23). Both series are often connected at the lower end by a median bridge (U-shaped placenta, Leinfellner 1951) (Fig. 2.24) and very rarely also at the upper end (O-shaped placenta, Leinfellner 1969a, as in some Winteraceae).

In both free carpels and in syncarpous gynoecia a placenta may contain more than one longitudinal series of ovules (diffuse as opposed to linear placenta) (Fig. 2.23). If it is situated on the extended carpel flanks, it is

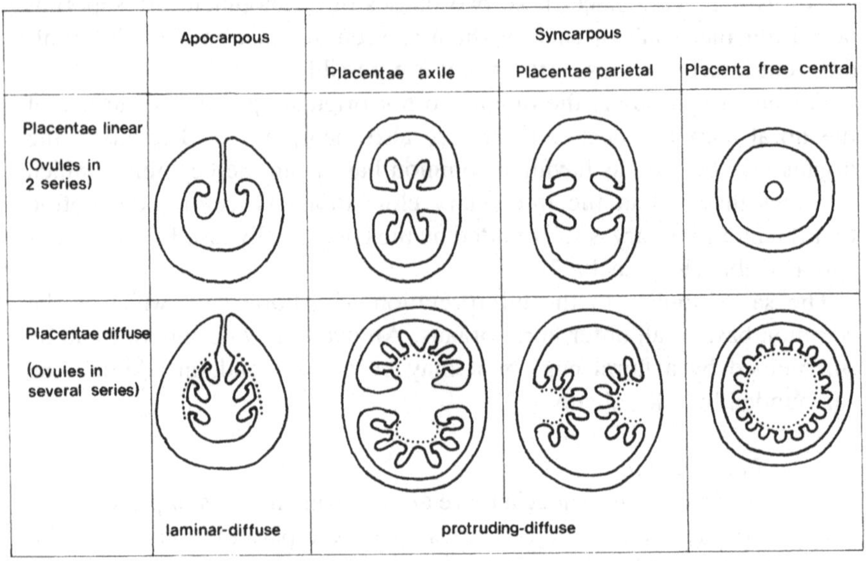

	Apocarpous	Syncarpous		
		Placentae axile	Placentae parietal	Placenta free central
Placentae linear (Ovules in 2 series)				
Placentae diffuse (Ovules in several series)				
	laminar-diffuse	protruding-diffuse		

Figure 2.23. Major forms of placentae as seen in transverse sections of ovaries.

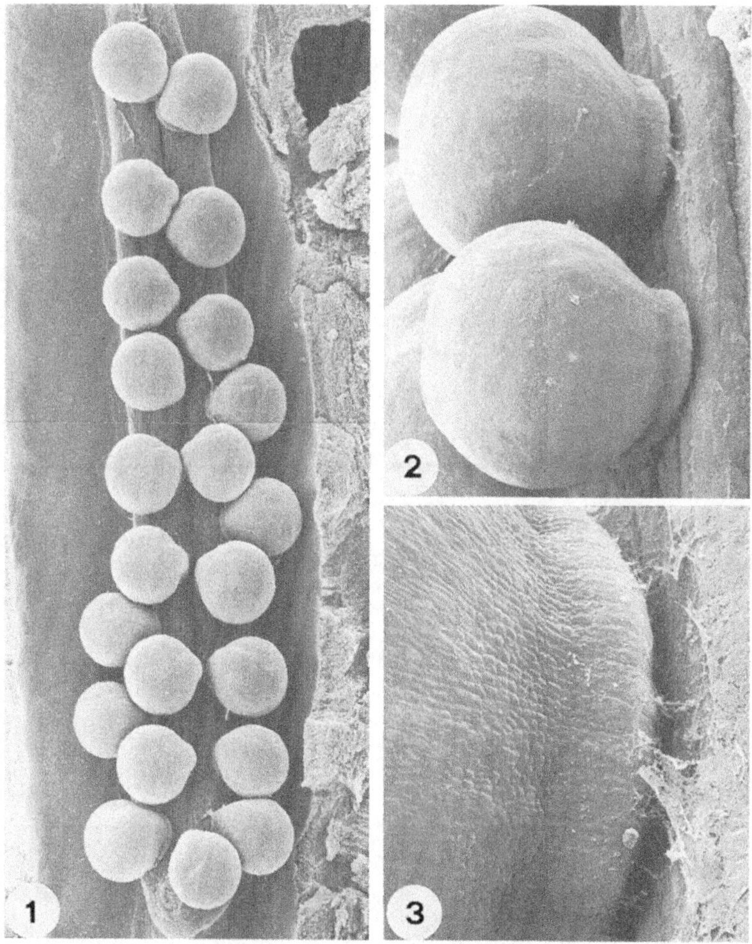

Figure 2.24. *Strelitzia reginae*. 1. Linear U-shaped placenta at anthesis (× 13). 2. Two ovules with micropyle appressed to placenta (× 45). 3. Contact between micropyle and placenta by secretion (× 150).

called a laminar-diffuse placenta (e.g. some Nymphaeaceae, Lardizabalaceae, Butomaceae, Hydrocharitaceae, Cyclanthaceae). In other groups the extended field of ovules is at the juncture of two carpel margins (either of the same carpel or of two adjoining carpels) on a thickened, protruding-diffuse placenta (e.g. some Passifloraceae, Saxifragaceae, Gentianaceae, Solanaceae, Gesneriaceae, Scrophulariaceae, see Fig. 2.25).

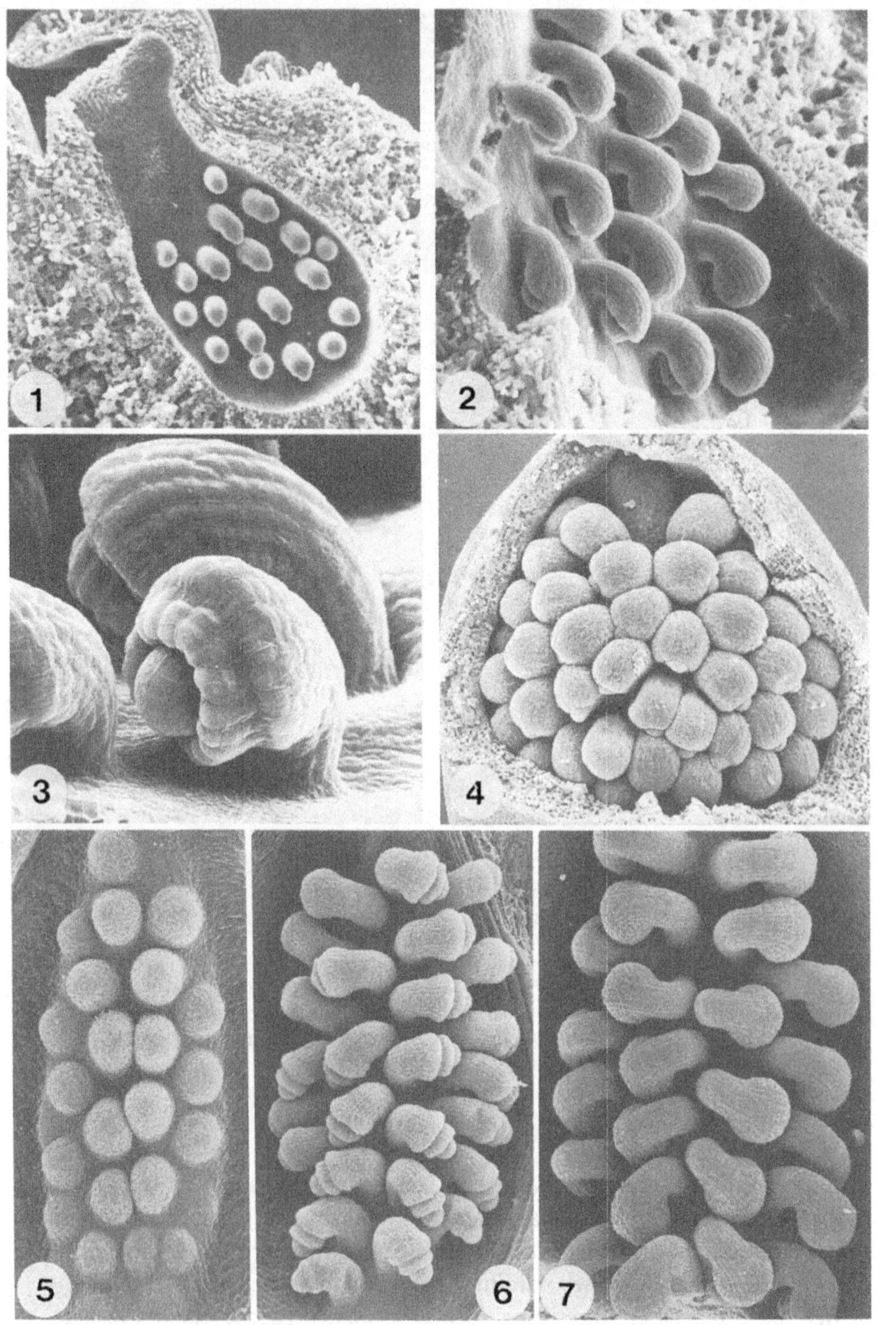
1
2
3
4
5
6
7

Stigma

Stigma differentiation is diverse among the angiosperms. At the morphological level, the stigma may attain many shapes. It is compact or subdivided into lobes. The lobes may be the free distal parts of the carpels in a syncarpous gynoecium or further subdivided parts of them. An example is Euphorbiaceae with generally three carpels and variously subdivided stigmatic lobes (Fig. 2.26).

At the histological level, a major distinction is between 'wet' and 'dry' stigmas, since they are also associated with different compatibility control systems (J. Heslop-Harrison 1976b) (see section 6.5). Wet stigmas have an abundant, free-flowing surface secretion; in dry stigmas the secretion forms a thin superficial pellicle. During stigma differentiation several waves of secretion may occur (Sethi & Jensen 1981, for *Gossypium*, Malvaceae); the secretion may be spontaneous or induced by pollination (J. and Y. Heslop-Harrison 1985). In addition, particles other than pollen may induce stigmatic secretion (Sedgley & Blesing 1982, for *Citrullus*, Cucurbitaceae). Wet stigma secretions contain lipids, lipoproteins and viscous polysaccharide mucilages, the pellicles of dry stigmas are proteinaceous and lying over a discontinuous cuticle. Dry stigmas may be selective in pollen retention, while this is not the case in wet stigmas (J. & Y. Heslop-Harrison 1985).

The site of stigmatic secretion may be the epidermis alone (e.g. in dry stigmas) or additionally subepidermal tissue (whereby the secretion is released through intercellular spaces; e.g. *Lycopersicon*, Solanaceae; Dumas *et al.* 1978). Whether it may also be secreted in deeper zones of a transmitting tract, if a hollow style is present, remains doubtful (Rosen 1971), especially since there is mostly at least a short zone of morphological closure of the canal just below the stigma. The composition of the secretions is different in the style and on the stigma (J. & Y. Heslop-

Figure 2.25. Forms of diffuse placentae. 1–3. *Nymphaea tetragona*. Laminar-diffuse placenta. 1. Median longitudinal section of carpel with young ovules curved downwards, peripheral ovules retarded (× 50). 2. Slightly older stage, from the side (× 60). 3. Young ovule with outer integument just initiated (× 270). 4. *Lycopersicon esculentum* (Solanaceae). Protruding-diffuse placenta. Ovules curved in diverging directions (× 150). 5–7. *Passiflora holosericea*. Protruding-diffuse placenta. 5. Young ovules, peripheral ovules retarded. 6–7. Older stages, ovules curved in diverging directions (× 90).

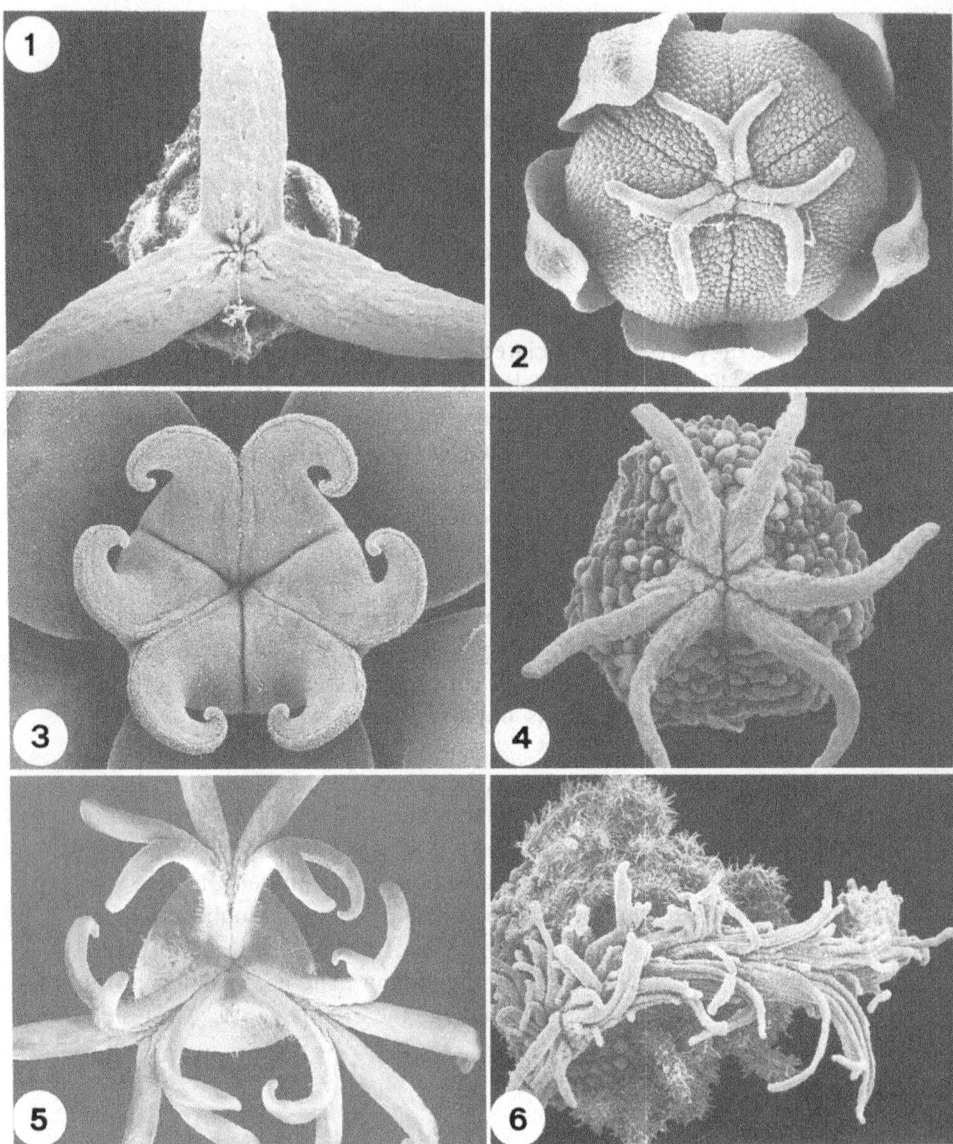

Figure 2.26. Diversity but uniform organization of stigmatic lobes in tricarpellate gynoecia of Euphorbiaceae. 1. *Alchornea thozetiana*. Stigmatic lobes entire (× 18). 2. *Phyllanthus niruri*. Stigmatic lobes forked (× 45). 3. *Sauropus macranthus*. Stigmatic lobes forked (× 15). 4. *Ricinocarpus pinifolius*. Stigmatic lobes forked (× 18). 5. *Cnidoscolus urens*. Stigmatic lobes triply forked (× 13). 6. *Macaranga involucrata*. Stigmatic lobes reduced to one that is multiply subdivided (× 15).

Harrison 1982b, for *Trifolium*, Fabaceae), the most obvious being the lipophilic components on the stigma that are lacking in the style.

More detailed classifications of stigma forms with morphological, histological and physiological aspects have been proposed by J. Heslop-Harrison (1976b), Y. Heslop-Harrison & Shivanna (1977), Y. Heslop-Harrison (1981) and Schill *et al.* (1985).

The main parameters are the amount of stigmatic secretion, the relief of the stigmatic surface, and the structure of stigmatic papillae. Thus, Y. Heslop-Harrison (1981) distinguishes the following types (with the last one added by Calder & Slater 1985):

- Surface dry (Fig. 2.27)
  - Receptive cells dispersed on a plumose stigma (D Pl) (e.g. Poaceae)
  - Receptive cells concentrated in ridges, zones or heads
    - Surface non-papillate (D N)
    - Surface distinctly papillate
      - Papillae unicellular (D P U)
      - Papillae multicellular
        - Papillae uniseriate (D P M Us)
        - Papillae multiseriate (D P M Ms)

- Surface wet (Fig. 2.28)
  - Receptive cells papillate; secretion moderate to slight, flooding interstices (W P)
  - Receptive cells non-papillate; secretion usually copious (W N)
  - Receptive area a depression with viscid secretion (matrix) containing detached, suspended cells (W Dc) (Orchidaceae).

Schill *et al.* (1985) add to the dry and wet stigmas a third type, where the receptive surface is more or less hidden in a tube ('Röhrennarben', tubular stigmas). This is an interesting modification but is not directly parallel to the above mentioned categories.

Wet stigma types seem to have two main secretory mechanisms: holocrine secretion in lipophilic exudates (e.g. *Petunia*, Solanaceae) and granulocrine secretion in hydrophilic exudates (e.g. *Lilium*) (Knox 1984, Dumas *et al.* 1988). In dry stigma types, the outermost coating of the stigma cells, the proteinaceous pellicle, is hydrophilic.

The stigma type may be constant (e.g. Bignoniaceae, type W P) or variable (e.g. Liliaceae, types D Pl, D P U, D P M Us, W P, W N) within a larger taxonomic group (Y. Heslop-Harrison & Shivanna 1977). It will be interesting to work out how such types are correlated with particular breeding systems or pollination modes. Y. Heslop-Harrison & Shivanna (1977) point to the association of sporophytic self-incompatibility systems with dry, papillate stigmas and of most gametophytic systems with wet stigmas. Further, trinucleate pollen tends to be associated with dry stigmas, while wet-stigma taxa tend to have binucleate pollen; but binucleate pollen occurs with both wet and dry stigmas. Tubular stigmas are present in some 'buzz-pollinated' groups, but not exclusively so.

The size of the stigmatic surface is not considered in these classifications, although it is also of interest in floral biology. Wind-pollinated flowers tend to have relatively large stigmas. In other pollination types a wide spectrum of sizes occurs. Flowers with very high numbers of ovules tend to have large and wet stigmas, the type W Dc (wet with detached cells), which is only known from orchids, being an extreme case.

Figure 2.27. Dry type stigmas. 1. *Passiflora suberosa*. Papillate, multicellular (seemingly unicellular) (× 70). 2. *Pisonia umbellifera* (Nyctaginaceae). Papillate, multicellular (× 130). 3–4. *Oryza sativa* (Poaceae). Plumose, with dispersed receptive areas. 3. Entire gynoecium (× 20). 4. Part of a stigmatic plume (× 200).

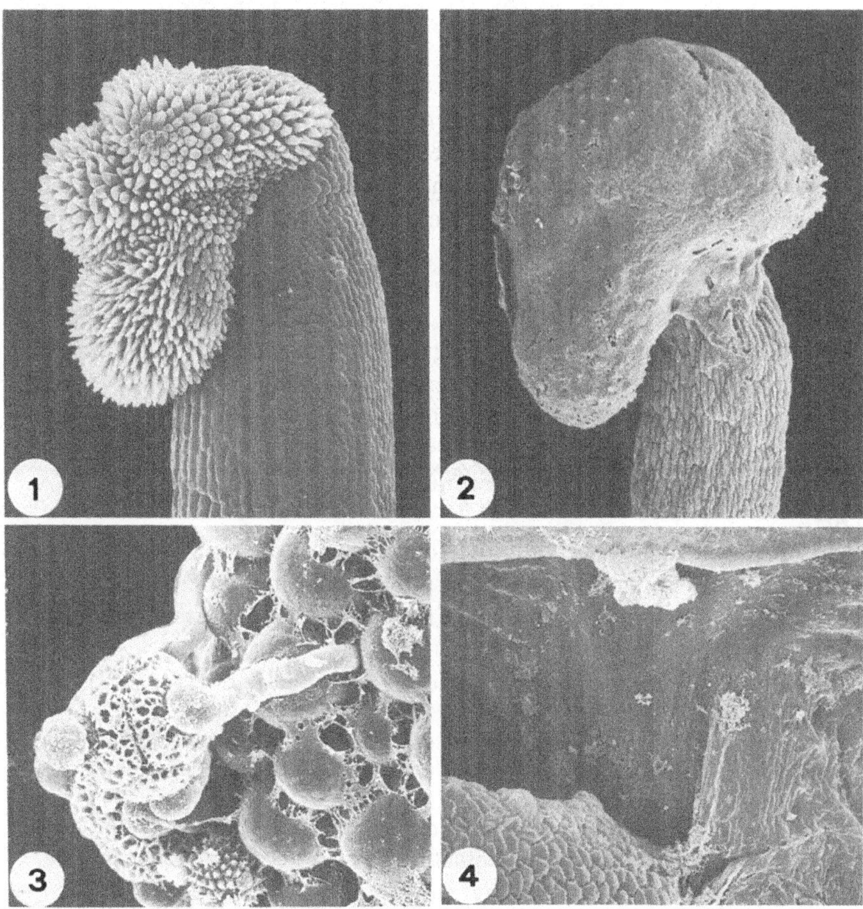

Figure 2.28. Wet type stigmas. 1–2. *Lantana camara* (Verbenaceae). Papillate, unicellular. 1. Dry stage, before anthesis (× 110). 2. Secretory stage, at anthesis (× 90). 3. *Drimys winteri* (Winteraceae). Papillate, unicellular. A germinating pollen tetrad with pollen tubes growing into the stigma (× 500). 4. *Ctenanthe* sp. (Marantaceae). Non-papillate (concave) (× 180).

From my experience the most widespread stigma form in angiosperms is wet and papillate (W P). Wet and non-papillate stigmas (W N) are rare. They are largely restricted to concave receptive surfaces (e.g. some Caesalpiniaceae, Zingiberales).

Pollen tube transmitting tract

The diversity of the stigma is in some respects paralleled by that of the pollen tube transmitting tract in the style and ovary (Behrens 1875). Capus (1879) showed the development of the tract as originating from the surface of the stylar canal, Dalmer (1880) discussed some functional correlations between morphology and histology within the tract, and Guéguen (1901) brought a comparative survey for a number of angiosperm families.

An early classification of the pollen tube transmitting tracts in the style had been proposed by Hanf (1936). He distinguished between 'open', 'half-closed' and 'closed' transmitting tracts. Open tracts have a lumen that is surrounded by a secretory (transmitting) epidermis; half-closed tracts have a lumen that is surrounded by a secretory (transmitting) tissue several cell layers thick; closed tracts are without a lumen and have a secretory (transmitting) tissue several cell layers thick (see also Knox 1984).

At present, a somewhat more detailed classification seems appropriate. Two aspects are important: (A) the histological differentiation of the tissue providing the tract and the amount of secretion; (B) the form of the inner morphological surface of the carpels.

Therefore, the following classification is proposed here (Fig. 2.29):

(A) (1) Epidermis strongly secretory, path of pollen tubes external in the ample secretion that fills the lumen of the stylar canal and ovary (Fig. 2.30.1).
(2) Epidermis weakly secretory, path of pollen tubes superficial on the epidermis (transmitting tissue 1 cell layer thick) (Fig. 2.15.1–2).
(3) Epidermis weakly secretory, path of pollen tubes internal in the cell walls of the transmitting tissue (transmitting tissue several cell layers thick) (Fig. 2.30.2).

(B) (1) Stylar canal with large lumen and large inner surface.
(2) Stylar canal with small lumen but large inner surface (in that the surface is folded).
(3) Stylar canal with small lumen and small inner surface.

Numbers (1), (2) and (3) of (A) and (B) often seem to be correlated (Fig. 2.29). A further feature correlated with the kind of differentiation of the pollen tube transmitting tract is the ovule number per gynoecium. A gynoecium with very many ovules tends to have a several-layered tract, a gynoecium with a single ovule rather a single-layered one (but: many-layered in Lamiaceae with 4 ovules, Rudall 1981, and Geraniaceae with 10 ovules, Sauer 1933).

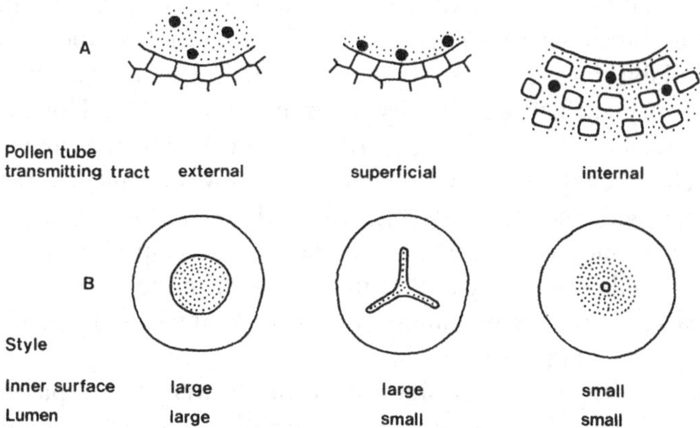

Figure 2.29. Forms of pollen tube transmitting tract (A) and correlation with inner morphological surface of style (B) as seen in transverse sections.

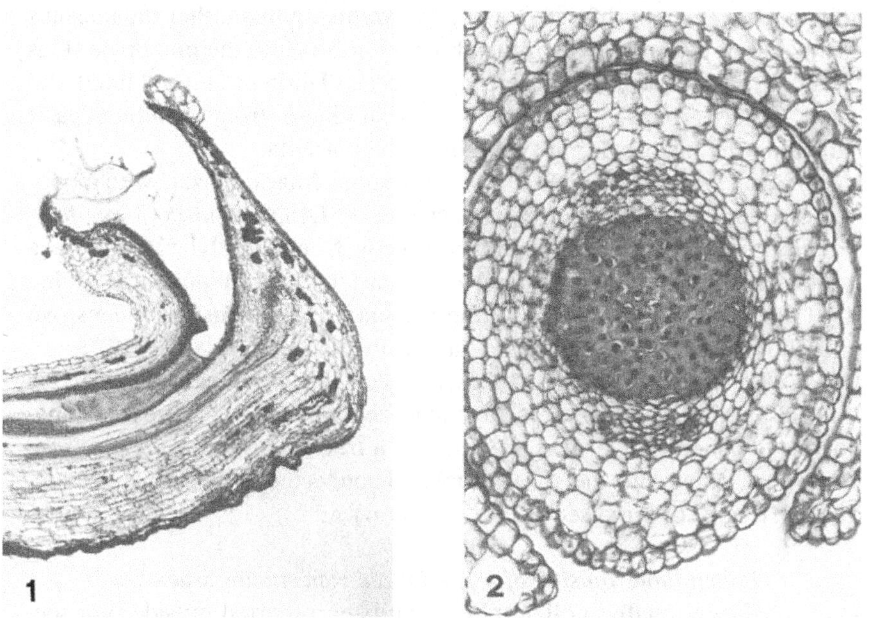

Figure 2.30. Forms of pollen tube transmission tract. 1. *Maranta leuconeura* (Marantaceae), longitudinal section. External tract (style hollow, filled with secretion) (× 45).
2. *Jacobinia pohliana* (Acanthaceae), transverse section. Internal tract (inner stylar surfaces postgenitally fused) (× 230).

The special differentiation of the stigma and of the pollen tube transmitting tract are somewhat correlated. They both depend on the number of ovules available in a gynoecium. A multiovulate gynoecium often has a large secretory surface and a massive transmitting tract. However, the amount of secretion may differ greatly in stigma and style; as an example, *Trimezia* (Iridaceae) has a dry stigma but a hollow transmitting tract completely filled with secretion (Bystedt 1990). In some Liliaceae secretions of the stylar canal pass onto the stigma (J. & Y.Heslop-Harrison 1982a).

Lloyd & Wells (1992) argue that the similarity in the differentiation of both regions is due to the evolutionary derivation of the angiosperm stigma from the transmitting tissue (see section 9.3).

A critical region is the transition from the ovarial tissue (placenta) to the ovule. In addition to the presence of secretions various morphological devices have also been developed to bridge this gap or even to omit the placenta:

(1) In Euphorbiaceae the nucellus is in direct contact with an obturator, a transmitting plug of the placental region. Both parts, the obturator and the nucellus, may proliferate to varying extents, in that either the nucellus grows out of the micropyle or the obturator grows into the micropyle (Figs 2.31.1–3). Obturators also occur in Liliaceae (Tilton & Horner 1980) and some other groups, where they may originate from different parts adjoining the micropyle, not only from the placenta.

(2) Basal anatropous ovules as occur in some Anacardiaceae with monomerous or pseudomonomerous gynoecia (*Anacardium*, *Mangifera*, *Pistacia*) may receive pollen tubes directly from the stylar canal via a protrusion of the funiculus (Joel & Eisenstein 1980, Wannan & Quinn 1991) (Fig. 2.31.4). In omitting the placenta this protrusion bridges two morphologically distant but topographically adjacent parts.

(3) The bridge for pollen tubes between the stylar canal and a free central-basal placenta (e.g. in Myrsinaceae and Lentibulariaceae) is analogous to condition (2). It is provided by a free extension of the placental summit reaching into the stylar canal and connected with it by direct contact (*Myrsine*) or via a secretion (*Aegiceras*).

Pollen tubes outside of a specialized transmitting tract

Exceptionally, pollen tubes have been reported outside of a specially differentiated pollen tube transmitting tract. In some cleistogamous flowers, not only the floral buds, but also the anthers remain closed; the pollen germinates in the anther and pollen tubes grow through the anther wall to the stigma. Extreme cases are some Malpighiaceae (*Gaudichaudia*, *Janusia*), where in highly reduced cleistogamous flowers pollen germinates

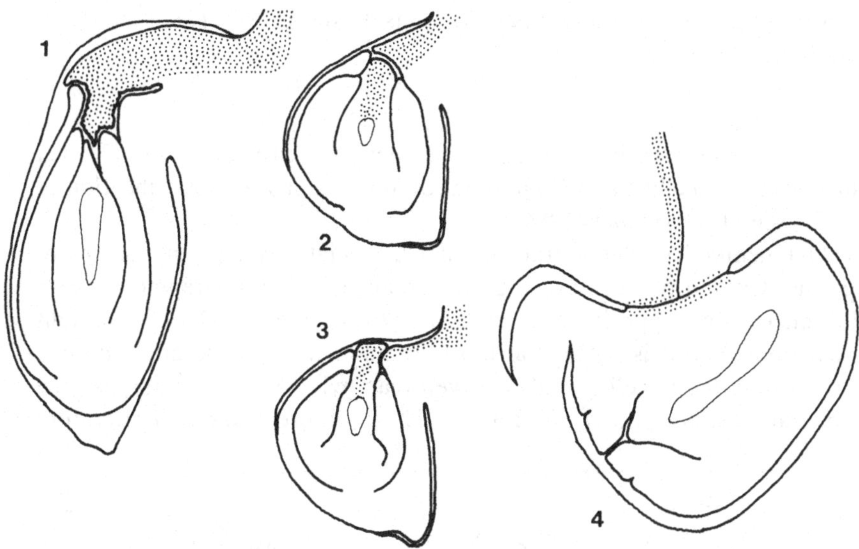

Figure 2.31. Different forms of bridges between placenta and micropyle for pollen tubes as seen in median longitudinal sections of carpels and ovules; pollen tube transmitting tract stippled (× 45). 1. *Pedilanthus tithymaloides* (Euphorbiaceae). Obturator protruding into micropyle. 2. *Dalechampia spathulata* (Euphorbiaceae). Nucellus apex partly protruding out of micropyle. 3. *Dalechampia triphylla* (Euphorbiaceae). Nucellus apex protruding out of micropyle. 4. *Anacardium occidentale*. Thickening on dorsal side of ovule in contact with lower end of stylar canal.

inside the anther and pollen tubes grow down in the filament and through the floral base into the ovary (Anderson 1980), and species of *Callitriche* (Callitrichaceae) where the same happens between male and female flowers (Philbrick 1984).

This is also the case in chalazogamous taxa where pollen tubes grow through the funicle and chalaza of the ovule before they reach the upper end of the embryo sac. Except for *Casuarina*, where chalazogamy was first described (Treub 1891), this is found mainly in some extratropical Hamamelididae, where there is a long delay between pollination and fertilization due to immaturity of ovules at anthesis (Luza & Polito 1991).

Even growth through free spaces has been reported (Nicholls & Cook 1986, for *Typha*). Wilms (1980) found in spinach (*Spinacia oleracea*, Chenopodiaceae) that the first pollen tubes grow intercellularly in the transmitting tract but later ones grow inside the cells between cell wall and plasmalemma.

Ovules

Ovules vary in three major respects: (1) curvature of ovule; (2) thickness of nucellus; and (3) number of integuments. For classification of ovules see also Bouman (1984).

(1) Uncurved ovules (orthotropous) are relatively rare in the angiosperms. Curved ovules have the attachment region topographically near the micropyle, opposite the chalaza. Mostly the nucellus is straight (anatropous); this is by far the most common form in the angiosperms. More rarely the nucellus is also curved (campylotropous); weakly curved forms are hemitropous (Fig. 2.32.1). These different kinds of curvature

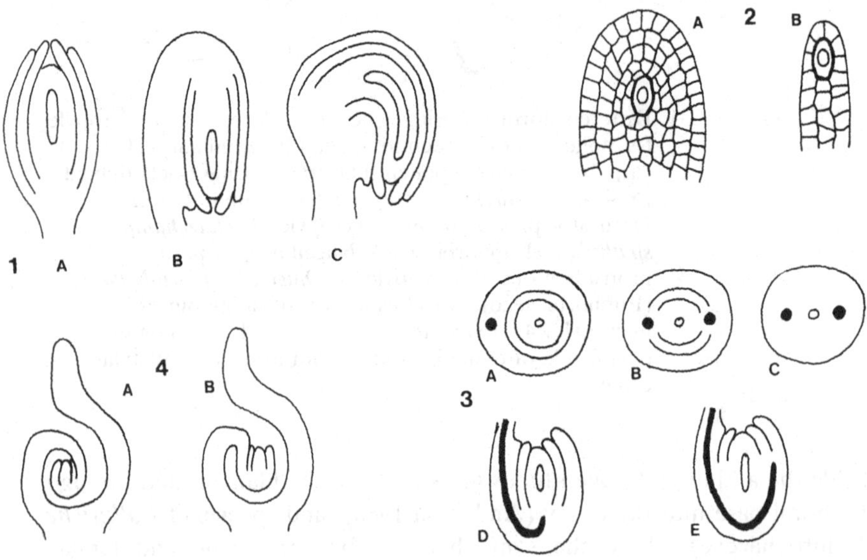

Figure 2.32. Diversity of ovule forms. 1. Curvature. A. Orthotropous. B. Anatropous. C. Campylotropous. 2. Nucellus thickness; meiocyte with thick contour. A. Crassinucellar. B. Tenuinucellar. 3. Chalaza; black, vascular bundle. A–C. Transverse sections. D–E. Median longitudinal sections. A,D. Ovule with unextended chalaza. B,E. Perichalazal ovule. C,E. Pachychalazal ovule. 4. Direction of ovule curvature in relation to carpel curvature as seen in median longitudinal section of carpel and ovule. A. Syntropous. B. Antitropous.

are better understood if viewed in the context of the architecture of the ovarial cavity and with respect to seed structure (see next section and section 2.4.6).

(2) The nucellus may be thick and the meiocyte(s) surrounded apically and laterally by more than one cell layer (crassinucellar) or it may be thin and the meiocyte surrounded apically and laterally only by the epidermis (tenuinucellar) (Fig. 2.32.2). Crassinucellar ovules exhibit a broad range of nucellar thickness, while tenuinucellar ovules are more narrowly defined and exhibit only one extreme end of the spectrum.

(3) The ovule has one integument (unitegmic) or two (bitegmic). The presence of three integuments (in a few Annonaceae) or the lack of integuments (in some Santalales, some Gentianaceae) is very rare. Transitions between bitegmic and unitegmic ovules occur, e.g., in *Impatiens* (Balsaminaceae) (Boesewinkel & Bouman 1991).

By far the most common combinations are: anatropous-crassinucellar-bitegmic (e.g. most Magnoliidae, Rosidae, Dilleniidae) and anatropous-tenuinucellar-unitegmic (e.g. Asteridae, Orchidaceae). However, most other combinations also exist and they are of much systematic interest (Philipson 1974, 1977).

The vascular bundle that serves the ovule usually ends in the chalaza. However, it may continue into one of the integuments on the opposite side of the raphe and it may even branch out into several strands around the entire ovule (surveys in Kühn 1928, Corner 1976, Boesewinkel & Bouman 1984). In bitegmic ovules the vascular bundles are either in the inner (e.g. *Dipterocarpus*) or, more often, in the outer integument (e.g. Annonaceae) (Corner 1976). Both cases may occur in the same family (e.g. Euphorbiaceae, Wunderlich 1967). Integumentary vascular bundles tend to occur more often in ovules that develop into large seeds.

Other ovule modifications are in the chalazal region (Fig. 2.32.3). (1) Perichalazal ovules: the chalaza is more extended in the median plane than in the transversal plane. Nucellus and integuments are congenitally fused in the median plane for some distance, while they are already free from each other in the transversal plane (Fig. 2.32.3.B,E) (Corner 1976). They occur, for example, in Annonaceae and Austrobaileyaceae (Endress 1980c). (2) Pachychalazal ovules: the entire chalaza is considerably extended with respect to nucellus and integuments (Fig. 2.32.3.C,E) (Corner 1976). A consequence is that the embryo sac may extend far down into the chalaza ('inferior embryo sac'), such as, for example, in Lauraceae (Endress 1972a). There is some correlation between pachychalazal ovules and large seed size (von Teichman & van Wyk 1991).

Correlations between ovule structure and ovary structure

It seems plausible that the curved forms of ovules (anatropous, campylotropous) predominate in angiosperms because in this type the micropyle comes to lie very close to the ovule attachment. Thus, the ovule can take up the pollen tube immediately from the placenta.

The relatively rare occurrence of uncurved ovules is always correlated with some extreme gynoecium construction and can also be functionally explained. They occur under the following circumstances.

(1) The gynoecium contains only one ovule, which is attached at the base of the ovary, and the micropyle is directed immediately towards the stylar canal, where it can take up a pollen tube. This occurs in several unrelated groups (Piperaceae, Polygonaceae, Juglandaceae, Myricaceae, Araceae) (Fig. 2.33.2).

(2) The ovary has parietal placentation and each ovule has its micropyle directed toward the neighbouring placenta and touches its pollen tube transmitting tissue (some Flacourtiaceae, Mayacaceae, Saururaceae, Fig. 2.33.1).

(3) The ovary wall secretes large amounts of mucilaginous substances which may completely fill the ovary cavity. The mucilage is commonly produced by special pads of hairs on the placentae. Pollen tubes may grow through this mucilage and can also easily enter uncurved ovules. This is the case in several water plants, but also in other ones, again in unrelated groups (*Hydrocharis*, Hydrocharitaceae; *Pistia*, Araceae; *Akebia*, Lardizabalaceae; *Barclaya*, Nymphaeaceae) (Fig. 2.33.3).

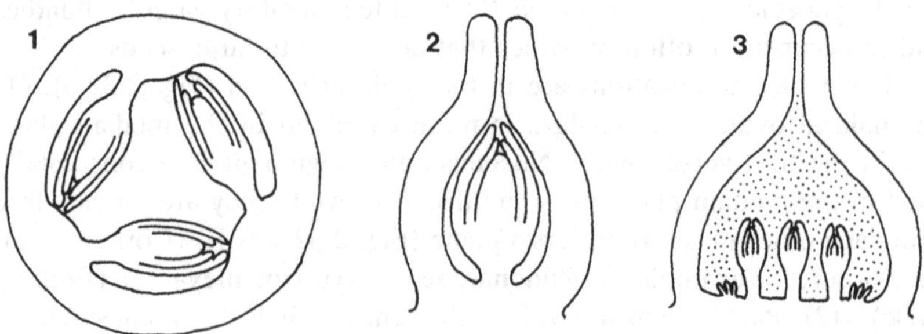

Figure 2.33. Orthotropus ovules in different gynoecium architectures. 1. Parietal placenta, micropyles touch neighbouring placenta (e.g. *Casearia*, Flacourtiaceae). 2. Basal placenta, single ovule, micropyle touches lower end of stylar canal (e.g. *Piper*). 3. Basal placenta, several ovules, ovary filled with secretion as pollen tube transmitting medium (stippled) (e.g. *Pistia*, Araceae).

Correlations between ovule orientation and carpel margin

As described above, the ovules are mostly formed at or near the carpel margin (marginal or submarginal placenta). However, since carpels have a certain thickness, the margin is not a well-defined line but rather a more or less broad zone. In free carpels the ovules are usually curved away from the carpel margin. Thus, if the carpel is seen as an involute structure, the ovules continue in the direction of the carpel involution by their own curvature. The result is that, as in the carpel, the outer side of the ovules shows stronger development than the inner, which is inhibited (Rohweder 1967, van Heel 1983) (Fig. 2.14.2).

This is also the case in laminar-diffuse placentae, such as, in Nymphaeaceae, Lardizabalaceae and Gentianaceae (Gopal & Puri 1962, Padmanabhan *et al.* 1978, Shamrov 1990) or Butomaceae, where all ovules of one placenta face the same direction, i.e. away from the carpel margin (Fig. 2.25.1–2).

Accordingly, in an axile or parietal protruding-diffuse placenta, the ovules face in two opposite directions, away from each other, because two (hypothetical) margins are involved, in an axile placenta by a single carpel (Hallé 1967) (Fig. 2.25.4), in a parietal placenta by two adjacent carpels (Fig. 2.25.6).

This most common behaviour of ovules, to continue the direction of carpel involution by their own curvature, is reversed in a few groups. There, the ovule curvature goes in the opposite direction to the carpel curvature. This difference seemed so remarkable to Engler (1931) (and earlier authors) that he distinguished the two orders Sapindales and Geraniales, mainly based on this difference. The existing terminology of these ovular features is confusing. Björnstad (1970) calls the first case, depending on the position of the ovule in the ovarial cavity: epitropous-dorsal or hypotropous-ventral or pleurotropous-dorsal, the second case: epitropous-ventral or hypotropous-dorsal or pleurotropous-ventral. They correspond to some extent with the earlier terms apotropous and epitropous (Warming 1913). Björnstad's terminology may be useful for mere descriptive purposes. It is, however, not practical in a comparative evolutionary sense. Therefore, I propose the following terminology: the ovule is incurved (syntropous) if curved in the same direction as the carpel, but recurved (antitropous) if curved in the other direction, irrespective of its apical, basal or lateral position (Fig. 2.32.4). The two forms of ovule curvature need further comparative developmental study.

Correlations between gynoecium and fruit structure

In some groups the initial stages of special differentiations of fruits are already apparent at anthesis. In Lauraceae the unicarpellate gynoecium has a uniform organization and uniform size at anthesis, but the fruits are very different in size. In all gynoecia at anthesis there is a dorsal and a ventral vascular bundle. However, in those that form big fruits there are already additional vascular bundles at anthesis (e.g. *Persea*), while they are lacking in the small-fruited ones (e.g. *Cinnamomum*) (Endress 1972a) (Fig. 2.34). In burr-fruits with hooks at the surface, these hooks may be initiated before anthesis and be present in the flowers, although they have no function yet (e.g. *Rulingia*, Sterculiaceae) (Fig. 2.35). In *Soliva* (Asteraceae) the style tips become hard and pointed in fruit and act as trample-burr devices. The styles are already stout in flower (Fig. 2.35). Relationships between fruits and flowers are also discussed by Primack (1987).

2.4.5 Additional functions

In some angiosperm groups the gynoecium plays an important role as a display for secondary pollen presentation (Asteraceae, Campanulaceae, Goodeniaceae, some Rubiaceae, Polygalaceae, Fabaceae, Proteaceae, Marantaceae; e.g. Brantjes 1982, Vogel 1984, Robbrecht 1988, Thiele 1988, Bamert 1990, Leins & Erbar 1990, Classen-Bockhoff 1991, Puff 1991) (see sections 8.6.3 and 8.9.6). These flowers are protandrous. The anthers open in bud, and pollen is deposited on the style, where it is presented to pollinators before the stigma becomes receptive. In Asteraceae and certain Campanulaceae the anthers are postgenitally fused into a tube that firmly surrounds the style. They may form a highly elabor-

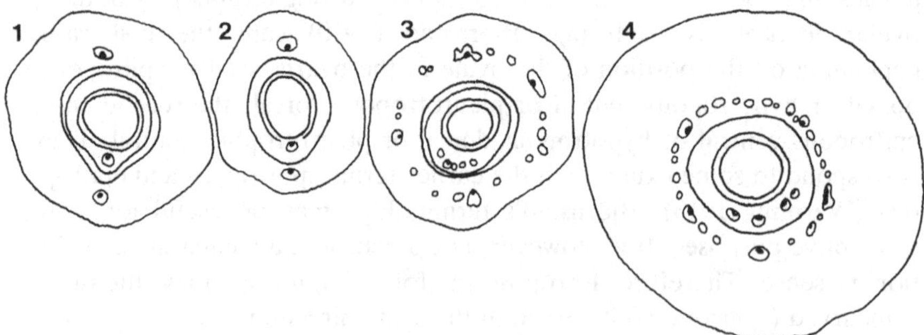

Figure 2.34. Unicarpellate, uniovulate gynoecia of Lauraceae at anthesis (transverse sections) with uniform organization but different vascular pattern according to later fruit size (× 32) (modified after Endress 1972a). 1. *Cinnamomum camphora*. 2. *Litsea glaucescens*. 3. *Persea americana*. 4. *Persea botrantha*.

ate dispenser mechanism (Brantjes 1983a, Thiele 1988, Leins & Erbar 1990). Still more complicated are the flowers of Asclepiadaceae and Orchidaceae where the attaching devices of the pollinaria are formed by highly elaborated parts of the gynoecial apex (see sections 8.7.1 and 8.10.1).

Nutritious stigma secretion may be taken by pollinators (e.g. *Tasmannia*, Winteraceae, see section 8.1.5; *Aristolochia*, Aristolochiaceae, Baker *et al.* 1973). In *Omalanthus*, Euphorbiaceae, a nectary is situated on the carpel immediately below the stigma.

In unisexual pollen- or nectar-flowers, in which female flowers do not offer any reward, the stigma may mimic anthers in shape and colour (e.g. Begoniaceae, Cucurbitaceae) or it may even mimic perianth parts (Caricaceae) (see section 5.10). The stigma in some bisexual flowers may be large and distinctly coloured (e.g. *Eustigma*, Hamamelidaceae, Endress 1989).

2.4.6 Evolutionary aspects

There are two major evolutionary trends in the angiosperm gynoecium among a multitude of 'oscillating' (minor) evolutionary changes. The first is the advent of syncarpy from apocarpy; the second, the evolution of tenuinucellar, unitegmic ovules from crassinucellar, bitegmic ones. These trends are major because the derived features are both predominant in the higher advanced angiosperms, but very rare in the 'lower' (primitive) groups. It seems, therefore, that, once acquired, these characters were no longer easily relinquished.

From apocarpous to syncarpous gynoecia

There are probably several advantages of a syncarpous over an apocarpous gynoecium, the most obvious being the presence of a compitum, i.e. the possibility of an even distribution of pollen tubes among the carpels, and the occurrence of centralized pollen tube selection in a single transmitting tract (Endress 1982). This may be of almost the same magnitude as the step from gymnospermy to angiospermy, whereby pollen tube selection was also greatly enhanced (D. Mulcahy 1979).

Other (minor) advantages of syncarpy over apocarpy may be the greater versatility of fruit behaviour at maturity and greater material economy (Endress 1982). Apocarpy prevails in Magnoliidae and Alismatidae. It still occurs in some Rosidae but is almost absent in the higher advanced groups. Concomitant with the advent of syncarpy there is a canalization of carpel number to (mostly) 2–5, because there is an advantageous gynoecium architecture with this range of numbers.

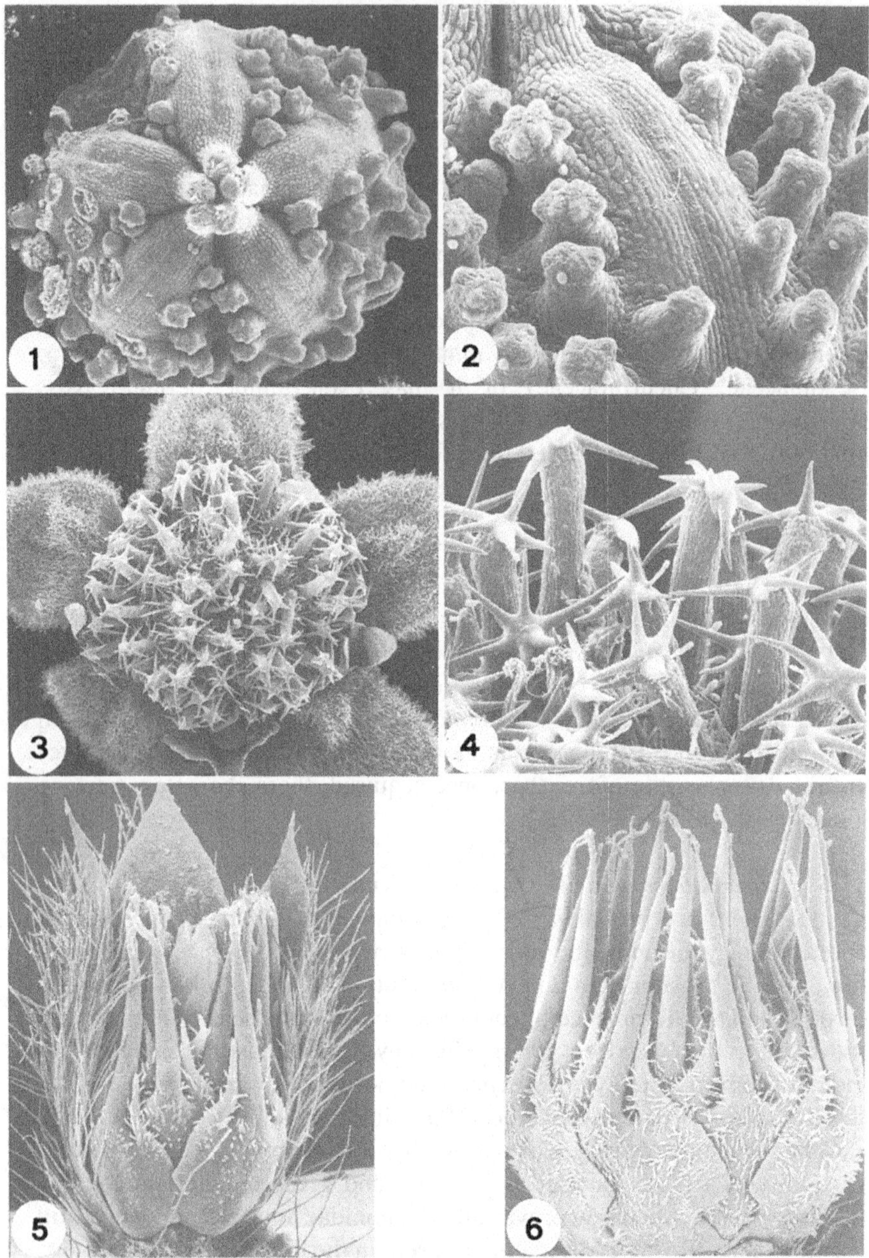

Figure 2.35. Gynoecia with differentiations at anthesis that act only in the fruiting stage. 1–4. *Rulingia pannosa* (Sterculiaceae) with burrs containing anchor-like protrusions. 1–2. Slightly

From superior to inferior ovaries

It has long been supposed that inferior ovaries are derived from superior ovaries. This is suggested by the more complicated development of inferior ovaries and their more prominent presence in higher advanced taxonomical groups. Inferior ovaries are rare in Magnoliidae but are predominant in groups such as Rubiales, Campanulales, Zingiberales and Orchidales. An additional, ecological viewpoint is that inferior ovaries tend to occur in bird-pollinated flowers as an evolutionary response to the vulnerability of the ovary to damage by bird bills (V. Grant 1950a). Bird-flowers are often derived from bee-flowers (see section 4.1.6).

However, evolutionary reversals from inferior to superior ovaries may also occur. This is apparent from detailed studies of erratic taxa with a superior ovary within groups where inferior ovaries predominate (Eyde & Tseng 1969, *Tetraplasandra*, Araliaceae).

From crassinucellar, bitegmic to tenuinucellar, unitegmic ovules

Tenuinucellar, unitegmic ovules are smaller than crassinucellar, bitegmic ones. An advantage is, therefore, that they require less space in the ovary and can be produced in greater quantities. Therefore, plants with this ovule form are more versatile as to ovule number. Exceptionally large numbers (millions) occur in some Orchidaceae, but they may also be reduced to a single ovule (consistently in Asteraceae). Most plants with this ovule form have relatively small seeds but the potential to produce bigger seeds has not been lost (*Helianthus*, Asteraceae; Convolvulaceae, Kühn 1928, Corner 1976). Tenuinucellar, unitegmic ovules are highly characteristic of the Asteridae, and they also occur in Orchidaceae, thus in the highest evolved groups of the dicots and monocots. Only in rare cases can the transition between the two ovule types be followed within the narrow limits of a genus (e.g. *Impatiens*, Balsaminaceae, Bouman & Boesewinkel 1991).

Ecologically, in parasitic or mycotrophic plants the ovules tend to be reduced and increased in number, since the germinating seeds may depend on their hosts from the beginning of development and, therefore, do not

Fig. 2.35 (*cont.*)
before anthesis with early development of the protrusions. 1. (× 60). 2. (× 130). 3–4. Young fruit with prominent protrusions. 3. (× 13). 4. (× 45). 5–6. *Soliva pterosperma* (Asteraceae) with trample-burrs containing thickened styles with hard, pointed ends. 5. Inflorescence at anthesis, styles already thickened (× 25). 6. Fruiting stage, styles thick and pointed (× 18).

need to build up large endosperms or perisperms or to form a large embryo in the seed (e.g. Rafflesiaceae, Orchidaceae). In some mycotrophic Gentianaceae the ovules even lack a distinctive integument and fail to become anatropous (Oehler 1927, Bouman & Devente 1986).

From anatropous to campylotropous ovules

Another, less prominent, trend is from anatropous to campylotropous ovules. Bocquet (1960) suggests that there are different evolutionary pathways to campylotropy. Advantages of campylotropous ovules over other ovule forms are discussed by Takaso & Bouman (1984). Campylotropous seeds can contain embryos up to twice as long as the seed itself or even longer, which may be advantageous for seedling establishment under certain conditions. Indeed, the majority of campylotropous seeds have long embryos reaching from micropyle to chalaza (Takaso & Bouman 1984, Bouman & Boesewinkel 1991).

From one to several longitudinal series of ovules on a placenta

It can be assumed that the primitive state in pluriovulate carpels is a placenta with one longitudinal series of ovules. From this state the laminar-diffuse placenta may be derived by broadening of the placental region and the protruding-diffuse placenta by thickening of the placental region. In the most primitive group of the Magnoliidae (Magnoliales) only the first condition occurs among pluriovulate carpels. The second condition occurs in some Nymphaeales and Ranunculales. The third condition is common in higher advanced groups, such as Rosidae and Asteridae.

From syncarpy to secondary apocarpy with postgenital fusion of free carpel apices

An unusual trend from syncarpy back to apocarpy has occurred in a few scattered families, most obviously in the Apocynaceae and Asclepiadaceae. Here, the most highly evolved group, the Asclepiadaceae and the Apocynoideae of the Apocynaceae, is apocarpous, in contrast to most other Gentianales (Fig. 8.40). But here, the two carpels are postgenitally fused in the apical region so that, at least in some groups, a compitum is still present (Endress *et al.* 1983, Fallen 1986, Kunze 1991). However, in the special case of some Asclepiadaceae, this kind of specialization makes a compitum less advantageous than in other groups (see section 8.7.2). Other groups with perhaps secondary apocarpy but postgenital fusion of the carpel apices providing a compitum occur in some Rutaceae, Sterculiaceae and Simaroubaceae (Jenny 1985, 1988, Ramp 1988) (Fig. 2.36). There are also groups that are syncarpous at the base with postgenitally fused carpel apices, where there is probably no basal compitum but an

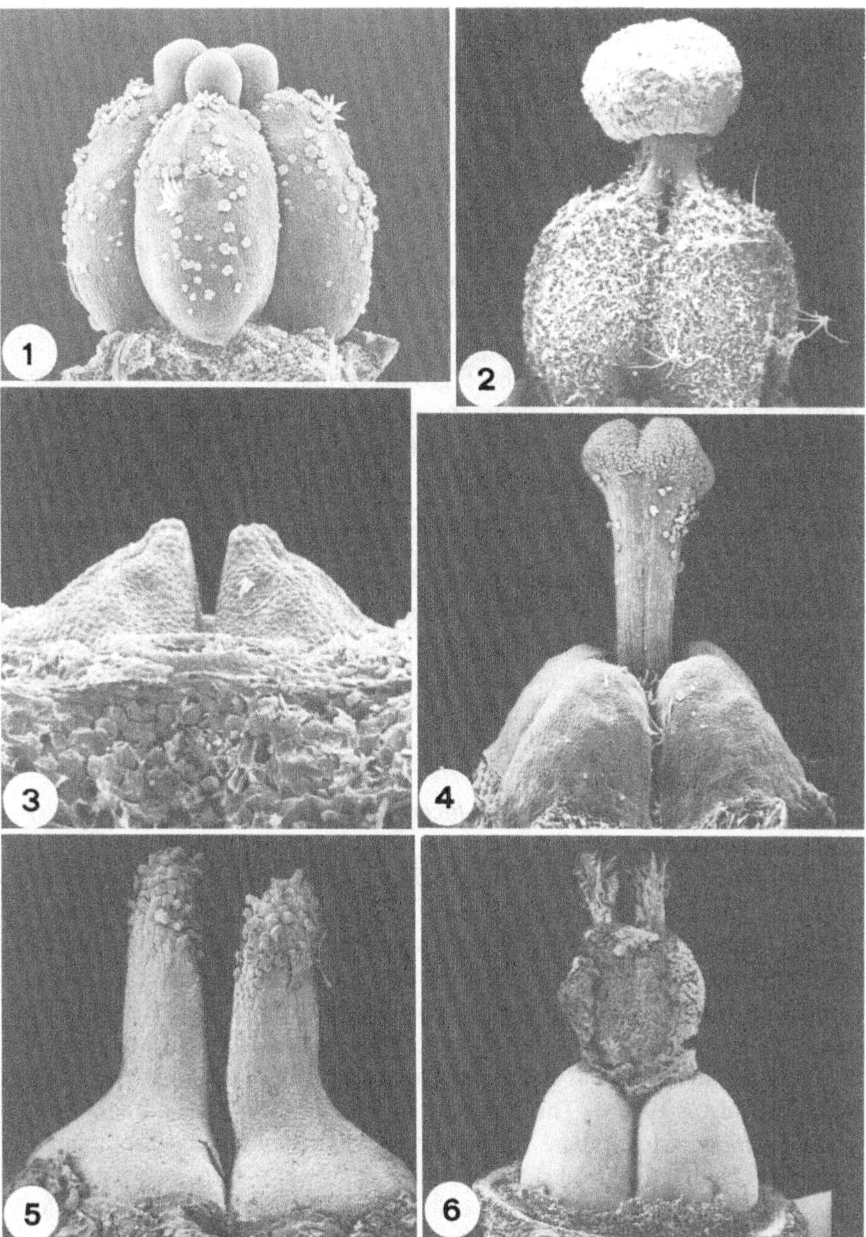

Figure 2.36. Apocarpous gynoecia with postgenitally united apical parts of carpels. 1–2. *Brachychiton australe* (Sterculiaceae). 1. Young stage with free carpels (× 50). 2. At anthesis (× 12). 3–4. *Zieria smithii* (Rutaceae). 3. Young stage with free carpels (× 160). 4. At anthesis (× 50). 5–6. *Mortoniella pittieri* (Apocynaceae). 5. Young stage with free carpels (× 140). 6. At anthesis (× 27).

apical one (some Loganiaceae, Staphyleaceae; Endress *et al.* 1983, Ramp 1987).

A number of Liliidae and most Zingiberidae have septal nectaries, which are formed by differential postgenital union of the carpels (Hartl & Severin 1981, van Heel 1988). In contrast to the dicot groups mentioned above, the carpels unite along their entire length, not only in their upper part. Also, in these monocots a compitum is formed by this postgenital union. It has not been explored, however, whether these forms are derived from syncarpous or from apocarpous ancestors.

A common feature of all these groups is that formation of the style tends to be retarded in development and postgenital carpel union is synchronous with initial style elongation (Hartl & Severin 1981, Endress *et al.* 1983) (Fig. 2.36).

Internalization of the pollen tube transmitting tract

An obvious evolutionary trend is the decrease in the morphological inner surface of the style and concomitant transgression of the transmitting tissue also to subepidermal layers, whereby a compact tube of circular diameter arises (internalization of transmitting tract). This feature is perhaps more frequent in Asteridae than in less advanced dicots (Vasil & Johri 1964, Bhatnagar & Uma 1969).

From unicarpellate to multicarpellate, syncarpous (paracarpous) gynoecia

In the Magnoliidae with prevailing apocarpous and unicarpellate gynoecia there are a few cases with a strongly syncarpous gynoecium. Perhaps surprisingly, the ovary is then paracarpous: it has a single cavity with parietal placentae. In each case, in the most closely related groups, there occur unicarpellate gynoecia. This has led to the hypothesis that, in these cases, syncarpy may have arisen not by fusion of the carpels of apocarpous ancestors but by internal multiplication of an originally single carpel (Endress 1990b). This can even be seen in statu nascendi in *Monodora* (Annonaceae), where the gynoecium starts ontogeny like a single carpel (in having a ventral slit) but results in a several-carpellate paracarpous structure (in developing several parietal placentae with the ovules diverging, and not directed away from the primary margins of the gynoecium primordium as in the case of laminar-diffuse placentae). In the dispute over whether this is a unicarpellate (Leins & Erbar 1980, 1982) or multicarpellate-syncarpous gynoecium (Guédès & Le Thomas 1981, Deroin 1985), therefore, both opinions are probably partly right but do not elucidate the particular evolutionary situation.

Other groups of such pairs of closely related paracarpous and unicarpellate forms are: Canellaceae/Myristicaceae; Papaveraceae/Berberidaceae; Saururaceae/Piperaceae (Endress 1990b); perhaps also *Takhtajania*/other Winteraceae (Leroy 1977, Vink 1978, but see Tucker *et al.* 1979).

A similar change from unicarpellate to paracarpous can also be seen in teratological cases of normally unicarpellate gynoecia where a second carpel has been formed (*Parkinsonia*, Caesalpiniaceae, Nair & Kahate 1961) or a change from apocarpous to paracarpous in the centre of a multicarpellate gynoecium (*Eupomatia*, Endress 1977b).

From eusyncarpous to paracarpous gynoecia

The common evolutionary pathway toward paracarpous gynoecia is via eusyncarpous forms. This evolutionary direction is evident in some clearly homogeneous groups with constant carpel number, because it is concomitant with an increase in ovule number. This step occurs with the transition from Scrophulariaceae to Orobanchaceae, from Scrophulariaceae to Gesneriaceae, from Ericaceae to Pyrolaceae/Monotropaceae, or from Apostasioideae/Cypripedioideae to monandrous Orchidaceae (Burtt 1970).

Unicarpellate and pseudomonomerous gynoecia

In this context the pseudomonomerous gynoecia also should be discussed, a topic that has not found a new comparative study after the extensive treatment by Eckardt (1937). Pseudomonomerous gynoecia appear superficially monomerous; they have only one fertile carpel but have traces of additional formations on it that are interpreted as additional reduced carpels. For Lauraceae it has been shown ontogenetically and by comparison with Monimiaceae that the allegedly pseudomonomerous gynoecia are in reality truly monomerous (Endress 1972a). Other cases such as Chrysobalanaceae or some Icacinaceae still have to be developmentally and comparatively investigated.

In truly pseudomonomerous cases the gynoecium starts development of at least two carpels. All but one cease development in a more or less early stage and persist as rudiments at anthesis. If they stop very early they may be only little solid knobs on one side of the fertile carpel. If they stop somewhat later they may develop small ovarial cavities but lack ovules. Another possibility of pseudomonomerous development is that the gynoecium commences like a single carpel and later, retarded primordia of additional carpels begin very limited growth. The curious gynoecium of *Monodora* (Annonaceae) and related cases, as discussed above, could be interpreted in a similar manner of sequential initiation. Many groups with pseudomonomerous gynoecia have not been studied developmentally as yet so that it is not known which type of development is the most common.

It is generally assumed that the pseudomonomerous condition has evolutionarily arisen by partial reduction of a syncarpous gynoecium with several carpels (D. Müller-Doblies, 1970, for *Typha*; Philipson, 1985, for Poaceae; Barabé *et al.* 1987, for Araceae). This may be the case generally but one should also bear in mind the possibility of irregular secondary addition by the expression of more structural units; an experimental example is gynoecium hyperplasia caused by heat stress in rice (Takeoka *et al.* 1991).

Extreme reduction in gynoecium structure

The parasitic family Balanophoraceae has extremely reduced female flowers and gynoecia. In *Balanophora* they resemble a moss archegonium more than an angiosperm gynoecium (Fig. 3.4.1). They are compact, and an internal morphological surface is lacking. Therefore, there is no ovule. An embryo sac and, after fertilization, an embryo develop in the thickened basal part of the gynoecium. Thus, there is the paradoxical situation of a seed plant without seeds but with normal sexual reproduction.

This extreme condition may well be derived from less reduced stages of various degrees in several families of the Santalales: Viscaceae, Loranthaceae, Santalaceae, Olacaceae (Zweifel 1939, Fagerlind 1948, Johri & Bhatnagar 1961). In this general trend of reduction the ovules at first fail to differentiate and the embryo sacs develop in the solid base of the ovary; in a second step the ovarial cavity fails to develop concomitant with reduction in size of the gynoecium.

2.5 Arrangement patterns of the structural units (floral phyllotaxis)

2.5.1 Floral phyllotaxis and floral diagrams

The floral organs or structural units are arranged in certain patterns, the floral phyllotaxis, which can be represented by floral diagrams as the arrangement of the rooms in a house is by an architectural ground plan. A floral diagram is the vertical projection of the base of the floral organs onto a plane. A transverse section of a flower with firm organs almost automatically provides a floral diagram; an example can be found as early as in Sprengel (1793, Tab. XIII, Fig. 21) with a drawing of *Persea indica*. However, the first comparative presentation of a few floral diagrams was by A. Braun (1831), and Eichler (1875, 1878) in his classic work *Blüthendiagramme* gave the first account on a larger scale throughout the angiosperms.

The floral phyllotaxis is generally in some way regular or symmetrical. This pattern can often be recognized most easily in the youngest developmental stages of the structural units and can be better understood by studying the course of floral development. Therefore, floral phyllotaxis will be discussed here in an ontogenetic context.

2.5.2 Simple patterns

The two most simple patterns in flowers are spiral (helical) and whorled (cyclic), but there are also more complicated and irregular patterns (Endress 1987a). Obviously, the most complicated, most elaborate flowers are the most regular ones in terms of symmetry, because high symmetry is an important precondition for synorganization of floral parts (Endress 1990a).

Spiral phyllotaxis

The geometrically most simple pattern is the spiral one, although it is more difficult to describe with words and more difficult to perceive at once than the whorled pattern (Fig. 2.37.7). Ontogenetically subsequent organs are positioned in regular angles, the divergence angles (formed together with the floral centre), and they are initiated in regular time intervals, the plastochrons.

In many spiral flowers the divergence angles are about 137°, as is the case in the vegetative region of plants with spiral foliage leaf phyllotaxis. This is also called the limit(ing) divergence or Fibonacci divergence or golden angle. In the famous Fibonacci sequence each number is the sum of the two preceding numbers: 1 1 2 3 5 8 13 . . . The limiting divergence is the limiting value of a sequence of ratios of subsequent Fibonacci numbers: 1/2 1/3 2/5 3/8 5/13 . . . This sequence converges to a limit of 137.5°, if the ratios are fractions of 360° (Hirmer 1931, Mitchison 1977, Richter & Schranner 1978). An example in flowers is *Austrobaileya* (see section 8.1.4) (Fig. 2.38.1–2).

Another angle that occurs more rarely in spiral phyllotaxis of flowers is 99.5°. This is the Lucas divergence. It is also the limiting value of a sequence generated by ratios of Fibonacci numbers: 1/3 1/4 2/7 3/11 5/18 . . . of 360° → 99.5°. An example in flowers is *Drimys winteri* (Winteraceae) (Hiepko 1966).

The prevailing occurrence of the 'limiting divergence' in spiral phyllotaxis (floral and vegetative) can also be explained by its relation to an exponential function $y=2^{-x}$. This is interesting in view of the hypothesis of an inhibitor-activator mechanism in pattern formation (Meinhardt 1982). The inhibitory influence of newly formed primordia determines the position of the following primordia. If we assume exponential decay of the inhibitor, the inhibiting effect of preceding primordia correspondingly decreases exponentially. In the exponential function *y* is thereby the concentration of the inhibitor, and *x* is the time interval given in plastochrons. Subsequent grades of inhibitor concentration show the same ratio as the 'limiting divergence' with its complementary angle to 360° (Richter & Schranner 1978).

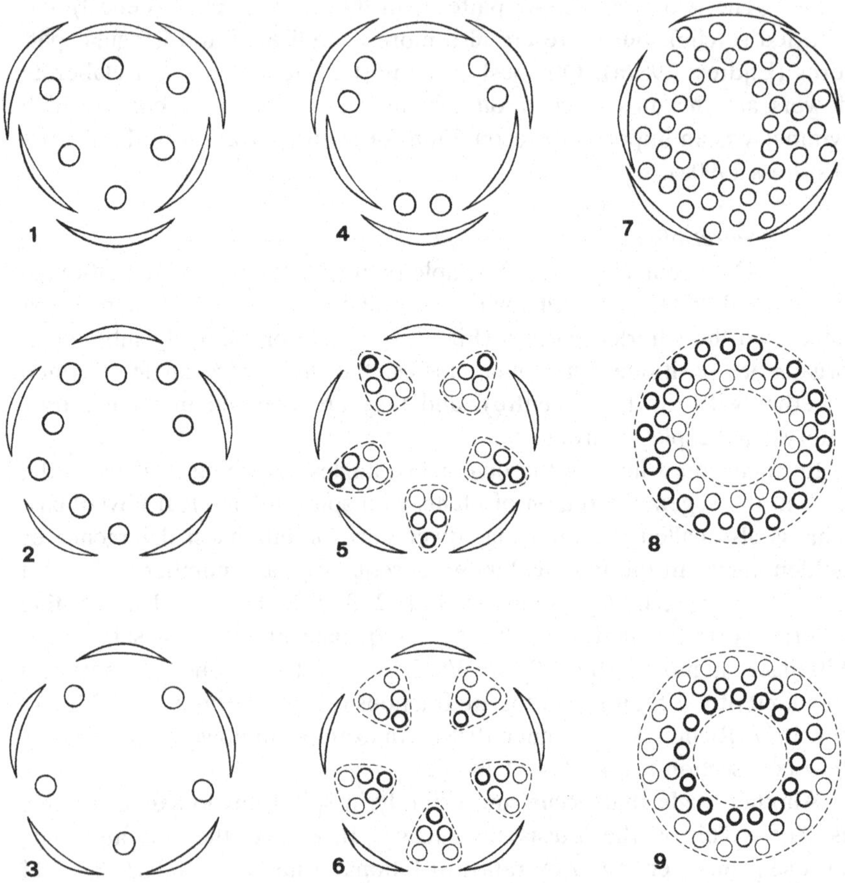

Figure 2.37. Patterns of stamen position and initiation with regard to perianth. 1–3. Stamens in alternating whorls. 1–2. Two whorls. 3. Single whorl. 4–6. Stamens in whorled groups. 4. Double position. 5. Centripetal initiation within a group. 6. Centrifugal initiation within a group. 7. Perianth organs and stamens in a spiral. 8. Stamens arising centripetally on ring primordium. 9. Stamens arising centrifugally on ring primordium.

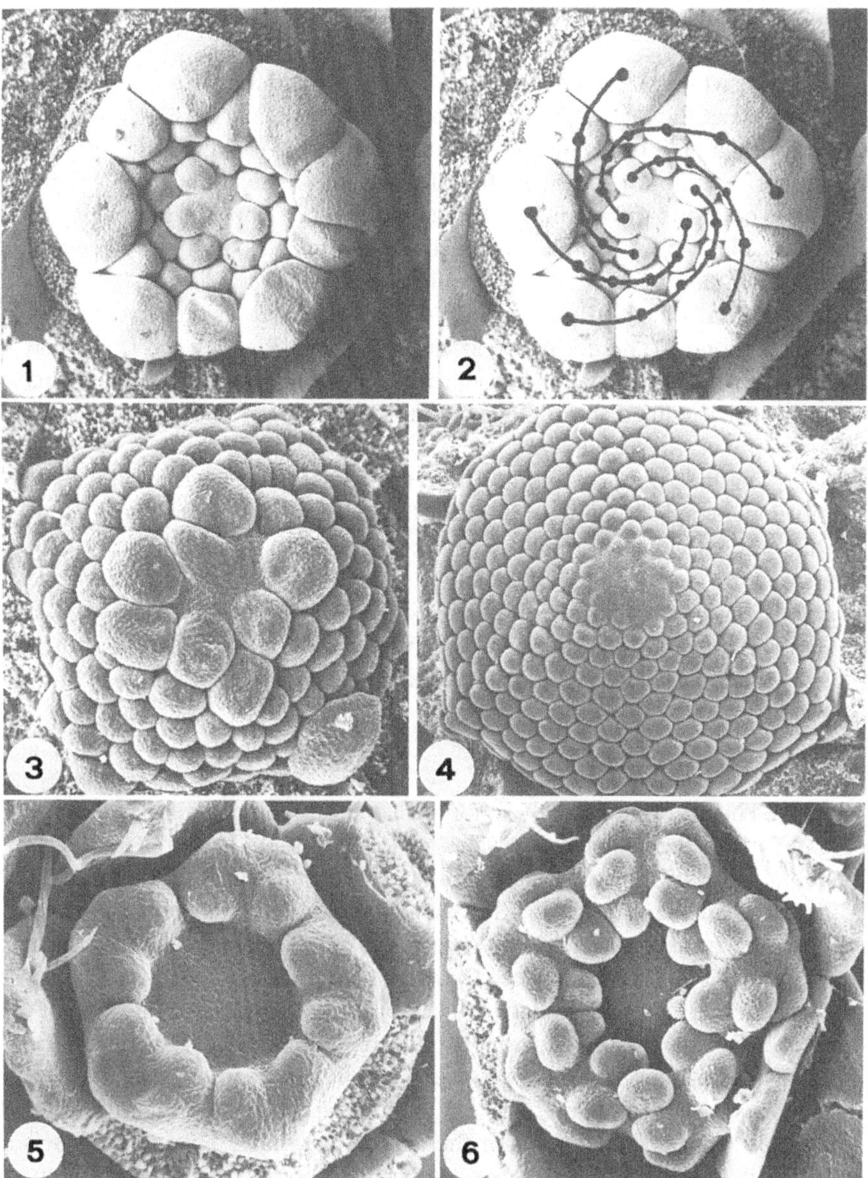

Figure 2.38. Patterns of stamen position in polyandrous flowers. 1–2. *Austrobaileya scandens*; regularly spiral, in 2 the set of five parastichies is indicated (× 140). 3. *Exospermum stipitatum* (Winteraceae); irregular (× 70). 4. *Annona cherimolia*; irregular (× 50). 5–6. *Kitaibelia vitifolia* (Malvaceae). 5. Five primary primordia with paired secondary primordia (× 130). 6. Five primary primordia with several paired secondary primordia, centrifugally initiated, in a slightly older stage (× 80).

In other words: structural units that arise around a growing apex tend to originate at the farthest possible distance from already existing ones, for instance by an inhibiting mechanism in which a newly formed primordium produces an inhibitor substance that diffuses into its surroundings. If there is as yet only one primordium, the second will be opposite. For the third the situation is more complicated in that it is influenced by both preceding ones. The inhibitor substance of the first one has already decayed to some extent so that the third primordium will arise nearer to the first than to the second one. If the decay is exponential, as one would expect, and the primordia arise in constant plastochrons, the result will be a spiral. If this pattern is maintained throughout organ initiation of the flower, the result is a completely spiral flower.

The regular spiral resulting from the subsequent primordia is the (onto)genetic spiral. Concurrent with this pattern there are sets of other spirals, the parastichies. They are easier to perceive at once because they are steeper than the ontogenetic spiral. They link all primordia which are distant from each other at a Fibonacci number. The number of spirals of a set is also a Fibonacci number. The two sets of 5 and of 8 are often the easiest to perceive (Fig. 2.38.2), depending on the outline of the individual organs.

Whorled phyllotaxis

The whorled pattern is somewhat more complicated. Here, the organs are in regular whorls (cycles) (Fig. 2.37.1–3). The organs of subsequent whorls alternate with each other. Rarely are they opposite to each other, i.e. on the same radii; this usually occurs only between whorls of different organ categories. It may be taken as a rule that new primordia appear in the largest free spaces available (Hofmeister 1868, Leins 1964). This is commonly the case between the organs of the preceding whorl. A critical case are obdiplostemonous flowers. Here, the stamens of the second of the two stamen whorls have smaller base areas than those of the first whorl, and, consequently, the free space is still slightly larger between the first stamens, when the carpels originate. Thus, the carpels are on the same radii as the inner stamens and not alternating with them. Diplostemonous and obdiplostemonous flowers may also occur in the same family (e.g. Rutaceae); Zygophyllaceae are obdiplostemonous, Liliaceae are diplostemonous (Eckert 1966).

All organs of a whorl are initiated more or less simultaneously. Between the last organ of a whorl and the first of the subsequent whorl a larger plastochron is intercalated. Thus, not single organs but whole groups of organs (the whorls) are initiated in regular plastochrons and divergence angles. On closer inspection, especially as revealed by the SEM, in many whorled flowers the organs of a whorl also arise in very rapid spiral sequences. The sequential appearance may be more pronounced under lower temperatures (Lyndon 1979). However, organ position in these flowers should not be called spiral because after each whorl there is an abrupt change in the divergence angle and plastochron. If there are small differences in organ position at inception due to the spiral initiation of each whorl, they are equalized in later development.

For convenience, initiation of organs is often equated with appearance of primordia at the surface. However, initiation seems to occur well before appearance (see also section 9.2.1).

In mature flowers no trace of the spiral initiation sequence of each whorl remains, except for the calyx with its frequent quincuncial aestivation. As an exception, some Convolvulaceae may be mentioned, where the stamens have different lengths according to the position in the whorl (Wagner 1989).

The role of prophylls (bracteoles) and calyx in spiral and whorled phyllotaxis

Going back now to the explanation of pattern formation for the spiral pattern: the first two organs are more or less opposite and with the subsequent organs the spiral becomes apparent (Fig. 8.19.2). This is exactly what happens in many dicot flowers, not only in ones that are spiral, but even in ones that are whorled: they start with two opposite (lateral) prophylls and then the 5-merous calyx is formed in a spiral, which results in a quincuncial-imbricate aestivation. If the two prophylls are lacking, correspondingly, the first two sepals have a lateral position (Breindl 1934). Now, the floral apex becomes somewhat 5-angled by the broadening of the 5 sepals and this sets the geometrical framework for the following whorls of often 5 petals, 5 or 5+5 stamens, and 5 (or often fewer) carpels (Figs. 8.25, 8.38).

The prophylls are not counted as floral organs. They are usually not so intimately integrated into the floral architecture as are the sepals. Often they remain small and sometimes they are early-caducous. They can also bear a flower in their axil. Although they seem unimportant in older buds and at anthesis, they have two important functions in early floral ontogeny: (1) a morphogenetical function: as the two first organs at the floral axis they mediate the onset of the spiral of the calyx; (2) an ecological function: they are protective organs for the sepal primordia.

In the monocotyledons with their prevailing 3-merous whorls usually only one, adaxial, prophyll is present. A spiral onset of the floral parts is less obvious (Schumann 1890, Sattler 1973).

2.5.3 More complicated patterns

More complicated patterns often arise when the floral organ number is fairly high and the organs of different categories have bases of rather different breadth.

Whorls with doubling ('dédoublement') and halving of organ positions

In whorled flowers at the transition from the corolla to the androecium there is often a doubling of organ positions (Figs 2.37.4, 2.39). At the expected place between the petals, rather than single stamens, collateral pairs of stamens develop (Čelakovský 1895, Murbeck 1914, Lundblad 1922, Endress 1987a). In other words, the whorl following the petals has twice the number of organs. Doubling occurs more often in 2- and 3-merous than in 5-merous flowers. This doubling of organ positions is perhaps enhanced by the narrow bases that stamens usually have compared with those of the perianth parts. It may also be correlated with a relatively long plastochron between the last perianth whorl and the first androecial whorl so that the circumference of the floral apex may widen considerably during the plastochron. This is especially the case in flowers where petals are lacking. In the gynoecium the number may decrease (halve) again.

If there are several whorls of stamens the number of stamens may double or halve again in subsequent circles. This apparently depends on the rate of further expansion of the floral apex (Karrer 1991, for Papaveraceae).

The pattern of organ pairs alternating with single organs occurs equally in families with otherwise isomerous whorls and in families with otherwise more extensive subdivision of single primary primordia. The first is the case in, e.g. Zygophyllaceae (*Nitraria*), the second in Sterculiaceae (*Theobroma*). These cases are morphologically similar but the evolutionary pathway of their origin may be different.

A doubling or irregular increase in organ number may also occur concomitant with organ reduction. In *Bauhinia galpinii* (Caesalpiniaceae) narrow staminodes occur in this pattern (Fig. 8.23.3); in some *Thunbergia* species (Acanthaceae) the reduced sepals occur in increased number (Fig. 8.50.3). In addition, the often doubled disk scales may be derived from small staminodes (e.g. in Cunoniaceae, Hamamelidaceae). The corona in Passifloraceae may also be interpreted as staminodial with this pattern (see section 8.5.1).

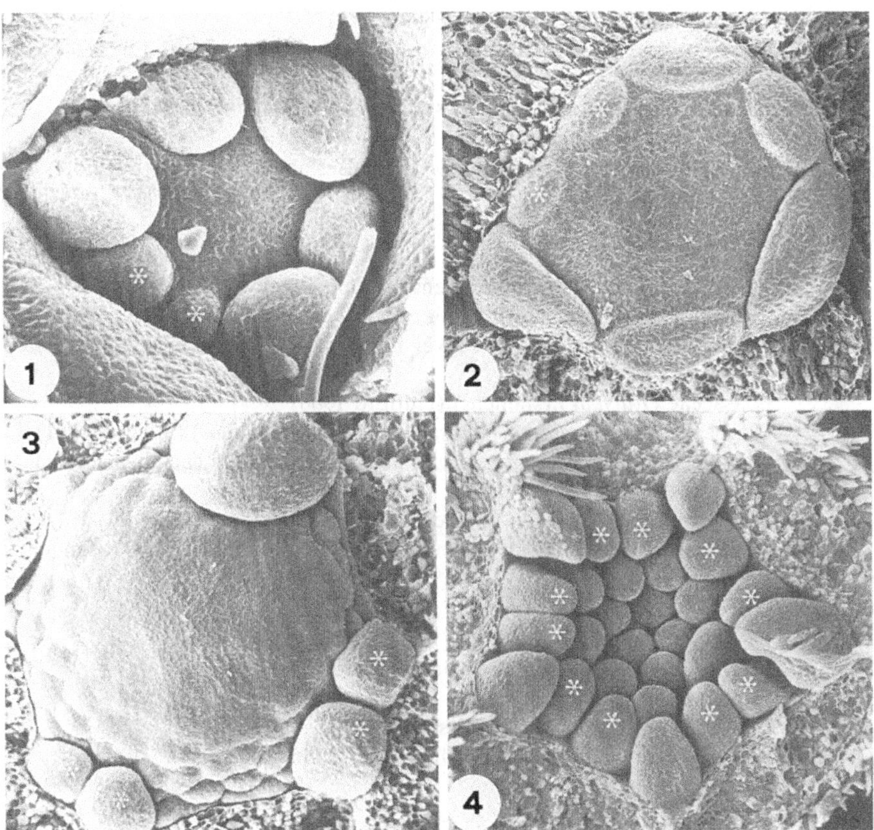

Figure 2.39. Examples of double positions in inner perianth region or outer androecium region (marked with asterisks). 1. *Artabotrys hexapetalus* (Annonaceae); in a sector of the inner perianth whorl (× 180). 2. *Caloncoba echinata* (Flacourtiaceae); in a sector of the inner perianth whorl (× 180). 3. *Exospermum stipitatum* (Winteraceae); two positions of innermost perianth (× 100). 4. *Hedycarya angustifolia* (Monimiaceae); second whorl of androecium (× 65).

Examples of doubling are predominantly in Magnoliidae, at the middle evolutionary level of the dicots, and in Alismatidae.

Magnoliidae:

Annonaceae (tepals, tepals–stamens) (Endress 1986a) (Fig. 2.39.1)
Magnoliaceae (tepals–stamens) (Erbar & Leins 1981)
Winteraceae (tepals) (Vink 1970) (Fig. 2.39.3)
Monimiaceae (tepals, tepals–stamens) (*Hedycarya*, *Wilkiea*,

Tambourissa) (Endress 1980a) (Fig. 2.39.4)
Cabombaceae (*Cabomba*) (petals–stamens) (Rohweder & Endress 1983)
Aristolochiaceae (petals–stamens) (Leins & Erbar 1985)
Ranunculaceae (petals–stamens) (Schöffel 1932)
Papaveraceae (Eschscholzioideae) (petals–stamens, stamens) (Endress 1987a, Karrer 1991)

Rosidae:

Rosaceae (petals–stamens, stamens) (Kania 1973)
Caesalpiniaceae (stamens–staminodes) (*Bauhinia galpinii*, personal observation; *B. malabarica*, Tucker 1988b: staminodes in female flowers) (Fig. 8.23.3)
Loasaceae (*Eucnide*) (stamens) (Hufford 1988)
Olacaceae (*Minquartia*, *Ochanostachys*, *Coula*) (stamens) (Stauffer 1961)
Geraniaceae (*Sarcocaulon*, *Monsonia*, *Hypseocharis*) (petals–stamens) (Rama Devi 1991)
Zygophyllaceae (*Nitraria*) (stamens) (Ronse Decraene & Smets 1991a)
Myrtaceae (*Chamaelaucium*) (petals–stamens) (Ronse Decraene & Smets 1991b)

Dilleniidae:

Flacourtiaceae (*Caloncoba*) (petals) (Fig. 2.39.2)
Capparaceae (stamens) (Karrer 1991)
Brassicaceae (stamens) (Endress 1992b)
Begoniaceae (stamens) (Ronse Decraene & Smets 1990a)
various Malvales (petals–stamens) (van Heel 1966, Bayer & Hoppe 1990)

Caryophyllidae:

Amaranthaceae (*Achyranthes*) (tepals–stamens) (Joshi 1932)
Polygonaceae (tepals–stamens) (Galle 1977)

Asteridae:

Oleaceae (sepals–petals) (Torgaard 1924)

Alismatidae:

Alismataceae (petals–stamens) (Sattler & Singh 1978)
Butomaceae (petals–stamens) (Sattler & Singh 1978)
Limnocharitaceae (petals–stamens) (Sattler & Singh 1978)

Irregular (chaotic) patterns

In textbooks, Magnoliidae are often characterized as having flowers with many organs that are spirally arranged; it is also acknowledged that some groups of the Magnoliidae have whorled flowers. However, it has largely been overlooked that in a large number of Magnoliidae the arrangement of the floral organs is neither spiral nor whorled but irregular (chaotic) (Figs. 2.38.3–4). Irregular patterns also occur in many other groups with higher numbers of floral organs. In some cases floral phyllotaxis starts with a regular pattern. This, however, decays as more primordia are initiated. The preconditions for an irregular pattern seem to be the following.

(1) Small size of single primordia compared with floral apex. Since stamen primordia are narrower than sepal and petal primordia, this happens most often in the androecium.

(2) Very rapid sequence of initiation of a larger number of organs (short plastochrons).

(3) Flowers with a reduced perianth seem more often to have irregular patterns, since the perianth does not establish a symmetrical framework for the further organ initiation (see Endress 1990a, Tucker 1990, 1992c, Lehmann & Sattler 1992, Scribailo & Tomlinson 1992).

Rutishauser (1983) showed, for vegetative shoots, an interesting correlation between phyllotaxis patterns and the size of the organ primordia and of the apex of the entire shoot. A corresponding comparative study for flowers is still lacking.

Flowers with androecia exhibiting secondary subdivision of the primordia (secondary polyandry)

In a number of groups of the dicotyledons the number of stamens in a flower is much larger than that of the outer organ categories (review by Ronse Decraene 1992 and Ronse Decraene & Smets 1992). Often there are as many primordia in the androecium as there are petals, but then there is secondary subdivision of these primordia (Fig. 2.37.5–6). Thus, one can distinguish between primary and secondary primordia. Only the secondary primordia will give rise to stamens. Each primary primordium may eventually yield a high number of stamens. A distinctive consequence of such an androecium initiation pattern may be the presence of groups of stamens that are basally fused (stamen bundles) in the mature flower. This is especially prominent in some Clusiaceae (e.g. *Montrouziera*, *Symphonia*) (Fig. 2.40) or Myrtaceae (e.g. *Beaufortia*, *Cladothamnus*, *Tristania*, *Melaleuca*). In flowers with primary and secondary androecium primordia the vascular bundles of the individual stamens tend to unite into 'trunk bundles', before they join the vascular tissue of the floral stele (Tucker 1972, Stebbins 1974a). The number of these trunk bundles corresponds to the number of primary primordia. This vascular pattern reflects the ontogenetic sequence of primary and secondary primordia.

Instead of distinct primary primordia an entire ring-mound may form, which will then give rise to many stamen primordia (Fig. 2.37.8–9). This behaviour occurs in many different families mainly of the Dilleniidae (Dilleniaceae, Lecythidaceae, Capparaceae) and Rosidae (Myrtaceae). There are different patterns by which stamens are initiated on the primary primordia. Usually there are irregularities because of the high stamen number in these flowers.

Morphologists have distinguished between centripetal patterns (outer secondary primordia formed before inner ones) and centrifugal patterns (inner secondary primordia formed before outer ones). The occurrence of centrifugal patterns is especially noticeable, as they are opposed to the developmental direction in the flower. The existence of centrifugal patterns has been known since Payer (1857) and Hofmeister (1868), but only Hirmer (1918) and still more so Corner (1946) pointed to their systematic significance. Cronquist (1957, 1968) evaluated it for the first time in the framework of the entire dicots or angiosperms. Cronquist emphasized that the centrifugal pattern seemed to be characteristic for larger systematic entities. Since then many species have been investigated, recently mainly with the SEM. Examples where a ring-like mound forms first and then the secondary primordia arise in a centrifugal direction are in Dilleniaceae (Corner 1946), Capparaceae (Leins & Metzenauer 1979, Karrer 1991), Lecythidaceae (Hirmer 1918), Cactaceae (Ross 1982, Leins & Schwitalla 1986), Limnocharitaceae (Leins & Stadler 1973, Sattler & Singh 1973) (see section 8.4.1, Figs 8.12, 8.15). Examples where a ring-like mound forms

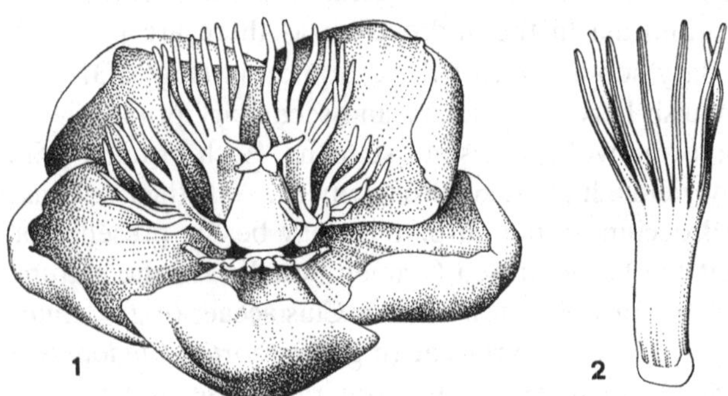

Figure 2.40. *Montrouziera gabriellae* (Clusiaceae). 1. Flower with five stamen bundles on the same sectors as the contort petals (× 0.3). 2. Stamen bundle from dorsal side with the long anthers (× 0.7).

first and then the secondary primordia arise in a centripetal direction are in Papaveraceae (Karrer 1991) and Myrtaceae (Ronse Decraene & Smets 1991b). In flowers with a ring-like androecial mound the gynoecium appears before the individual stamens or at least before the last stamens, irrespective of whether the initiation pattern is centrifugal or centripetal. Thus, this pattern may be seen as a delay in the initiation of individual stamens while the floral apex expands, so that the androecial part of the floral apex is delimited from the gynoecial part by a furrow and proliferates in the form of a ring-mound, before the stamens are initiated. The latest stamens may appear long after gynoecium initiation. In flowers with several separate primary primordia centripetal (*Eucalyptus*, Myrtaceae; Drinnan & Ladiges 1989) and centrifugal (*Stewartia*, Theaceae; Erbar 1986) patterns also occur.

In addition to the more uniform ones, there are also very polymorphic groups with centripetal and centrifugal patterns (e.g. palms, Uhl & Moore 1980, Uhl & Dransfield 1984, Uhl 1988). Therefore, as always, it is not an absolutely reliable feature for the delimitation of larger taxa.

As yet, only a very small range of polyandrous androecia have been analysed. It seems premature to characterize the Dilleniidae by a prevalence of the centrifugal pattern. The flowers should also be studied in other biological respects of their androecia for a meaningful evaluation of the pattern.

2.5.4 Evolutionary aspects

The notion of spiral versus whorled floral phyllotaxis characterizing primitive versus advanced flowers is much too simplistic. It has been shown that in the Magnoliidae the diversity of patterns is extensive. There are perhaps fewer taxa with regularly spiral flowers than expected. Families with especially flexible patterns are Ranunculaceae (Schöffel 1932, C. Wagner, personal communication), Menispermaceae, Monimiaceae (Endress 1980a,b), Nymphaeaceae (M. Wolf, personal communication). In all these groups spiral and whorled flowers coexist in closely related taxa, sometimes even in the same species; more complicated and chaotic patterns also occur.

The occurrence of primary and secondary primordia in the androecium is puzzling. Hagemann (1984) and Leins & Erbar (1991) discuss the possible antiquity of this pattern. However, the complexity of this pattern and its lack in the early fossil record make its primitive status at the level of the entire angiosperms very unlikely.

Recent comparative studies of organ initiation patterns within larger taxonomic groups of a uniform floral groundplan have shown that the initiation patterns are not as stable as one would think. In particular, the studies by Tucker (1987, 1989) on Fabales have revealed considerable variation. Variation also seems to be present in grasses (Clifford 1987) (see also section 9.2.1).

3

Floral construction (architecture, gestalt)

It has long been known that flowers of a superficially similar form may occur as a result of convergence in very different taxonomic groups, owing to architectural constraints. Thus, a similar form (construction) is not necessarily congruent with a similar organization. The German terms 'gestalt' and 'bauplan' are often also used in English to express this contrast (Arber 1937). Troll (1928a) has analysed this for flowers and flower-like structures. Since entire inflorescences or, on the other hand, parts of flowers may form flower-like structures, the term 'blossom' is often used to distinguish such formations from flowers as defined by bauplan. In this sense 'flower' is related to the organization, and 'blossom' to the gestalt. The flower-like cyathium of a *Euphorbia* or the flower-like third part of a flower of certain Iridaceae (e.g. *Moraea*) are blossoms, not flowers.

Architectural constraints on flowers have the effect that organs are not able to attain any imaginable shape and size. They have much to do with the phenomenon that flowers appeal to us because of their harmonious and aesthetic forms. Detailed mechanical analyses of flower constructions have rarely been carried out. Brantjes (1981a, 1983b) studied the mechanical tuning of flowers.

Here an elaborate classification is not given. It may suffice to show some prominent forms. Other classifications are to be found in Delpino (1868–74), Vogel (1954), Werth (1956a), Kugler (1970), Leppik (1972, 1977), Koepcke (1974) and Faegri & van der Pijl (1979).

3.1 Large and small flowers, allometry

Flowers of the same organization, with the same number and arrangement of floral parts, may differ in size by several orders of magnitude. In the same order Scrophulariales, *Oroxylum* (Bignoniaceae), with massive, bat-pollinated flowers (10 cm in diameter) and *Micranthemum* (Scrophulariaceae), with minute cleistogamous flowers (0.1 cm in diameter), may have a mass difference of the order of ten or a hundred thousand to one. Comparative studies on this phenomenon are largely lacking. Studies of allometry or scaling and its influence on changes in the internal structure in flowers are also still in their infancy. Niklas (1992) gave a general framework for biomechanics in plant evolution.

It is evident that the problem of mechanical stability is quite different in small and big flowers. An obvious feature is that the largest flowers have united perianth organs. The union of organs provides stability to the entire floral architecture. The free upper parts of the perianth are then either relatively short (e.g. *Aristolochia* p.p.; *Rafflesia*; *Brugmansia*, Solanaceae; *Fagraea*, Loganiaceae) or if they are long, then thay are lax and pendulous (*Aristolochia* p.p.; *Paphiopedilum*, Orchidaceae).

Homologous parts in small and large flowers may have different thicknesses because of different numbers of cell layers. Larger parts also tend to have more vascular bundles. At organ initiation the floral size may be relatively similar in small and large flowers; as a rule, at first a simple median vascular bundle is formed in each organ. Later, secondary bundles may develop in smaller or larger numbers. Stability may also be increased

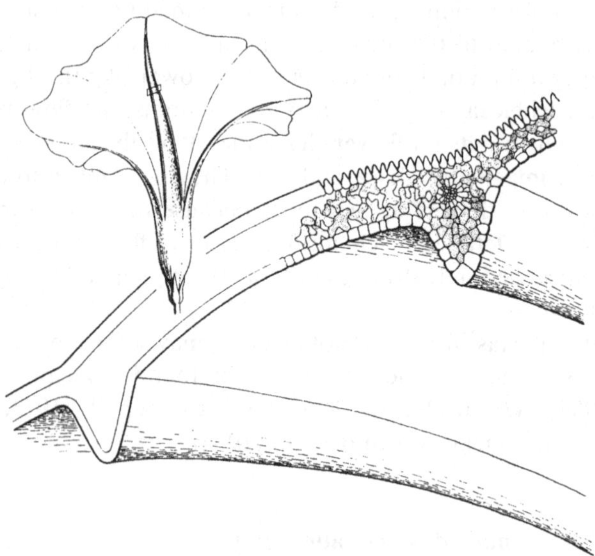

Figure 3.1. *Ipomoea tricolor* (Convolvulaceae). The large corolla has a weak texture because of the thin cell walls and large intercellular spaces but is reinforced by its architecture: the petals are united into a funnel-shaped structure with 10 longitudinal ribs (from Matile 1978, with permission).

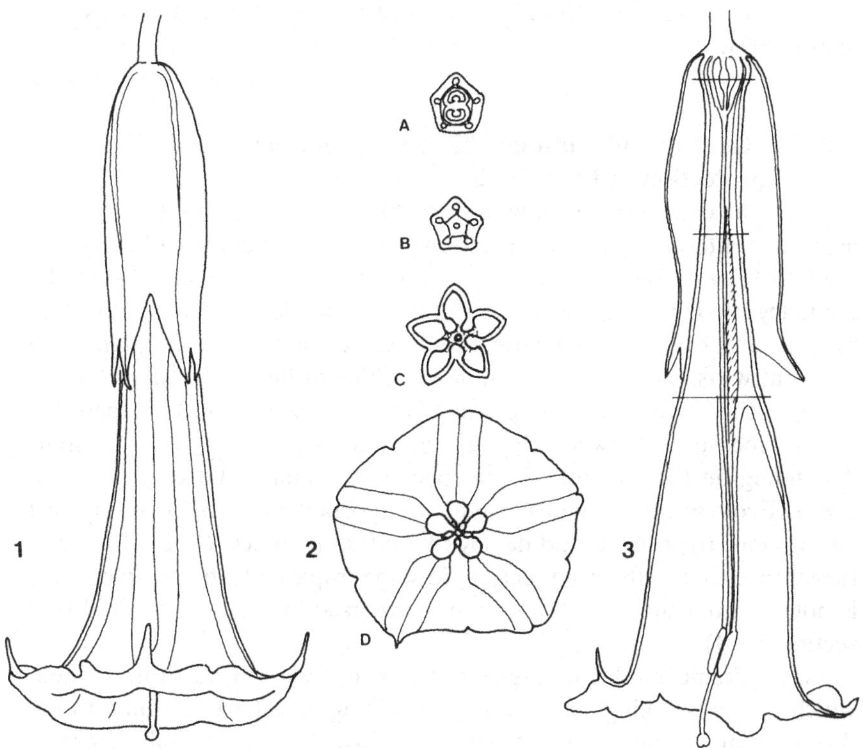

Figure 3.2. *Brugmansia sanguinea* (Solanaceae) (× 0.5). 1. Flower from the side. 2. Transverse section series; levels A–C indicated in 3; D is a view from outside into the flower, showing 15 longitudinal ribs. The stamens together with the corolla form five canals through which nectar may be reached by long-billed hummingbirds. Nectar is secreted at level A around the ovary.

by longitudinal ribs (Wainwright 1988), mostly over vascular bundles. This often occurs in perianth organs (Figs 3.1, 3.2). In widely open flowers with unfused perianth organs, longitudinal ribs may occur even in small-sized flowers (see section 3.3).

The fruits of Lauraceae have the same organization but very different sizes. In large fruits there are more vascular bundles than in small ones. Formation of additional vascular bundles may begin before anthesis, although there is no difference in ovary size at that stage (Endress 1972a) (see section 2.4.4).

Sprengel (1793) found that small-flowered species of *Malva* are devoid of a nectar cover ('Saftdecke') in contrast to large-flowered species (see section 3.4).

3.2 Stacking and stuffing: the struggle for space and protection in floral buds

The comparative study of floral bud architecture has largely been neglected, although it has repercussions on the understanding of particular shapes of floral organs at anthesis. In buds there is a conflict between the tendency of single organs to expand and the tendency of the entire bud to keep the organs in a restricted space. A compromise is attained in different ways. There is always dense packing in bud with minimal space between the organs. In some groups, hairs are used as 'packing material' in the free space between the organs, and they may be long or short depending on the amount of free space (e.g. Hamamelidaceae, Endress 1989). Ecologically, a tight packing is important as a protection against desiccation, frost, heat, and damage by infesting insects (Berg 1990a). In the strange case of the water calyces in some tropical plants, organ packing is not tight but the young organs are immersed in a protective fluid (see section 2.2.4).

Two extreme modes of organ packing in buds can be distinguished, which I term 'stacking' and 'stuffing'. Stacking buds have a regular three-dimensional arrangement pattern of organs. They commonly occur in flowers with a stable organ number and regular whorled phyllotaxis (Fig. 3.3.1–2). Stuffing buds have an irregular three-dimensional arrangement pattern of organs. They occur especially in flowers with variable and high organ numbers and spiral or irregular phyllotaxis (Fig. 3.3.3–6). In stuffing buds the organs shape each other by mutual pressure and thus attain irregular outlines. The individual form is superimposed by the influence of contiguity (Endress 1975). In later development, when the organs become free from the constraints in bud, the irregularity is either retained or it may be lost again. There are prominent examples, where the irregular shape is retained at maturity: in Annonaceae carpels and stamens are irregularly angular, since they are totally contiguous in bud (Fig. 8.1.9), and the sterile anther apices form a protective shield by their extreme contiguity. In Austrobaileyaceae the shape of the inner staminodes is influenced by their neighbours and by the stamens that surround them (Endress 1980c).

In *Balanophora* the extremely reduced, perianthless female flowers are densely packed in large numbers. They are covered and supported by clavate organs that fill the space between the flowers (Fig. 3.4). The clavate

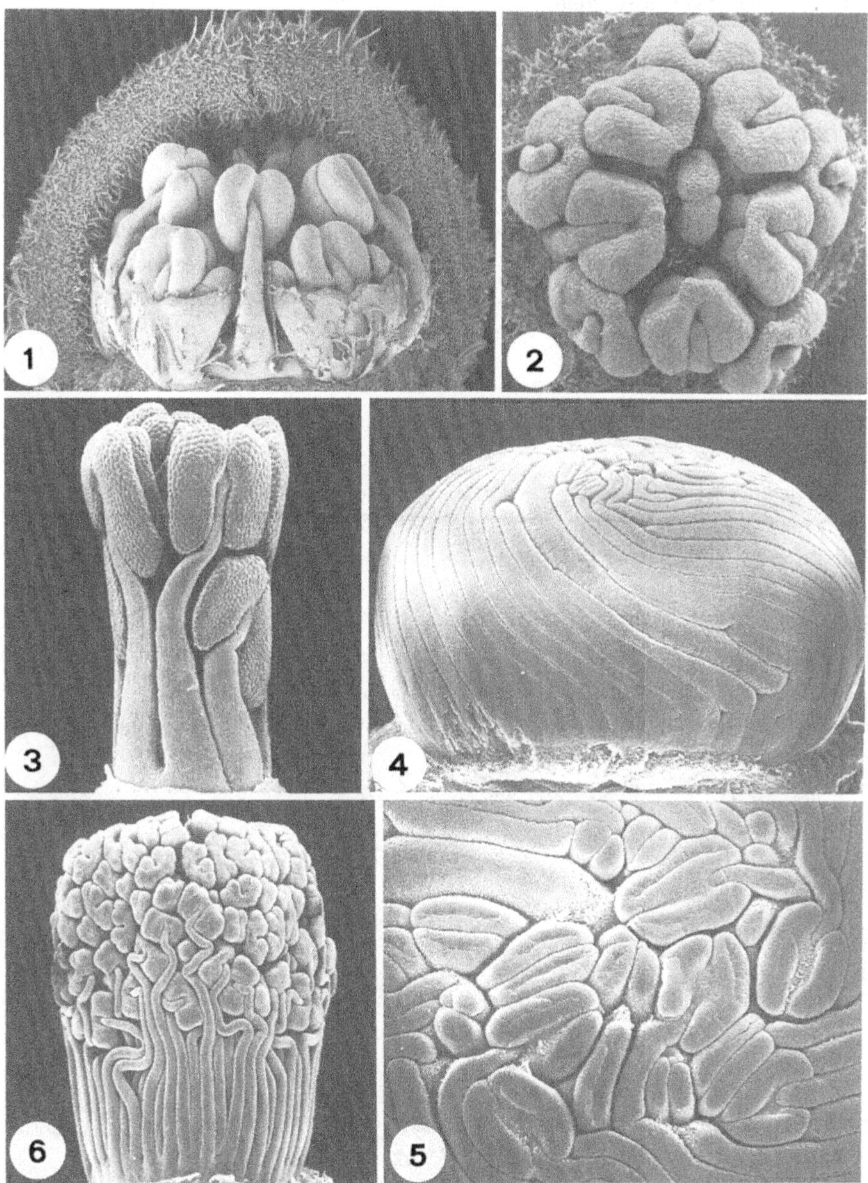

Figure 3.3. Stacking (1–2) and stuffing (3–6) of stamens in floral buds. 1. *Gillbeea adenopetala* (Cunoniaceae); anthers in two tiers, from the side (× 25). 2. *Caldcluvia* sp. (Cunoniaceae); anthers in two tiers, from above (× 60). 3. *Pisonia umbellifera* (Nyctaginaceae); from the side (× 10). 4–5. *Barringtonia samoensis* (Lecythidaceae). 4. From the side (× 15). 5. From above (× 25). 6. *Archidendron vaillantii* (Mimosaceae) (× 17).

organs may represent bracts or sterile female flowers that have attained protective functions. A similar situation is to be found in the female inflorescence of *Typha*.

3.3 Flowers with open access to the centre and to rewards

In many flowers access to the floral centre is not restricted. They are mostly choripetalous but there are also sympetalous forms (e.g. *Swertia*, Gentianaceae).

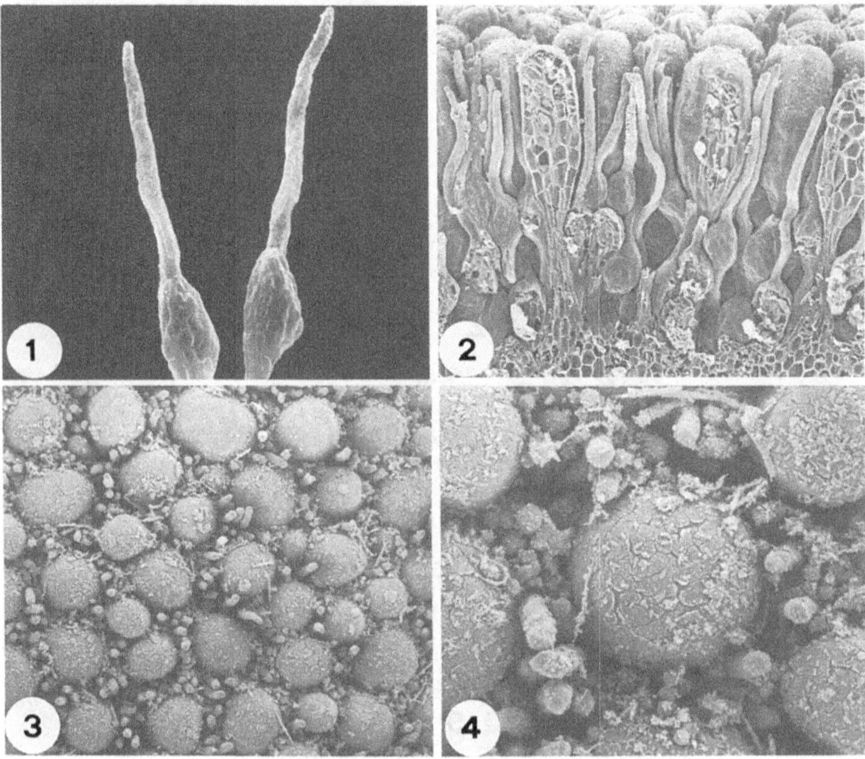

Figure 3.4. Stacking of female flowers in the inflorescence of *Balanophora fungosa*. 1. Two isolated female flowers at anthesis (× 65). 2. Longitudinal section of female inflorescence with stacking of flowers and club-shaped protective organs (× 35). 3. Surface view of female inflorescence with flowers and protective organs (× 35). 4. Magnification of 3 (× 110).

Disk- and bowl-shaped flowers

In many flowers the sepals, petals and stamens are spreading at anthesis and the floral centre is open. The flowers are flat or bowl-shaped. Nectar, if present, is readily accessible. Such flowers are often visited by a wide spectrum of insects, usually by taxa with a short proboscis. Insects with a long proboscis may have difficulty in expanding the proboscis in such flowers. Most pollen-flowers belong to this type (e.g. Dilleniaceae, Ochnaceae, Begoniaceae). There is also a large number of nectariferous flowers of this shape (e.g. Anacardiaceae, Celastraceae, Araliaceae, some Euphorbiaceae, some Liliaceae). Often the nectary is a flat disk and it may in itself be a nectar guide by its contrasting coloration.

In flat flowers the spreading perianth parts that form the peripheral region of the flat construction are often reinforced at the base by longitudinal ribs on the upper surface. These organs are part of the platform for pollinators, and since they are not fused in most flat flowers a reinforcement may be important. This should be comparatively investigated (Fig. 5.5.3).

The capitula of most Asteraceae may be seen as disk-shaped blossoms, although nectar is included in a short floral tube. But since the individual flowers are very small, nectar can also be reached by short-tongued insects.

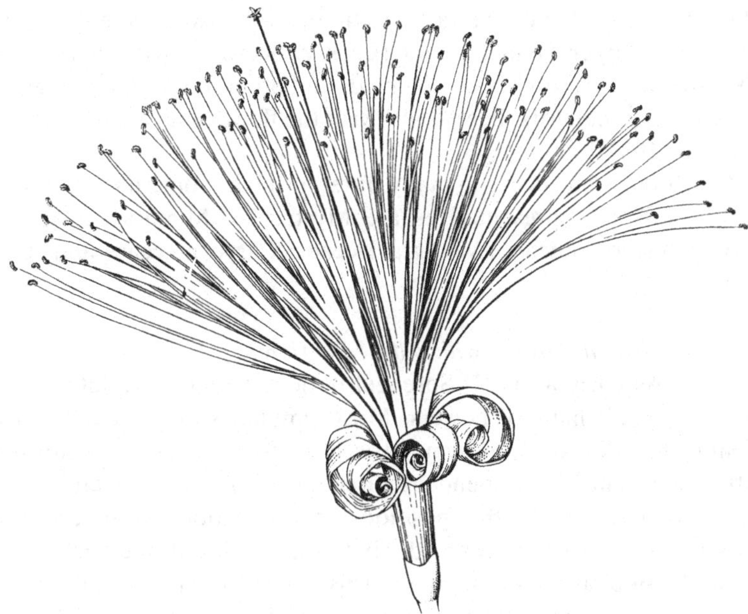

Figure 3.5. Brush-flower: *Pachira aquatica* (Bombacaceae) (× 0.4).

Brush-flowers

Brush-blossoms are flowers or inflorescences where the stamens dominate the other floral organs in their appearance. The stamens are relatively long and showy. If the brush-blossoms are single flowers, the stamen number usually is greatly increased. If they are entire inflorescences, the flowers are densely aggregated and the stamen number is often also increased. Brush-flowers are either mere pollen-flowers or they produce nectar. Large brush-flowers often have basally united stamens, e.g. *Barringtonia* (Lecythidaceae), *Pachira* (Bombacaceae), and *Archidendron* (Mimosaceae) (Figs 3.5, 8.30). This is ontogenetically derived from common primary primordia and may be seen as enhancing stability of the long filaments.

Since pollen and nectar are in relatively easy reach, brush-flowers are often exploited and pollinated by a wide range of animals. *Barringtonia calyptrata* (Lecythidaceae), a tree in coastal regions of tropical Queensland, is an example. The white brush-flowers are in pendulous inflorescences. In the evening they emit a sweetish-musty smell and are visited by bats. In the morning lorikeets and bees are frequent visitors (personal observation). *Syzygium cormiflorum* (Myrtaceae), also from tropical Queensland, is successfully pollinated by bats, birds and various insects (Crome & Irvine 1986). *Syzygium syzygioides* is visited by many short-tongued Hymenoptera and Diptera but also by some Lepidoptera and some larger Hymenoptera (Lack & Kevan 1984). In *Inga brenesii* (Mimosaceae) various bees, sphingids and other moths, hummingbirds and bats act as pollinators; beetles also visit the flowers but are poor pollinators (Koptur 1983).

Brush-blossoms occur in many different taxa. Some families with many brush-blossom representatives are Myrtaceae, Mimosaceae, Lecythidaceae–Planchonioideae, Marcgraviaceae, Caryocaraceae and Capparaceae.

Pendent flowers with reflexed perianth

As early as 1793, Sprengel noticed certain correlations between direction and shape of flowers. Most familiar is the predominantly horizontal direction of monosymmetric flowers (see below). A form that is often correlated with pendent direction is a widely open flower with reflexed perianth. In this position the pollination organs are relatively protected against rain. Since the petals are reflexed, they are also visible from the side, as are the stamens. This flower form is especially well suited for buzz-pollination (see section 5.1). But there are also large flowers with this form that are adapted to various large pollinators. In these a peculiar

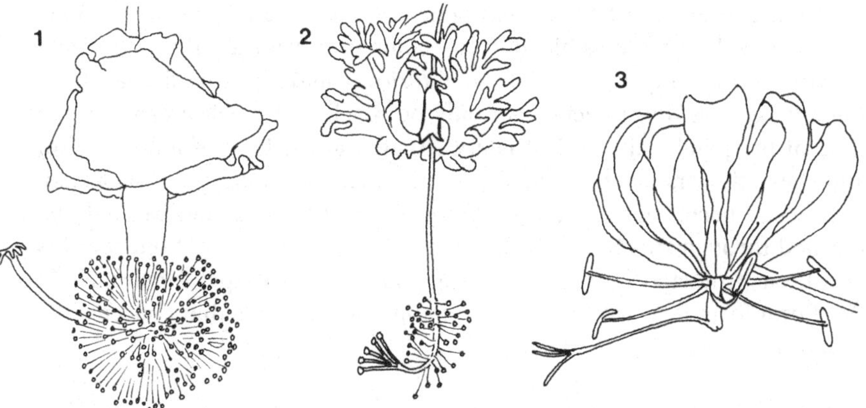

Figure 3.6. Large, pendent flowers with reflexed perianth and stigma sharply curved to one side. 1. *Adansonia digitata* (Bombacaceae) (× 0.4) (after Porsch 1935). 2. *Hibiscus schizopetalus* (Malvaceae) (× 0.8). 3. *Gloriosa superba* (Liliaceae) (× 0.4).

change in the architecture may occur. While in the small flowers the style is directed downwards, in some large flowers it is sharply curved to one side so that the stigma is touched by an animal that approaches the flower from the side and moves around the flower to completely exploit it (revolver flower, see section 3.4). Examples are *Gloriosa superba* (Liliaceae; butterflies), *Adansonia digitata* (Bombacaceae; bats), *Hibiscus schizopetalus* (Malvaceae; butterflies or birds?) (Fig. 3.6) (see also Baker 1986a).

3.4 Flowers with canalized access to the centre or to other sites with rewards

In the majority of flowers the general access to the floral centre is impeded in such a way that only certain groups of potential pollinators may reach nectar (or other rewards). Nectar (or pollen) may also be protected from draughts or rain. The most common device is a narrowed tubular part, mostly formed by the corolla, between the floral centre and the outer world. Often the narrow part is, in addition, provided with hairs (Fig. 5.13.1). This had been noticed by Sprengel (1793), who interpreted it as protection against raindrops ('Saftdecke' or nectar cover). Later interpretations rather stressed the protection against nectar thieves (first by Delpino, 1890) and against nectar evaporation. Nectar may also be stored in spurs. But even in flowers that appear to be widely open, nectar may be concealed and accessible only to pollinators with a sufficiently long proboscis. In many Liliaceae (e.g. *Gloriosa*) and Caesalpiniaceae (e.g. *Caesalpinia*, *Delonix*) nectar can only be reached through a canal in the base of the petals, although it is not situated at the base of a floral tube, and these flowers are pollinated by butterflies with long proboscides. In Malvaceae nectar is secreted on the inside of the sepals and can only be reached through slits between the base of the petals; in addition, the slits are often protected by stiff hairs originating from the petal margins. In this floral form it is just the wide spreading of the petals that conceals the nectar.

Tubular flowers

Tubular flowers have a more or less long narrow part, which canalizes the access to the floral centre where the nectar is usually situated. There are different possibilities for the construction of tubular flowers. In most cases the tubular structure arises by the congenital fusion of the petals (sympetaly) and intercalary elongation of the fused region. The tubular part can also arise by the incomplete opening of a contort choripetalous corolla (e.g. *Malvaviscus*, Malvaceae) (Fig. 4.2.1), or by postgenital fusion of free petals (e.g. *Oxalis tubiflora*, Oxalidaceae; *Correa*, Rutaceae; *Stackhousia*, Stackhousiaceae, Hartl 1957; Loranthaceae, Feehan 1985). A floral tube can also be provided by the calyx, which, by its own fusion or simply by its stiffness, holds the free petals together in a tubular manner (e.g. *Wormskioldia*, Turneraceae).

Hence, tubular flowers have diverse organizations, but mostly they have congenitally united petals. This is apparently the easiest way to achieve this form. All other ways are more complicated and therefore less likely to occur.

Salverform flowers

Most flowers with a tube have an upper part, which is spreading and so directs floral visitors to the floral centre. The spreading part often consists of the free petal apices. If the tubular part goes relatively abruptly into the spreading part, the flowers are called salverform. Salverform flowers are common in Gentianales (Gentianaceae, Loganiaceae, Apocynaceae), Rubiaceae and Solanaceae, but they also occur in choripetalous flowers, e.g. Pittosporaceae and Escalloniaceae. Salverform flowers are especially prominent among those pollinated by Lepidoptera, but there are also many bee-pollinated ones. The entrance into the narrow part of the salver may be partly covered by scales (ligules) at the inner side of the perianth organs and so canalize the way to the nectar (e.g. Erythroxylaceae).

In many bat-pollinated flowers, the upper, extended part is bell-shaped rather than flat, e.g. some Bignoniaceae, *Solandra* (Solanaceae) (Vogel 1957, 1968, 1969a,b). According to different proportions and shapes of the upper part or of both lower and upper part, more types are often distinguished (Werth 1956a, Kugler 1970).

Spurred flowers

A spur is a sac-like appendage of a floral organ, mostly of a petal. It originates by differential dilatation of a flat organ. Usually, a spur contains nectar, which is secreted in the spur itself or near the entrance of the spur. A flower mostly contains a single spur and is, therefore, monosymmetric. Many orchids have spurs (e.g. *Angraecum*) (Fig. 8.64.3); in Balsaminaceae they are a family character. Spurs are rare in groups with predominantly tubular flowers. Rarely, each petal of a flower has a spur (e.g. *Halenia*, Gentianaceae, with four spurs). *Diascia* (Scrophulariaceae) has two spurs that contain oil. It is common for oil-flowers to have the elaiophores in pairs, since the bees collect oil with their legs and not with the proboscis (see section 5.3).

Usually spurs are visible from the outside of a flower; however, this is not the case for the 'inner' spurs. They arise in highly monosymmetric floral cups. The inside of the floral cup is extended downwards on one side by differential intercalary growth. It looks then as if the ovary sits on one side of the wall (or on top) of the floral cup and a canal leads down to the nectary on the other side. Inner spurs occur, for example, in *Pelargonium* (Geraniaceae), *Bauhinia*, *Amherstia* (Caesalpiniaceae), *Dactyladenia* (Chrysobalanaceae) and *Epidendrum* (Orchidaceae) (Fig. 8.21).

Spur length is highly plastic, and this plasticity is often used for different pollination biological modes within genera (outer spurs, e.g., in various orchids, Nilsson 1988; *Impatiens*, Grey-Wilson 1980; inner spurs, e.g., in *Pelargonium* and *Bauhinia*, Vogel 1954).

Flowers with tubular anthers

In some pollen-flowers, the anthers are large and tubular and open by a pore at the upper end. Pollen is only available if the anthers are buzzed, thus only to a limited array of floral visitors: certain groups of bees (Roubik 1989). Pollen is also protected against rain in this way. There is a large number of tropical angiosperm groups with this specialization (see section 5.1).

Flowers with concave (hidden) stigmas

Concave stigmas occur particularly in flowers with precision mechanisms in the pollination apparatus, flowers with pollinaria (Asclepiadaceae,

Orchidaceae) (Fig. 8.68.6), or flowers where the anthers and stigmas are in close proximity (Zingiberaceae, Costaceae, Marantaceae) (Fig. 8.62).

They also occur in other cases. Flowers that are pollinated by larger animals, even in a relatively imprecise manner, such as birds or Lepidoptera, may have peculiar stigmas: the receptive surface is in a dimple surrounded by a raised whorl of stiff, unreceptive hairs like a comb (*Delonix*, Caesalpiniaceae; *Alpinia*, *Globba*, Zingiberaceae) (Figs 8.22, 8.26.4, 8.62). It may be that the hairs, while touching the pollinator's body or wings, generate electrostatic forces that attract adherent pollen or that they hold a drop of secretion that is squeezed out of the stigmatic chamber or hollow style through its movement by the pollinator's body (*Globba*, Zingiberaceae, Müller 1931). In other flowers of this kind the stigma is hemispherical (capitate), sticky and has a much larger diameter than the style that bears it (*Bauhinia*, *Peltophorum*, Caesalpiniaceae; *Orthosiphon grandiflorum*, Lamiaceae, van der Pijl 1972; *Stachytarpheta*, Verbenaceae) (Fig. 8.22.1).

Revolver flowers

Some flowers have more than one access towards the nectariferous part. To exploit the entire nectar content a visitor has to probe each of them. This usually involves its rotation around the flower axis and efficient (abundant) contact with the pollination organs. The pollinator commonly has a radial orientation on the flower. The number of channels for the mouth parts of a pollinator often matches the number of stamens or petals, since these organs are often synorganized to provide the separate entrances. Examples are Apocynaceae and Asclepiadaceae (Fig. 8.36) or *Brugmansia* (Solanaceae) (Fig. 3.2).

Roundabout flowers

Some large flowers have the reward arranged in a circle. To exploit it the pollinators (usually bees) have to move around the entire flower with the body tangentially oriented (in contrast to revolver flowers). Pollination is nototribic. Examples are species of *Passiflora* with nectar as reward or species of *Dillenia* with pollen as reward (personal observation) (Fig. 8.16). In *Rafflesia*, female flowers are roundabout flowers (see section 8.3.2).

3.5 Monosymmetric and polysymmetric flowers

Monosymmetric (zygomorphic) and polysymmetric (actinomorphic) flowers (see section 2.1) are mentioned here again. The reason is that monosymmetry has different causes. It may be genetically deeply rooted and be a feature of the groundplan, or it may be more superficial. In any case, it is closely connected with the 'gestalt' of flowers (Mair 1977, Endress 1992b). Monosymmetry may also arise in different phases during ontogeny. A century ago physiologists were interested in the mechanical causes of monosymmetry. They tried to induce it artificially in polysymmetric flowers and, vice versa, to produce polysymmetric flowers out of naturally monosymmetric ones. Vöchting (1886) showed that in some groups gravity induces monosymmetric flowers (positional monosymmetry), in others flower form is not influenced by gravity (constitutional monosymmetry), and in still others both gravity and constitution play a role. Only positional monosymmetry occurs in groups with largely polysymmetric flowers; it is attained only late in ontogeny (e.g. *Schlumbergera* of Cactaceae, see Fig. 4.2.3). Constitutional monosymmetry occurs in groups with predominantly monosymmetric flowers; it is attained early in ontogeny (e.g. Scrophulariales, see Fig. 8.55, Wunderlin 1992).

That gravity plays some role, directly or indirectly, in this context, may also be seen in the fact that in most monosymmetric flowers the median plane is the symmetry plane. The natural position of monosymmetric flowers is mostly such that the symmetry plane is vertical. This coincides with the direction of the symmetry plane of the pollinating animals, which are also monosymmetric.

In addition, naturally occurring deviant polysymmetric flowers among species that normally have monosymmetric flowers, the pelories, were studied (e.g. Peyritsch 1872, W. & A. Bateson 1891). They occur as terminal flowers of racemes with normal flowers or, in some aberrant specimens, all flowers may be replaced by polysymmetric ones. Pelories have often been interpreted as atavistic forms. However, they are rather monstrosities because they may have five spurs (e.g. in Antirrhineae, Scrophulariaceae). Normal spurred flowers among angiosperms (with very few exceptions) have a single spur and the presence of a spur is only a consequence of monosymmetry.

3.6 Flag-flowers (papilionate flowers) and lip-flowers

There is a large number of monosymmetric (zygomorphic) flowers that are oriented to the side and are differentiated into a lower and an upper lip. The pollination organs, anthers and stigma, are positioned near the median plane and pollen is often more or less protected by the upper or lower lip. The lower lip functions as a landing platform. This floral form may be based on quite different floral groundplans: the lips are mostly formed by the corolla but they may also consist of sepals or even of parts of the stamens. A large number of such flowers are pollinated by Hymenoptera. From their systematic distribution it may be assumed that the evolutionary advent of such flowers was largely triggered by Hymenoptera. The pollination organs may be protected by the upper lip and the bees pollinate with their back (nototribic) or by the lower lip and the bees pollinate with their underside (sternotribic) (in slightly asymmetric flowers also with their flanks: pleurotribic). The first form is known as 'lip-flowers', the second as 'flag-flowers' ('papilionate flowers'). The gynoecium tends to be embraced by the united stamen filaments or by a single elaborate stamen, in flag-flowers from below, in lip-flowers from above (Fig. 3.7). Both kinds are concentrated in certain orders and families of the flowering plants, but the opposite form is always also present in low number, showing that evolution from one to the other is relatively easy. Lip-flowers are centred around Scrophulariales, Lamiales, some Campanulales, Zingiberales and Orchidales; flag-flowers mainly occur in Fabales and Polygalaceae. In both forms there are special elaborations leading to complete hiding of the pollination organs and mechanical application during a pollinator's visit in various ways. Application may be possible several times or restricted to the first visit, when the flower opens explosively. Increased

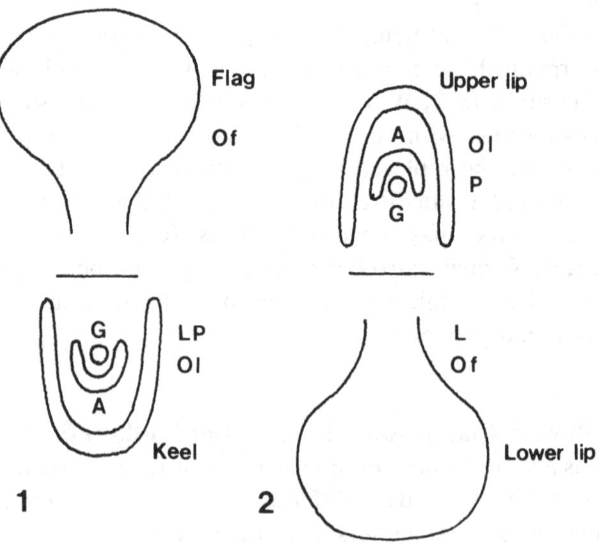

Figure 3.7. Analogies in elaborated flag-flowers (1) and lip-flowers (2) (diagrams). L, landing place; P, pollination organs (A, androecium, G, gynoecium); O, optically attractive organs (f, frontally exposed, l, laterally exposed).

synorganizations between corolla, stamens and style helps to keep the precision mechanisms also in enlarged flowers. There is a great diversity in these elaborations (see chapter 8).

Evolution has also led to larger, bird- or bat-pollinated flowers, which are prominent in a number of these groups. An extreme case are the large bird-pollinated flowers of *Strelitzia reginae* (Zingiberales), flag-flowers with the stamens and style included in a keel formed by the two lower petals. This keel is robust and elastic and is also subtended by a very stout inflorescence bract. It opens under the weight of a perching bird and closes again when relaxed (see section 8.9.2) (Figs 8.57, 8.59).

In some rare cases several flowers together form a kind of a lip-flower (lip-blossom), e.g. in *Mimetes* (Proteaceae) (Vogel 1954, Classen-Bockhoff 1990).

3.7 Trap-flowers

Trap-flowers contain a chamber which pollinators actively or passively enter and in which they remain for a shorter or longer period. The exit is in some way impeded or even made impossible for a while. Pollinators are predominantly Diptera. Some trap-flowers are also pollinated by beetles or bees. Trap-flowers are large compared with their pollinators, and the largest flowers in the world (*Rafflesia* and *Aristolochia*) are trap-flowers. The large floral size is due only to the trap, commonly formed by the perianth, not to the pollination organs, which may be rather small.

Some groups, particularly many Araceae, have trap-blossoms, in which the trap is formed by a large bract (spathe) and the pollination part consists of an entire inflorescence of reduced flowers concealed in the bract.

In the most elaborate trap-flowers (blossoms) the trap is a long tubular part (pitcher trap), which contains the pollination organs in its often widened base. Especially characteristic are window-panes (Yeo 1972) in the trap wall around the pollination organs. The window-panes are wall areas that lack pigments and an optical tapetum so that light from the outside can enter freely through the wall. In contrast, less light enters from the side of the floral entrance, since the trap tube is often dark, narrow and sometimes curved (*Aristolochia*, Aristolochiaceae; *Ceropegia*, Asclepiadaceae, *Arisaema*, Araceae). Thus, the light from the 'wrong' side impedes the navigation of the visitors in the trap but directs them to the pollination organs (McCann 1943, Troll 1951, van der Pijl 1953).

The entrance of the flower often exhibits contrasts between light and dark pigmented lines or spots. It is often directed sideways or is protected by a roof-like extension. The entrance may also contain osmophores. In extreme cases, where carrion is mimicked, the floral scent is obvious. Scent is produced by the perianth (e.g. *Ceropegia*, *Aristolochia*) or by the appendix of the inflorescence in Araceae (Vogel 1963a). The entrance may also be covered by wax glands, the secretions of which are involved in precipitation of the visitors into the trap (*Ceropegia*, Asclepiadaceae; some Araceae) (Vogel 1961).

The tubular part of the trap may contain narrow regions constricted by diaphragm-like structures that limit the size range of floral visitors (e.g. *Aristolochia*, *Rafflesia*, *Ceropegia*) and also impede a rapid exit (Figs 8.7, 8.8, 8.37). Furthermore, stiff hairs that are directed towards the floral centre may reinforce this effect (*Ceropegia*, *Aristolochia*, Araceae) (Vogel 1961).

Trap-flowers or trap-blossoms tend to be protogynous. This may enhance outcrossing rather than protandry, since the floral visitors remain in the traps for some time.

Less elaborate (fly) trap-flowers lack long floral tubes. They occur, for example, in *Abroma*, *Theobroma*, *Herrania*, and *Sterculia* (Sterculiaceae) (Young 1984) (Fig. 3.8), *Tacca* (Taccaceae) (van der Pijl 1953) and *Trechonaetes* (Solanaceae) (Vogel 1954). They tend to have several lateral entrances; this often occurs by a failure of the sepal or petal tips to separate as the flower opens (*Hydnora*, Hydnoraceae; *Sterculia*, Sterculiaceae; *Cer-*

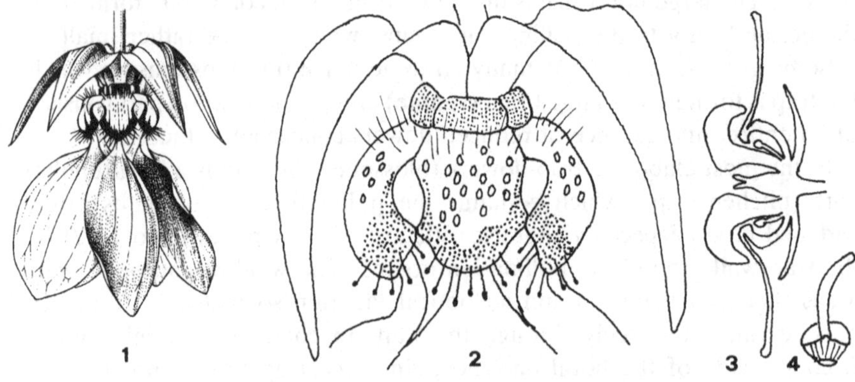

Figure 3.8. Trap-flowers in Sterculiaceae (in natural position). 1–2. *Abroma augustum*. 1. Flower from the side (× 1). 2. Trap formed by petal bases, showing lateral openings between petals, numerous window-panes in the petal bases, dark markings and stiff hairs (× 3.3). 3–4. *Herrania albiflora*. 3. Flower in longitudinal section, with trap formed by petals and inner staminodes (× 1). 4. Petal from ventral side, basal part with longitudinal lamellae (× 0.7).

opegia p.p., Asclepiadaceae; *Cryptophoranthus*, *Bulbophyllum*, *Satyrium*, Orchidaceae); in *Abroma* and *Theobroma* (Sterculiaceae) lateral windows arise by the firm and concave structure of the petal base (Fig. 3.8), in *Trechonaetes* (Solanaceae) by the incurved stamen filaments.

Bee trap-flowers are different but also contain some of the elements of the fly (and beetle) trap-flowers. They are mainly known from some orchids (Cypripedioideae, Stanhopeae). The flowers have a sliding zone with an oily surface. In the lady's slippers and in *Coryanthes* the flower visitors fall into a container. They can leave it only through a narrow path, in which they touch the pollination organs. In *Stanhopea* they touch the pollination organs directly on the slide (see section 8.10.1). In *Cypripedium*, window-panes occur in the trap (Troll 1951). However, in contrast to the fly trap-flowers, they are not deceptive but real guides towards the exit.

4

Floral adaptation to different pollinators (mode, style)

That wind and certain insects, mainly bees, flies, butterflies and moths, play an important role in pollen transfer and pollination has long been known and was impressively demonstrated two hundred years ago by Sprengel (1793). The role of birds and bats as pollinators was only discovered at the end of the last century in North America and, mainly, in the tropics. Only recently were some other mammals, such as certain primates, rodents and marsupials, found to act as pollinators in some regions. The history of these discoveries is summarized in Schmid (1975) and Baker (1980, 1983).

Not all floral visitors are pollinators. There are illegitimate visitors that are able to take nectar or pollen from a flower without pollinating it. They are also called nectar or pollen thieves (Inouye 1980, Neill 1987). There are also more efficient and less efficient pollinators. Potential pollinators may not pollinate each time they visit a flower. Thus, there are all kinds of intermediate situations between pollination and mere theft (see extensive discussion in Roubik 1989). The pollination efficiency of a certain pollinator depends on many factors (Stiles 1978b).

Many flowers have a large spectrum of different pollinators (polyphilic flowers or blossoms, see Faegri & van der Pijl 1979). More rarely are flowers adapted to a very narrow group of pollinators (oligophilic flowers), in extreme cases to a single species (monophilic flowers). Conversely, some animals visit many different flowers (polytropic visitors, see Faegri & van der Pijl 1979), others visit only a few (oligotropic visitors) or even a single species (monotropic visitors). On both sides the more specialized cases have found more attention in studies of pollination biology than the generalized, opportunistic ones. In the tropics opportunistic pollination syndromes are especially widespread in the canopy of wet forests (Frankie 1975, Gentry 1982).

In many cases pollinators transport pollen on particular parts of their body, depending on their behaviour on the flowers and on the floral shape.

From this part of the body, pollen is also deposited on the stigma. It is practical to distinguish (since Delpino 1868–74) three kinds of pollination: nototribic (by the back), sternotribic (by the underside), and pleurotribic (by the flanks).

Flowers that are pollinated by specific animals (zoophilous flowers) or abiotic agents (anemophilous and hydrophilous flowers) may show a characteristic syndrome of features, which is often recognizable at once. These syndromes may be termed the floral mode or style (Delpino 1868–74, Vogel 1954, 1980, 1990c, van der Pijl 1960, 1961). The term 'mode' may be preferred, because 'style' is also used in floral morphology to describe a part of the gynoecium. A particular mode may encompass very different floral organizations and also different floral architectures. The floral mode concept is practical. However, limits against gestalt or bauplan are not always clear (see section 9.4.2).

The recognition of pollination biological 'syndromes' is a powerful way of prediction for the design of studies in species whose pollination biology is unknown (e.g. Vogel 1968). The pollination mode of *Rafflesia* by carrion flies was predicted by Delpino (1873) and established more than a hundred years later (Beaman *et al.* 1988, Bänziger 1991). It is self-evident that the other side of the coin exhibits the dangers of false intuition, which may divert from an agnostic, flexible approach (e.g. Wyatt 1983).

Baker & Hurd (1968) have hypothesized that in pollinator–plant coevolution the initial selective force is created by the pollinator. The strongest evolutionary response is by the plant in the form of a suite of adaptations to attract the visitor consistently enough that it becomes a reliable pollinator. A specific floral construction could thus be evolutionarily modulated into a floral mode, a syndrome of characteristic features. This has been supported by other authors (e.g. Stiles 1981).

Rather than single species, entire guilds of species (see section 7.6) may be involved in coevolution (Nilsson *et al.* 1987, for hawkmoth-pollinated orchids; Neill 1987, for hummingbird-pollinated *Erythrina*, Fabaceae). Since there is commonly no strict one-to-one relationship between interacting plant and animal populations or species, this seems to be a common evolutionary situation ('diffuse coevolution') (Ehrlich & Raven 1964, Janzen 1980). It has been discussed for flower–pollinator relationships in several contributions in Gilbert & Raven (1975), and later by Paulus (1978, 1988), Feinsinger (1983a) and Schemske (1983b). The evolutionary plasticity of floral biological modes is an important facet of adaptive radiation (e.g. V. & K. Grant 1965) (see section 9.4.2).

4.1 Flowers pollinated by animals

4.1.1 Hymenoptera

Hymenoptera, and especially bees, are the pollinators par excellence, and a large part of the diversity of the angiosperms may be due to their coevolution with bees. Flowers adapted to bee-pollination are called melittophilous flowers. Generally, bees feed on nectar and the females also collect pollen for their own needs and for their larvae. Pollen collection is by buzzing (see section 5.1) in many tropical bee groups, especially Euglossini, many Anthophoridae, some Apidae (*Melipona* and *Bombus*), most Colletidae and Oxaeidae, some Andrenidae, Melittidae, Stenotritidae and Halictidae. Most pollen-flowers are pollinated by bees. Thus, bees play a predominant role as pollinators in such a large family as the Melastomataceae (Renner 1984a, 1989b). Some specialized bee groups also collect oil (some Anthophoridae), scent (Euglossini) or resin (several groups) (see chapter 5). The predominant, but not exclusive, colours of bee flowers are blue, yellow and ultraviolet. Nectar guides are common. The early morning is a major foraging time for bees. Therefore, many bee flowers open at that time or earlier, during the night (e.g., *Passiflora foetida*, sections 7.3 and 8.5.1) (Roubik 1989, Armbruster & McCormick 1990).

Bees are diverse in body size, length of proboscis and many other aspects of their biology (Roubik 1989). Social bees are especially versatile in their flower exploitation methods and they are also able to exploit a broad range of different floral forms. Although they are especially polytropic they often show individual flower constancy (V. Grant 1950b, Manning 1957, Waser 1986). For these reasons it is convenient to discuss some groups separately. Gross subdivision into larger and smaller bee groups is not only practical with regard to different floral forms but also with regard to different position of flowers in the vegetation: large bees prefer to forage in the canopy of trees, while smaller bees prefer lower regions in the forest (Perry 1984, Bawa *et al.* 1985b, Frankie *et al.* 1990).

Xylocopa (Anthophoridae)

Large bees, especially the carpenter bees (*Xylocopa*) play an important role in the pollination of tropical plants. The genus *Xylocopa* is not restricted to, but has its centre in, the tropics. Often, *Xylocopa*-flowers are large and striking and have, therefore, found much attention. A seminal paper to show the syndrome of *Xylocopa*-flowers and to discuss many biological aspects of *Xylocopa*-pollination was that by van der Pijl (1954). Cammerloher (1931) had already briefly discussed features of flowers pollinated by *Xylocopa*.

Xylocopa bees are large and strong; correspondingly, flowers pollinated by them may have an especially strong architecture. Examples include Caesalpiniaceae: *Bauhinia purpurea*, *Cassia*; Fabaceae: *Canavalia*, *Centrosema*, *Vigna* (Fig. 8.33); Melastomataceae: *Bellucia*; Asclepiadaceae: *Calotropis*; Acanthaceae: *Thunbergia grandiflora* (Figs 8.48, 8.49); Costaceae: *Costus speciosus*) (van der Pijl 1941a, 1954, Hurd 1978, Arroyo 1981, Gottsberger *et al.* 1988, Renner 1984a, 1987, 1989b). A number of them are shown in chapter 8. Females and males forage for nectar with their long proboscides. Females also collect pollen by buzzing (e.g. *Cassia*, Dulberger 1981; *Dillenia*, personal observation).

Further features of *Xylocopa* flowers include: hidden nectar that can only forcefully be reached, or, in pollen flowers, poricidal anthers; pale, unsaturated colours; fresh odours; nototribic (or pleurotribic) pollination; one-day flowers and often steady-state flowering strategy, since *Xylocopa* bees are important trapliners (mainly after van der Pijl 1954). Many plants pollinated by large bees are self-incompatible (Bawa 1974, Frankie *et al.* 1983).

Euglossini (Apidae)

The euglossines are an entirely tropical group of larger bees, confined to tropical America. They contain five genera, three of which are important flower pollinators (*Eufriesea*, *Euglossa*, *Eulaema*). They have striking metallic, iridescent colours. They inhabit mainly the understorey of wet forests. Because of their peculiar biology related to flowers they have attracted much attention and are therefore better known than other tropical bees. Male and female euglossines collect nectar. Females also collect pollen by buzzing. Females have a large foraging area and may trapline over distances of more than 20 kilometres (Janzen 1971) (see section 7.5). This seems to be the case for males as well (Williams & Dodson 1972). However, some individuals remain in smaller areas (Dressler 1982). Individuals, males included, have a relatively long life span, which may be more than two months (Ackerman & Montalvo 1985). This may be important in the context of traplining. The successive flowers of plant individuals on certain constant profitable routes may then be exploited for a longer time by an individual bee.

Important food sources for nectar are Costaceae, Marantaceae (Williams 1982), Lecythidaceae (Prance 1976), Apocynaceae (Janzen 1971) and Gesneriaceae (Wiehler 1978) (Fig. 8.45.1). Because of their long proboscides they can take nectar from flowers with long tubes. They often visit flowers that are also pollinated by butterflies (Vogel 1966a). However, since their proboscis is not coiled when not in action, but held straight under the body, nectar uptake is perhaps not possible from blossoms such as those of Asteraceae (Zucchi *et al.* 1969). Pollen is collected from flowers with many stamens and especially from buzz-flowers with poricidal anthers (Dressler 1982). Examples are *Cassia* and *Swartzia* (Fabales) (Janzen 1971). Janzen (1971) estimates that 'any given species at any given locality at any specific time of year usually has one or two primary pollen hosts and perhaps twice as many nectar hosts'.

Of special interest is pollination of 'perfume-flowers' by male euglossine bees, who collect scent substances that are produced abundantly by the flowers. They use it in some way in their sexual behaviour. The perfume is brushed by the forelegs from the substrate and transferred to the hind tibial organs where it is stored in special chambers. They also mix labial gland lipids with the perfume, which may increase the efficiency in retaining fragrance. It may be comparable with the 'enfleurage' technique of the perfume industry (Whitten *et al.* 1989).

This symbiosis based on the sexual behaviour of the bees and the scent of flowers occurs in many orchids and also in various other families but in no case as extensively as in orchids (see sections 5.5 and 8.10.1). There are still wide gaps in our knowledge of the detailed function of the scent substances in the sexual cycles of these bees (Williams 1982, Williams & Whitten 1983, Dressler 1982, Ackerman 1983).

The picture is made more complicated by the fact that the male euglossines also collect scented material from other sources, especially from the sap of wounded trees, rotting wood or rotting fruits. Further, although many orchid species have their particular perfumes, the mutualism between euglossine species and orchid species may not be very specific, at least in some cases (Ackerman 1989) (see sections 5.5 and 8.10.1).

In long-term studies carried out in Panama, Roubik & Ackerman (1987) found that over 20% of the euglossine species were never seen with orchid pollinaria, and thus probably seldom visited orchids. While many orchids completely depend on euglossines for pollination, many euglossines may survive without orchids (or other perfume flowers). This notwithstanding, diffuse coevolution between euglossines and perfume flowers is likely (Roubik & Ackerman 1987).

Other large bees

The other principal large bees in the tropics are *Bombus* and *Centris*. The widely distributed genus *Bombus* has been studied more in temperate than in tropical regions (Roubik 1989). *Centris* exhibits behavioural traits similar to those of *Xylocopa* (Frankie *et al.* 1990). However, it has not yet been determined whether there are specific floral adaptations connected with *Centris* pollination, perhaps with the exception of the oil-flowers (Vogel 1974). The few species of *Apis* are important, versatile pollinators in tropical Asia, also for many crop plants (Crane 1991). They have also been introduced in the Neotropics.

Smaller bees

Major tropical groups of smaller bees are Halictidae, small Apidae (*Melipona*, *Trigona*) and small Anthophoridae (*Exomalopsis*) (Roubik 1989). Flowers pollinated by them are small (less than 1 cm), white to cream, generally short-tubed and polysymmetric. Dioecy often occurs. Nectar production is low or, in many species, absent. This is based on a survey in Costa Rica by Frankie *et al.* (1983). More species exhibit a mass flowering than steady-state strategy, since small bees do not trapline. However, foraging distances of *Trigona* (Apidae) may cover 1 km (Appanah *et al.* 1986). *Trigona* bees often exploit pollen by biting off pieces of the anthers (Renner 1983) without pollinating the flowers.

Wasps

Wasps visit flowers for nectar. In an extensive comparative study carried out in Costa Rica, Heithaus (1979b) showed that wasps preferred flowers with shorter corollas and shorter floral cups than do bees. Flower colours are preferentially green, yellow-green or white. This flower type is 'opportunistic' and also visited by a number of other insects groups (bees, beetles, butterflies, flies) (Heithaus 1979b). Wasps also seem especially attracted to inflorescences with a large number of flowers. A special symbiosis has evolved between figs and chalcid wasps (see section 5.9).

4.1.2 Diptera

Flies are the second most important insect group of pollinators, although a majority of them are not dependent on flowers for food. They may be archaic pollinators (e.g. Lloyd & Wells 1992). A relatively large number of Magnoliidae are fly-pollinated (Endress 1990b). Many flies take nectar (or stigmatic exudate) from open, unspecialized flowers, which occur in so many plant families. These flowers may be effectively pollinated by the flies. Glistening nectar or pseudonectaries seem to play an optically attractive role for flies as well as for Hymenoptera (Kugler 1955). Some flies also feed on pollen. Syrphids and other flies with an elongate proboscis or 'snout' are able to exploit tubular flowers. These flowers are usually polyphilic and are pollinated by other insects as well (e.g. Kevan & Baker 1983).

Other more specialized flowers are more specifically associated with flies (my(i)ophilous flowers). They usually have dull colours, brown or green, often with dark spots or stripes. Often, the perianth organs or other parts are tail-like. They often emit musky smells. Odours are the primary attractants, while the specific colours and forms secondarily enhance attractiveness (Kugler 1956). More elaborate flowers may include lateral entrances and vibratile hairs. Trap-flowers with lateral entrances and with window-panes at the base of the trap are predominantly myophilous (see section 3.7) (van der Pijl 1953).

A number of myophilous flowers work by deceit. They may be subdivided into two categories. Sapromyophilous flowers mimic carrion or dung by their smell, colour, texture and surface patterns. They are visited by flies such as Calliphoridae, Sarcophagidae, Muscidae and Drosophilidae. The flies lay their eggs in the flowers and thereby act as pollinators. The larvae usually cannot develop. Thus, the plants parasitize the flies. Sapromyophilous flowers occur in primitive as well as highly advanced angiosperms (e.g. *Austrobaileya* and *Bulbophyllum*, respectively). It is noteworthy that the largest flowers in the world are sapromyophilous: *Rafflesia arnoldii*, *Aristolochia grandiflora*, and possibly *Paphiopedilum sanderianum* (Figs. 8.7, 8.9, 8.10). Additionally, sapromyophilous flowers are sometimes the largest ones in a given family: *Stapelia*, up to 30 or 40 cm diameter (Asclepiadaceae) (Fig. 8.37), *Sapranthus*, up to 15 cm long (Annonaceae) (Janzen 1983a; Olesen (1992) found beetles as floral visitors of *Sapranthus*). If the word 'flower' is replaced by 'blossom', one may also add some Araceae. However, the largest representative of Araceae, *Amorphophallus titanum*, is not pollinated by flies but by carrion beetles (van der Pijl 1937). In sapromyophilous flowers the perianth is generally the largest part and exhibits the osmophores and dark brown colours. Is the large size of these flowers an optical component of the mimicked model or is it mainly important for the flowers to have a large volatile emitting surface? Some sapromyophilous blossoms are thermogenic, especially in the Araceae (Meeuse & Raskin 1988).

A second category of elaborate fly-flowers working by deceit are the fungus gnat flowers (Vogel 1978b). They mimic the fruiting bodies of mushrooms by scent, lamellate or porate surface patterns, and position near the substrate. Diptera (e.g. Mycetophilidae, Sciaridae) that lay their eggs in mushrooms do the same in such flowers and pollinate them. In this case also, the larvae usually cannot develop. This syndrome was first found in *Asarum* and later in species of *Aristolochia*, some Araceae (*Arisarum*, *Arisaema*) and Orchidaceae (*Dracula*) (Vogel 1978b). It is to be expected that a number of uninvestigated tropical plants are sapromyophilous. For gall midges as pollinators, see section 8.2.2.

Interestingly, myophilous flowers of different categories tend to be present in certain plant groups (e.g. Aristolochiaceae, Asclepiadaceae, Araceae, Orchidaceae). The phylogenetic relationships among these taxa are poorly known.

Another group may be the Sterculiaceae–Buettnerieae. *Abroma augustum* seems to be sapromyophilous (van der Pijl 1953 found Milichidae as visitors). *Theobroma cacao* is pollinated by small midges (Ceratopogonidae and Cecidomyiidae) (Young *et al.* 1986). *Herrania* species are pollinated by Phoridae (Young 1984). *Herrania albiflora* with fungose, lamellate petals and mushroom-like scent (personal observation) should be tested for fungus mimesis (Fig. 3.8).

4.1.3 Coleoptera

Like flies, beetles have been considered to represent archaic pollinators. Beetles are important pollinators in some larger-flowered Magnoliidae (Gottsberger 1977, 1988). However, their prominent position as the first pollinators in early angiosperm evolution has become doubtful, since the more basic groups of the Magnoliidae are not beetle-pollinated (Gottsberger 1974, Crepet *et al.* 1992, Lloyd & Wells 1992). There are also some more highly evolved angiosperms with beetle pollination. Cyclanthaceae are even exclusively beetle-pollinated (Gottsberger 1991).

Beetles are uncouth pollinators. They feed on pollen, various floral tissues, nectar and other floral exudates. Their activity on flowers has been called 'mess and soil pollination', because they often destroy floral organs by their coarse mouth parts. Flowers adapted to beetle-pollination (cantharophilous flowers) produce relatively large amounts of pollen and sometimes special nutritive tissues in perianth parts or stamens (Schatz 1987, Thien *et al.* 1990). Flower colours are generally dull (cream, greenish, brownish). Scent acts as primary attractant. A variant may be exhibited by *Amphicoma* (Scarabaeidae), which seem to be primarily attracted by red colours (Dafni *et al.* 1990).

Mutualism or parasitism of flowers by beetles may be closely associated (Silberbauer-Gottsberger 1990 for palms). Conversely, some plants work by deceit and beetle eggs deposited in the flowers cannot develop. Such beetles normally oviposit on fruits or on carrion. *Amorphophallus* and some other Araceae have impressive carrion beetle blossoms (van der Pijl 1937). *Hydnora* (Hydnoraceae) is a carrion flower pollinated mainly by dermestid beetles (Musselman & Visser 1989). *Eupomatia* (with a fruity scent) is perhaps the only example studied where the flowers seem to be the regular brood substrate for the larvae (Armstrong & Irvine 1990) (see section 5.9).

Some beetles (e.g. Chrysomelidae, Staphylinidae, Cantharidae,

Cerambycidae) take part in the pollination of relatively unspecialized, polyphilic flowers, sometimes together with flies, bees, and butterflies. Such flowers are generally small, relatively open and aggregated in dense inflorescences (e.g. *Flindersia*, Rutaceae; *Alphitonia*, Rhamnaceae; Irvine & Armstrong 1988, 1990). Some beetles have elongated mouth parts and may also exploit tubular flowers (Cerambycidae, Nemognatha) (Kevan & Baker 1983). However, some flowers are pollinated exclusively by beetles, and there are some highly specific floral adaptations associated with particular beetle groups (especially of the Curculionidae and Scarabaeidae), which are best treated as separate syndromes. It has been suggested that beetles may also play an important role as pollinators at the community level, as found in Australian rain forests (Irvine & Armstrong 1990).

Curculionid beetles

The flowers of some tropical plants are pollinated exclusively by small weevils (Curculionidae). These flowers range from small to large and provide shelter by narrow entrances into an inner space where the weevils may stay for a long time. Thermogenicity may occur (Silberbauer-Gottsberger 1990), but this is less well investigated than in the larger dynastid-pollinated blossoms. In the Magnoliidae, Eupomatiaceae and Degeneriaceae are exclusively weevil-pollinated (e.g. Armstrong & Irvine 1990, Thien 1980). Other examples with weevil-pollination include *Myristica* (Myristicaceae) (Irvine & Armstrong 1990), *Zygogynum* and *Exospermum* (Winteraceae) (Thien 1980, Thien *et al.* 1990), *Annona* sect. *Atta* (Annonaceae) (Gottsberger 1989b), Carludovicoideae (Cyclanthaceae) (Gottsberger 1991) and various palms (Greathead 1983, Silberbauer-Gottsberger 1990).

Nitidulid beetles visit and pollinate similar flowers.

Dynastine scarab beetles

Dynastine scarab beetles (Scarabaeidae: Dynastinae) are striking, large beetles that pollinate a number of plant groups (perhaps 900 species) in the Neotropics (Schatz 1990). The spectacular dynastine scarab beetle pollination syndrome was first recognized by detailed studies on *Victoria* (Nymphaeaceae) by Prance & Arias (1975) and *Nymphaea* (Prance 1980a), and later also established for palms and Cyclanthaceae (Beach 1982, 1984), Araceae (Gottsberger & Amaral 1984, Gottsberger & Silberbauer-Gottsberger 1991, Young 1986, 1990) and Annonaceae (Gottsberger 1989a,b, 1990).

The blossoms (large single flowers in Magnoliidae, inflorescences of numerous small flowers in Arecidae) are nocturnal and light in colour. They are strongly odoriferous (fruity or musty) and thermogenic (see section 5.5). They form hot chambers, either by the large and robust perianth (Magnoliidae) or by a spathe around the inflorescence (Arecidae), where the beetles congregate, mate and breed. The nocturnally active beetles are primarily attracted by the strong scent. Protogyny is common to all scarab-pollinated taxa (Schatz 1990).

A similar relationship may exist between cephaline scarab beetles and certain taxa of the same families of plants (Pellmyr 1985).

Evolutionary relationships

As in fly-pollinated flowers, different cantharophilous syndromes, including curculionid and dynastine pollination, are present in some plant groups, such as Annonaceae (Gottsberger 1989a,b), Cyclanthaceae (Gottsberger 1991) and palms (Silberbauer-Gottsberger 1990). The more specialized dynastine syndrome has probably evolved from curculionid or other beetle syndromes in these groups (Gottsberger 1991). Thermogenicity, a component of the dynastine (and curculionid) syndrome, is found in a few genera of four subfamilies in the palms and it must have repeatedly arisen in the evolution of the family (Silberbauer-Gottsberger 1990).

4.1.4 Lepidoptera

Three major lepidopteran groups are important pollinators: moths (Noctuidae), hawkmoths (Sphingidae), and butterflies (Rhopalocera). According to their different biology, floral adaptations to these groups are quite elaborate and distinctive. A common feature of many of these flowers is that nectaries are concealed in narrow tubes or spurs, since most Lepidoptera have a long proboscis. The anthers may be included in the floral tube or be exserted, and pollen is deposited on the proboscis or wings and body, respectively (Haber & Frankie 1989). Many sphingids and butterflies cover long distances rapidly and are efficient pollinators for rare plants. A fourth group, the small, pollen-eating Micropterigidae, has been less well studied but suggests that primitive Lepidoptera may have played a role in early angiosperm evolution (Thien *et al.* 1985, Pellmyr *et al.* 1990, Pellmyr 1992).

Hawkmoths

Most hawkmoths (sphingids) are active at dusk or at night, when they visit flowers for nectar. Accordingly, flowers pollinated by them (sphingophilous flowers) often open at dusk; they are white or at least light in colour so that they can also be recognized in the darkness. Deeply and narrowly lobed or fringed blossom contours may enhance direction of the pollinators towards the floral centre (Vogel 1954). Sphingids have the longest probosces among the insects and the corresponding flowers have the longest tubes or spurs. The flowers have a strong, sweet fragrance, which is often stronger at night than during the day (Brantjes 1973). This indicates a rhythmic diurnal change in the production of volatiles, which has been analysed by Matile & Altenburger (1988) for *Stephanotis floribunda* (Asclepiadaceae) and some other species. The scent often has nitrogen-containing compounds (Nilsson *et al.* 1985) and may then be reminiscent of rotten vegetables. Long-distance attraction by the fragrance and short-distance visual orientation has been reported (Nilsson *et al.* 1985).

To what extent tropical hawkmoths display floral constancy is unknown; however, a certain degree of constancy has been observed in some species under certain circumstances (Nilsson *et al.* 1987). In experiments with temperate species flower constancy was based on scent (Brantjes 1973).

In contrast to other Lepidoptera, hawkmoths generally do not alight on the flowers, so no landing platform is necessary. Thus the flowers with their long tubes do not need special reinforcements in their architecture. Flowers are often slender and gracile, sometimes in pendulous inflorescences (*Clerodendrum wallichii*, Verbenaceae). The sphingids are hairy and a large, exposed stigma seems to be adapted to relatively imprecise pollination (Nilsson & Rabakonandrianina 1988).

Hawkmoth-flowers occur in many parts of the tropics (and temperate regions). Madagascar is famous for its especially long-tongued hawkmoths and long-spurred orchids. At least 70 species (10% of the orchid flora) have spurs of 8 cm and more, in contrast to continental Africa with a much lower percentage. The diversity of long-tongued hawkmoths is especially high in Madagascar. The classic example is the orchid *Angraecum sesquipedale* with a spur often exceeding 30 cm in length, for which Darwin (1862) predicted a hawkmoth with a proboscis of similar length. Only later were a few species with tongues of more than 20 cm length detected. These or similar sphingids are potential pollinators of *Angraecum sesquipedale*, although pollination has never been observed (see Nilsson *et al.* 1985, 1987). The spurs are generally slightly longer than the proboscis of the pollinators. This induces the visitor to press against the pollination organs, at least near the end of nectar extraction. The spurs are curved in angraecoid orchids and the flowers pendulous and movable. By insertion of the proboscis into the spur the flowers attain a position suitable for pollination. Examples of hawkmoth-pollinated flowers with inner spurs include *Bauhinia* (subgen. *Bauhinia*, sect. *Macrosiphonia*,

Caesalpiniaceae) (Silberbauer-Gottsberger & Gottsberger 1975) and *Pelargonium* (Geraniaceae) (Vogel 1954).

A large number of sphingophilous groups occur in the Asteridae because of the ready evolution of long floral tubes in sympetalous corollas (Apocynaceae, Rubiaceae, Solanaceae, Campanulaceae–Lobelioideae) but they have also evolved in some other families, such as Liliaceae, Caricaceae, Combretaceae (Eisikowitch & Rotem 1987), Nyctaginaceae (V. & K. Grant 1983b, Martinez del Rio & Burquez 1986) and Pittosporaceae. Prominent examples with especially long tubes are species of *Hymenocallis*, *Crinum* (Liliaceae), *Brugmansia*, *Datura*, and *Nicotiana* (Solanaceae) (Kugler 1971, V. & K. Grant 1983a), *Posoqueria* and *Oxyanthus* (Rubiaceae) (Figs 3.2, 4.1).

Some sphingophilous flowers are open brush-flowers. Since the stamens obstruct the access to the nectary for larger animals a long proboscis is also necessary; examples are some Capparaceae (Werth 1942, Vogel 1954) and *Pithecellobium* (Mimosaceae) (Haber & Frankie 1989).

In tropical montane forests, hawkmoth-pollination is limited by lower temperatures. Cruden *et al.* (1976) suggest that, unlike in temperate regions, in tropical montane forests a shift from nocturnal to diurnal activity of hawkmoths is inhibited by competitive interactions with hummingbirds.

Long-spurred orchids and long-tongued hawkmoths are interesting systems in which to study coevolutionary relationships (Nilsson *et al.* 1987). There appears to be an evolutionary trend to form longer and longer flowers and proboscides. Darwin (1862) suggested it to be the result of a kind of race between flowers and pollinators. This hypothesis was tested by Nilsson (1988) on some orchids with deep flowers. He found that moths insert their probosces no further than necessary to obtain nectar. In order to assure contact of the pollinator with the pollination organs the floral tube (or spur) should, therefore, be longer than the proboscis of the moth. It was shown that an experimental reduction in flower depth reduced both the male and female components of fitness in the orchids, because the moths then did not contact the pollination organs. In natural populations there is a correlation between flower depth and female fitness as measured by fruit set.

Noctuid moths

Noctuid moths have relatively short proboscides. Accordingly, flowers adapted to moth-pollination (phalaenophilous flowers) have relatively short tubes. As in sphingid-flowers, scent is the primary attractant,

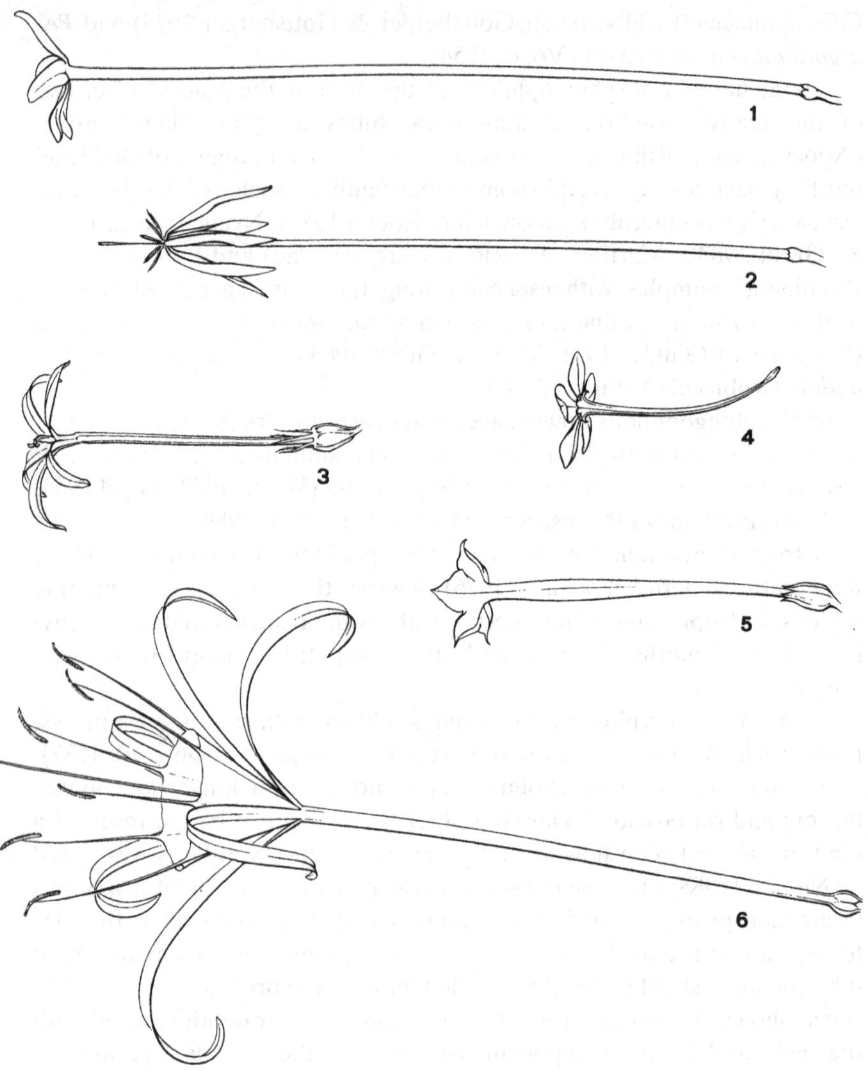

Figure 4.1. Hawkmoth-pollinated flowers with long tubes (× 0.5). 1. *Posoqueria latifolia* (Rubiaceae). 2. *Oxyanthus formosus* (Rubiaceae). 3. *Isotoma longiflora* (Campanulaceae). 4. *Quisqualis indica* (Combretaceae). 5. *Nicotiana sylvestris* (Solanaceae). 6. *Hymenocallis littoralis* (Liliaceae–Amaryllidoideae).

but noctuids react less specifically (Brantjes 1978). The flowers tend to be light-coloured, yellowish, greenish, purplish, but not purely white (Vogel 1954). *Thuranthos macranthum* (Liliaceae) is an example of a noctuid-pollinated plant (Stirton 1976).

Butterflies

In contrast to hawkmoths, butterflies not only have shorter proboscides, but they are active during the day, and they commonly alight on the flowers to take nectar. Accordingly, flowers adapted to butterfly-pollination (psychophilous flowers) are open during the day (opening in the morning), have shorter tubes or spurs and provide a landing platform. They are brightly coloured, often orange, red or pink, often also with several contrasting colours, in which red plays an important role. A small number of butterfly groups have also been found to be able to eat pollen (Gilbert 1972, DeVries 1979). Two flower forms in butterfly-pollinated plants are especially prominent:

(1) The flowers are relatively large. Each individual flower provides a landing place. The stamens and styles are long and pollen is deposited imprecisely on the wings or the body (e.g. *Delonix regia*, *Caesalpinia pulcherrima*, Caesalpiniaceae; *Globba*, Zingiberaceae). The flowers are polysymmetric, salverform, with a narrow tube, or monosymmetric with stamens and styles exserted far out of the flower. In non-tubular flowers a nectar tube may be separate in a special fold of a petal (*Delonix*, *Caesalpinia*, *Globba*) or of several petals (*Gloriosa*, *Lilium*, both Liliaceae) or in a spur (Vochysiaceae) (Figs 1.1, 8.28). The anthers are very versatile (seesaw-mechanism). The style is also elastic. The stigma is tubular, surrounded by a basket of stiff, non-receptive bristles, whose function may be to hold copious stigmatic secretion or to scratch wings or body surface to electrostatically attract pollen (?). In *Globba*, a pollinator visit mechanically causes a drop of stylar fluid to appear in the bristle basket to take up pollen grains from the visitor's body (Müller 1931).

(2) The flowers are relatively small. Numerous flowers together form a landing platform, often by a dense, umbel-like arrangement. Examples include *Lantana camara* (Verbenaceae), *Asclepias curassavica*, *Epidendrum radicans* (Orchidaceae) (Fig. 7.2); *Bauhinia integrifolia*, Caesalpiniaceae (Fig. 8.29); *Stachytarpheta* (Verbenaceae) (DeVries 1987) (Fig. 7.1); *Cordia* (Boraginaceae) (Opler *et al.* 1975, Opler 1983); and *Bougainvillea spectabilis* (Nyctaginaceae) (Werth 1942).

In some butterfly-pollinated flowers red 'flags' formed by sepals or bracts or postanthetic remaining flowers occur (e.g. *Delonix*, *Bauhinia*, Caesalpiniaceae; *Mussaenda*, *Warscewiczia*, *Carphalea*, Rubiaceae).

4.1.5 Thysanoptera

Thrips often occur as pollen (and probably also nectar) feeders in flowers (Kevan & Baker 1983, Kirk 1984). Their role as pollinators, however, has rarely been established. Usually they are regarded as minor, accessory pollinators. In *Lantana camara* (Verbenaceae), which is primarily butterfly-pollinated (Schemske 1976), successful thrips-pollination has been found in the absence of butterflies (Mathur & Mohan Ram 1986). However, some cases of predominant thrips-pollination have recently been described. In the mass flowering *Shorea* (Dipterocarpaceae) thrips seem to play the main role as pollinators (Appanah 1981, Appanah & Chan 1981, Ashton *et al.* 1988), although bees seem to be the major pollinators of other genera of the family (Ashton 1988, Dayanandan *et al.* 1990). The short-lived thrips may build up large populations within a short time and are therefore particularly apt pollinators of mass flowering plants. Thrips are perhaps also the major pollinators of *Belliolum* species (Winteraceae) (Thien 1980, Carlquist 1983, Pellmyr *et al.* 1990, personal observation) and some Myristicaceae (Bawa *et al.* 1985b). Flowers adapted to thrips-pollination provide shelter by narrow entrances towards the floral centre. They either have narrow tubes or, in more open flowers, the stamens are crowded and provide narrow slits through which the thrips can enter the protected floral centre. Scent is a primary attractant, and white colours are especially attractive in conjunction with scent (Kirk 1985). Pollen is a major food source. Nectar may be present or absent. Nutritive tissues of floral organs may also be present and used as food for larvae or adults (e.g. *Belliolum*, Endress 1986a). Flowers may be used as breeding places (e.g. *Mollinedia*, Monimiaceae; Gottsberger 1977). Apparently, the syndrome of flowers adapted to thrips pollination somewhat resembles that of flowers pollinated by small beetles.

4.1.6 Birds

Bird-flowers (ornithophilous flowers) and flower-birds have long excited naturalists, since both are equally spectacular because of their often bright colorations. Extensive comparative investigations on bird pollination were first made by Porsch (e.g. 1924a), and later, based for the first time on broad comparative field studies, by K. & V. Grant (1968).

Flower-pollinating birds take nectar from the flowers. Besides the classical flower birds, the hummingbirds (Trochilidae), in the New World, there are also important flower birds in the Old World, sunbirds (Nectariniidae), honeyeaters (Meliphagidae), honeycreepers (Drepaniidae) and lorikeets (Psittacidae). Since hummingbirds usually hover during their flower visits, while the other flower birds need a perch, the architecture of the inflorescences and flowers of Old World and New World bird-pollinated plants is often somewhat different (van der Pijl 1937). However, there are deviations from these patterns; some non-hummingbirds are also able to hover and may pollinate flowers without a perch. This habit is even not unusual in nectarinids (Westerkamp 1990). Further, in the Neotropics other flower-birds also occur: honeycreepers (Drepaniidae) and orioles (Icteridae), and a few hummingbird species regularly perch while feeding (Feinsinger & Colwell 1978).

Bird-flowers are diurnal and are scentless. Since many flower-birds have long beaks and tongues, bird-flowers often have a relatively long and tubular part, commonly with protruding stamens and stigmas. The length of bird beaks and flower tubes is evolutionarily plastic, which is reflected by the diversity in both groups (Stiles 1978b). Flowers may be polysymmetric and simply tubular. In contrast, if combined with monosymmetry and horizontal position, the lower perianth parts are sharply reflexed ('dogfish-flowers', Proctor & Yeo 1973) (Figs 4.2.3, 8.55). Bird-pollinated flowers produce relatively large amounts of dilute nectar of low viscosity and with low amino acid content (Baker 1978, Stiles 1981); thus they tend to have large nectaries. They also tend to have robust, sclerified tissues at the periphery of floral organs, especially at the sites where damage by beaks or feet would be fatal for the plants (Pass 1940). This is more pronounced in flowers pollinated by larger birds, thus more common in Old World than in New World plants. Obvious examples are some Proteaceae in Australia and Africa with robust, wiry flowers, or Myrtaceae with robust, calyptrate floral envelopes (*Eucalyptus*). A bias towards inferior ovaries in bird-flowers may also be seen in this context, since ovaries in this position may be less vulnerable to bird bills (V. Grant 1950a).

Bird-flowers are often bright red, sometimes combined with yellow and other colours. Nectar guides also occur. In contrast to bee-flowers the ultraviolet component of nectar guides is weak (Kugler 1966), although hummingbirds are able to see ultraviolet (see section 5.7.1). The predominance of red has been interpreted as exploitation of an ecological niche with regard to the phylogenetically older class of bee-pollinated flowers (K. Grant 1966, Raven 1973). Red is at the long-wave end of the visible spectrum of electromagnetic radiation. Bees and some other insects are not able to see red (see section 5.7.1). The evolutionary path to red bird-flowers is therefore particularly open. Once established, the effect may be reinforced by the presence of many red bird-flowers, which renders red a general 'flag' for birds, which may be especially efficient in ecosystems with migratory hummingbirds (Johnsgard 1983). One could therefore say that, paradoxically, the predominance of red may be at the same time a

Figure 4.2. Hummingbird-pollinated flowers (× 1). 1. *Malvaviscus arboreus* (Malvaceae); flower upright, polysymmetric. 2. *Podandrogyne brachycarpa* (Capparaceae); flowers oriented to the side, monosymmetric, organ number low and fixed. 3. *Schlumbergera truncata* (Cactaceae); flower oriented to the side, monosymmetric, organ number high and variable.

cryptic colour to avoid visits by insects that could act as pollen robbers. The same could be true for the lack of ultraviolet nectar guides. Non-red bird-flowers occur especially at the lower and upper end of the gamut of corolla tube lengths (Baumberger 1987).

Bird-pollination is prominent in montane forests of some regions (Stevens 1976, 1985). There is a correlation between epiphytism and ornithophily, as illustrated by Gesneriaceae, Ericaceae and Rubiaceae (Wiehler 1983).

Hummingbirds

Flowers adapted to hummingbird-pollination are often solitary or loosely clustered in a horizontal or pendent position and have flexible pedicels. Since most hummingbirds hover while visiting flowers, landing platforms are lacking. Pendent flowers can be easily reached from below and lifted up into a more horizontal position by the hummingbirds (Toledo 1974). Flowers are often red or yellow (K. & V. Grant 1968). Hummingbirds have no innate preference for red (Bené 1941), but they learn to associate particular colours with nectar sources. The preference for certain colours may thus change during the seasons of a year (Wagner 1946).

As in the case of the Madagascan orchids and their hawkmoths there is a case of striking coevolution in a hummingbird:flower pair. The Andean swordbill (*Ensifera ensifera*) has a bill that may attain a length of more than 10 cm, and is the pollinator of *Passiflora mixta* (subg. *Tacsonia*), which has a floral tube (including calyx and corolla) of up to 15 cm length (D. & B. Snow 1980) (see section 8.5.1). Its nectar is perhaps inaccessible to all other hummingbirds. The swordbill is a trapliner and visits a large number of individuals over a relatively wide area. Pollination by hummingbirds is more prominent in the montane than in the lowland rain forests (Cruden 1972, Feinsinger 1983a, Bawa 1990), although Renner (1989a) found a predominance of bee-pollination (buzz-pollination) at high altitudes in Venezuela. The two main groups of hummingbirds, the hermits and the non-hermits, differ in distribution and behaviour. The hermits predominantly occur in the lowlands. They are associated with large monocot herbs, especially Zingiberales, also with some Acanthaceae (Stiles 1981, McDade 1992). The non-hermits prefer middle elevations and are also present at higher altitudes. They are dependent on flowers of many dicots and of bromeliads (Stiles 1981).

Some tropical ornithophilous genera may have followed hummingbirds to extratropical regions (e.g. Acanthaceae) with the result that ornithophilous species also originated in predominantly bee-pollinated extratropical genera (e.g. *Delphinium*, Ranunculaceae) (K. & V. Grant 1968). For the evolution of North American hummingbird-flowers K. & V. Grant (1968) and Stebbins (1989) showed that the most easily modified floral shape is either broadly tubular or narrowly bell-shaped. Such flowers are usually pollinated by relatively large bees, which enter the flowers when seeking nectar. A further modification may involve backward-turning of the petal lobes, especially the lower lip ('dogfish-flowers', see above).

Flower-pollinating mites live in tubular hummingbird-pollinated flowers and carry pollen from flower to flower within an inflorescence. Thus they may effect autogamy and geitonogamy. They are transported to other inflorescences in the nostrils of hummingbirds (Colwell 1983). Two genera of such mites (Gamasida: Ascidae), with *ca.* 200 species, are known (Heyneman *et al.* 1991). Each mite species depends on one or very few host plant species that provide a year-round supply of flowers. Floral specificity is assumed to be based on olfactorial cues. They show less specificity with their hummingbird hosts. However, mite species of short-tubed flowers tend to travel on short-billed birds and vice versa (Heyneman *et al.* 1991). More than a hundred plant species have been found with these mites in the flowers (Colwell 1983). An investigated example is *Heliconia* (Dobkin 1984, 1987). Mite pollination in Old World plants has also been suggested but not studied in detail (Stevens 1976, Smith *et al.* 1992).

Other birds

Most birds, with the exception of hummingbirds, tend to alight on the plant while taking nectar. Some plants have obvious adaptations in this respect. In *Phygelius capensis* (Scrophulariaceae) the flowers are curved back on the inflorescence and resupinate, so that they can be reached from the floral axis (Fig. 8.55). In *Spathodea* and *Deplanchea* species (Bignoniaceae) or *Etlingera* (Zingiberaceae) the flowers are curved upwards so that they can also be reached from the centre of the inflorescence (Weber & Vogel 1984, Westerkamp 1990, Classen 1987) (Figs 8.47, 8.60). Often the flowers are clustered in dense partial inflorescences and the clusters may be exploited from the inflorescence axis (*Agave*).

An extreme condition is that birds not only sit in or near the flowers but also pollinate the flowers with their feet. This is the case in *Strelitzia* (S. & P. Frost 1981) and in *Norantea* (Marcgraviaceae; Perry 1990). In these flowers the gynoecium is extremely robust and pollen and stigma are very sticky.

In Australia the diversity of bird-flowers is especially high (Ford *et al.* 1980), probably owing to the large size differences and different behaviour of pollinators, which are mainly Meliphagidae and Psittacidae; the latter also eat pollen in addition to nectar. In addition to tubular and 'dogfish-flowers' (e.g. *Eremophila*, Myoporaceae), there are particularly many brush-flowers, usually in dense inflorescences. Two prominent families are Myrtaceae (e.g. *Eucalyptus*, *Callistemon*, *Syzygium*) and Proteaceae (e.g. *Banksia*, *Grevillea*). Many of these genera are prominent components of the vegetation. In the American tropics passerine nectarivores are important as pollinators in seasonally dry areas, when hummingbirds are scarce (Stiles 1981).

4.1.7 Bats

Bat-pollination is an exclusively tropical phenomenon. That certain tropical bats feed on nectar in flowers was reported as early as in the eighteenth century (Dobat & Peikert-Holle 1985). However, it was not until a hundred years later (Moseley 1879) that it was assumed that such bats pollinated the flowers. Porsch (1931) was the first to describe bat pollination and flower adaptations for *Crescentia* (Bignoniaceae) in its natural habitat. Comparative investigations were later carried out in Eastern Asia by van der Pijl (1936, 1941a, 1956), in Africa by Jaeger (e.g. 1954), and in South America by Vogel (1957, 1968, 1969a,b) who considerably extended the systematic range of known bat-flowers. As in the case of flower-birds, the Neotropics and Palaeotropics have different bat groups among the pollinators. Old World flower-bats belong to the Megachiroptera, New World flower-bats to the Microchiroptera. Nevertheless, the syndromes of bat-flowers (chiropterophilous flowers) are similar in both regions (Vogel 1957). Dobat & Peikert-Holle (1985) list 64 angiosperm families containing bat-flowers. Some families have a relatively high proportion of bat-flowers (Bignoniaceae, Bombacaceae, Cactaceae, Caesalpiniaceae, Campanulaceae–Lobelioideae); in other families they are rather exceptional.

Bats take nectar and pollen and are nocturnally active; they fly long distances (Start & Marshall 1976). Bat-flowers produce copious mucilaginous nectar and often copious pollen and they often open at dusk. They may be white but often have dull colours, such as greenish or brownish (e.g. *Kigelia*, *Oroxylum*, both Bignoniaceae; *Mucuna*, Fabaceae). Most characteristic is the smell, often of fermented material. As an extreme, the flowers of *Fagraea racemosa* (Loganiaceae) (Fig. 7.1) smell pungently of vinegar (personal observation). Smells reminiscent of rotting fruit are also common (e.g. *Barringtonia calyptrata*, Lecythidaceae). Vogel (1983) argues that the dull colours are not a means of attraction for bats but rather connected with the heavy scent production of the petal surfaces. They may also be cryptic for other animals that are not pollinators. Single flowers or inflorescences are strong enough to carry a bat, so often the largest flowers within a family are bat-flowers. Brush-flowers (e.g. *Adansonia*, *Ochroma*, *Durio*, Bombacaceae, Jaeger 1954, Soepadmo & Eow 1976; *Marcgravia*) or widely open bowl-shaped flowers (e.g. the above-mentioned Bignoniaceae) are common, where the bats can insert their head and tongue (Fig. 8.47). The bowls of some bowl-shaped flowers are especially thick, since the bats hold themselves with their claws. Some Neotropical (Glossophaginae) and Palaeotropical bats (Macroglossinae) hover and may use smaller flowers without strong stems (Heithaus 1982). Bat-pollinated plants are mostly trees; some are epiphytes; rarely they have a giant herbaceous habit (e.g. *Louteridium*, Acanthaceae). Among Cactaceae bat-flowers are only found in tall species (Vogel 1969b). The flowers or inflorescences are often produced in a relatively open space so that they can be easily reached by bats. They often stand (*Oroxylum*, Bignoniaceae) or hang (*Kigelia*, Bignoniaceae) out of the foliage or are sessile on a tree trunk (cauliflory) (*Crescentia*, Bignoniaceae) (Marshall 1983).

4.1.8 Non-flying mammals

It is a relatively recent insight that some of the most bizarre flowers and inflorescences are apparently adapted to pollination by non-flying mammals. Porsch (1936) discussed the potential significance of marsupials for the pollination of certain Australian Proteaceae and Myrtaceae and of lemurs for the Madagascan *Symphonia* (Clusiaceae). This was later elaborated by Sussman & Raven (1978). *Ravenala* (Strelitziaceae) seems to be especially adapted to lemur pollination (Kress *et al.* 1992). Grünmeier (1990) observed visits of prosimians to the otherwise bat-pollinated African *Parkia* (Mimosaceae). Rourke & Wiens (1977) showed apparent convergence in Australian and South African Proteaceae (especially *Dryandra*, *Banksia*, and *Protea*) that are pollinated by mouse-like mammals: marsupials in Australia and rodents in Africa. These plants have the inflorescences near the ground, and the floral organs are highly lignified, often with a wire-like texture. Rourke & Wiens (1977) speculate that the syndrome has evolved from bird-pollinated plants in response to fires in the sclerophyllous vegetation where they occur. Arboreal marsupials also visit the flowers on trees of other Proteaceae and some Myrtaceae. More detailed studies further substantiated the importance of these mammals as pollinators of various Proteaceae (Collins & Rebelo 1987, Goldingay *et al.* 1991). In New Guinea Hopkins (1992) found nocturnal possum pollination in a *Mucuna* species (Fabaceae).

More recently, non-flying mammals were found to pollinate flowers in the Neotropics also. In Costa Rican cloud forests, arboreal rodents take nectar from three species of the otherwise bee-pollinated genus *Blakea* (Melastomataceae) (Lumer 1980, Lumer & Schoer 1986). The nocturnal flowers seem to be adapted especially to these visitors. They are campanulate, pendent, green with purple anthers and have copious nectar, which is sucrose-rich (as the nectar of corresponding Proteaceae). They are scentless to humans but apparently not to the rodents. The species grows at high altitudes under windy conditions, where bats are less active. Marsupials may be pollinators of *Mabea* (Euphorbiaceae) (Steiner 1981) and *Pseudobombax* (Bombacaceae) (Gribel 1988). Further, monkeys feed on nectar but are highly destructive of flowers (Garber 1988, for *Symphonia*, Clusiaceae). However, it is likely that monkeys also function as pollinators in the Neotropics, as is supposed for species of *Combretum* (Combretaceae) (Prance 1980b), and several Bombacaceae (Janson *et al.* 1981).

Special adaptation of some Australian and African Proteaceae to non-flying mammals as pollinators is now well established (Turner 1982, Wiens *et al.* 1983). Porsch (1936) speculated that the woody tissues in proteaceous flowers of sclerophyllous regions evolved as devices for water storage and were, then, a preadaptation for pollination by birds, and – even more – by non-flying mammals. It would certainly be interesting to do a comparative investigation of the impact of primates in pollination biology, especially as compared with the contribution of bats (we all poke our noses into flowers and my little daughter likes to pollinate tulips with her nose).

4.2 Flowers pollinated by abiotic agents

4.2.1 Wind

Wind is an omnipresent but unfaithful pollinator. Wind-pollination occurs in places where animal pollinators are rare and where pollen transport is not obstructed by dense vegetation. Wind-pollination has mainly been studied in temperate regions but it is not negligible in the tropics either (Regal 1982). It is more prominent in dry or seasonally dry environments, since rain hinders the dispersal of pollen in the air. It is particularly important in open grassland (most Poaceae), savannahs (e.g. *Ateleia*, Fabaceae, Janzen 1989), sclerophyll forest (*Casuarina*), semi-deserts (*Amphipterygium*, Anacardiaceae), above the timberline (Arroyo *et al.* 1990), and in many water plants (Cook 1988). In dense vegetation it is more restricted to emergent trees with free canopy (*Araucaria*, conifers), on mountain crests (*Balanops australiana*; *Nothofagus*, Fagaceae) and along rivers and coastlines (*Pandanus*, Pandanaceae, Cox 1990; a few palms, Henderson 1986, Silberbauer-Gottsberger 1990). But even in the understorey of wet forests wind pollination may not be completely absent, as shown for *Trophis involucrata* (Moraceae) with explosive pollen release during the drier hours of the day (Bawa & Crisp 1980). On the other hand, in the grasses, a wind-pollinated family par excellence (Connor 1980), some species of the rain forest understorey have switched to insect pollination (Soderstrom & Calderon 1971). Wind-pollination is favoured in species-poor vegetation where individuals of a species are not too widely spaced. Synchronous mass flowering is also correlated with wind-pollination. In grass species a short daily pulse of pollen release at a specific hour occurs (Kerner 1905, Dowding 1987). This daily staggered flowering helps in the partitioning of pollen availability between several anemophilous species, as present in grasslands.

The pollen:ovule ratio of plants adapted to wind-pollination (anemophilous flowers) is very large. Abundant pollen production is achieved by enlarged anther size, or increased stamen number per flower or male flower number per plant (Endress 1977c, 1986b). Pollen is dry, powdery and distributed in single grains. The grains are relatively small but not too small (predominantly between 20 and 40 μm diameter), since this provides the best properties for the capture of grains by the stigma (Whitehead 1969, 1983, Crane 1986a). The pollen surface is smooth, and pollen is especially resistant to desiccation. Commonly the anthers, or, if the flowers are unisexual, the entire male flowers or inflorescences are pendulous, having lax stalks. In some groups the thecal halves do not gape widely after dehiscence but only form a more or less narrow slit, through which the pollen may be released little by little by gusts of wind (Figs 2.7.1, 2.7.4). There may also be pollen-arresting mechanisms outside the anther, e.g. on bracts, where pollen collects until the mass is taken by a gust of wind. In some other groups, on the contrary, pollen is released at once by an explosive opening of the anther under dry conditions (some Moraceae, Urticaceae, *Ricinus* of Euphorbiaceae). Pollen dispersal distances are limited by the longevity of pollen grains and their ability to be carried aloft.

In order to catch the randomly distributed pollen grains from the air, stigmas are well exposed and stigma surfaces are large and subdivided into smaller parts. An extreme form are stigmatic lobes with numerous secondary branches, as they are commonly found in grasses and related families (Fig. 2.27). Stigmas of anemophilous plants are generally of the dry type. This may be related to the action of electrostatic forces in pollen capture (Niklas 1985). Most wind-pollinated plants have a single ovule per gynoecium or, if there is more than one, a single seed will develop, while the other ovules degenerate.

The perianth is reduced, since visual attractants are not necessary or rather deleterious in view of pollen thieves and since free access of wind for pollen transport should not be hindered. However, Niklas (1985, 1987) showed that also in cases where the stigmas of anemophilous plants are not freely exposed, the airflow is channelled toward the stigmas by the specific architecture of the blossoms. There is a trend towards unisexual flowers, and the male and female flowers may considerably diverge in their form. This is plausible for two reasons: (1) anthers and stigmas do not need to be close together in the same flower and there is no necessity for optical similarity of male and female flowers, so that they may be shaped solely by their divergent physical needs; and (2) the dissociation of the sexes allows a greater production of male flowers compared with female ones to attain a higher pollen:ovule ratio. Mast-fruiting is also known to be associated with wind-pollination in some plants (e.g. bamboos and some temperate trees) (Smith *et al.* 1990).

Wind-pollination is evolutionarily secondary at the level of the angiosperms. This is suggested by evidence from different fields: (1) the fossil record of early angiosperms and pollinating insects (Crepet *et al.* 1991, Pellmyr 1992); (2) the almost total lack of wind-pollination in the primitive extant Magnoliidae (Endress 1990b); (3) the occurrence of wind-pollination in many different, more specialized angiosperm groups: in some larger families (e.g. grasses), or in a few genera (e.g. Euphorbiaceae, palms, Pandanaceae) or species in otherwise zoophilous groups (Cox 1991a).

4.2.2 Water

Most water plants raise their inflorescences above the water surface and are pollinated by animals or by wind (Cook 1988). However, there are some curious adaptations of tropical and other water plants to pollination at the water surface or even under water (hydrophilous or hydatophilous flowers). Most of these plants are representatives of the Alismatidae and the majority are marine plants. A great amount of work to elucidate the mechanisms of water-pollination has been done in the past two decades (reviews by Ducker & Knox 1976, Pettitt *et al.* 1981, McConchie 1982, Pettitt 1984, Cox 1988).

In hydrophilous pollination water is the vector in pollen transport. The pollen itself may or may not come into direct contact with water. Pollen remains dry if anthers or flowers carry pollen across the water surface (Cox 1988). The syndrome exhibits some common traits with wind-pollination, since both are wasteful abiotic methods of pollen transport. These include small, inconspicuous, unisexual flowers with a reduced perianth, single carpels with single ovules but abundant pollen production resulting in a high pollen:ovule ratio, a large receptive surface sometimes with secondary branching of stigmas, and mass flowering (Markgraf 1936, Tomlinson 1982, McConchie *et al.* 1982, Les 1988). However, since the water surface is a two-dimensional space, pollen transport may be more precise than in the three-dimensional space under water (or in the air). Cox & Tomlinson (1988) found that pollen dispersion also tends to be two-dimensional underwater.

It is useful to distinguish three categories of water-pollination: (1) pollen transported above the water surface, (2) pollen transported on the water surface, (3) pollen transported beneath the water surface (Ernst-Schwarzenbach 1944, Cox 1988). Usually, in category (1) plants, the underwater male flowers abscise and float to the water surface. Here, they open, the sepals bend backwards and the anthers are raised in the air and function as sails. Female flowers also float on the water surface but they remain attached to their long pedicels. Male flowers are blown about on the water surface and may so reach a female flower. They tend to aggregate around female flowers. Pollen may be directly deposited from the anther to the stigma (*Nechamandra*, *Lagarosiphon*), or the unwettable pollen may be released to the water surface before it reaches the stigma (*Elodea*), or the male and female flowers, whose inside is unwettable, may become enclosed in an air bubble by a wave, they tip over and pollination occurs (*Vallisneria*), or pollen may be explosively released and transported to a nearby female flower (Cook 1982). Category (1) plants occur mainly in the Hydrocharitaceae.

In category (2) pollen is transported directly upon the water surface, where it forms extended rafts ('search vehicles', Cox 1983) that reach the floating, filamentous stigmas. The search vehicles may consist of filiform pollen (*Halodule*, *Thalassodendron*, Cymodoceaceae), or of pollen contained within mucilaginous tubes or mats (*Halophila*, *Thalassia*, Hydrocharitaceae; *Lepilaena cylindrocarpa*, Zannichelliaceae); other examples of category (2) include *Elodea* (Hydrocharitaceae), *Ruppia* (Ruppiaceae), *Cymodocea* (Cymodoceaceae) (Tomlinson 1969, Cox & Knox 1988, 1989, Cox 1988, 1991b, Cox & Tomlinson 1988).

In category (3), submersed pollination, pollen is transported to the stigma below water surface. This category is the most difficult to study. In *Thalassia testudinum* (Hydrocharitaceae) flowering is at spring tides (Phillips *et al.* 1981). Male flowers open at night. Pollen is released underwater in strings of mucilaginous slime along the substrate surface and collides with the stiff papillate stigmas of the female flowers (Cox & Tomlinson 1988). In other plants stamens abscise under water, float to the surface and release the pollen grains, which then slowly sink (*Ceratophyllum*; *Lepilaena bilocularis*, Zannichelliaceae) (Cox 1988). It has been suggested that in some plants of category (3) water surface pollination (category 2) may also occur at low tides (*Syringodium*, Cymodoceaceae) (Tomlinson & Posluszny 1978, Cox *et al.* 1990).

Incompatibility mechanisms have not been reported in plants with underwater pollination, perhaps because of constraints imposed by the water-liability of recognition substances and because of reduction of the exine in hydrophilous pollen (Pettitt & Jermy 1975, Les 1988). In *Posidonia australis* (Posidoniaceae) pollen germinates precociously in seawater if a stigma is nearby, pollen adhesion is not specific, and the sites of pollen adhesion and pollen tube penetration on the stigma may be distant (McConchie & Knox 1989). This suggests that the stigma releases a germination factor triggering pollen tube growth and that the pollen grains may regulate water uptake without stigma contact (McConchie & Knox 1989). Precocious pollen tube development is also known from other underwater plants.

An impressive range of possibilities of animal-, wind- and water-pollination has been explored even in closely related groups, as demonstrated by the Hydrocharitaceae (Cook 1982). Still more surprisingly, even within a single species, *Groenlandia densa* (Potamogetonaceae), pollination in the air, on the water surface and underwater occurs (Guo & Cook 1990). In *Althenia filiformis* (Zannichelliaceae) pollination at the water surface or underwater in air bubbles (bubble autogamy) has been recorded (Cook & Guo 1990). For *Potamogeton*, also, air and bubble pollination have been recorded (Philbrick 1988). All these finds make an evolutionary origin of water-pollination from wind-pollination most likely (Philbrick 1988, Cook & Guo 1990, Guo & Cook 1990). The systematic distribution of categories (2) and (3) suggests that each of these pollination modes has evolved independently several times.

5

Special differentiations associated with pollinator attraction

A multitude of devices in flowers to attract pollinators (except for pollen) are only loosely associated with certain floral organs. There is a great deal of positional flexibility in these differentiations, which manifest themselves mainly at the histological level. It has long been known that flowers display colours and scent and provide pollen and nectar for their pollinators (Kölreuter 1761, Sprengel 1793). In contrast, previously unknown flower classes, which offer 'perfume', oil and resin to special hymenopteran groups, which actively collect these floral secretions, were major discoveries in the field of ecology of the last few decades.

5.1 Pollen and pollen-flowers

Pollen-flowers provide only pollen to the pollinators or rarely small quantities of nectar in addition. The main pollinators of such flowers are beetles and bees in the widest sense, while flies are less prominent. Many pollen-flowers occur in taxa with polyandrous androecia, but representatives of a large number of families with oligandrous androecia also have specialized pollen-flowers.

Different forms of pollen-flowers

Vogel (1978a) distinguishes three types of pollen-flowers: (1) the *Magnolia* type having numerous stamens, sometimes in trap- or shelter-flowers, with sticky pollen, that is often shed on the perianth and eaten by pollinating beetles (mainly in Magnoliales, Nymphaeaceae and Araceae); (2) the *Papaver* type also with many stamens, usually in simple, open bowl-shaped flowers, but often with more powdery pollen that is either shed on the perianth or collected by bees directly out of the (sometimes poricidal) anthers (mainly in Ranunculaceae, Papaveraceae, Dilleniales, some Violales, Tiliaceae; or in single genera of otherwise nectariferous families, such as *Rubus*, *Mimosa*, *Portulaca*, *Opuntia*, Vogel 1978a); sometimes beetles are also pollinators of such flowers (Dafni *et al.* 1990); (3) the *Solanum* type (see also van der Pijl 1954) with oligandrous androecia, often in widely open flowers facing downwards or sideways with the petals often reflexed, with powdery pollen in often poricidal, large, optically attractive anthers (Figs 5.1, 5.2.1). This type occurs in many different groups of the angiosperms, e.g. *Solanum* (Solanaceae), *Ramonda* (Gesneriaceae), *Exacum* (Gentianaceae), *Brackenridgea* (Ochnaceae), *Ardisia* (Myrsinaceae), *Keraudrenia* (Sterculiaceae), *Dichorisandra* (Commelinaceae), *Dianella* (Liliaceae), *Xerophyta* (Velloziaceae) and *Cyanastrum* (Cyanastraceae). Its form is so peculiar that it has been termed a special type several times: 'shooting star flower form' by Corbet

Figure 5.1. Buzz-pollinated flowers with poricidal anthers.
1–2. *Solanum luteum*; flower pendent, polysymmetric.
1. From the side (× 13). 2. To the inside (× 18).
3–4. *Lycianthes rantonnei* (Solanaceae); flower oriented to the side, monosymmetric. 3. From the side (× 15). 4. To the inside (× 13).

et al. (1988), 'solanoid flower' by Faegri (1986) and 'tipo boragineo' by Delpino (1867).

Pollen can be collected by bees in different ways (Thorp 1979, Roubik 1989). Commonly it is done simply by contacting the floral parts and then grooming pollen adhering on the body. Another, outstanding way is by vibrating the stamens, so that dry pollen is loosened out of the anthers.

Pollen as a reward

Pollen is rich in nutrients, such as amino acids, polysaccharides, and lipids, and sometimes vitamins (Baker 1978). H. & I. Baker (1979) examined the occurrence of starch and oil in correlation with pollen size and pollinators. Starch is a less compact store of energy than oil. Therefore, if pollen grain size is a limiting factor, oil will be stored rather than starch, even if the pollinators do not demand it. In flowers with pollen as the only reward, pollen tends to be smaller so that enough grains can escape consumption. Many of these flowers have poricidal anthers, and small pollen can also be more easily released through the small anther pores. In pollen collected by Hymenoptera and Diptera there is selection of oil-containing pollen. In autogamous and anemogamous species there are high proportions of starchy pollen. In lepidopter- and bird-pollinated flowers there is selection for large starchless (or very large starchy) pollen, because pollen tubes must grow through long styles. Since pollen tubes are dependent on their own reserve substances in the first phase of development (see section 2.4.1) they need more reserves in long-styled flowers.

Buzz-pollination and poricidal anthers

The most fascinating evolutionary response of flowers to pollen collection by vibration is the reduction of anther dehiscence to apical pores and at the same time elongation of the anthers. This has happened in many different groups of the dicots and monocots, especially in Dilleniidae, Rosidae, Asteridae and Liliidae. Harris (1905) had noted taxa with poricidal anthers for 33 families, and in the meantime more have turned up; there may be such taxa in about 50 families. In the list given by Buchmann (1983) there is a still higher number but not all of them fall in the class of pollination by vibration. However, poricidal anthers are notably lacking in the Magnoliidae in spite of its many pollen-flowers, probably because flies and beetles, and not Hymenoptera, are the main pollinators in the majority of the Magnoliidae (Endress 1990b). The Araceae are perhaps similar, although slimy pollen may be released through pores.

In these flowers the bees cling to the tubular anthers while shivering their flight muscles; pollen is sprayed out of the anther pores by the vibrations and is caught in hairs over the surface of the bee's body (Roubik 1989). Electrostatic forces may play a role in the adherence of the powdery pollen to the pollinator's body and in the transfer to the stigma (Erickson 1975, Thorp 1979, Corbet *et al.* 1982, 1988, Erickson & Buchmann 1983). The usually small, pointed stigma may also be advantageous in the context of electrostatic forces (Niklas 1985). This method of pollen gathering was termed buzz-pollination by Michener (1962) for *Cassia* where the long, tubular, poricidal anthers are especially conspicuous, and Buchmann & Hurley (1978) discussed the properties of such flowers. However, observations on pollen collecting by vibrations are much older (e.g. van der Pijl 1939, in Melastomataceae; even Sprengel, 1793, observed in *Solanum* that bees were striking the anthers). Buchmann & Cane (1989) showed that bees are able to assess pollen returns while buzzing *Solanum* flowers.

Such stamens tend to have large, showy, often bright yellow anthers. In many groups the apical parts of the filament are also enlarged and brightly coloured, sometimes with the help of tufts of moniliform hairs (Fig. 5.13), so that the entire complex simulates extra large anthers (some Liliaceae; *Keraudrenia*, Sterculiaceae; *Cassia javanica*, Caesalpiniaceae; Commelinaceae; some *Cleome* species, Capparaceae, Iltis 1959, Kers 1969). There is also a tendency for the anthers to form a more or less postgenitally united tube (e.g. *Lycopersicon*, Solanaceae) (Figs 5.2, 5.3). The filaments are

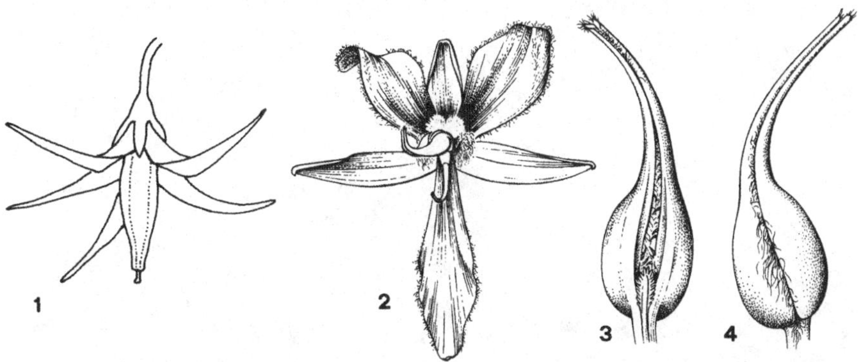

Figure 5.2. Highly elaborate buzz-pollinated flowers with anther openings hidden. 1. *Lycopersicon esculentum* (Solanaceae); flower pendent, anthers postgenitally united into a tube, anther openings inside tube (× 2.5). 2–4. *Cochliostema odoratissimum* (Commelinaceae); flower oriented to the side. 2. Flower (× 1). 3. Two hose-like cuculli formed by outgrowths of stamen filaments enclosing the anthers from ventral (3) and dorsal (4) side (× 3).

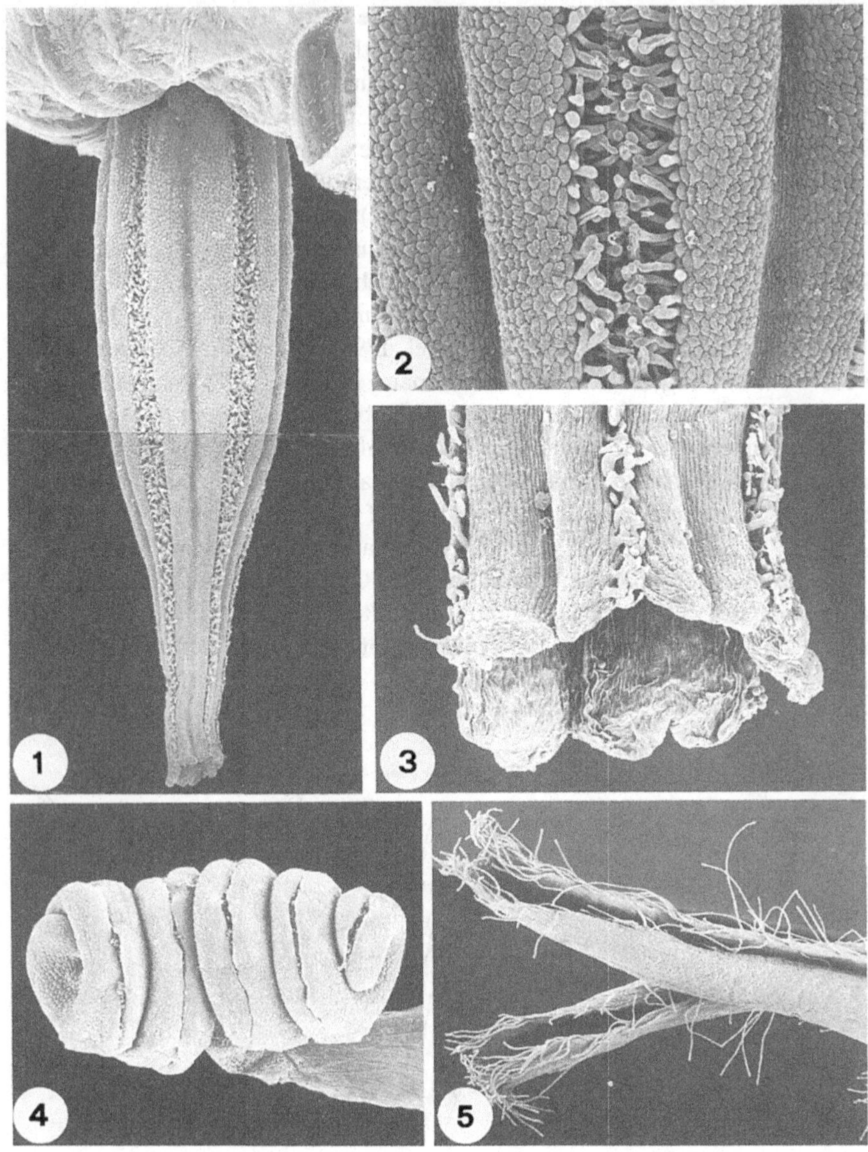

Figure 5.3. Highly elaborate buzz-pollinated flowers with anther openings hidden. 1–3. *Lycopersicon esculentum* (Solanaceae); pendent flower from the side. 1. Androecium (× 13). 2. Postgenitally united flanks of anthers (× 45). 3. Distal end of hose formed by the five united anthers (× 70). 4–5. *Cochliostema odoratissimum* (Commelinaceae). 4. Anther with helically twisted thecae (× 20). 5. Distal end of the two hose-like cuculli enclosing the anthers (× 13).

short and stout or, if elongated, they are thick and strong but elastic so that they may take up the vibrations of the bees. Single anthers or the entire androecium may form a resonating chamber (Corbet *et al.* 1988). In monosymmetric, horizontally oriented flowers the filaments are often arching (e.g. Melastomataceae; *Swartzia*, Fabaceae; *Cassia*, *Senna*, Caesalpiniaceae; *Cleome*, Capparaceae) (Fig. 8.20.3–4).

This plasticity in the morphological structure of the anthers is paralleled at the histological level. In some (but not all) taxa with poricidal anthers that are buzz-pollinated, an endothecium with partial cell wall thickening is not differentiated but the thecal walls may be reinforced by evenly thickened cell walls, sometimes in the epidermis instead of the hypodermis. Such deviations from the usual pattern have been reported for representatives of Dilleniaceae (Rao 1961: wall thickenings only in the epidermis), Ericaceae, Melastomataceae (Matthews & Knox 1926, Matthews & Maclachlan 1929), Gesneriaceae (Huber 1953), Mayacaceae and Commelinaceae (Gerenday & French 1988). Conversely, it is noteworthy that a normal endothecium is also absent in certain taxa that are nonporate but closely related to porate groups (Xyridaceae, Gerenday & French 1988; Epacridaceae, Endress & Stumpf 1990). Are these evolutionary reversals or is buzz-pollination, rather than poricidal dehiscence, influencing the histological change?

The entire biological situation makes it probable that the use of vibrations for pollen collecting came first, since it is widely used in flowers with dry pollen (Buchmann 1985, Proença 1992), not only in those with tubular anthers where it had been first described. The differentiation of long, tubular, poricidal anthers was a response to this method of pollen collecting. The better protection of the pollen and portionwise, parsimonious release out of the anthers must have been an advantage for the plants. Renner (1989a) found a particularly high proportion of buzz-pollinated plants on the high tepuis of Venezuela.

Heteranthery

Another evolutionary response to buzz-pollination is the advent of heteranthery. Since H. Müller (1883) compared heteranthery in several pollen flowers (and Darwin 1877 had emphasized the general evolutionary significance of different floral morphs in the same species) the phenomenon has attracted much attention and was again comparatively treated by Vogel (1978a). It is now known from about twenty families but often from only a few members within a family (Dilleniaceae: *Dillenia*, Lecythidaceae, Capparaceae, Tiliaceae: *Clappertonia*, *Sparmannia*, Cochlospermaceae, Fabaceae: *Swartzia*, Caesalpiniaceae; *Cassia* s.l., Anacardiaceae,

Malpighiaceae, Lythraceae: *Lagerstroemia*, Melastomataceae, Gentianaceae, Solanaceae, Scrophulariaceae, Commelinaceae, Pontederiaceae, Haemodoraceae, Liliaceae) (Vogel 1978a; personal observation). Heterantherous flowers contain stamens of different shapes and colours. A striking pattern is the presence of smaller but optically more attractive (often yellow) stamens where pollinators actively collect pollen (feeding stamens) and a lower number of larger but more cryptic (often bluish or red) stamens. In extreme cases only a single larger stamen is present (e.g. *Solanum* sect. *Androceras*, Whalen 1978; *Rhynchanthera* species, Melastomataceae, Renner 1990; *Pyrrorhiza neblinae*, Haemodoraceae, Simpson 1990). Pollen from the larger stamens is somehow transmitted to the pollinator's body (mostly the back), at places where it is not collected (groomed) but may later be transferred to stigmas, and, therefore, constitutes the main pollinating pollen (pollinating stamens). The release of the pollinating pollen to the bee's body may occur because of the vibrations of the bee, which are transmitted in some way to the pollinating stamens although the bee does not cling to them (e.g. *Dillenia*, *Cassia*, *Swartzia*) (Fig. 8.20.3–4). In many groups, pollen from both feeding stamens and pollinating stamens is fertile (Renner 1989b).

Heterantherous flowers may be polysymmetric or monosymmetric. However, they tend to be monosymmetric in some groups of otherwise mainly polysymmetric flowers (e.g. in Commelinaceae, Lecythidaceae, Gentianaceae, *Solanum*). One of the most striking groups is the Lecythidaceae where the pollinating stamens surround the gynoecium in a more or less regular manner, while the feeding stamens form a one-sided extension of the androecium (Prance 1976, Ormond *et al.* 1981). Deviating from the usual pattern, the feeding stamens are larger than the pollinating stamens in Lecythidaceae. In extreme cases, within the Lecythidaceae, the feeding stamens form an incurved hood over the pollinating stamens, protecting them, and instead of pollen nectar may be provided to pollinators from evolutionarily new nectaries (*Lecythis* sp., *Bertholletia*, *Eschweilera*, *Couratari*); in addition, there is an evolutionary shift to bat-flowers in *Lecythis poiteaui* (Mori *et al.* 1978) (see section 8.4.1).

Direction of the pollen jet to the pollinator's body may be canalized by special floral devices. In certain Ochnaceae the corolla does not expand but remains half-closed and forms a hose-like structure (Kubitzki & Amaral 1991), or flat staminodes around the stamens may have the same effect (Fig. 1957 in Goebel 1933). In *Dillenia* the open end of the poricidal anthers is curved towards the pollinator. In some *Chamaecrista* species (Caesalpiniaceae) the banner petal is incurved to form a 'cucullus', which acts as a kind of hose guiding the pollen stream (Gottsberger 1988). The most elaborate example is perhaps *Cochliostema* (Commelinaceae) (Vogel 1978a), where of the three fertile stamens the two lateral ones produce two hose-like cuculli, in which all three anthers are enclosed at maturity (Masters 1873) (Fig. 5.2). These two hoses are of about the same length as the style and with the exit near the style. In addition, the three anthers are long and helically twisted, which may be a special elaboration for prolonged pollen supply (Fig. 5.3). Interestingly, in the two last genera enantiomorphy (see section 6.3) additionally occurs. In all three genera mentioned, heteranthery is present. Pollen protection and parsimony is further enhanced by these elaborations. Spirally twisted anthers, as in *Cochliostema*, also occur in the buzz-pollinated *Orpheum* (Gentianaceae, Johnson 1992) and the pollen-flowers of *Philydrum* (Hamann 1966).

Pollen heteromorphy

In parallel with heteranthery, pollen heteromorphy has been found in several groups of pollen-flowers. Pollen from the feeding anthers is often sterile and also its structure may differ from that of the pollinating anthers. In *Couroupita guianensis* (Lecythidaceae) fertile pollen has a smooth, sterile pollen a verrucose surface (Mori *et al.* 1980). In *Swartzia* (Fabaceae) the sterile pollen is only 2/3 the size of the fertile pollen (Baker 1978). In *Tripogandra* (Commelinaceae) both pollen morphs develop under normal meiosis. However, in the sterile pollen meiosis is delayed (2 days) and after meiosis it degenerates (Mattsson 1982, Gamerro 1986).

Sterile pollen is evolutionarily released from all those constraints that act on fertile pollen between pollination and fertilization and one could, therefore, expect 'heteromorphy' also in the supply of reserve substances. Preliminary studies by Buchmann (1986) did not reveal significant differences among the two stamen morphs in heterantherous *Solanum* species. However, he found very high values for nitrogen and protein. H. & I. Baker (1979) showed, that, as expected, buzz-pollinated flowers with poricidal anthers have on average very small pollen.

5.2 Nectaries, nectar and nectar-flowers

Nectar-flowers

Nectar-flowers are the most important flower class because most pollinator groups, from the smallest bees and flies up to the largest bats, are nectar consumers. Therefore, the diversity of nectar-flowers is especially large. In many nectar-flowers pollen is also collected by the pollinators.

Nectaries

Nectaries are glands that secrete nectar, a liquid rich in assimilates, the main components being water and various sugars; amino acids are commonly present in low quantities. Nectaries are not confined to flowers. They also occur on stems and leaves, this predominantly in tropical plants, and not only in angiosperms but also in certain ferns. Their primary function in an evolutionary context is perhaps a physiological one, namely the elimination of superfluous assimilates (Frey-Wyssling 1933). Only secondarily did nectaries acquire various ecological roles.

The occurrence of nectar in flowers has been known for many centuries (Lorch 1978). But only Sprengel (1793) paved the way for an understanding of the ecological role of nectar in flowers and Darwin (1859) added the evolutionary dimension.

The ecological role of nectaries on vegetative organs was recognized still later. Delpino (1873) and Belt (1874) detected the association of nectaries with ants that protect the plants and Janzen (1967b) gave experimental confirmation for the efficiency of this symbiosis in *Acacia*, as did Schemske (1980) for *Costus*. Comparative studies on structure and function of nectaries on vegetative organs and on the outside of sepals were made by Nieuwenhuis – von Uexküll-Güldenbandt (1907), Cammerloher (1929a), Bentley (1977) and Elias (1983).

The delimitation between nectaries in flowers and on vegetative parts and between nectaries with pollination and ant guard function is not always congruent (discussion in Vogel 1977, Schmid 1988). Therefore, different terms are in use for the distinction. On the topographical level, floral and extrafloral nectaries (situated on floral organs vs. not on floral organs) are distinguished (Caspary 1848), or, more recently, reproductive and non-reproductive nectaries (situated within the inflorescence vs. outside the inflorescence) have been proposed (Schmid 1988). On the functional level, nuptial and extranuptial (involved in pollination biology vs. not involved in pollination biology) are used (Delpino 1873). All three pairs of terms may be useful in certain contexts, whereby the first set is the least ambiguous.

Nectaries are not confined to particular floral organs. They may occur anywhere in flowers. Nevertheless, there are certain systematic patterns (Brown 1938, Fahn 1954, Smets 1986). The two most prominent types are the disk nectaries in the middle and higher evolutionary level of the dicots and the septal nectaries in the monocots. Disk nectaries are mostly situated at the floral base between androecium and gynoecium and form an often conspicuous ring (Rao 1971) (Figs 5.4, 5.5). Septal nectaries are within the ovary septa of the gynoecium and have exits at the ovary surface (Fig. 5.6). While disk nectaries have long been known, septal nectaries were

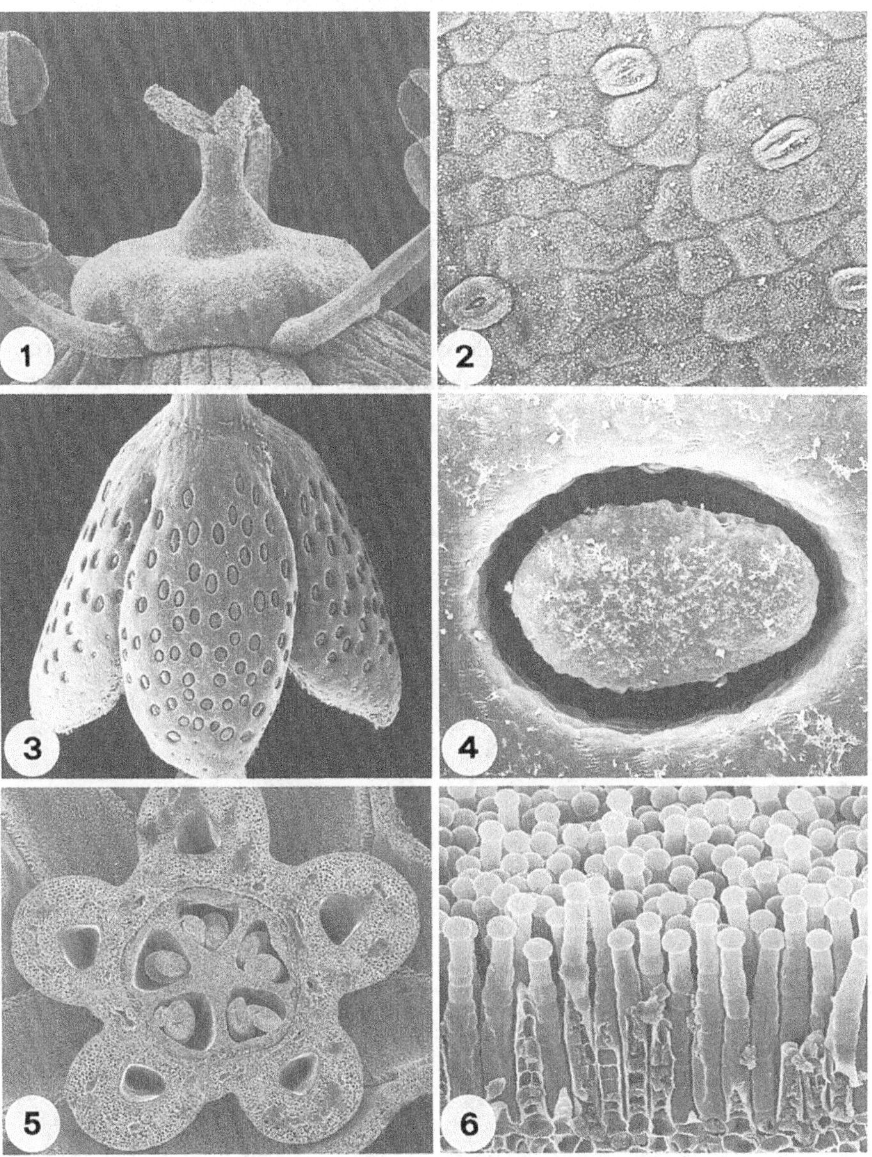

Figure 5.4. Forms of floral nectaries. 1–2. *Maytenus emarginata* (Celastraceae); mesophyll nectary forming a disk. 1. Floral centre from the side (× 20). 2. Surface view of nectary with water pores. 3–4. *Nepenthes* sp.; epithelium nectary on upper side of tepals. 3. Reflexed tepals with nectaries (× 13). 4. A nectary (× 250). 5–6. *Abutilon megapotamicum* (Malvaceae); trichome nectaries on upper side of sepals. 5. Floral base with upper gynoecium parts removed to show the five nectarial hair pads on the five sepals (× 13). 6. Group of multicellular hairs of the nectarium (× 250).

not detected until 1854 by Brongniart (see reviews by Daumann 1970, Schmid 1985, Böhme 1988). Other sites of nectar production are the outer or inner surface of sepals (*Thunbergia*, Acanthaceae, see section 8.8.4; Malvales), the inner surface of petals (*Gloriosa*, Liliaceae; *Halesia*, Styracaceae; Triuridaceae, Rübsamen-Weustenfeld 1991), stamens (Lauraceae), carpels (*Omalanthus*, Euphorbiaceae); the stigmatic secretion may also be copious, sugar-rich and function as nectar (e.g. several Araceae, Daumann 1931, Ramirez & Gomez 1978, Croat 1980).

Further, there is a considerable diversity in the histological structure of floral nectaries. Vogel (1977) distinguishes three main types (already

Figure 5.5. Forms of disk nectaries. 1–2. *Phyllanthus hypospodius* (Euphorbiaceae); male flower, disk as five separate lobes between petals and stamens, each water pore in a groove. 1. Flower (× 15). 2. Disk lobe (× 70). 3–4. *Mangifera indica* (Anacardiaceae); bisexual flower, disk as five separate lobes between petals, with convoluted surface. 3. Flower (× 13). 4. Surface of disk (× 130).

characterized by Behrens 1879 and Fahn 1952) (Fig. 5.4): (1) mesophyll nectaries: nectar is secreted in the mesophyll and reaches the nectary surface through intercellular spaces and modified stomata (water pores) (see also Zandonella 1977); (2) epidermis (epithelium) nectaries: nectar is secreted by the epidermis; (3) trichome nectaries: nectar is secreted by tufts of hairs.

Within flowers type (1) seems to dominate, especially in disk nectaries. This type may be evolutionarily derived from hydathodes, as they occur on foliage leaves (not from extrafloral nectaries!) (Vogel 1990a for Cucurbitaceae). Type (2) is more frequent in extrafloral nectaries, where often the epidermis is differentiated as a pronounced epithelium. Within flowers it occurs in the septal nectaries of the monocotyledons (Böhme 1988), in the equally slit-like elaborated nectaries of Asclepiadaceae (Christ & Schnepf 1985, 1988), on the upper surface of the tepals in *Nepenthes* (Nepenthaceae) (Fig. 5.4), or, without a pronounced epithelium, on the inner tepals of *Asimina triloba* (Annonaceae). It is also common in the extrafloral but nuptial nectaries of *Euphorbia* and related genera (Rauh 1984). The presence of type (1) in flowers and type (2) on vegetative parts among the same plant has been shown, for example, in *Passiflora* (Durkee *et al.* 1981, Durkee 1982), *Thunbergia* (Acanthaceae) (Zandonella & Piolat 1982), and *Ixonanthes* (Ixonanthaceae) (Link 1992). And finally, types (1) and (2) may even occur intermixed at the same place, such as in the nectariferous spur of *Tropaeolum majus* (Rachmilevitz & Fahn 1975). Type (3) seems relatively rare but is very characteristic of Malvales and Dipsacales, and also occurs in Periplocoideae (Asclepiadaceae) (Schick 1982b), in a few Scrophulariaceae (Vogel 1977), Cucurbitaceae (Vogel 1981b), and in *Avicennia* (Verbenaceae) (Werth 1922). A kind of mixed form between type (2) and (3) are disk-shaped, short trichomes with an apical multicellular epithelium. They occur, for example, on the calyces of *Thunbergia* (Acanthaceae) and of many Bignoniaceae. In *Nepenthes* this type occurs intermixed with the broader epidermis nectaries already mentioned (Daumann 1930). Nectaries may also originate at the abscission layers of early caducous flower buds (*Apios*, *Canavalia*, Fabaceae), or on undeveloped floral buds (*Sesamum*, Pedaliaceae) (Vogel 1977).

The modified stomata (water pores) in mesophyll nectaries seem to have lost the capacity to close their pores by contraction of the guard cells. Only at the end of the secretory phase may the pores become occluded. However, few taxa have been studied in detail, e.g. *Vicia faba*, Fabaceae (Davis & Gunning 1992).

Nectaries have a rich vascular supply, and the phloem is relatively well developed, compared with other vascular bundles in the flower. Frei (1955) studied the different kinds of vascular supply of floral nectaries in connection with sugar concentration in the secreted nectar, which was later found to be of interest for diverse pollinating syndromes (H. & I. Baker 1975, 1983). In general, nectaries that produce nectar with a relatively high sugar content have relatively more phloem elements in their vasculature than those that produce relatively watery nectar. Nectarial tissue, like other glandular tissues, is rich in cytoplasm and sometimes contains starch, which is used during nectar production (e.g. *Passiflora*, Durkee *et al.* 1981).

A distinction, interesting from the evolutionary and ecological point of view, is whether or not nectaries fall off after anthesis (nectaria caduca vs. nectaria persistentia, Smets 1986). This distinction had also been made by Delpino. The first category is mainly on petals and stamens, the second on disks and carpels or perhaps also on persistent sepals. From the evolutionary point of view the persistent nectaries are concentrated in the medium and higher evolved groups of the dicots (disk nectaries) and in monocots (septal nectaries). An ecological aspect of the persistent nectaries is that they may be able to function far beyond anthesis and attain new functions, e.g. in the attraction of ant guards for the developing fruits (e.g. *Mentzelia nuda*, Loasaceae, Keeler 1981; *Ruellia radicans*, Acanthaceae, Gracie 1991). In addition, extranuptial nectaries on the undersurface of the sepals may attract ant guards for the flowers against attacks by other organisms (e.g. *Thunbergia grandiflora*, Acanthaceae, Krebs 1990) or against nectar robbing by regular pollinators (van der Pijl 1954). Persistent nectaries may also be able to reabsorb nectar (Daumann 1930, for *Nepenthes*; Cruden *et al.* 1983, Burquez & Corbet 1991).

Some flowers with copious nectar production have special compartments, where nectar is stored (Sprengel 1793 distinguished it as 'nectar holder' from the 'nectar gland'). These parts are often the base of a floral tube in sympetalous flowers or spurs (see section 3.4). The same effect is produced by free petals that are directed forwards and with overlapping flanks (e.g. *Quassia amara*, Simaroubaceae, or *Malvaviscus arboreus*, Malvaceae) (Fig. 4.2) (Opler 1983).

The most complicated nectaries are the septal nectaries already mentioned of many Liliidae and Zingiberidae. They arise at the flanks of the originally free carpels (Fig. 5.6). By differential postgenital fusion of the three carpels either a continuous canal system or three canals are formed, which lead to the surface of the ovary at a certain level. The epidermis of the canals is secretory and nectar may be stored partly in the canals. In some bird- and bat-pollinated taxa with copious nectar production the

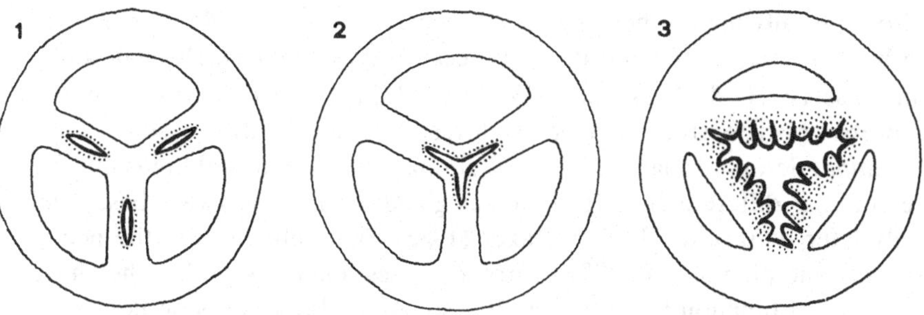

Figure 5.6. Forms of septal nectaries as seen in transverse sections of ovaries. 1. Separate (e.g. *Beschorneria*, Agavaceae). 2. Confluent-simple (e.g. *Heliconia*). 3. Confluent-convoluted (e.g. *Pitcairnia*, Bromeliaceae).

surface of the secretory epidermis is much enlarged by convolutions (Vogel, 1977, calls it a 'nectar kidney'). This is the case, for example, in *Pitcairnia* (and other Bromeliaceae, Böhme 1988), or in *Musa* (Fahn & Benouaiche 1979) (Fig. 5.6). In *Mucuna* (Fabaceae) the disk-nectaries are highly lobed (see section 8.6.3).

Nectar

Apart from the omnipresent sugars (sucrose, glucose and fructose), water and amino acids, nectar may contain several other compounds, such as other sugars, lipids and antioxidants, and – as potential deterrents – alkaloids, phenolic substances and glycosides (H. & I. Baker 1983). The content and concentration of sugars and amino acids may be different depending on the requirements of the particular flower pollinators or in correlation with different flower forms or with their evolutionary history. Hummingbird-flowers have high sucrose:hexose ratios, while other bird-flowers have low ratios. H. &. I. Baker (1983, 1990) suggest a historical reason: hummingbird-flowers are often derived from bee-flowers, which are characteristically sucrose-rich; in other flower-birds a 'taste' for hexose sugars may have developed from the common use of fruit juices. Although sugar concentration may be affected by environmental conditions, there is a general correlation with the pollinator types. Sugar concentration is low in lepidopter-flowers because nectar of high viscosity could not be taken up by the long, narrow mouth parts. This is also true for hummingbird- and microchiropteran bat-flowers, since they spend only a second or less at a flower (H. &. I. Baker 1983). Amino acids seem to be important for insects that have no other source of protein-building materials except nectar, especially Lepidoptera. Amino acid concentration is also especially high in flowers visited by carrion and dung flies. It is lower in bird- and bat-pollinated flowers, because these animals also eat pollen and insects (H. &. I. Baker 1973, 1983). Lipids in nectar were found in flowers of some trees in tropical forests (Baker 1978). In some cases (*Catalpa*, Scrophulariaceae; some Bignoniaceae) nectar and lipids may be secreted at different sites (disk versus glandular hairs at ovary base) and only later be mixed. H. & I. Baker (1983) found alkaloids in the nectar of 12% and phenolics in 49% of tropical forest plants. Whether they have a deterrent function has not been investigated. Hawkmoths on flowers of *Datura meteloides* (Solanaceae) behaved as if they were dizzy, perhaps because of alkaloids in the nectar, but seemed to like it (V. & K. Grant 1983a).

Physiological aspects of nectar production have often been studied. The transport and transformation of precursors from the phloem to the surface of the nectary and the secretion of the nectar cannot be discussed here (reviews: Findlay 1982, Fahn 1979, 1988).

5.3 Elaiophores, floral oil and oil-flowers

Oil-flowers

Vogel (1969c) made an important discovery when he reported the existence of flowers that offer oil, and not nectar, to their pollinators. He also demonstrated (1974) the occurrence of oil-flowers in a large number of angiosperms, among them many tropical taxa. The pollinators are some specialized groups of bees, which use the floral oil to nourish their larvae and perhaps also for lining the nest walls. In the meantime, knowledge of the phenomenon has been greatly expanded (e.g. Simpson & Neff 1981, 1983, Simpson *et al.* 1990, Buchmann 1987, Vogel 1988, 1989, 1990a, Vogel & Machado 1991, Cocucci 1991). At present, oil-flowers are known from more than 70 genera out of 8 angiosperm families (Malpighiaceae, Krameriaceae, Cucurbitaceae, Primulaceae, Solanaceae, Scrophulariaceae, Iridaceae, Orchidaceae) (Vogel 1988).

Elaiophores

The floral oil is secreted by special glands, the elaiophores (Vogel 1974). Both arrangement and histological structure of the elaiophores differ in some way from that of nectaries.

In contrast to nectar, the bees do not take up the floral oil with their mouth parts but with their legs. The two front legs, and sometimes also the middle pair of legs, have special devices to collect oil (Vogel 1974). Consequently, the elaiophores often occur in pairs; they are often situated not towards the centre of the flower but more laterally where they can be reached with the legs. This is obvious in the double spurs of *Diascia* or *Angelonia* (Scrophulariaceae), the bilateral arrangement of the elaiophores on the sepals of Malpighiaceae, on the petals in some Cucurbitaceae, or on the lip of *Oncidium* (Orchidaceae) or near the gynoecium in Krameriaceae (Vogel 1974, Steiner & Whitehead 1990) (Figs 5.7, 5.8). In other groups they have a circular arrangement around the entire flower, such as in some Iridaceae or Cucurbitaceae (Vogel 1974, 1990a). Another method of oil-collecting is abdominal 'oil-mopping'. It was found in the bees (Ctenoplectridae) that pollinate Cucurbitaceae (Vogel 1990a).

Two prominent kinds of morphological and histological differentiation occur, which have been extensively studied by Vogel (1974, 1986, 1990a) (Fig. 5.8). (1) The elaiophores occur in the form of more or less raised knobs or groups of knobs or ridges, which are obvious on the flowers. The epidermis forms a pronounced epithelium with strongly elongated, dark staining cells. The epidermis is covered by a very thick cuticle. In the centre of the elaiophore the cuticle separates from the cell walls and oil is secreted into this interspace. As the flowers are visited by their bees the oil is squeezed out through a weak part of the cuticle, where it easily ruptures when manipulated. This kind of elaiophore is present in Malpighiaceae, Krameriaceae and a few Orchidaceae (e.g. *Oncidium ornithorrhynchum*). (2) The elaiophores are dense mats of unicellular or multicellular hairs (trichome elaiophores). They are less striking than the first kind. If the hairs are multicellular they have a multicellular head on an uniseriate stalk, whereby usually one stalk cell has cutinized cell walls. If the hairs are unicellular, they may also be capitate (Iridaceae). Mats of multicellular hairs occur in Scrophulariaceae (*Calceolaria*, *Diascia*, *Angelonia*), Primulaceae (*Lysimachia*) and Cucurbitaceae (*Momordica*, *Thladiantha*). Mats of unicellular hairs occur in several Iridaceae, and *Zygostates* (Orchidaceae).

As compared with nectaries, these two types of elaiophore histology correspond to the epidermis (epithelium) type and the trichome type. In contrast to nectaries, the mesophyll type, where the secretion is released through stomata, seems to be absent in elaiophores.

In some instances elaiophores may be evolutionarily derived from nectaries. In Malpighiaceae elaiophores occur only in New World representatives, while in Old World groups the same glands on the outer side of the sepals function as nectaries (e.g. *Hiptage*). It has been debated whether these Old World taxa still represent the ancestral condition or have secondarily reversed to it (Vogel 1974, 1990b, Anderson 1990). Vogel (1990b) assumes that the ancestral condition in the family was to have pollen-

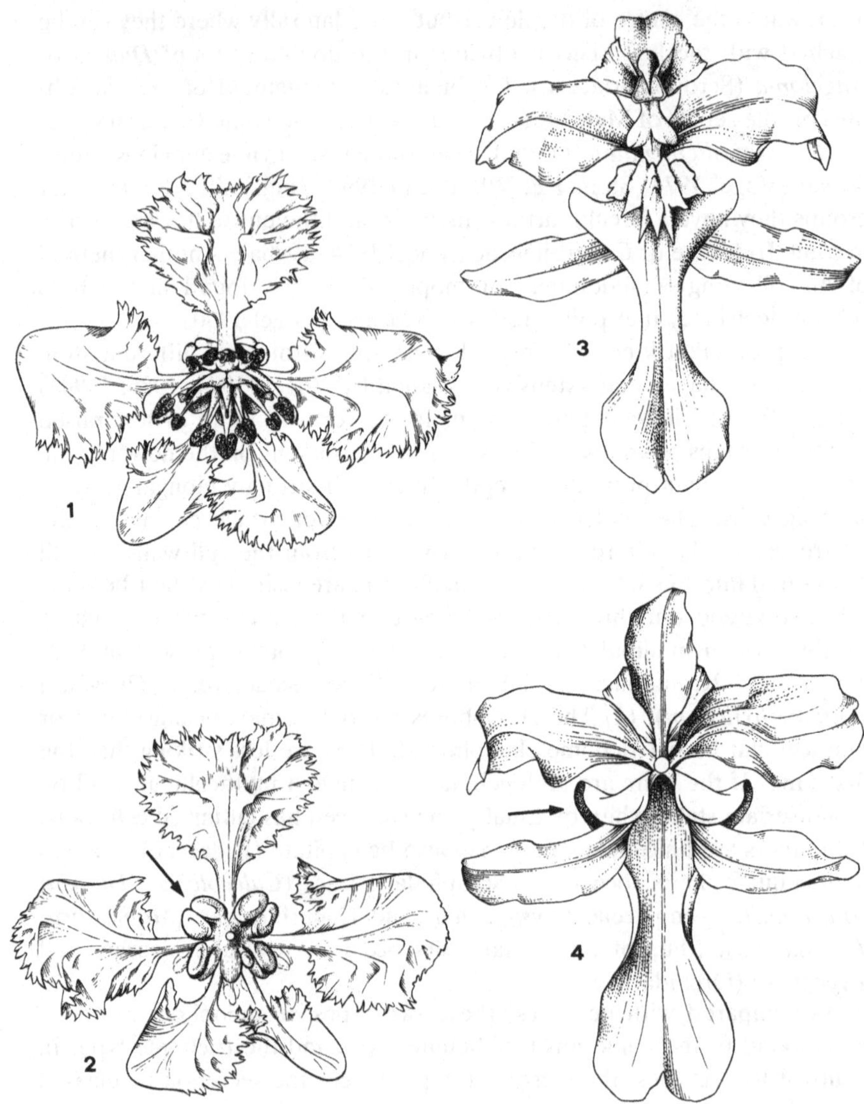

Figure 5.7. Oil-flowers of two different families with paired elaiophores on outer side of flower (arrows) (× 3). 1–2. *Malpighia glabra*. 1. From inner side. 2. From outer side. 3–4. *Oncidium ornithorrhynchum* (Orchidaceae). 3. From inner side. 4. From outer side.

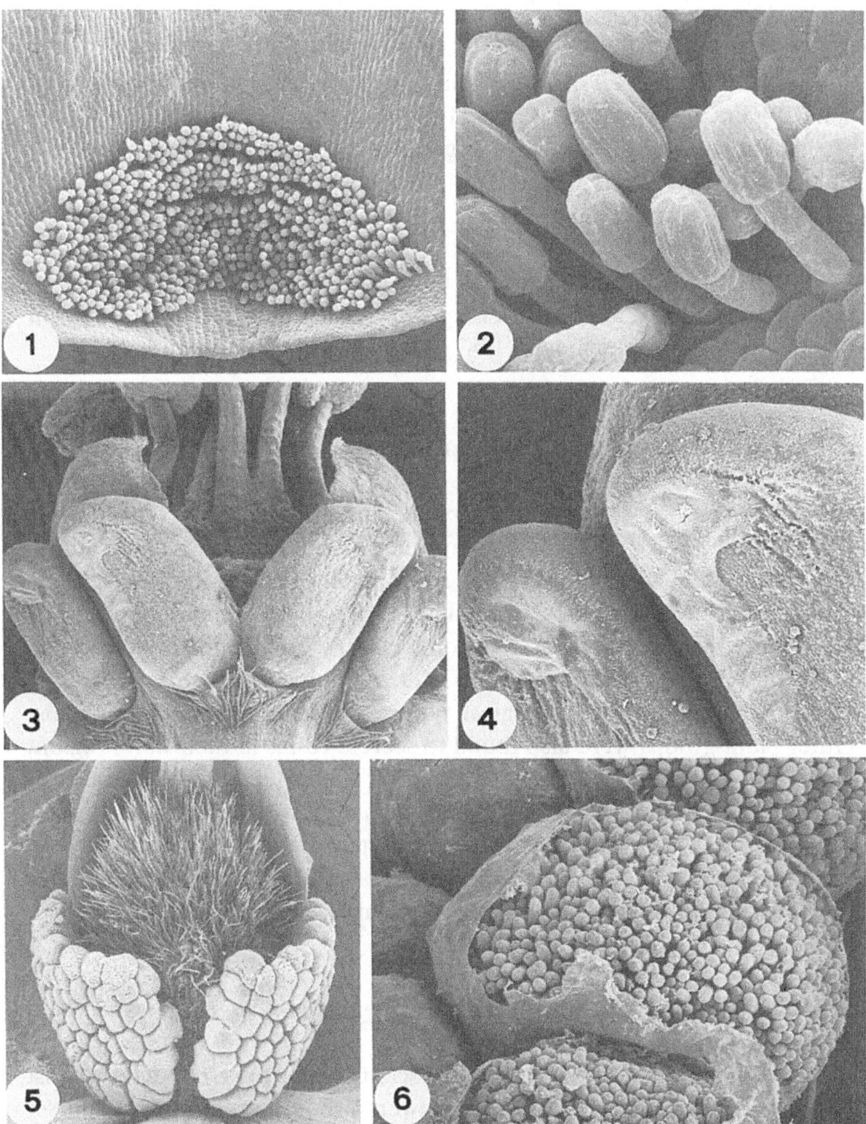

Figure 5.8. Forms of elaiophores. 1–2. *Calceolaria scabiosaefolia* (Scrophulariaceae). 1. Trichome elaiophore pad on distal part of lower lip (× 40). 2. Multicellular hairs of elaiophore pad (× 300). 3–4. *Malpighia glabra*. 3. Epithelium elaiophores, two pairs on outside of sepals (× 18). 4. Two elaiophores, showing thick cuticle forming oil-filled blisters. 5–6. *Krameria grayi*. 5. Epithelium elaiophores, one pair flanking ovary base, subdivided into numerous portions (× 13). 6. One portion, with cuticle partly removed (× 150).

flowers (see also Baker 1978 for other groups). Elaiophores consisting of mats of hairs may rather be derived from similar hairs that secrete resinous or oily substances on floral and vegetative parts, secretions that are not collected by pollinators (Vogel 1974, Steiner 1985).

Floral oil

Floral oils are specifically fatty oils, mono- and diglycerides of monohydroxy fatty acids; 3-acetoxy fatty acids were found in different taxonomic groups (Buchmann 1987, Vogel 1990a).

5.4 Resin glands and resin-flowers

Many insects collect plant resins for use in nest construction, especially bees of the families Apidae and Megachilidae (Armbruster 1984). These resins are usually taken from wounds of various plants. Early reports of bees collecting resins from inflorescences in *Dalechampia* (Euphorbiaceae) were by H. Müller (1879) and Cammerloher (1931). However, only recently was the floral biology analysed more closely in this group (Armbruster & Webster 1979, Armbruster 1992). Froebe *et al.* (1983) studied the morphological nature of these (extrafloral) glands. Other plant groups were also found to secrete resins in their floral region, which are collected by bees. Skutch (1971) reported it from stamens of *Clusia*, Armbruster (1984) also from staminodes of the female flowers. Armbruster (1984) lists *Dalechampia* (15 species) and *Clusia* (8 species). Simpson & Neff (1981) also mention the use of floral waxes of some species of *Maxillaria* (Orchidaceae) by bees for similar purposes.

5.5 Osmophores, floral scent and perfume-flowers

Scented flowers

In plants the ecological role of scents seems to be restricted almost entirely to flowers and fruits. That flowers produce scent has, of course, long been known; Delpino (1868–74) provided a classification of floral odours. However, only in the second half of this century was the localization (Lex 1954, von Aufsess 1960) and structure and function of floral scent glands comparatively studied (Vogel 1963a, 1990d).

In many of the popular fragrant flowers scent comes from the entire surface of the petals, such as in *Polianthes* (Agavaceae) or *Citrus* (Rutaceae). In a number of primitive flowering plants, where petals are lacking, scent is produced by the stamens and staminodes, e.g. Chloranthaceae, Austrobaileyaceae, Eupomatiaceae (Endress 1980c, 1984a,b,

1987b). Scent production by anthers also occurs in other groups, e.g. *Solanum* (D'Arcy *et al.* 1990). In some specialized flowers and inflorescences the scent-producing parts are also morphologically prominent, like a kind of antenna, as shown by Vogel (1963a) for some Orchidaceae, Burmanniaceae, Araceae, Asclepiadaceae, and Aristolochiaceae, where he distinguished 'flat-', 'whip-', 'paint-brush-', and 'palp-osmophores' (Fig. 5.9).

Pollen grains may also be carriers of scent. They may influence which flowers bees visit by signalling the presence of pollen reward (Dobson 1989). The volatiles are mainly terpenoids. They seem to be constituents of the pollenkitt produced by the tapetum (Dobson 1988).

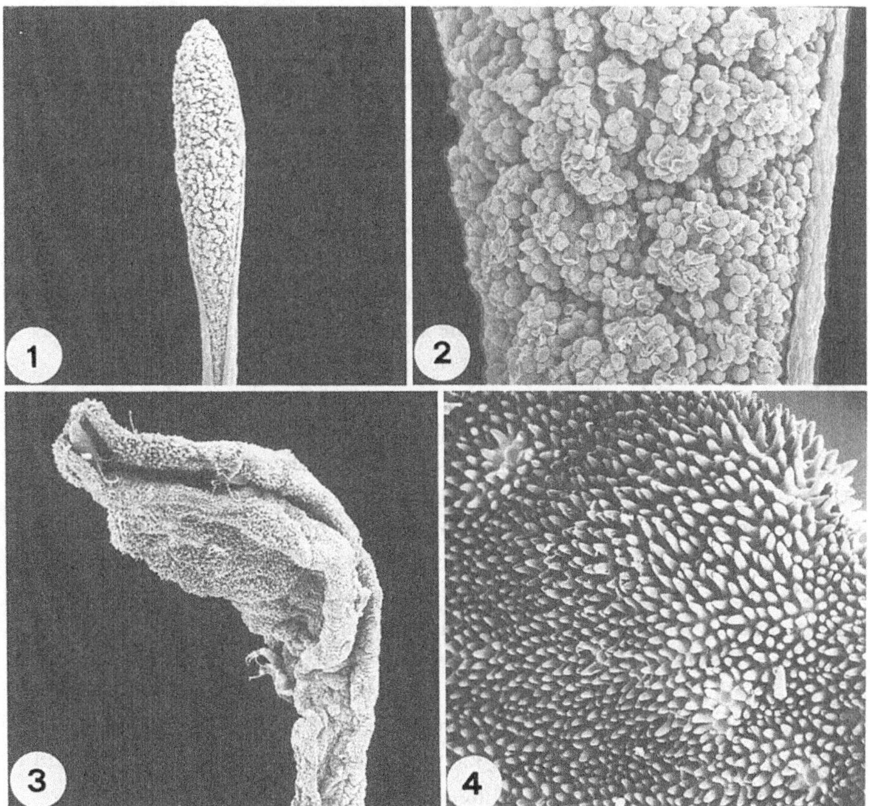

Figure 5.9. Osmophores. 1–2. *Restrepia elegans* (Orchidaceae). 1. Distal part of median sepal (× 13). 2. Papillate surface (× 80). 3–4. *Austrobaileya scandens*. 3. Distal part of inner staminode (with two drosophilid eggs in furrow) (× 14). 4. Papillate surface (× 70).

Sometimes the scent-producing region is restricted to the floral centre in the form of scent guides. This was found by Lex (1954) in experiments with bees. Vogel (1963a) showed that scent guides are often also 'negatively ultraviolet' guide marks (see section 5.7.2). In contrast to the minute organs of insects, our clumsy noses are able to localize scent-producing regions only in very big flowers (*Rafflesia*, Bänziger 1991; *Aristolochia grandiflora*). We also have to bear in mind that we are not able to perceive all the volatiles that may be apparent to pollinators. Most animal-pollinated flowers are typically scented, the least scented are bird-pollinated flowers. Scents are an important means by which oligo- or monophilic flowers address their pollinators (Vogel 1963a).

Osmophores

The scent glands, osmophores, in flowers have a particular histological structure (Vogel 1963a, 1990d). Besides the general features of secretory tissues, such as rich vascularization, richness in cytoplasm, large size of nuclei, and reduced cuticle, osmophores typically show richness in starch (except for the secretory epidermis), and either a conspicuously papillate epidermis with the large nuclei lying in the papillae or, more rarely, numerous stomata. Thus, there seem to be two different histological types: (1) epidermis (epithelium) osmophores, where the volatiles are secreted by the epidermis, and (2) mesophyll osmophores, where volatiles are secreted in the mesophyll and released through stomata.

The presence of long or otherwise prominent papillae on the secretory epidermis of type (1), which is in contrast to other gland types in flowers, suggests that the enlarged surface is a means of efficient emission of volatiles. The same effect may be at work, at the morphological level, in the often antenna-like structure of osmophores. Examples of type (1) are *Ceropegia* (Asclepiadaceae) (Vogel 1961, 1963a), *Austrobaileya* (Endress 1980c), and especially *Restrepia* (Orchidaceae) with bizarre mushroom-like papillae (Vogel 1963a, Pridgeon & Stern 1983) (Fig. 5.9). Type (2) occurs in the oft-studied appendices of Araceae inflorescences (Vogel 1963a). The 'food-bodies' of *Sauromatum*, perhaps modified sterile flowers, are also osmophores and a variant of type (2): the volatiles are released through numerous small lacunae between the epidermis cells and not through stomata (Meeuse *et al.* 1984).

Contrary to other gland types in flowers, the secretions of osmophores are usually not visible, since they are highly volatile. Therefore, it is more difficult to recognize and delimit an osmophore by superficial observation. A quick test is staining with neutral red solution. All types of glands readily take up neutral red solution in the living state, because their cuticles are reduced. By exclusion of other types of secretion osmophores can be found in cursory tests (Vogel 1963a, Stern *et al.* 1986).

Floral scents

A great number of compounds of floral scents are known. Vogel (1983) provides a principal classification: (1) components of 'agreeable' smells: methyl-esters of fatty acids; aliphatic and cyclic mono- and sesquiterpenes, diterpenes; compounds with benzene rings; phenylpropanes; (2) components of 'unpleasant' smells: hydrocarbons; fatty acids; nitrogenous volatiles (ammonia, indole, cadaverine, putrescine, skatole) (see also Williams 1983).

Rhythmicity of scent emission

Many flowers change the intensity or quality of their scent during anthesis. This is especially prominent in many hawkmoth-pollinated flowers that become heavily fragrant in the evening but are almost scentless during the day. In *Hoya carnosa* (Asclepiadaceae) the physiological mechanism of this nocturnal scent production is an endogenous circadian rhythmicity: scent production continues for two to three cycles under constant illumination (Altenburger & Matile 1988). In *Stephanotis floribunda* (Asclepiadaceae) there is a rhythmic change of fragrance composition. The maxima of the compounds of the floral scent do not coincide in time: the maximum of methylbenzoate and linalool occurs at midnight but that of 1-nitro-2-phenylethane at noon (Matile & Altenburger 1988). An ecological explanation may be the advantage of a more cryptic behaviour during the day when appropriate pollinators are lacking.

Scent production coupled with heat production

Scent emission may be a very rapid process. In a number of plant groups, scent production is coupled with conspicuous heat production (thermogenicity). The best studied examples are in the Araceae (Meeuse & Raskin 1988). Thermogenicity has also been reported from the related families Cyclanthaceae and Arecaceae, and among dicots from Annonaceae, Aristolochiaceae and Nymphaeaceae (Raskin *et al.* 1987). Temperature rise seems to be primarily a mere accompanying effect of scent production, but secondarily it may efficiently promote scent emission and may also be a pollinator attractant by itself. In all these cases flies and beetles are involved in pollination, and a number of cases are trap-blossoms. The floral odours involved often contain aliphatic amines, indole and skatole.

Salicylic acid induces heat production in Araceae (Raskin *et al.* 1987).

In *Sauromatum* salicylic acid is produced in the primordia of the male flowers. From there it moves into the appendix where it exerts its action after a lag-time of about a day (Meeuse & Raskin 1988). The thermogenic process is under tight biological control, as the same ceiling temperature is always reached, regardless of ambient temperature, e.g. 42–44 °C in *Xanthosoma robustum* and *Philodendron selloum* (Meeuse & Raskin 1988).

Perfume-flowers

Since the experiments by K. von Frisch (1919) it has been known that traces of scent are used by honey bees to advertise a promising pollen or nectar source to other worker bees in the hive. Scent may be carried directly on the body surface or on pollen grains taken up (Dobson 1989). Conversely, bees mark with scent and reject freshly visited flowers (Giurfa & Nunez 1992). It was a relatively recent find, however, that male euglossine bees in the Neotropics actively collect floral scents and apparently use them as precursors of sex pheromones to attract females to a mating site. The existence of such 'perfume-flowers', which secrete scents in large quantities, and their biological functioning was discovered in the orchid *Stanhopea* and elucidated by Vogel (1960, 1963c, 1966a), Dodson & Frymire (1961), Dressler (1968) and Dodson *et al.* (1969). They are now best known in orchids, where all species of the subtribes Stanhopeinae and Catasetinae are pollinated exclusively by male euglossine bees. They also occur in Gesneriaceae (*Gloxinia*; Vogel 1966a), Bignoniaceae (*Saritaea*; Dressler 1982), Solanaceae (some *Cyphomandra* species; M. Sazima & Vogel 1989), Euphorbiaceae (species of *Dalechampia*, Armbruster & Webster 1979), Araceae (*Spathiphyllum*, *Anthurium*; Dressler 1968), Haemodoraceae (*Xiphidium*; Dressler 1982). In the orchids more than 600 species have perfume-flowers and more than 200 bee species may act as floral visitors (Ackerman 1985, 1986b).

Owing to their excessive scent production, the osmophores of perfume-flowers have large surfaces; sometimes, as in *Stanhopea* (Orchidaceae), the surface has irregular ridges and groves (Stern *et al.* 1987). As in other osmophores the tissue cells have large nuclei, fairly large vacuoles and contain lipid droplets and large amounts of starch. Surprisingly, the starch disappears only at the end of anthesis and shows no change during the active phase of scent production.

Earlier, these special perfume osmophores had been mistaken for 'food tissue' because of their rich content in assimilates and also because of the striking behaviour of the flower visitors. They manipulate the glands with their forelegs and carry the perfume in leg cavities. It had been noticed by Darwin (1859) that the bees behaved as if drugged when entering the flowers of *Coryanthes* (Orchidaceae).

The main components of the floral fragrance in certain orchids and in *Dalechampia* (Euphorbiaceae) were found to be identical (Whitten *et al.* 1986). Comparative surveys of scent compounds of perfume-flowers are given by Williams & Whitten (1983) and Gerlach & Schill (1991). They contain chiefly monoterpenes, sesquiterpenes, and phenylpropanes.

5.6 Evolutionary relationships between pollen-flowers, nectar-flowers, oil-flowers, perfume-flowers and resin-flowers

It seems that certain angiosperm groups have acquired a special versatility in producing different kinds of floral secretions; in some families, such as Gesneriaceae, Solanaceae and Orchidaceae, pollen-flowers, oil-flowers and perfume-flowers occur together, sometimes in closely related groups.

In *Cyphomandra* (Solanaceae) with '*Solanum* type' pollen-flowers the anther connective is scented (D'Arcy *et al.* 1990) and in some species scent is collected by euglossine bees instead of pollen; by the scraping movements of the bees pollen is released out of the anthers without any buzzing (Sazima & Vogel 1989). A similar relationship between pollen-flowers and perfume-flowers seems to occur in *Xiphidium* (Haemodoraceae) (Buchmann 1980). In *Dalechampia* (Euphorbiaceae) pollen-flowers, resin-flowers and perfume-flowers occur (Armbruster 1992). It may be noted that constituents of both resin and perfume are terpenes.

Many oil-flowers may have evolved from nectar-flowers (Vogel 1974, 1988, Baker 1978). This is suggested by the similar but more complicated structure of elaiophores compared with nectaries and by the systematic and biological context of nectar-flowers. The oil-flowers of *Nierembergia* (Solanaceae) probably evolved from nectar-flowers by the loss of the nectary disk and differentiation of trichome elaiophores (Cocucci 1991).

However, for Malpighiaceae Vogel (1990b) assumes an origin of the oil-flowers from pollen-flowers with extrafloral nectaries, as they still occur in Old World genera. In *Krameria* the poricidal anthers may also point to former pollen-flowers. This evolutionary direction is also suggested for the oil-flowers among the genus *Lysimachia* (Primulaceae), which still have the shape of *Solanum*-type pollen-flowers, as they occur in other *Lysimachia* species and in many genera of the sister family Myrsinaceae (Vogel 1986). In *Diascia* (Scrophulariaceae) with predominantly purple oil-flowers (Vogel 1974) a few species have yellow flowers and pollen is the only floral reward (Steiner 1992). They may have reverted from oil-flowers to pollen-flowers. A trend from oil-flowers to pollen-flowers is also seen in *Calceolaria*, where a quarter of the species are visited for pollen, although some still have rudimentary elaiophores (Molau 1988).

The transition from nectar-flowers to pollen-flowers and vice versa has occurred many times. An example for the first direction are Solanaceae with *Solanum* (Vogel 1978a); examples for the second are Melastomataceae (Renner 1989b, Stein & Tobe 1989) and Lecythidaceae (Prance 1976).

5.7 Optical displays, visual cues

Different colours and shapes make flowers immediately attractive to us, and they are also properties of prime importance in pollination biology. In contrast to the other floral attractants discussed above, optical displays are not rewards for pollinators. That petals are colourful organs is commonplace, but all other floral organs may also be involved in colour display. Even organs outside the flowers, such as floral bracts, may be the main flags (e.g. Zingiberaceae; Heliconiaceae; *Poinsettia*, Euphorbiaceae; *Bougainvillea*, Nyctaginaceae). Leppik (1977) used the term 'semaphylls' for the optically attractive floral parts. The floral contour ('dissection') may also be important as an optical display (Hertz 1935, Leppik 1953, Vogel 1954, Barth 1991).

5.7.1 Colours

The visual part of the electromagnetic spectrum, the 'spectrum' in its original sense, is what we perceive as light. Visible light is a particular range of wavelengths, and subsections of it appear as different colours. Our visual spectrum is between 400 and 750 nm wavelength, encompassing violet–blue–green–yellow–orange–red. The spectrum of honeybees (and other bees) is between 250 and 650 nm, thus including ultraviolet at the lower end but excluding red at the upper end. Hummingbirds and some butterflies may perceive both red and some ultraviolet (Chen *et al.* 1984).

Humans are trichromatic; we have three types of colour receptor, each of which is sensitive to a particular section of the spectrum. These sections are blue, green and red. Honeybees are also trichromatic but with emphasis on different sections: ultraviolet, blue, and yellow (Autrum & von Zwehl 1964, Menzel & Backhaus 1991); many mammals and beetles are only dichromatic (Burckhardt 1977, Silberglied 1979). For bee-pollinated flowers it has been shown that sharp steps in the reflection spectra occur precisely at those wavelengths, where the pollinators are most sensitive to spectral differences (Chittka & Menzel 1992).

It has long been known that flowers may exhibit ultraviolet patterns, but their ecological meaning was recognized only later (Kugler 1947, Daumer 1958). Most likely, the primary function of ultraviolet receptors in insects was that of detectors of polarized light, an important tool in orientation. Only secondarily were these receptors used for colour vision, thus becoming involved in floral biology (Wehner 1976).

Yellow flowers are often visited by a large variety of insects. Unspecialized insects (Diptera) and more primitive butterfly groups seem to prefer yellow flowers (Kevan 1978). Specialized bee-flowers are often blue. Blue is also preferred by more specialized Diptera (Bombyliidae) (Knoll 1921, Kevan 1978). This parallels the observation that blue flowers are concentrated at the middle and higher evolutionary level of the angiosperms (see below: Gottsberger & Gottlieb 1981). Bright red flowers are often associated with hummingbirds and higher advanced groups of butterflies (K. & V. Grant 1968, Kevan 1978). Nocturnal flowers are frequently white or pale, light colours being better visible in the darkness (Baker 1961).

Cytological level: pigments

Flower colours are manifold. The entire visual spectrum is displayed in flowers; such displays involve many different kinds of molecules (Scogin 1983). At the cytological level three types of pigments may be distinguished: (1) they may be dissolved in the vacuoles (Harborne 1993); (2) they may be in the cytoplasm as chromoplasts of diverse structures (Sitte *et al.* 1980) or chloroplasts; (3) they may (rarely) be in the cell walls (*Leonotis*, Lamiaceae, Geitler 1934).

Type (1) is made up of flavonoids containing anthocyanins (blue, violet, red); anthoxanthins (yellowish, white, ultraviolet); anthochlor (chalcones and aurones) (yellow and ultraviolet) (e.g. many Asteraceae; *Antirrhinum*, Scrophulariaceae); betalains (pink or red) (restricted to Caryophyllidae). Type (2) is made up of chlorophylls (green) and carotenoids, containing carotenes (red) and xanthophylls (yellow). The pigment of type (3) seems to be unknown.

Thus, similar colours may be attained by different combinations of pigments. This is important for evolutionary considerations. Red may be attained by anthocyanins, betalains or carotenes, yellow by xanthophylls, anthoxanthins or anthochlor, ultraviolet by anthochlor and other flavonoids (Thompson *et al.* 1972). Black may be effected by concentrated anthocyanins or by the cooperation of different chromophores. White may be attained by certain flavonoids, but in most cases it is a 'structural' colour, generated at the histological level by light reflection in the mesophyll and not by molecular pigments (see below). Iridescent blue as a structural colour, in contrast to many animals, is not known for flowers as yet. Among reproductive organs of angiosperms it has been reported only from fruits of *Elaeocarpus* (Lee 1991), where it is caused by the particular ultrastructure of the outer epidermal cell walls.

The different floral colours seem to exhibit a random distribution among the higher angiosperm groups with the exception of blue. Blue anthocyanins seem to be absent in the more primitive angiosperms (most Magnoliidae, Alismatidae) but more concentrated in the Rosidae, Asteridae, Commelinidae, and Liliidae (Gottsberger & Gottlieb 1981). There is also a correlation of blue flowers with predominantly hymenopteran pollination and herbaceous habit.

Pollen is mostly yellow because of the accumulation of carotenoids, but other colours and pigments also occur (Wiermann *et al.* 1981). Blue pollen is present, for example, in many Malvaceae.

Histological level: colour qualities and intensities

In flowers, anthocyanins, ultraviolet-absorbing flavonoids, and some carotenoids are often localized in the epidermis. Only in a few taxa have they been found mainly in the mesophyll; carotenoids and ultraviolet-absorbing flavonoids are often in both the epidermis and the mesophyll (Kay et al 1981). This is in contrast to the vegetative parts of the plants, where pigments are commonly in the mesophyll and not in the epidermis. The predominant epidermal position of floral pigments may enhance the brilliance of the coloration.

Colour qualities are not only dependent on the participating molecules (pigments) but also on their distribution in the tissue and on other properties of the tissues. Organ surface structures and the presence of intercellular spaces in the mesophyll may greatly influence colour effects.

If a larger amount of incoming light is reflected inside a coloured organ it appears brighter and the colours more vivid. Light is reflected from cell walls, and especially from surfaces between tissue and air spaces. Therefore, more light is reflected in organs with extended intercellular spaces than in compact tissues. Attractive flower parts often have a layer of conspicuous intercellular spaces in the mesophyll below the coloriferous layers (Fig. 5.10.6). This reflection layer is also called an 'optical tapetum' (Troll 1951, Vogel 1961). Such parts appear bright and not transparent, in contrast to parts without a tapetum, which are matt and transparent. Light passes through the pigmented layer, is reflected on the tapetum and passes back again through the the pigmented layer. The colours appear bright and saturated. Organs without pigments but with an optical tapetum appear bright white. To this group belong white petals and stamens with inflated filaments that appear white. In trap-flowers the trap walls often have an optical tapetum, but near the pollination organs in the base of the trap there is a 'window-pane' that lacks a tapetum (*Aristolochia*; *Ceropegia*, Asclepiadaceae) (Troll 1951, Vogel 1961, Yeo 1972) (see section 3.7).

Additive colours and subtractive colours

Some colours are attained by the cooperation of more than one pigment. As shown by F. & S. Exner (1910) there are two principal possibilities for pigment distribution. Pigments may lie side by side in different cells of the epidermis (in the manner of pointillism). They are then perceived as a mixture of both components (additive colours). Pigments may also lie one on top of another, either in different cells or even in the same cell. Often one pigment is in the papillate tip and another in the base of an epidermal cell. This is a means to attain dark colours (subtractive colours). Studies on additive and subtractive colours were done on several temperate species (F. & S. Exner 1910, Vogel 1950, Meeuse 1961, Kugler 1970) but are largely lacking for tropical plants.

5.7.2 Surface effects and colour patterns

Optically attractive surfaces are often not smooth but have a variety of specific regular surface patterns that may efficiently influence deepness of colour and gloss or lustre effects (Fig. 5.10). Such effects were first comparatively studied by F. & S. Exner (1910), and more recently by Barthlott & Ehler (1977), Kay *et al.* (1981) and Kay (1988).

A simple smooth surface effects a reflecting lustre. The reflected light is parallel to the incident light, to a large extent. This is the case, for example, in pseudonectaries (see below). Smoothness is mostly produced by an even, flat epidermis surface or, more rarely, by a film of an oily secretion on the epidermis (fatty lustre) (Knoll 1922).

A papillate surface produces a velvety effect with saturated, warm colours. This is a frequent property of petals (Fig. 5.10.1–3). Only a minor fraction of the incoming light is directly reflected at the surface. The largest proportion is dispersed on the oblique parts of the papillae and does not fall back but is finally absorbed in the tissue. If the surface is viewed from the direction of the incoming light, a velvety gloss appears via a small point of reflection on each papilla. The common situation is that each epidermal cell is projected as a papilla (e.g. *Tibouchina*, Melastomataceae). Very rarely, the epidermal cells are elongate and each has several aligned papillae: the cells are 'multiple-papillate' (e.g. some Caryophyllaceae, Kay *et al.* 1981).

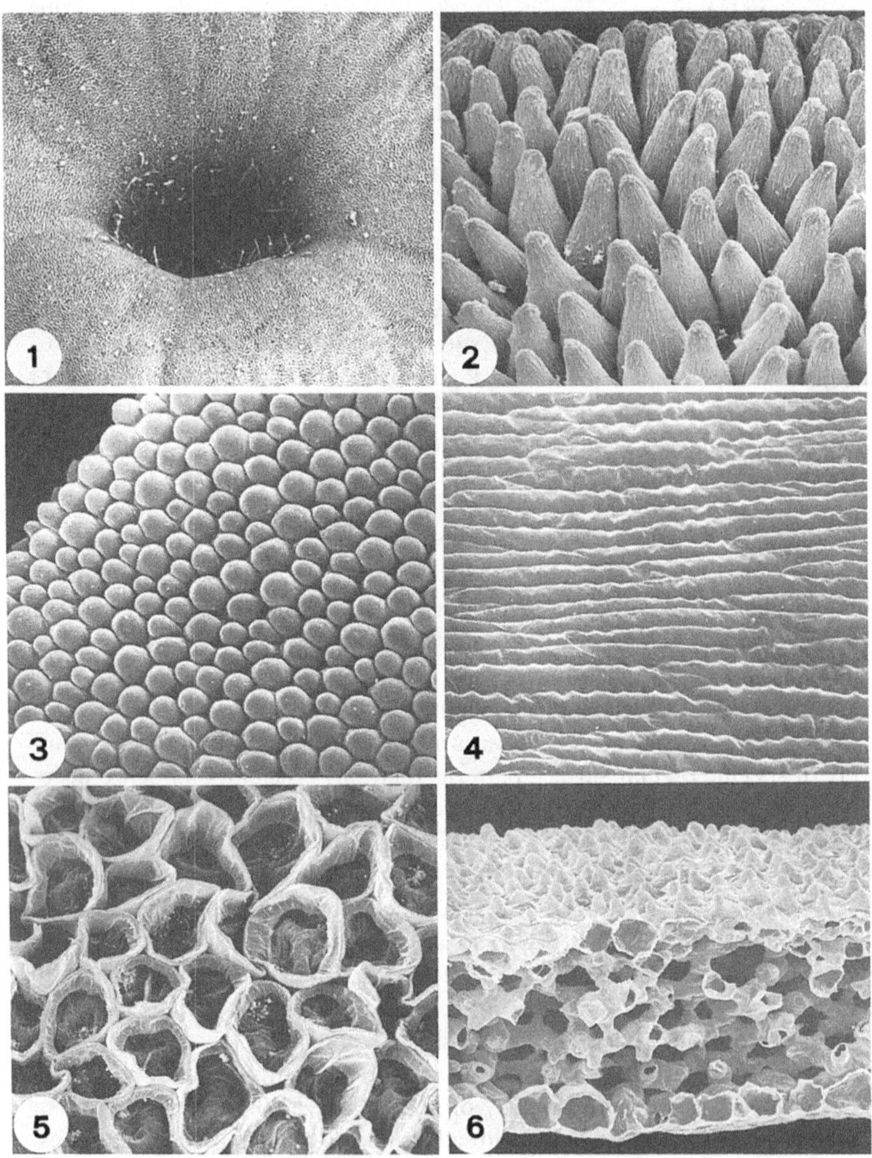

Figure 5.10. Epidermis differentiations of floral organs with different optical effects. 1–2. *Lantana camara* (Verbenaceae); long, conical papillae on petals. 1. Centre of expanded, papillate part of corolla (× 18). 2. Papillae (× 450). 3. *Saintpaulia ionantha* (Gesneriaceae); rounded papillae on petals (× 100). 4. *Schlumbergera bridgesii* (Cactaceae); elongate parallel ridges on petals (× 100). 5. *Saintpaulia ionantha* (Gesneriaceae); grooved surface on anthers (× 200). 6. *Tibouchina semidecandra* (Melastomataceae); transverse section of petal showing papillate surface and spongy parenchyma (× 200).

Another kind of velvety effect is produced by concave epidermis cells. This occurs in the large yellow anthers of *Saintpaulia* (Gesneriaceae) (Fig. 5.10.5).

A surface in which the epidermis cells are elongate and form parallel ridges has a silky lustre (*Schlumbergera*, Cactaceae, Fig. 5.10.4) (Mathä 1936). Light is reflected only when the surface is viewed from an angle perpendicular to the elongate ridges. The direction of epidermal cell elongation is usually that of preferential petal elongation during development. Thus, the silky lustre effect appears if the flower is viewed from any direction.

Another pattern of epidermal cells is reversed-papillate (e.g. *Roscoea*, Zingiberaceae) or reversed multiple-papillate (e.g. *Setcreasea*, Commelinaceae). In these cells the upper surface is more or less flat but the lower cell walls are protracted like papillae (Kay *et al.* 1981).

Lenticular epidermal cells occur in *Lophospermum atrosanguineum* (Scrophulariaceae).

Many epidermal surfaces on floral organs have a variety of cuticular patterns (Barthlott & Ehler 1977). In addition to optical properties they may also function as hydrophobic surfaces (Corbet 1990) or be linked with scent emission or function as tactile cues (see section 5.8). Especially complicated surface patterns appear on the highly evolved flowers of Asclepiadaceae (Ehler 1975) and Orchidaceae (Ehler 1976).

Nectar guides

Many flowers display contrasting colour markings at the entrance towards the nectar source. Yellow and negative ultraviolet (ultraviolet-absorbing markings against an ultraviolet-reflecting periphery) are the predominant colours. In general, most nectar guides reflect longer wavelengths than does the floral periphery (Kevan 1978). It was again Sprengel (1793) who discussed the significance of nectar guides for the first time. From the prevailing yellow colour he concluded that yellow was more attractive to bees than red. A guiding effect on bees and bumblebees was later experimentally confirmed (Manning 1956). Nectar guides occur in many different floral forms, the least so in brush-flowers (Kugler 1966). Nectar guides are most prominent in bee-pollinated flowers and are especially elaborated in monosymmetric flowers (lip- and flag-flowers), where the anthers and the floral centre are more or less hidden (Müller 1876, Kugler 1963).

Osche (1983) discusses the hypothesis that yellow nectar guides mimic anthers with pollen, stamens, or entire androecia. Therefore, they are seen as an evolutionary reminiscence of the colour of pollen, since pollen was an important floral reward in early flower evolution. The original function of the yellow pigmentation of pollen is seen in its protection against ultraviolet radiation. Although many insects, such as Hymenoptera and Lepidoptera, readily learn which colour patterns are rewarding, a standard signal, such as yellow markings, may enhance the efficacy of the lure. Bumblebees show innate reactions to floral markings of high spectral purity (Lunau 1990, 1991).

Colour change

Certain flowers change colours during anthesis. Sprengel (1793) mentioned this for *Aesculus* (Hippocastanaceae), for example. Today it is known from representatives of at least 74 families (Weiss 1991). In some groups it is more common than in others (e.g. Verbenaceae, Boraginaceae, Rubiaceae, Caesalpiniaceae, Sapindaceae). In many flowers colour change is more or less restricted to the nectar guides. Colour change may also affect the ultraviolet part of the spectrum (Eisner *et al.* 1973).

Fritz Müller (1877) observed colour change in the flowers of *Lantana camara* (Verbenaceae) where it is perhaps better known than anywhere else. He also found that old, non-functional flowers increase the optical attractiveness of the inflorescence and at the same time become distinctive from the young functional flowers that provide rewards to pollinators (Fig. 7.2). This was later confirmed for this species by many other authors (Vogel 1950, Barrows 1976, Kugler 1980, Mathur & Mohan Ram 1986, Weiss 1991). Weiss (1991) also studied *Phyla* (Verbenaceae) and four temperate genera with colour change and found the same pattern, as did the work by Vogel (1950) on mostly temperate plants. Studies in some Australian Myrtaceae and Proteaceae also yielded the same result (Lamont 1985). In *Quisqualis indica* (Combretaceae) the colour changes from white to red; white flowers of the first night are pollinated by hawkmoths, older, red flowers by various other animals (Eisikowitch & Rotem 1987). In the hummingbird-pollinated flowers of *Malvaviscus arboreus* (Malvaceae), in contrast, the colour changes from scarlet in bud to lighter red at anthesis, by the loss of certain pigments (Gottsberger 1971).

A variant method of colour change is found in many Caesalpiniaceae where the flag petal folds after the first phase of the flower. Since the nectar guide is only on the ventral side of the flag petal, it is no longer visible in the folded condition (Jones & Buchmann 1974; section 8.6.1).

The collapse or abscission of entire corollas, shortly after pollination, may also result in a colour change of the flower, and thus be favourable for both plant and pollinator (Gori 1983). These changes are inducible by pollination in certain groups. In other groups they are not inducible and part of a fixed temporal programme. Both types may occur in the same genus, e.g. *Lupinus* (Fabaceae) (Gori 1983).

Pseudonectaries

Some flowers display glistening bodies, which have been observed to attract insects (mainly flies and bees) in a few cases. They are interpreted as mimicking nectar drops. They are best known in the temperate genus *Parnassia*, where they have been termed 'pseudonectaries'. However, similar structures also occur in genera of tropical and subtropical regions, such as *Solanum* (Buchmann 1983), *Lopezia* (Onagraceae) and *Nemacladus* (Campanulaceae) (Fig. 5.11). They are situated on the petals or stamens. They are completely dry, and there is no sign of secretory activity (personal observations). However, in many cases they are situated near the real nectaries, which are also present but in a more hidden position. Pollinators may be led to the nectar source by the pseudonectaries. In fly-pollinated *Lopezia* species (Onagraceae) a snapping mechanism of stamens and stigma is released if a sensitive part near the pseudonectary is touched (Eyde & Morgan 1973). A similar association of pseudonectary and snapping mechanism (although different in details) is also present in *Stylidium* (Stylidiaceae) (Erbar 1992), where fly-pollination is also common (Armbruster & Edwards 1992).

The glistening effect is due to the convex and completely smooth surface of the pseudonectary. In contrast, the adjacent regions often have a papillate surface (Fig. 5.11.2). The gadgets of *Nemacladus* are bizarre. Two stamens have a protruding socket at the filament base. Each socket bears three or more reflexed clavate giant cells, which cause the glistening effect (Fig. 5.11.5).

As a variant, *Memecylon* (Melastomataceae) may be mentioned. Here, on top of the anthers there is a glistening area, which is a small oil-secreting gland (Fig. 5.12.1–3). When it was detected, the flowers were interpreted to be oil-flowers (see Buchmann 1987). However, Renner (1989b) showed that the minute quantities of oil are not collected by oil-bees but that this gland rather seems to function as a pseudonectary in providing a glistening film. Similar minute glands of unknown function occur on the anthers of Bonnetiaceae (Kubitzki 1978). A combination of convex, smooth bodies with a secretory area at the concave apex was found at the two tips of the crescent-shaped petals of *Gillbeea* (Cunoniaceae) (personal observation) (Fig. 5.12.4–6).

In the fly-pollinated *Stelis* (Orchidaceae) extracellular oxalate raphids are present on the upper surface of the lip and petals. They produce a

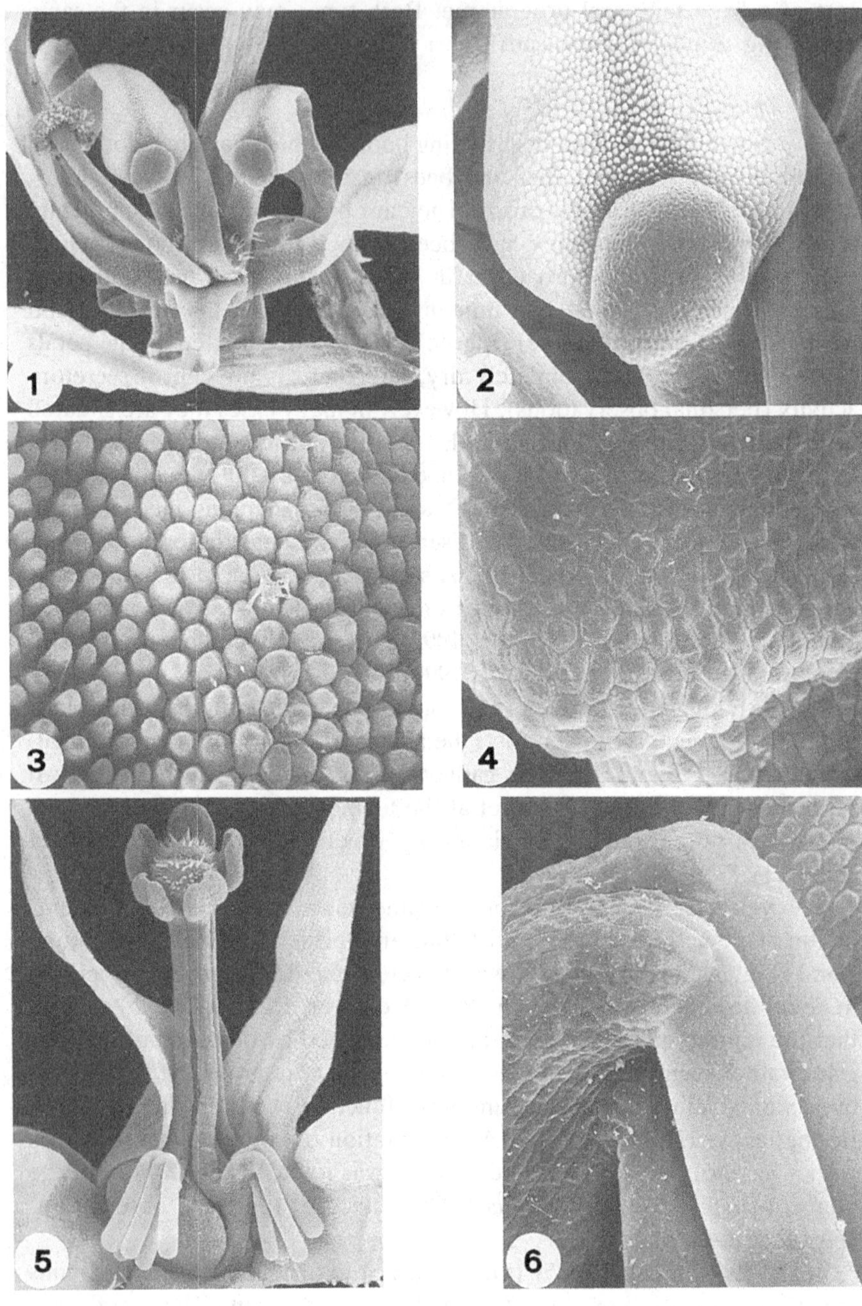
1
2
3
4
5
6

glistening effect by their refractile properties. They have also been interpreted as pseudonectaries (Chase & Peacor 1987).

Pseudopollen

As mentioned above, the (predominantly yellow) nectar guides may have been evolved as anther mimics (Osche 1983). In some cases not only the colour but also the specific granular surface pattern suggests a direct pollen mimicry. This has especially been described for several orchids. In species of *Eria*, *Maxillaria* and *Polystachya*, special hairs produce granular structures that had been interpreted as pseudopollen (van der Pijl & Dodson 1966, Ehler 1976, Goss 1977, Dafni 1984). Most other examples described among the orchids are not tropical. In some Gratioleae (Scrophulariaceae) an entire anther with pollen seems to be simulated by a basal outgrowth of the stamen filament with a granular surface (Magin *et al.* 1989, Endress 1992b) (see section 8.8.5). It has not been studied in most cases to what extent this pseudopollen is collected and whether it contains nutrients that can be used by potential collectors (Ackerman 1986a).

5.7.3 Evolutionary aspects

Since petals were probably not present in the first angiosperms, the floral semaphores were at first stamens (or bracts or tepals). This is reflected by some primitive relic extant angiosperms, where stamens are the most conspicuous floral organs (e.g. Austrobaileyaceae, Chloranthaceae). This is paralleled by the location of scents (see section 5.5). Thus, a primary evolutionary step in angiosperms was transference of the semaphore role from stamens to petals. A secondary evolutionary step was diversification by the potential of transference of this function to all other floral organs, either more inside: stamens and carpels, or more outside: tepals and even bracts (Table 2.2).

Figure 5.11. Forms of dry pseudonectaries. 1–4. *Lopezia racemosa* (Onagraceae). 1. Flower with two pseudonectaries (tubercles) on upper petals (× 12). 2. Petal with pseudonectary (× 40). 3. Papillate surface of petal (× 160). 4. Smooth surface of pseudonectary (× 160). 5–6. *Nemacladus glanduliferus* (Campanulaceae). 5. Floral centre showing two groups of clavate appendages at filament base (× 25). 6. Group of clavate appendages with smooth surface (× 180).

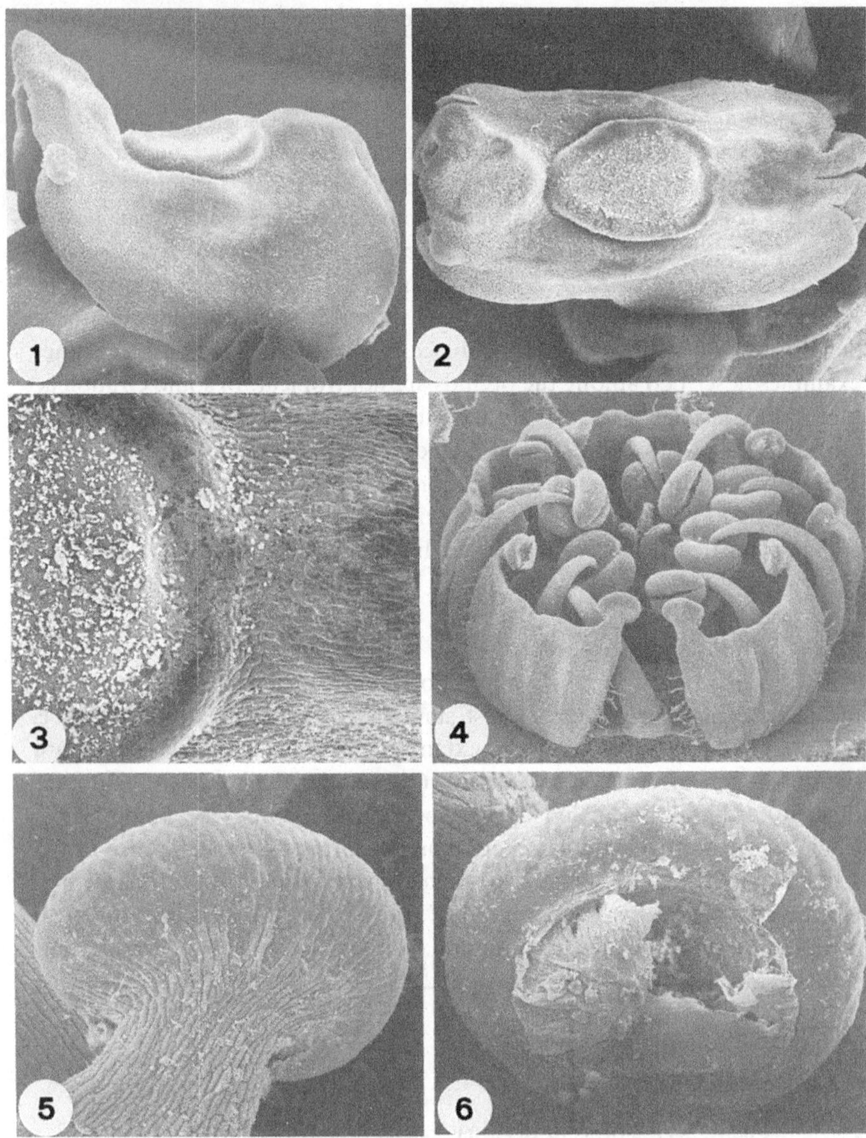

Figure 5.12. Forms of secretory pseudonectaries. 1–3. *Memecylon floribundum* (Melastomataceae). 1. Anther from the side with pseudonectary on top (× 35). 2. Anther from above (× 40). 3. Part of pseudonectary with secretion (× 150). 4–6. *Gillbeea adenopetala* (Cunoniaceae). 4. Floral centre with ten pseudonectaries, two on each petal (× 18). 5. Pseudonectary from outside (× 150). 6. Pseudonectary from inside showing secretion (× 180).

5.8 Tactile guides

The surface of the same floral organs serving as holding devices for pollinators may be extremely diverse within a larger taxonomic group. This is shown for wings and keels of Fabaceae (Stirton 1981, Tewari & Nair 1984). Here, the surface structure, folds and ridges, is at the supracellular level. Many lip-flowers have transverse wrinkles as holding devices on the lower lip, e.g. *Jacaranda* (Bignoniaceae) (Morawetz 1982) and *Thunbergia grandiflora* (Acanthaceae) (van der Pijl 1954) (Fig. 8.51.4). This is also prominent in personate flowers, where the pollinators have to forcefully open the flowers and holding devices are useful, e.g. *Linaria* (Scrophulariaceae) (Müller 1929).

Petal microtexture at the cellular level may also be significant. Kevan & Lane (1985) showed that honeybees are able to distinguish between different textures: between different species as well as between the lower and upper part of the same petals.

In the region where pollinating bees settle in flowers, there is often a peculiar kind of hair. The hairs are made up of a chain of more or less spherical cells, and are therefore called moniliform hairs. Such hairs are found in flowers of many different angiosperm families but they are uncommon in vegetative organs. They exhibit different patterns of surface sculpture (Fig. 5.13). A comparative study of such hairs is lacking. More rarely, unicellular hairs are moniliform (in some Scrophulariaceae, Raman 1989, 1990, 1991; *Monophyllaea*, Gesneriaceae, Weber 1976a). A moniliform pattern may also occur at a higher structural level: in *Sparmannia africana* (Tiliaceae) the entire staminode is moniliform and bright yellow, while the stamen filaments are smooth and red (cryptic for bees?).

It has been supposed that such moniliform structures in flowers are tactile cues and stimulate the bees to carry out pollen-collecting movements (Vogel 1978a). In the cases where the moniliform hairs (and staminodes of *Sparmannia*) are brightly coloured (often yellow) they may also mimic the presence of larger amounts of pollen (Osche 1983). The hairs may also serve to retain pollen dislodged from the anthers (Renner in Faden 1992).

5.9 Flowers providing brood places for pollinating insects

Ficus

Ficus (Moraceae) is the most intensively studied genus involving a symbiosis and coevolutionary patterns between a plant and its pollinator (Galil 1977, Janzen 1979, Wiebes 1979, Bronstein & McKey 1989). Figs have urceolate inflorescences (the 'figs' or syconia), ranging from 3–4 mm

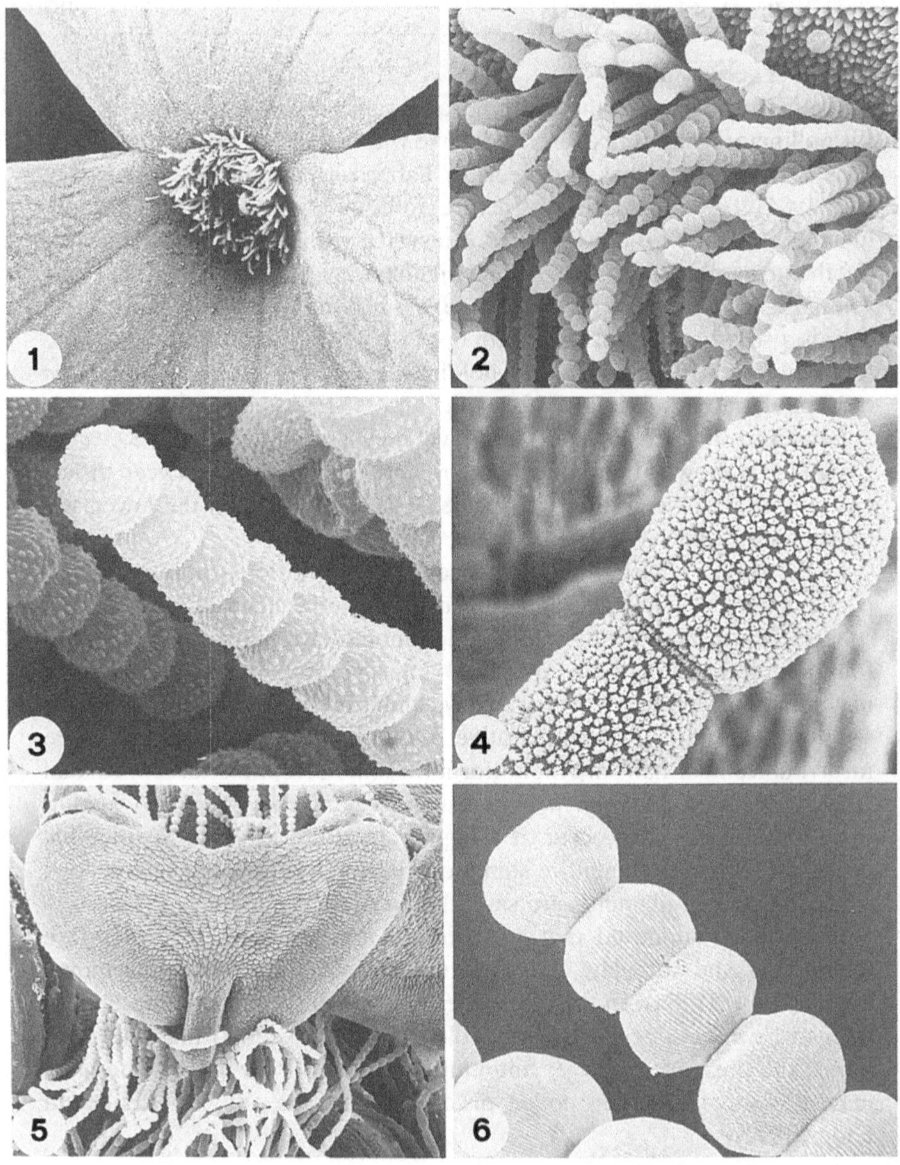

Figure 5.13. Moniliform hairs near pollination organs. 1–3. *Verbena peruviana*. 1. Floral entrance surrounded by moniliform hairs (× 13). 2. Group of moniliform hairs (× 100). 3. Moniliform hair with surface sculpture (× 450). 4. *Thunbergia alata* (Acanthaceae); distal part of moniliform hair from stamen with surface sculpture (× 450). 5–6. *Rhoeo spathacea* (Commelinaceae). 5. Stamen surrounded by moniliform hairs (× 35). 6. Distal part of moniliform hair with surface sculpture (× 450).

up to 10 cm in diameter and containing 50–7000 unisexual (male and female) flowers (Verkerke 1989) (Fig. 5.14). There are always more female than male flowers. The flowers are small and have a simple, reduced perianth. The male flowers generally have 1–3 stamens. The female flowers have a unilocular ovary with a single ovule and a laterally inserted style. There are monoecious and gynodioecious species.

Figs are pollinated by agaonid wasps, which lay their eggs within the syconia. The larvae feed on the developing seeds. Fig species are almost always associated with a single wasp species (Bronstein & McKey 1989). There are over 700 independently interacting pairs of figs and pollinators (Bronstein & McKey 1989). Syconia are highly synchronous on an individual tree. There is strong protogyny. The male flowers may start anthesis 2–6 weeks after the female flowers. Female wasps transport pollen in special pockets on the thorax or on the forelegs (Galil 1977). When they enter the syconia, they deposit pollen on the stigmas and oviposit in some of the stigmas. Wasp larvae develop and the new wasps copulate in the fig. Females then collect pollen from newly open male flowers and leave the syconium and search for other trees with younger syconia, to which they are probably attracted by long-distance olfaction (Bronstein 1989).

In monoecious figs there is extensive variation in style length of the female flowers (e.g. 0.6 – 2.3 mm in *Ficus religiosa*) (Verkerke 1989).

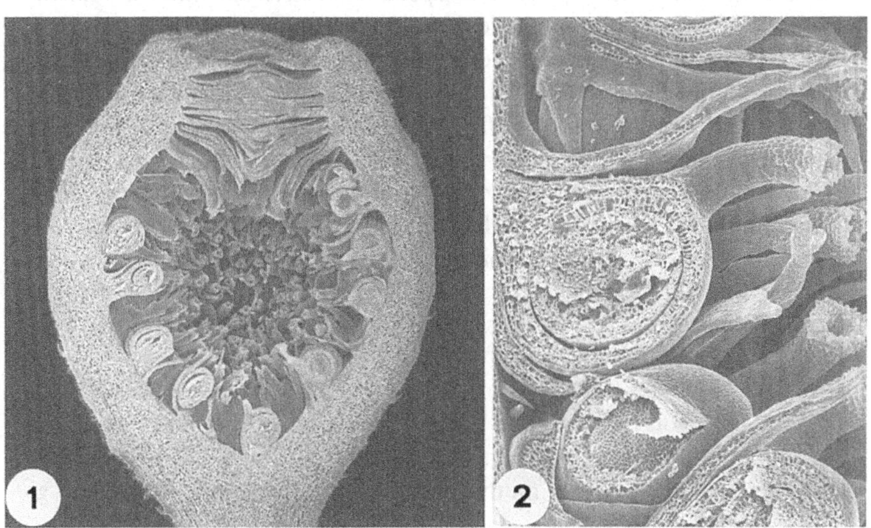

Figure 5.14. *Ficus montana* (Moraceae). 1. Syconium in longitudinal section; inside lined with female flowers and bracts at the entrance (× 13). 2. Female flowers (× 70).

There is reciprocal variation in pedicel length. Thus all stigmas are at the same level, but the ovaries are at different levels, which allows denser packing. In addition, flowers with longer styles are less likely to receive wasp eggs than those with shorter styles. Although the wasp's ovipositor is long enough for the longest styles, there seem to be difficulties in reaching the ovule through the stylar canal. Therefore, more of the more deeply buried ovaries develop seeds than do the upper-level ovaries (Verkerke 1989). Neighbouring stigmas are contiguous; all stigmas form a platform ('synstigma'), on which the wasps move. In *Ficus sur* there is evidence that pollen tubes can grow over the synstigma surface and enter neighbouring flowers (Verkerke 1989).

Gynodioecious species have specialized figs for wasp production with male and female flowers ('gall figs') and for seed production with only female flowers ('seed figs'). All styles are short in gall flowers, while they are long in seed flowers (up to 10 mm), with the effect that in seed flowers the ovules cannot be reached by the wasp's ovipositor. In the gall flowers fertilization of the ovary is probably favourable, but not essential, for endosperm development (Berg 1990b).

The syconia of *Ficus* probably evolved from discoid to cup-shaped inflorescences, as they occur in several Moraceae (Berg 1990b). In the predominantly anemophilous genus *Artocarpus* of the Moraceae, with head-like inflorescences, an insect-pollinated species, *A. heterophyllus*, is known, where the larvae of pollinating flies develop in the giant inflorescences (van der Pijl 1953). In contrast to other *Artocarpus* species the pollen is sticky. Strong scent production is shared with some other species. Nectar secretion from the perianth base in male flowers is known from *A. altilis*, in which nectar and pollen are taken by *Trigona* bees (Brantjes 1981c).

The inflorescences of Moraceae, with their densely arranged small flowers and numerous potential hiding places between them, are apt candidates as brood places for certain insects. But there are also many structural devices developed by the inflorescences and flowers to prevent attacks by insects, as was impressively shown by Berg (1990a).

Other cases

The second well-known case of an exclusive symbiosis involving pollination and brood place in the flowers is *Yucca* (35–40 species, Agavaceae) and the yucca moths, which, however, is not tropical (e.g. Baker 1986b).

Two other, tropical, examples have only recently been discovered and are less well known: *Eupomatia* and *Siparuna*, both in the Magnoliidae.

The flowers of the isolated relic Eupomatiaceae (Magnoliales) (see section 8.1.2) are pollinated by curculionid beetles of the genus *Elleschodes* (Hamilton 1898, Hotchkiss 1959, Armstrong & Irvine 1990). The beetles are attracted by the strong fruity fragrance of chiral esters produced by the staminodes (Endress 1984a, Bergström *et al.* 1991). Stamens and staminodes are united into a massive, nutritious synandrium. It provides shelter and a brood place for the pollinators. After abscission of the synandrium in *Eupomatia laurina* the larvae develop in it on the ground and then pupate in the earth (Armstrong & Irvine 1990). The adult beetles hatch only two weeks after oviposition. Since the beetles contain copious fat deposits in their bodies they may be in an inactive state between subsequent flowering seasons of *E. laurina* (Armstrong & Irvine 1990). Thus, *Eupomatia* and *Elleschodes* possibly constitute a highly specific symbiosis, if not only *Eupomatia* completely depends on *Elleschodes* but also vice versa. The flowers are outstanding in structure and behaviour.

In *Siparuna* (Monimiaceae) the urceolate flowers are pollinated by gall-midges (Cecidomyiidae), which use them as a brood place (Feil 1992) (see section 8.2.2).

5.10 Mimicry in flowers

Flowers may mimic flowers of the other sex of their own species, or flowers of other taxa, or even animals. The orchids are notorious for this behaviour (see section 8.10.1) (Little 1983, Dafni 1984, 1986). Sapromyophilous flowers of various groups mimic carrion or dung; fungus gnat flowers mimic mushrooms (see section 4.1.2). Guilds may also contain mimics (see section 7.6).

A less far-fetched mimicry is that found in unisexual flowers with differential rewards, where female flowers mimic male flowers. It is obvious that unisexual flowers that are animal-pollinated are not free to become too different from each other, in contrast to wind-pollinated flowers. Both morphs of unisexual flowers are more likely to be visited the more similar they are. In a mimicry system the models are more numerous than the fakes. This is also the case in the following examples. In all of them the stigmas are elaborated and mimic parts of the male flowers.

Begonia (Begoniaceae) has unisexual pollen-flowers. Thus, the female flowers do not provide any reward. However, they are visited by pollen-collecting bees, since their stigmas mimic anthers with pollen. They are branched and contorted, papillate and bright yellow (Vogel 1978a, Agren & Schemske 1991).

Ecballium (Cucurbitaceae) has unisexual flowers, the male flowers with nectar, the female ones nectarless. They resemble each other to a large extent. Both are visited by nectar-collecting solitary bees (Dukas 1987). Other cucurbits also exhibit this kind of mimicry (G. & U. Shrivastava 1991).

Jacaratia (Caricaceae) has unisexual, sphingid-pollinated flowers, again with the female flowers nectarless. The stigmas apparently mimic the petals of male flowers (Bawa 1980c). Cases of mimicry or 'mistake pollination' may also occur in other Caricaceae (Baker 1976). Other, similar examples may also be found in Euphorbiaceae (Bawa 1980c) and other families (Willson & Agren 1989).

In species with nectar-flowers there is a tendency towards differential nectar production in a group of flowers. A smaller number of flowers produce only little reward ('empty flowers') (Bell 1986, Gilbert *et al.* 1991). They escape the costs of nectar production to some extent, without losing the benefits of pollination, since the 'cheaters' cannot be discerned from other flowers, if the nectar is concealed ('lucky hit' strategy, Southwick 1982; 'blank-bonanza' strategy, Feinsinger 1983b).

6

Special differentiations associated with the breeding system

There are various mechanisms to avoid exclusive extremes in reproduction. This is a chapter on compromises of conflicting goals: the compromise of quantity vs. quality in the production of gametophytes, seeds, and offspring (Willson 1983), and the compromise of precise mechanism of pollination vs. avoidance of interference (Lloyd 1982, Lloyd & Webb 1986). The result is a flexible balance between extreme conditions. The first insight into the significance of such a balance is probably due to Hermann Müller (1873). It was later much elaborated (e.g. Stebbins 1950, 1957, 1974a, Lloyd 1979, Bawa & Beach 1981, J. & L. Lovett Doust 1988, Barrett & Eckert 1990a,b, Lloyd & Schoen 1992).

The field of breeding systems is multidimensional. Different classifications are possible. The current terms belong to different levels: structural, developmental, and physiological. A uniform system that is applicable to all levels is not possible.

The phenomena of uni- and bisexual flowers and of heterostyly are basically structural. The phenomena of dichogamy and temporal dioecy are basically developmental. The phenomena of self-compatibility and self-incompatibility and of agamospermy (apomixis) are basically physiological. Further, the phenomenon of dichogamy can be viewed in a wide sense (Lloyd & Webb 1986), encompassing different viewpoints, (i.e. flower level or plant level or population level) and include, in addition to the developmental aspect, the structural aspect of heterostyly.

The flexible balance may also be seen in the fact that the evolution of complicated mechanisms is not unidirectional. They can decay again. This has been shown in many instances (Stebbins 1974a, Jain 1976, de Nettancourt 1977, Barrett 1988, 1989). And it can occur repeatedly within a smaller group (e.g. Onagraceae, Raven 1988).

6.1 Sex expression: bisexual and unisexual flowers, monoecy and dioecy

Bisexual (hermaphrodite, perfect) flowers are predominant in the angiosperms but there is also a large number of groups with unisexual flowers. There are also many unisexual genera or species in otherwise bisexual families or genera. Morphologically, there are all kinds of intermediate forms between bisexual flowers and such unisexual flowers as do not exhibit rudiments of the reciprocal organs. However, seemingly bisexual flowers may be functionally unisexual. At the lowest systematic level, in species with normally perfect flowers there are often genotypes with deficient development of ovules or pollen sacs. This is especially well known in species of crop plants as male or female sterility (Sedgley & Griffin 1989). In *Nicotiana* (Solanaceae) the development and patterns of gene expression in normal and sterile anthers have been analysed in detail (Koltunow *et al.* 1990). Sex expression of flowers in an inflorescence may follow a rigid, predictable pattern (e.g. *Euphorbia*), or it may be flexible and adjustable to particular conditions of an individual (e.g. *Solanum* species, Diggle 1991a,b).

Unisexual flowers of both sexes are either produced on the same individuals (monoecy) or on different individuals (dioecy). Dioecy is rarely absolute: in seemingly dioecious plants, flowers of the opposite sex may appear in low numbers as well (Policansky 1982). Furthermore, the individuals of some seemingly dioecious species may change sex from one vegetation period to the next depending on environmental conditions (sequential dioecy, 'sex choice') (e.g. *Arisaema*, Araceae) (Schlessman 1988). Unisexual flowers may also occur in combination with bisexual flowers: often male and bisexual on the same individual (andromonoecious) (e.g. *Solanum*; Anderson & Symon 1989) or female and bisexual together (gynomonoecious). Gynodioecious taxa (female and bisexual flowers on different individuals) seem to occur in some groups where the bisexual blossoms are high inbreeders (e.g. *Bidens*, Asteraceae; Sun & Ganders 1986). This is perhaps mainly associated with the advantages of outbreeding for females (Lloyd 1982). In contrast, clear androdioecious examples are almost absent; most cases mentioned in the literature seem to be 'cryptically dioecious': in the seemingly bisexual flowers the stamens are not functional (Mayer & Charlesworth 1991), but they may just function to provide (sterile) pollen as a reward to pollinators.

Unisexual flowers have evolved in many wind-pollinated groups ('higher' Hamamelididae, many Cyperaceae, *Acer negundo*), some water-pollinated groups (Hydrocharitaceae) and in some taxa with with relatively unspecialized biotic pollination (Myristicaceae, Menispermaceae,

Monimiaceae). Similarly, Bawa (1980a,b) found in the Neotropics that species with unisexual flowers tend to have small, relatively unspecialized, white or pale yellow or green flowers.

Many tree species in the tropics, especially in rain forests, are dioecious (Ashton 1969, 1976, Tomlinson 1974, Bawa 1974, 1980a, Croat 1979). Among the biotically pollinated tree species of a forest in Costa Rica more than half had unisexual flowers (Bawa & Opler 1975). In mangroves there are more species with bisexual flowers (Primack & Tomlinson 1980).

Unisexual flowers are largely absent in highly elaborated, animal-pollinated flowers (e.g. Fabales, Scrophulariales, Orchidales). Surprising exceptions are the highly complicated orchids *Catasetum* and *Cycnoches* with unisexual flowers (see section 8.10.1).

All kinds of combinations are likely to have certain advantages and disadvantages in an evolutionary sense, depending on the particular situation of other features of reproductive biology. Evolutionary pathways from bisexual flowers to separation of sexes have been extensively discussed in the past decade (Lloyd 1982, Baker 1984). Not only the adaptive significance of outbreeding versus inbreeding but also altered patterns of resource allocation in unisexual flowers (of monoecious species) or even unisexual individuals (of dioecious species) may play an important role in an evolutionary context (J. & L. Lovett Doust 1988) (see section 6.7).

6.2 Dichogamy

Kölreuter (1761) found that in perfect (bisexual) flowers of *Epilobium* (Onagraceae) the male and female parts are not functional at the same time. Sprengel (1793) detected the pervasiveness of this phenomenon in plants with perfect flowers and termed it 'dichogamy' (vs. homogamy). Stout (1928) and Lloyd & Webb (1986) discussed dichogamy in detail.

The two basic forms are flowers with a female–male sequence of function (protogynous) and flowers with a male–female sequence (protandrous) (Fig. 8.56). In many taxa the two phases are not completely separated in a flower, but there is some overlap (incomplete dichogamy). More rarely, the separation is complete, and there may even be a neutral interval between the two phases (e.g. *Trochodendron*; R. Glimmann, personal communication). The most extreme case of neutral intervals (at the level of the individual) is 'sex choice', where the same individual acts as male or female in different years (Lloyd & Bawa 1984, Poppendieck 1987, Schlessman 1988).

Protandrous and protogynous flowers are both widely distributed in the angiosperms. In animal-pollinated plants the ratio is almost 2:1 (Lloyd &

Webb 1986). In some larger taxa the mode is highly constant (Magnoliidae, Campanulales), but in many other groups there is polymorphism at the generic level.

Protogyny is highly dominant among Magnoliidae but the reason for this is not clear (Lloyd & Webb 1986, Bernhardt & Thien 1987, Endress 1990b, 1992a) (see section 9.3). It is also predominant in abiotically pollinated plants, in inflorescences with dense flower aggregation (e.g. Araceae, Piperaceae) (Fig. 6.1), and in fly trap- and brood-blossoms, where pollinators stay in a blossom for a longer period (Lloyd & Webb 1986).

Protandry is dominant among Campanulales because of the secondary pollen presentation on the style. It is also prominent in upright spikes with acropetally opening flowers, and where the flowers are visited from bottom to top by bumblebees or birds (Acanthaceae, Verbenaceae, Lamiaceae) (Percival 1965, Wyatt 1983).

Dichogamy has been interpreted as a means to promote outcrossing. However, as, for example, Lloyd & Webb (1986) point out, dichogamy is combined with other 'outcrossing mechanisms' in many taxa, so that an evolutionary interpretation may be more difficult. Another important evolutionary aspect may be avoidance of interference of stamens or pollen and stigmas.

The basic level of dichogamy is that of a bisexual flower (intrafloral dichogamy, Lloyd & Webb 1986). If flowers of an individual are asynchronous, neighbouring flowers are still likely to be pollinated with the plant's own pollen (geitonogamy). Therefore, if dichogamy is to be an efficient outcrossing factor, it must be reinforced by further temporal mechanisms at (1) the inflorescence, (2) the individual or (3) the population level. (1) Flowers of certain regions of an inflorescence are in synchrony (e.g. successive nodes in the inflorescences of Lamiaceae or Scrophulariaceae). This may increase the outcrossing rate, depending on the behaviour of the pollinators. (2) All anthetic flowers of an individual are synchronous. This has also been described as 'temporal dioecism' (Cruden & Hermann-Parker 1977). It is known from about 40 angiosperm families and is more common in the primitive subclasses Magnoliidae, Hamamelididae, and Alismatidae than in other subclasses (Cruden 1988). A similar effect results if, in a monoecious plant, the male and female flowers do not open at the same time (interfloral dichogamy). In *Cupania* (Sapindaceae) the sequence of a flush of unisexual flowers is male–female–male (Bawa 1977). (3) Synchrony and flower bursts between different individuals of a population are coordinated in such a way that flowers of reciprocal stages tend to be available only on different individuals (*Eupomatia laurina*, Endress 1984a). There may also be two distinct types of individuals which have different daily phenologies (e.g. some Laura-

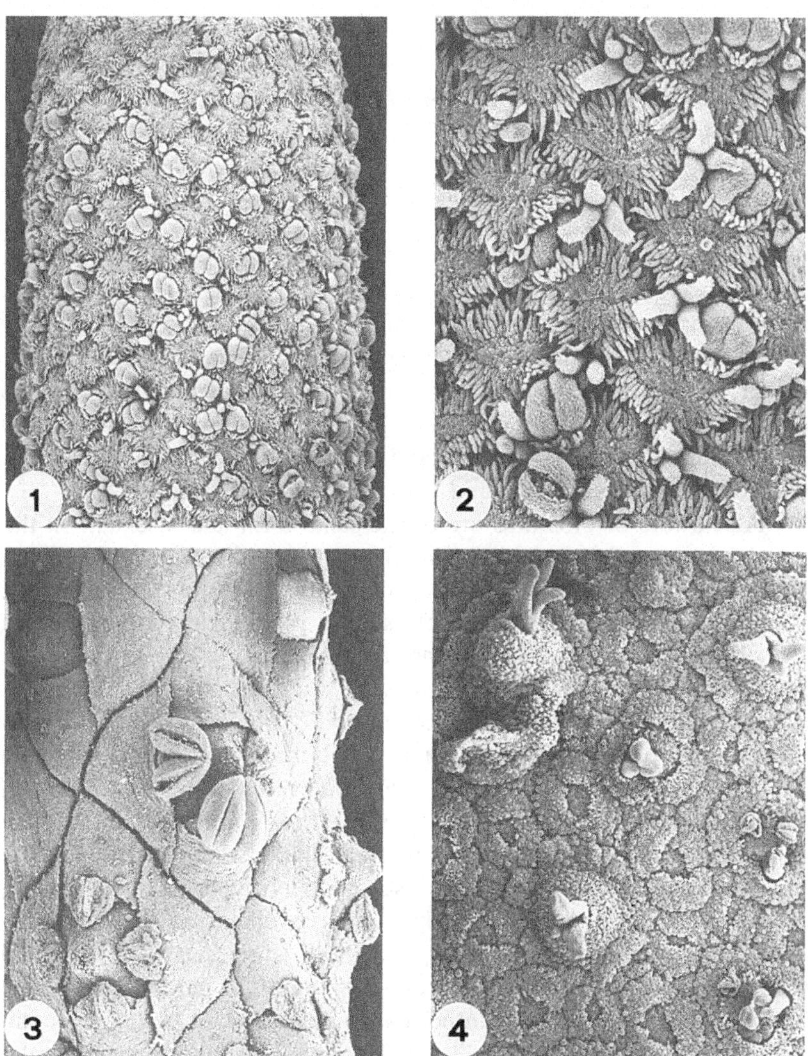

Figure 6.1. Dichogamy in densely arranged small flowers.
1–3. Protogyny in bisexual flowers. 1–2. *Lepianthes peltata* (Piperaceae). 1. Part of spadix (× 18). 2. Group of flowers in female and male stages (× 45). 3. *Anthurium scandens* (Araceae); part of spadix with flowers in female and male stages (× 18). 4. *Dorstenia psilurus* (Moraceae); unisexual flowers in different stages (× 13).

ceae; Stout 1927, Kubitzki & Kurz 1984). Another variant is that all flowers in a population are synchronous but there are protandrous and protogynous individuals (heterodichogamy) (*Trochodendron*; R. Glimmann, personal communication, Chaw 1992).

6.3 Herkogamy

Pollen presentation and pollen receipt may be spatially separated within blossoms or individuals. This is herkogamy. Thus, herkogamy and dichogamy are parallel phenomena at the spatial and temporal level. They may also occur in combination: within a flower the anthers move to the position of the stigmas and vice versa for the second phase of anthesis (*Clerodendrum*, Verbenaceae; *Eupomatia*) (Fig. 8.56). Webb & Lloyd (1986) discuss herkogamy in detail and distinguish three main classes: (1) homomorphic, (2) reciprocal and (3) interfloral herkogamy.

(1) All flowers or blossoms are of the same form and bisexual. In the least elaborate cases several pollinator contacts of anthers and stigma within a blossom are easily possible, although anthers and stigmas are not in the same position (e.g. brush-blossoms). In more elaborate cases a single ordered contact with pollen and stigmas occurs. In 'approach' herkogamy (Webb & Lloyd 1986) the stigmas are touched first, when the flower is approached. This is common and especially obvious in large flowers pollinated by large animals, such as birds, hawkmoths or bats (*Malvaviscus*, *Hibiscus*, Malvaceae; *Schlumbergera*, Cactaceae; *Gloriosa*, Liliaceae; *Adansonia*, Bombacaceae) (see section 3.3) (Figs 3.6, 4.2). In the reverse case, pollen is touched first. This is more rare and occurs in some tubular flowers with the stigmas included (*Dracophyllum*, Epacridaceae). In 'movement' herkogamy, movement of floral parts by the pollinators is necessary for a pollination. In species of *Thunbergia* (Acanthaceae), *Torenia* (Scrophulariaceae), and *Incarvillea* (Bignoniaceae) a horn on the anther has to be moved for pollen presentation (see section 8.8.5) (Figs 8.49, 8.51). In Fabaceae and some Polygalaceae (Arroyo 1981, Brantjes 1982, Bamert 1990) the flowers require tripping to make the stigma receptive (see section 8.6.3). The sensitive stigmas that close after being touched in many Scrophulariales are another example (see section 8.8.5).

(2) Herkogamy may also come about by two or more different floral morphs with reciprocal position of anthers and stigmas, as is the case in heterostylous flowers (see section 6.4). A curious case is enantiostyly, where the style is directed to the right or to the left in otherwise monosymmetric flowers. It is associated with pleurotribic pollination. Mostly these

two morphs occur on the same individual (e.g. *Solanum* sect. *Androceras*, Whalen 1978). However, in some Haemodoraceae they occur on different individuals (Ornduff & Dulberger 1978).

(3) In the widest sense unisexual flowers may also be regarded as herkogamous. This is herkogamy at the interfloral level. Andromonoecism, gynomonoecism, monoecism and dioecism would be steps of increasing herkogamy.

As in the case of dichogamy, herkogamy is sometimes combined with other means that promote outcrossing. Therefore, avoidance of self-fertilization is probably not the only selective force towards herkogamy. Avoidance of self-interference may play an important role as well.

Herkogamy is not parallel to dichogamy in every respect. Herkogamy makes more efficient use of floral visitors, since pollen removal and deposition are possible with a single visit, in contrast to dichogamy, where two visits are necessary (Baker & Hurd 1968, Webb & Lloyd 1986). Also expenditure on floral rewards may be lower. On the other hand, most kinds of herkogamy operate less precisely than dichogamy.

6.4 Heterostyly

The morphological phenomenon of heterostyly, the reciprocal placement of anthers and stigmas in two floral morphs within a species, has been detected several times since the 16th century (Ornduff 1992). Sprengel (1793) found it in *Hottonia* (Primulaceae). He could not explain it, but, for the first time, supposed it had certain unknown significance for pollination. Since the time of Ch. Darwin and F. Hildebrand, heterostyly has attracted the attention of many naturalists and later evolutionary ecologists and geneticists. It is found in about 25 angiosperm families (reviews in Ganders 1980 and Barrett 1992).

Apart from the position of anthers and stigmas at equivalent levels there are other features, mostly correlated with the two floral morphs. These are: (1) different structure of stigma and pollen; and (2) a sporophytically controlled, diallelic incompatibility system, which prevents or reduces self- and intramorph fertilizations. Heterostyly commonly occurs in the form of distyly with two corresponding morphs, more rarely as tristyly with three corresponding morphs in a species (e.g. *Eichhornia*, *Pontederia*, Pontederiaceae; *Oxalis*) (Barrett 1992, Dulberger 1992). Tristyly occurs only in flowers with two stamen whorls (Yeo 1975). The entire syndrome of morphological and physiological features is governed by a 'supergene', a set of tightly linked alleles (Lewis & Jones 1992).

Heterostyly has originated more than twenty times in the angiosperms.

Unlike in other polymorphic sexual systems, evolutionary reconstructions are almost impossible. Two potential pathways, summarized by Barrett (1992), are (1) from homostyly over self-incompatibility (because of inbreeding depression) to distyly (because of more efficient cross-pollination) (D. & B. Charlesworth 1979) and (2) from approach herkogamy over reciprocal herkogamy (because of more efficient cross-pollination) to distyly (because of pleiotropy and inbreeding depression) (Lloyd & Webb 1992). An evolutionary breakdown of heterostyly may result in autogamous and homostylous variants, as found in island populations of *Eichhornia paniculata* (Pontederiaceae) and *Turnera ulmifolia* (Barrett 1989).

Heterostyly is absent in Magnoliidae and Hamamelididae (Friis & Endress 1990) and in the most highly advanced groups (e.g. Campanulales, Orchidales) (Lloyd & Webb 1992). It requires flowers of a uniform but not too elaborate (polysymmetric rather than monosymmetric) architecture, as they occur especially at the middle evolutionary level of dicotyledons and monocotyledons. There are many heterostylous Rosidae (Saxifragaceae, Connaraceae, Erythroxylaceae, Linaceae, Lythraceae, Olacaceae, Santalaceae) and Liliidae (Pontederiaceae, Iridaceae, Liliaceae–Amaryllidoideae). Among Asteridae, heterostyly occurs in some groups with predominantly polysymmetric flowers (Gentianaceae, Menyanthaceae, Rubiaceae, Boraginaceae). It is also concentrated in insect-pollinated flowers of medium size (Dulberger 1992), where pollination is especially precise. Most heterostylous flowers have a stable and relatively low number of stamens. An exception is *Cratoxylum* (Clusiaceae) with numerous stamens (Lewis 1982, Barrett & Richards 1990).

6.5 Self-incompatibility and self-compatibility

In many angiosperms, pollen of the same individual (or of a similar genotype) as the gynoecium cannot lead to fertilization. Physiological processes impede the normal development of a pollen tube or at least of fertilization.

Such self-incompatibility is the principal device promoting outbreeding (Heslop-Harrison 1983), and it is a specific angiosperm invention, which is lacking in gymnosperms (Whitehouse 1950, Pandey 1960). Normally, incompatibility reactions are not directly between the gametes or gametophytes but between the male gametophyte (or factors derived from the male-acting sporophyte) and the female-acting sporophyte. The most common system among angiosperms is that in which the male gametophyte is involved (gametophytic self-incompatibility). The system in which factors of the male-acting sporophyte conveyed by the tapetum onto the pollen grain are decisive (sporophytic self-incompatibility) are more rare (Heslop-Harrison 1983).

Another distinction is made between homomorphic and heteromorphic self-incompatibility: the heteromorphic type is related to heterostyly. However, since not all heterostylous plants are self-incompatible (Barrett 1992), heterostyly is discussed separately (section 6.4).

For the genetic systems of both gametophytic and sporophytic self-incompatibility, one to four genes (rarely more) with many alleles are known (e.g. Lewis 1979). The study of genetic and molecular biological aspects of self-incompatibility of a few model systems is very advanced, e.g. *Nicotiana*, *Lycopersicon*, *Petunia*, Solanaceae (gametophytic), *Brassica* (sporophytic). Furthermore, there is a vast literature on genecological and evolutionary aspects. For surveys, the reader is here referred to Heslop-Harrison (1975, 1978, 1983), de Nettancourt (1977), Lewis (1979, 1980), Pandey (1980), Barrett (1988), Clarke *et al.* (1989), Haring *et al.* (1990).

Sporophytic and gametophytic self-incompatibility is expressed at different sites. In sporophytic systems it operates on the stigma surface: the pollen grains are not able to germinate or at least to penetrate the stigma surface. The message of the (male-acting) sporophyte is deposited on the outer pollen wall (the exine) and becomes active immediately on the stigma. Since all pollen grains of a plant express the same phenotype, matings are either fully compatible or fully incompatible. Half-compatible matings do not occur (see Knox *et al.* 1986).

In contrast, in gametophytic systems self-incompatibility operates only later inside the gynoecium: the pollen grains germinate but the pollen tubes are arrested during their course within the stigma (grasses; Heslop-Harrison 1980), style (in most cases), ovary or ovules (*Acacia*; Kenrick *et al.* 1985). Thus, the message of the gametophyte becomes active only during pollen tube growth (Heslop-Harrison 1975, 1983). An extreme case is *Theobroma cacao* (Sterculiaceae), where it acts only in the embryo, which aborts in the case of incompatibility.

Thus, in uninvestigated taxa, the nature of incompatibility can be surmised by the site of action in the gynoecium. There are other features highly correlated with one type or the other. Sporophytic systems have trinucleate pollen grains and often dry, papillate stigmas; gametophytic systems have binucleate pollen grains (except for the trinucleate grasses) and often wet stigmas (Y. Heslop-Harrison & Shivanna 1977).

Self-incompatibility is often not absolute; in both systems various levels occur. The proportion of seed or fruit set after self- and cross-pollination shows the occurrence and the degree of self-incompatibility. It can be measured by the index of self-incompatibility (Zapata & Arroyo 1978, Knox *et al.* 1986). It is calculated as the ratio of seed or fruit obtained by self-pollination to seed or fruit obtained by cross-pollination. On the other hand, self-pollination of predominantly outcrossing self-compatible species generally results in inbreeding depression (Schemske 1983a).

The determination of the degree of self-incompatibility may be technically difficult. The commonly used bagging experiments may adulterate the results: they increase the concentration of ethylene around the flowers, which can weaken the self-incompatibility response, as found in *Lycopersicon* (Williams & Webb 1987). Further, in some groups only a small amount of flowers regularly set fruit even under the best conditions (e.g. Mimosaceae, Caesalpiniaceae) (see section 8.6). Proportionately more tree species are self-compatible in high-altitude forests than in lowland forests (Sobrevila & Arroyo 1982, Tanner 1982, Bawa 1990).

Self-incompatibility is known from several hundred genera of more than 60 families and 30 orders of angiosperms (Knox *et al.* 1986, Barrett 1988, Gibbs 1988). It has been disputed whether self-incompatibility arose simultaneously with angiospermy and was a precondition for the successful rise of the angiosperms (Mulcahy 1979) or whether it only evolved when the early Magnoliidae already existed (Olmstead 1989). A monophyletic origin concomitant with the emergence of closed carpels was suggested by Whitehouse (1950). The present diversity of forms of self-incompatibility in angiosperms may then have arisen by superimposition of more complexity. Pandey (1980) shows possible evolutionary pathways for this and also for the evolution of angiosperm self-incompatibility from gymnosperm recognition systems, which do not operate against the same genotypes but against related species. In contrast, Bateman (1952) argued that self-incompatibility had arisen de novo several times in the angiosperms. There is only consensus in the view that gametophytic self-incompatibility is more primitive than sporophytic self-incompatibility in angiosperms (Barrett 1988), perhaps with the exception of Zavada & Taylor (1986).

6.6 Agamospermy

In 1841 J. Smith (see Baker 1980) detected the formation of viable seeds from entirely female plants in cultivation without the presence of pollen in the Australian *Alchornea ilicifolia* (Euphorbiaceae) and concluded that in this plant the ovule gave rise to new plants without the action of pollen. Agamospermy (apomixis) is the development of seeds without preceding fertilization; either an egg cell (or another cell) from an unreduced embryo sac or a nucellar cell (nucellar embryony) gives rise to an embryo. Cytological evidence for agamospermy has been gained mainly from several temperate plant groups (Rutishauser 1967, Nogler 1984, Asker & Jerling 1992) but indirect evidence points to its wide occurrence among tropical plants also (e.g. Kaur *et al.* 1978). Agamospermy has been reported from representatives of 38 orders of the angiosperms (Asker & Jerling 1992). The families Rosaceae, Asteraceae and Poaceae are especially predisposed to evolving agamospermy (Richards 1986, Haig & Westoby 1991a). The presence of nucellar embryony is suggested when several embryos develop in one ovule (polyembryony). Most tropical crop plants have races with polyembryony (e.g. Moncur 1988, Sedgley & Griffin 1989), which indicates how plastic a great number of species are in this respect and also that the potential for partial agamospermy must be common throughout the angiosperms. In the allopolyploid mangosteen (*Garcinia mangostana*, Clusiaceae), which is cultivated for its edible fruits, only plants with functionally female flowers are known, whose seeds are then always agamospermous (Richards 1986).

6.7 Allogamy and autogamy; outbreeding and inbreeding; chasmogamy and cleistogamy

Pollination takes place within a flower (autogamy) or between flowers (allogamy). Outbreeding (outcrossing, xenogamy) occurs by the transfer of pollen between flowers of different individuals of a species, whose genotypes are generally slightly different. Inbreeding (selfing) occurs by pollen transfer among flowers of a single individual. Two modes of inbreeding are possible: pollen transfer is within a single flower (autogamy) or between different flowers on the same individual (geitonogamy). Autogamy and geitonogamy may be different from a structural point of view, but their genecological effect is the same. However, if one takes into account that mutations may occur in different meristems of an individual (Klekowski 1988, Schmid 1992), a slightly enhanced variability of the offspring after geitonogamy compared with autogamy is likely.

Earlier it has been believed that the scattered occurrence of individuals of a species in tropical lowland rainforests would prevent substantial outcrossing rates. Experimental studies, however, showed that outcrossing is the major breeding mechanism in many instances (Ashton 1969, Bawa 1974, Bawa *et al.* 1985a, Arroyo 1976, 1979, Bullock 1985, Hamrick & Loveless 1989, Hamrick & Murawski 1990). Traplining is a most efficient method of long-distance pollen transfer (see section 7.5).

A precondition for allogamy to occur is that flowers open at anthesis

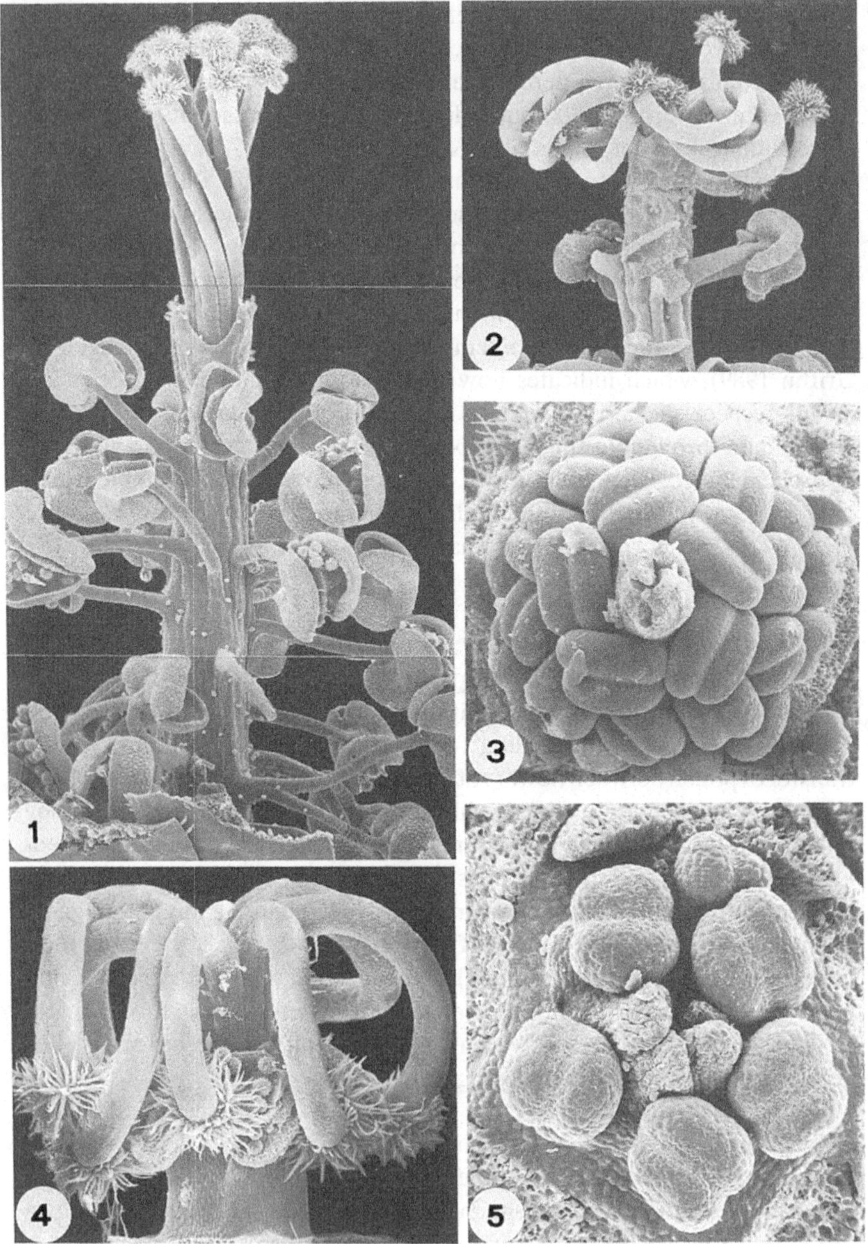

Figure 6.2. Chasmogamous (1–3) and cleistogamous (4–5) flowers in *Pavonia hastata* (Malvaceae). 1. Gynoecium and androecium of chasmogamous flower from the side, early

(chasmogamy). However, chasmogamous flowers per se do not guarantee allogamy; autogamy or geitonogamy may also occur, if there are no other means to prevent it. If flowers do not open at maturity (cleistogamy), only autogamy is possible. The transition from allogamy to autogamy is one of the most frequent evolutionary shifts. It may be due to a single change in a heterochronic gene (Guerrant 1989, Lord 1991) (see section 9.2.3). Thus switches between allogamy and autogamy occur frequently at low systematic levels. In the Onagraceae a switch to self-pollination is a characteristic feature in the evolution of 12 of the 16 genera (Raven 1988). At the species level there have been about 22 shifts in the pollination system during the course of evolution of the outcrossing species, but the switch from outcrossing to self-pollination has been much more frequent (Raven 1988). Facultative outcrossing is a mixed mating system in which the balance between self- and cross-pollination is a function of pollinator activity. These species are self-compatible, they are adapted for cross-pollination and have delayed autogamy (Cruden & Lyon 1989).

A good example to show different specializations for allogamy and autogamy, chasmogamy and cleistogamy is the family Malvaceae. In general, they have protandrous flowers. In the male phase the stigmas are still enclosed by the androecial tube. In the female phase the style elongates and the stigmatic branches are exposed. In many malvaceous flowers, if not pollinated, the free upper style branches curve downwards, to contact the upper anthers and take up pollen grains from them. If pollinated during downward curvature they may regain erect position and selfing does not occur (e.g. *Hibiscus trionum*, Buttrose *et al.* 1977). This is an example of facultative outcrossing, as defined above.

A further step is realized in the malvaceous *Pavonia hastata*. It produces chasmogamous and cleistogamous flowers. The chasmogamous flowers behave as described above (Fig. 6.2.1–3). The cleistogamous flowers remain closed and the style branches are curved towards the anthers and contact them in bud (Fig. 6.2.4–5). A further specialization in these dimorphic flowers of *Pavonia hastata* is different stamen numbers and size of androecium and gynoecium. In cleistogamous flowers only about

Fig. 6.2 (*cont.*)

female phase (× 13). 2. Late female phase with stigmatic branches curved (proximal part of androecium not shown) (× 13). 3. Floral bud of chasmogamous flower with numerous stamens (× 60). 4. Gynoecium and androecium of cleistogamous flower from the side (× 30). 5. Floral bud of cleistogamous flower with only five stamens and one reduced one (× 140).

5 stamens are produced against about 20–30 in chasmogamous flowers; furthermore, the androecial cylinder and the style remain much shorter, and the petals are reduced (Fig. 6.2.4).

Cleistogamy may occur as a rule in a species (constitutional cleistogamy) or it may occur only under adverse conditions (environmental cleistogamy) (Walter 1988). However, it may be generally stated that environmental factors play a role in the expression of every kind of cleistogamy. In most species with constitutional cleistogamy, chasmogamous flowers are also produced by the same individual (e.g. soybean, *Glycine max*, Fabaceae) (Robacker *et al.* 1988). Cleistogamy, therefore, is usually part of a floral polymorphism in a species. Cleistogamy has been reported for 287 species of 56 families of angiosperms; furthermore, in more than half of these species a dimorphism with cleistogamous and chasmogamous flowers exists (Lord 1981).

Cleistogamy has diverse forms. They have been reviewed by Uphof (1938), Lord (1981), Mohan Ram & Rao (1984), and in the meantime more extreme forms have been found. Often the corolla is reduced in size. Stamens may be reduced in number (*Pavonia hastata*, Malvaceae, Fig. 6.2; *Gaudichaudia*, Malpighiaceae, Anderson 1980). Pollen sac number per anther may be reduced (Endress & Stumpf 1991). Pollen number per anther may be reduced. Pollen grains may have thinner walls than in chasmogamous flowers. Number of carpels and ovules may also be reduced.

The relationships between stamens and carpels in cleistogamous flowers show a successive reduction. (1) Anthers may normally open and pollen grains be deposited on the stigma in bud (e.g. *Pavonia hastata*) (Fig. 6.2.4). (2) Anthers may normally open, pollen grains germinate in the anthers and pollen tubes grow towards the stigma. (3) Anthers may remain closed, pollen germinate within the anthers and pollen tubes grow through the anther wall towards the stigma. (4) Anthers may remain closed, pollen germinate within the anthers, grow down through the filament and floral base into the ovary (some Malpighiaceae) (Anderson 1980). (5) Anthers may remain closed, pollen germinate within the anthers, grow down through the filament and floral base into the inflorescence axis, from there into the base of a female flower and into its ovary (*Callitriche*) (Philbrick 1984).

In some cases the polymorphism is associated with amphicarpic behaviour. This occurs in mainly annual representatives of a number of families, among them some tropical Fabaceae, Poaceae, and Commelinaceae (survey in Cheplik 1987). In these species chasmogamous air-flowers and cleistogamous subterranean flowers are produced (Schoen & Lloyd 1984, Cheplik 1987). This results in a combined near and far dispersal strategy, with relative genetic stability at the established site and more genetic variation for exploring new sites.

6.8 Mate choice, male and female allocation

Mate choice, the selection of gametophytes, works at many levels including phenology, pollination biology, morphology, anatomy, and physiology. There are many mechanisms that affect maternal, paternal, and offspring fitness (Willson & Burley 1983, Marshall & Folsom 1991). A number of them have been discussed above.

Pollen:ovule ratio

Male gametophytes produced by an individual commonly by far exceed the number of female gametophytes. Many pollen grains are lost before they reach a stigma, and pollen tube selection in the style causes a further reduction of male gametophytes available for fertilization. Thus, reproductive offer and reproductive success may differ greatly (Urbanska 1989). As early as 1937, Pohl quantified the excessive pollen production of some wind-pollinated plants, in which the pollen:ovule ratio is extremly high. In a series of decreasing pollen:ovule ratios are plants that exhibit outcrossing, autogamy and cleistogamy, respectively. The range of pollen: ovule ratios in these modes of reproduction seems to be relatively constant. Therefore, pollen:ovule ratios may be used to predict the breeding systems of plants (Cruden 1977).

Pollen:pollen-tube ratio

Kölreuter (1761) found that in some plants more than one pollen grain was necessary for successful fertilization of one ovule. Indeed, in many groups there is a remarkable reduction in the number of microgametophytes from pollen grains produced to pollen tubes actually reaching the ovules. In self-incompatible plants more than one pollen grain may be necessary to release enough compounds to allow pollen germination (Brewbaker & Majumder 1961, Cruden & Miller-Ward 1981). Y. Heslop-Harrison *et al.* (1985) describe the successive reduction in number of pollen tubes from numerous to one at several critical sites of the pollen tube transmitting tract in *Zea mays*, and Cruzan (1986) for *Nicotiana glauca*. This is also correlated with the narrowing of the pollen tube transmitting tract towards the ovary (see section 2.4.1).

Pollen tube selection in the style

There is ample evidence that pollen of allogamous species is subject to intense competition (D. & G. Mulcahy 1983, 1987). This results not only in a decrease in number of pollen tubes eventually reaching the ovary, but also in a selection for the most vigorous tubes. This, in turn, has a positive effect on the properties of the sporophytes derived from these male gametophytes (D. Mulcahy 1979). Snow & Spira (1991) showed for *Hibiscus moscheutos* (Malvaceae) that in mixed pollinations faster-growing pollen tubes give rise to more progeny than do slower-growing tubes. A precondition for this pollen tube racing is the synchronous start of pollen germination (analysed by Thomson 1989). A synchronous start is made possible by the deposition of clumps of pollen, which is usually the case in pollination by animals, in contrast to wind-pollination. Evolutionary aspects of pollen tube growth rates at the population level are discussed by Walsh & Charlesworth (1992).

Compitum and pollen tube distribution and selection

In a pollen tube transmitting tract, pollen tubes tend to be regularly distributed. This is especially interesting if a gynoecium has free stigmatic lobes and not all lobes receive pollen grains. The compitum (see section 2.4.1) allows equal distribution to all carpels and ovules (e.g. Bystedt & Vennigerholz 1991b, for *Trimezia*, Iridaceae). The redistribution may need only very short distances, e.g. 0.5 mm (Jenny 1988, for *Brachychiton* and *Abroma*, Sterculiaceae; Ramp, 1987 for *Staphylea*).

Another aspect is that of centralized pollen tube selection. If pollen tube selection plays an important evolutionary role, as discussed above, centralized selection in a syncarpous gynoecium with a compitum is superior to non-centralized selection in an apocarpous gynoecium without a compitum. Given the same condition in a syncarpous and an apocarpous gynoecium, each with 4 carpels and 4 ovules and 40 pollen grains regularly distributed over the stigmatic surface, in an apocarpous gynoecium there will be 4 separate areas for competition and in each only the fittest of 10 pollen tubes will succeed, while in a syncarpous gynoecium with only one area for competition the 4 absolute best of the 40 male gametophytes will succeed. In other words, in a syncarpous gynoecium the probability that the 4 absolute best male gametophytes out of the 40 reach the 4 ovules is 1, but in an apocarpous gynoecium without a compitum it is only $1 \times 3/4 \times 1/2 \times 1/4 = 3/32$ (Endress 1982).

Selection of male and female gametophytes in the ovary, selection of seeds (seed:ovule ratio)

It is a common phenomenon that not all ovules of a carpel, not all carpels of a gynoecium, or not all gynoecia of an inflorescence or of an individual will reach maturity, even if all flowers have been pollinated (Stephenson 1981). Two main reasons may be resource limitation or a particular plant architecture that does not allow the development of every single ovule that has been initiated. Thus, the seed:ovule ratio may be highly affected. The seed:ovule ratio is higher in annuals than in perennials (Wiens 1984).

Such surplus initiation of ovules that are bound to abort from the onset may seem to be a waste of material. But there are other aspects to be considered. (1) It opens the possibility of selection of certain ovules provided with especially efficient male gametophytes. (2) It opens the possibility of maturing of the best reproductive units in case of partial predation. (3) The presence of 'surplus' flowers from the point of view of maximal seed yield may be important from the point of view of pollinator attraction. (4) The surplus production may be part of the bauplan and may, therefore, not easily be abandoned. Bisexual flowers whose gynoecium aborts may at least perform male functions.

How the regulation mechanisms work is not easy to establish. Experimental studies were carried out only in a few groups, e.g. *Cassia*, Caesalpiniaceae; *Asclepias* (Sutherland 1986, review in Lee 1988). In multiovulated pods (elongated fruits) pollen deposition patterns and stigmatic inhibition of pollen germination may regulate seed number (Shaanker & Ganeshaiah 1990). The position of fruit within an inflorescence and of seed within an ovary has an effect on abortion (Bawa & Webb 1984).

Incompatibility phenomena following particular crosses may be expressed only after fertilization ('seed incompatibility'). Embryo development may be disturbed and the seed aborts. It has been hypothesized that the cytogenetical status of the endosperm controls seed development. The relationship between the genomes of the (triploid) endosperm nuclei may play a decisive role (Rutishauser 1969).

7

The process of anthesis

7.1 Structural changes during anthesis; growth and movement of parts

In a flower where petals open and later drop or fade, the process of anthesis is obvious. In perianthless flowers the beginning and end of anthesis may be difficult to determine, and the duration of pollen availability and stigma receptivity have to be tested. Anthesis spans a more or less definite period of time. If anthesis is long, the flower may grow considerably in that time (e.g. *Trochodendron*, R. Glimmann, personal communication).

Flowers that open and close consistently at the same hour of the day, or that show a consistent rhythm of opening and closing over several days, have preferentially been studied because certain developmental processes may be exactly predicted for a given time, which facilitates investigation. Other phenomena studied include movements of parts induced by irritation (stimulation).

Detailed studies on the morphology, histology and physiology were carried out in *Ipomoea tricolor* (Convolvulaceae) by Matile (1978) and collaborators. The flowers open early in the morning and close in the late afternoon. In bud the corolla, although completely sympetalous, has a contort aestivation, like a folded umbrella. Each sector of the corolla that corresponds to a petal has two longitudinal ribs (Fig. 3.1). These ribs not only reinforce the floral architecture (see section 3.1) but also curve outwards and unfold the corolla in the beginning. At the end the corolla closes by rolling in of the entire margin, caused again by differential longitudinal growth of the ribs (see also Goldsmith & Hafenrichter 1932).

In *Nymphaea* species (Nymphaeaceae) anthesis spans several days (Schmucker 1932, Prance & Anderson 1977). The flowers are bisexual but they are functionally female at first, and later become functionally male. In addition, on each day the flowers open and close at the same time. Thus, a continuous development is superimposed by a rhythmicity. In the neighbouring genus *Victoria* the entire flowering process is compressed into two days (Valla & Cirino 1972, Prance & Arias 1975).

Among tropical flowers rapid and reversible movements following irritation occur in the androecium of *Sparmannia* (Tiliaceae) (Bünning 1959) and the stigmatic lobes in representatives of several families of the Scrophulariales (Newcombe 1922, 1924, von Guttenberg 1959, Hart 1990) (see section 8.8.5). The structural basis for the movement of the stamens is a contractile tissue as part of the epidermis on the ventral side of the filament base (Bünning 1959); the basis for the movement of stigmatic lobes is a contractile, collenchyma-like tissue on the inner side of the lobes (von Guttenberg 1959). The ecological function may be interpreted as effecting herkogamy and as a visual signal for pollinators indicating that a flower has been visited shortly before.

A structural basis for the coordinated behaviour of flowers may be the presence of vascular bundle connections in the floral base. Prominent 'girdling bundles' have often been mentioned (Sporne 1977). However, their function has not been analysed. It may be speculated that these bundle connections in the floral base form a relay system between all floral organs that may be involved in conveying certain signals (Deroin 1991a).

7.2 Flower longevity

Longevity of individual flowers is very diverse among the angiosperms. In tropical wet and dry forests flowers of trees are short-lived: 1–3 days, most of them just one day (Percival 1974, Primack 1985a). Many tropical herbs also have one-day flowers (Dobkin 1987). An extreme case are dipterocarp forests with almost exclusively one-day flowers (Chan & Appanah 1980, Ashton *et al.* 1988). Even entire inflorescences may be as short-lived as a night (*Brownea rosa-de-monte*, Caesalpiniaceae, Hudson & Sugden 1982; *Parkia*, Mimosaceae, Hopkins 1984). In a forest in the Venezuelan Guayana Highland a flower longevity of 12 hours was present in 63% of 55 species studied (Ramirez *et al.* 1990). However, mean flower longevity in forests in Costa Rica increases from 1 day in the lowlands up to 4–8 days in montane species, which may be interpreted as resulting from the higher unpredictability of pollination at higher altitudes (Stratton 1989). The same trend is exhibited by the family Melastomataceae, in which flower longevity is generally one day. However, in the high-altitude species *Graffenrieda fruticosa* the flowers last at least 7 days (Renner 1989b).

The generally short anthesis of tropical lowland flowers is also in contrast to temperate regions, where mean duration of anthesis is much longer. In many cases within a family there is a contrast between the short-lived flowers of tropical trees and and the more long-lived flowers of herbaceous or shrubby species (e.g. Fabales) (Primack 1985a). However, in some other families one-day flowers seem to occur exclusively (Commelinaceae, Convolvulaceae, Pontederiaceae, Turneraceae) (Primack 1985a). Bromeliaceae, Iridaceae and Lamiaceae have an anthesis duration of one to three days; Asclepiadaceae vary from two to seven days; Ericaceae, Saxifragaceae, Myrtaceae and Orchidaceae have rather long-lived flowers of between 4 and 19 days (Primack 1985a). An extreme case in orchids is *Paphiopedilum* with a flower longevity of 1–3 months (Cribb 1987). However, orchids are most versatile and contain extremes in both directions (see section 8.10.1).

At the species level, flower longevity may depend on whether or not a flower has been pollinated (Halevy 1986) (see section 2.2.2). This is best known for some orchids, where flowers fade as soon as pollination has taken place (see section 8.10.1). It may also depend on weather conditions (see section 7.3).

There is often a correlation between flower longevity and pollination mechanisms. Many grasses, as wind-pollinated plants, have flowers that are functional for only an hour (Kerner 1905). Flowers pollinated by trapliners (see section 7.5) are often short-lived but inflorescences may be long-lived, since only one or a few flowers are open at a time (e.g. *Thunbergia grandiflora*, Acanthaceae; *Dillenia suffruticosa*; *Passiflora foetida*; *Strelitzia*; *Heliconia*).

There is also a correlation between inflorescence structure and sequential flowering with only one or a few flowers (per inflorescence) open at a given day. Long racemes or spikes or monochasial inflorescences are especially apt for this behaviour. Examples for racemes are *Passiflora*, for spikes *Stachytarpheta* (Verbenaceae), for monochasia *Dillenia*, *Hibbertia* (both Dilleniaceae), *Hamelia* (Rubiaceae), *Strelitzia*, and *Heliconia* (Fig. 7.1). They all flower continuously or have extended flowering periods (*Passiflora*, *Strelitzia*, and *Heliconia*, see chapter 8; *Dillenia*, Corner 1940; *Hamelia*, Newstrom *et al.* 1991).

Figure 7.1. Inflorescences with sequential anthesis of one or few flowers at a time. 1. *Stachytarpheta cayennensis* (Verbenaceae); spike (× 0.7). 2. *Fagraea racemosa* (Loganiaceae); thyrse (× 1). 3. *Hamelia patens* (Rubiaceae); dichasium with monochasial parts (× 1).

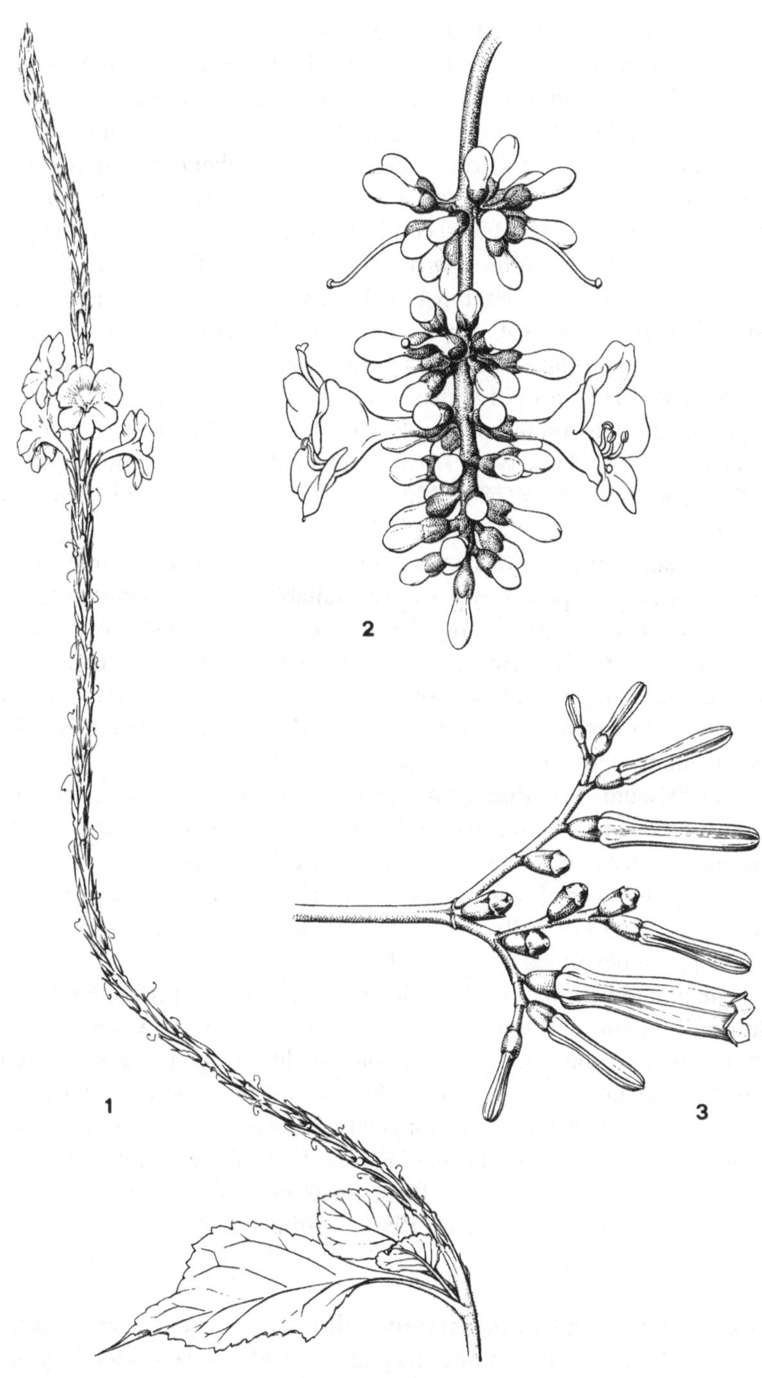
1
2
3

7.3 Rhythmicity and synchrony of anthesis

Many species exhibit a strong daily rhythmicity in flowering. Linnaeus (1751) was impressed by flowers that open and close at specific hours of the day and he devised a living 'flower clock' in the garden by planting particular species in the right sequence. The phenomenon is especially impressive in the tropics, and early naturalists were attracted if large flowers opened within a few seconds in the early morning or in the evening at dusk (e.g. *Ipomoea bona-nox*, Convolvulaceae (as *Calonyction*), Massart 1895, p. 9). van der Pijl (1930) published a list of some species in Malesia with different flower opening and closing times.

In several night-flowering species the functioning of a circadian clock has been shown. The onset of anthesis and fragrance production is triggered by a particular light quantity impulse on the preceding day (e.g. *Cestrum*, Solanaceae, Overland 1960; *Hoya*, Asclepiadaceae, Altenburger & Matile 1988). The particular sites of light perception seem to be unknown (Mohan Ram & Rao 1984).

'The daily timing of insect foraging represents a compromise between the constraints imposed by resource availability, daily fluctuations in temperature and thermal regulation costs, and risk of predation' (Armbruster & McCormick 1990). An analogous hypothesis may also be stated from the point of view of the flowers. Many flowers that are pollinated by daily active insects open in the early morning, while it is still dark, when many bees are already active (e.g. *Passiflora foetida*, see section 8.5.1). It seems that there is a selective pressure on such a protraction of time of flower opening and pollination. Floral rewards are higher and predator activity is lower the earlier floral visits occur.

A large number of tropical species open their flowers at dusk or at dawn. There are peaks of pollinator activity by many 'crepuscular' pollinators at these particular times of day (Baker 1961).

The phenomenon of rhythmicity in flowers is most prominent in tropical lowland regions and under steady climatic conditions. At higher altitudes or in other regions where favourable conditions for pollination are often limited and unpredictable, more flexible flowering strategies occur. The time of flower opening and longevity is directly dependent on weather conditions and may vary extensively (e.g. *Drimys brasiliensis*, Winteraceae, Gottsberger *et al.* 1980; *Stelis argentata*, Orchidaceae, Christensen 1992; see also Primack 1985b, Corbet 1990).

7.4 Flowering in populations, flowering periods, seasonality

There are many flowering patterns, and diverse kinds of rhythmicity.

There is also a wide range of intraspecific flexibility for these features (Frankie 1975, Primack 1985b). The life span of individual flowers, anthesis at specific times of the day and in specific seasons, flowering in bursts that last for only a short time, or with fewer flowers regularly distributed over longer periods; all this is very diverse, and each geographical region and each systematic category has its own particular set of diversity in this respect. Gentry (1974b) has demonstrated this diversity for Neotropical Bignoniaceae. He ordered the species around three phenological patterns: (1) 'Multiple bang species' and 'big bang species': species that produce many flowers suddenly and for short periods (mass flowering); (2) 'Cornucopia species': species that flower strongly over a period of 3–8 weeks and have flowers that last only one day; (3) 'Steady-state species': species that exhibit constant production of a few flowers each day (night) lasting only one day (or night). Surveys as extensive as that for Bignoniaceae do not exist for other families. These three kinds of behaviour also exist in other groups. In *Parkia* (Mimosaceae) all three types together occur at the generic level (Hopkins 1984). *Casearia praecox* (Flacourtiaceae) is an extreme case of mass flowering, where all individuals of a population flower at once, for a single day (Bawa 1983). Conversely, an individual of *Dillenia suffruticosa* flowers continuously with one or a few flowers every day for 50 years (Corner 1940). Deciduousness is not common in tropical rainforest trees, but in some mass-flowering species it is associated with flowering and so increases the conspicuousness of the flowers (e.g. *Bombax valetonii*, Appanah 1981). The tendency towards inbreeding by geitonogamy in mass-flowering may be lowered by 'anti-pollinators' (Gentry 1978), insect predators (or aggressive behaviour among the pollinators, Frankie 1976) that disperse pollinators between individual plants.

There are also plants that do not flower every year, extreme examples being bamboos with internal clocks (Janzen 1976) and dipterocarps with external triggers for flowering (Janzen 1974a, Ashton *et al.* 1988). Other examples are discussed by Appanah (1985, 1990) and Newstrom *et al.* (1991).

Eupomatia laurina (see section 8.1.2) is an example of synchrony at more than one level. (1) The flowers are protogynous one-day flowers and all open flowers are in the same phase. For newly opened flowers in the female phase, pollen is available from flowers that opened on the preceding day. (2) An individual plant tends to flower in flushes with at least one flowerless day in between, but the flushes are not synchronized between individuals (Endress 1984a). This pattern ensures pollination but excludes autogamy and greatly depresses the level of geitonogamy.

In tropical regions seasonality of flowering may be due to dry and wet periods. This may be seen in different populations of a single species whose range extends from humid to seasonally dry climatic conditions (Borchert 1980). Rainfall (Opler *et al.* 1976) or specific temperatures (Ashton *et al.* 1988) are triggers for flowering or for flower initiation (see also Augspurger 1982, 1985, Borchert 1983, and Rathcke & Lacey 1985). Another, indirect reason for seasonality of flowering may be the seasonal availability of migratory pollinators. In the Neotropics there are migratory hummingbird species, which are present only in winter. They are then territorial and defend flowers that provide nectar (Des Granges 1979). North American migratory birds play an important role in the pollination of the Neotropical *Norantea sessilis* (Marcgraviaceae) (Perry 1990).

7.5 Flowers that are pollinated by trapliners

More often than in temperate regions, tropical species of plant grow with relatively widely spaced single individuals. If pollinated by animals this requires long flight distances by the pollinators. This was first noticed by Fritz Müller (Ludwig 1897). The first detailed description and broader appreciation of the phenomenon was by Janzen (1968) for pollination in *Passiflora* and Janzen (1971) for the behaviour of euglossine bees. Aspects of this behaviour were also discussed for *Xylocopa* by van der Pijl (1954).

The phenomenon is now called traplining (Janzen in Baker 1973, Janzen 1974b, Frankie *et al.* 1974, Gentry 1974b), and several groups of animals are known as trapliners. Among the spectrum of pollinators trapliners are robust, relatively long-lived animals, such as birds, bats, hawkmoths and large bees (*Xylocopa*, euglossines) (Janzen 1983b). These trapliners search for foraging routes that include scattered but predictable high-yield locations, which they regularly exploit over a relatively long time. The behaviour of many less scattered plants that produce only single flowers at a time is perhaps a secondary adaptation to the presence of such traplining animals. Pollination by trapliners is an effective means of outcrossing. Traplined plants show a 'steady-state' flowering pattern (see section 7.4).

Plants that are pollinated by trapliners usually occur in the forest understorey. They are herbs, vines, shrubs or small trees. They also include a number of epiphytes (Ackerman 1986a).

Among hummingbirds there are wandering species that visit several habitats each year and follow seasonal blooms of flowering plants. They move about foraging on different species (Des Granges 1979). Species of *Erythrina* sect. *Erythrina* typically form low-density populations and are pollinated by traplining hummingbirds in Central America (Neill 1987). In contrast to many other trapline-pollinated plants and to hummingbird-pollinated plants in general *Erythrina* forms canopy trees (Stiles 1978a, Neill 1987).

Traplining is also important in the context of conservation of tropical organisms, since it shows that large areas may be necessary even for small organisms (Janzen 1974b) (see section 10.2).

7.6 Flowering in communities, staggered flowering, guilds

Species of the same genus that occur in the same community tend to have staggered flowering, if they are pollinated by the same vectors. The 15 *Miconia* species (Melastomataceae) near Manaus (Amazonia) are pollinated by the same bees and show staggered flowering (Renner 1984b); 19 *Miconia* species in Trinidad (Snow 1965) and 19 other *Miconia* species in Colombia (Hilty 1980) show the same behaviour.

Many *Erythrina* species (Fabaceae), all bird-pollinated, show staggered flowering, and populations may contain flowers over five months or longer (Neill 1987). In sympatric *Inga* species (Mimosaceae) in Costa Rica there are different flowering strategies, but a large amount of overlap in flowering times (Koptur 1983, 1984) (see section 8.6.2).

An impressive example is *Shorea* (Dipterocarpaceae) in eastern Asia with the same trigger for mass flower development in several sympatric species but a slightly different developmental period of the flowers so that staggered mass flowering results (Ashton *et al.* 1988). The flowering sequence between species always remains constant in each flowering period (LaFrankie & Chan 1991). This behaviour within a monophyletic group of species with staggered flowering in a particular area suggests evolution from one ancestor by diversification of the flowering times. Wind-pollinated species, such as grasses, that occur together in the same community, tend to exhibit staggered flowering during each day. This is a means to lower pollen interference on stigmas (see section 4.2.1).

Another kind of staggered flowering of very different taxa in a certain region may be seen as an adaptive peak for an entire community (Stiles 1977). Concomitant staggered fruit maturity may also be an important aspect, especially if the fruits are important food sources for dispersers (Primack 1985b).

A guild is a group of species that exploit the same environmental resources in a similar way (Simberloff & Dayan 1991). In terms of flowers a guild may be a group of plants that use the same pollinators. Robertson (1895) was one of the first to recognize the role of this kind of coordination between species in an ecosystem.

A disputed example of a guild that uses the same pollinators is *Lantana*

1
3
2

camara (Verbenaceae), *Asclepias curassavica* and *Epidendrum radicans* (*E. ibaguense*) (Orchidaceae), three common species in secondary growth in forest gaps in the Neotropics. These three species have very different floral ground plans but they are superficially alike. The flowers are in dense umbel-like aggregations that are bicoloured: red and orange-yellow (Fig.7.2). The flowering season is extended. They are all pollinated by the same butterflies (e.g. monarchs) (Boyden 1980). Boyden (1980) suggests that they form a guild and that *Epidendrum radicans*, a nectarless species, is deceptive by mimicking the two other nectar-producing species. It has also been assumed that in terms of mimicry classification *Asclepias curassavica* and *Lantana camara* represent a pair of Müllerian mimics, whose resemblance would allow them to attract more pollinator visits through the combined impact of their flower displays than either would solely on the basis of its abundance; on the other hand, *Epidendrum radicans* is a Batesian mimic because of its resemblance to the nectar-producing model species (Bierzychudek 1981). Bierzychudek (1981) and Wolfe (1987) also showed the difficulties in testing this hypothesis, because experiments are only possible on a temporally and locally limited scale and not in larger ('evolutionary') dimensions.

Nilsson *et al.* (1987) demonstrate a guild of long-spurred angraecoid orchids of four genera in central Madagascar, all pollinated by the same hawkmoth species. They have probably interacted over a long time and reflect diffuse coevolution (Janzen 1980 and section 4.1.4).

Figure 7.2. Species of a guild, flowers in umbel-like inflorescences with yellow (shaded) and red (unshaded) parts (× 1). 1. *Asclepias curassavica*; yellow parts: corona, red parts: petals. 2. *Lantana camara* (Verbenaceae); yellow parts: petals of freshly open flowers, red parts: petals of older flowers and of floral buds. 3. *Epidendrum radicans* (Orchidaceae), the nectarless cheater; yellow parts: gynostemium and part of the lip, red parts: other organs.

8

Floral diversity and evolution of selected systematic groups in the tropics

In this chapter, a number of predominantly tropical taxa have been selected to show their characteristics and their diversity. Since an understanding of their evolution requires focusing at all levels of the systematic hierarchy, ideally from the populations up to the entire kingdom, the taxa selected here are from different systematic levels: species (e.g. *Passiflora foetida*), genus (e.g. *Passiflora*), family (e.g. Orchidaceae) and order (e.g. Fabales, Scrophulariales, Zingiberales). They are intended to show, in concentrated form, relevant traits of floral diversity and evolutionary trends. Many prominent tropical groups are not discussed in this chapter, but some extremes have been chosen, such as in Magnoliales and Orchidales, that show different sections of the spectrum of angiosperm diversity.

Each group has its own character and its own idiosyncrasies, and emphasis in the knowledge of each group is different. Therefore, they are not treated here in a uniform way. If not otherwise stated, species numbers for larger groups are taken from Mabberley (1987). The classification follows Cronquist (1988) with some modifications (see Appendix).

8.1 Magnoliales (Magnoliidae)

The Magnoliidae, and especially the order Magnoliales, contains a number of isolated, archaic relic families, which have survived in the western Pacific region, including Eastern Australia, New Guinea, New Caledonia and Fiji. A group of four families with a somewhat similar behaviour of the flowers are Degeneriaceae, Himantandraceae, Eupomatiaceae and Austrobaileyaceae (Endress 1984b, 1986a). They are usually associated in some way with the Annonaceae, the largest family of the order. Austrobaileyaceae is perhaps the most isolated family among the four. It has been classified with Laurales by several authors, although it shows a number of special similarities with the Annonaceae (Endress

1980c) and fits better in Magnoliales than in Laurales (Cronquist 1981, 1988). The Winteraceae is another prominent archaic group among the Magnoliales but is more distant from the group of families around the Annonaceae. Phylogenetic studies may eventually reorganize the classification of the Magnoliidae but there is no consensus as yet, since recent palaeobotanical and molecular biological studies of primitive angiosperms provide entirely new views of the group (e.g. Donoghue & Doyle 1989, Friis & Endress 1990, Doyle & Hotton 1991, Loconte & Stevenson 1991, Taylor & Hickey 1992, Qiu *et al.* 1993).

8.1.1 Annonaceae

The family Annonaceae contains some 2000 predominantly tropical trees. The diversity of floral shapes has been reviewed by van Heusden (1992). The flowers commonly have three perianth whorls with three organs in each, three sepaloid and six petaloid (Fig. 8.1). More rarely there are two organs in a whorl, e.g. *Disepalum* (Johnson 1989), or four, e.g. *Tetrameranthus* (Westra 1985), or they have an irregular position, e.g. *Toussaintia* (Le Thomas 1969), or a double position of the inner petals occurs, e.g. *Ambavia gerrardii* (Deroin & Le Thomas 1989) (see also Fig. 2.39.1). In some genera sympetaly occurs, e.g. *Disepalum* (Johnson 1989), *Papualthia* (W. Morawetz, personal communication), *Fusaea* and *Isolona*. Most groups have a large number (up to 400 in *Annona*) of small, sessile stamens in irregular arrangement, which is best seen in very young buds (Fig. 2.38.4). The extrorse anthers are densely packed and form a rounded platform in the floral centre. They are cuneiform and have a more or less thick sterile shield at the apex, which may be flat or pointed (Fig. 8.3.7). It contains tannins or is even lignified (Endress 1975). The carpels, one to many (up to 400) per flower, are mostly free. They are commonly epeltate (Leinfellner 1969a) (Fig. 2.18.2), or rarely weakly peltate (Deroin 1988a) and contain one to several ovules in two rows along the flanks. The ovule position is lateral, even for a single 'basal' ovule. The ovules are often perichalazal. In rare cases they have more than two integuments (Corner 1976). The stigma may be pronouncedly transversally bifid (e.g. *Goniothalamus*). Increase (and decrease) of carpel number per flower and of ovule number per carpel is part of the diversification in the family (Deroin 1991b).

In taxa with smaller flowers and a smaller number of floral organs, the stamen position may be more regular. They are then in trimerous whorls (*Orophea*, Kessler 1988), often with double positions resulting in six or nine organs in a whorl, e.g. *Exellia scammopetala* (Le Thomas 1969) or *Monanthotaxis whytei* (Ronse Decraene & Smets 1990b). Three or six

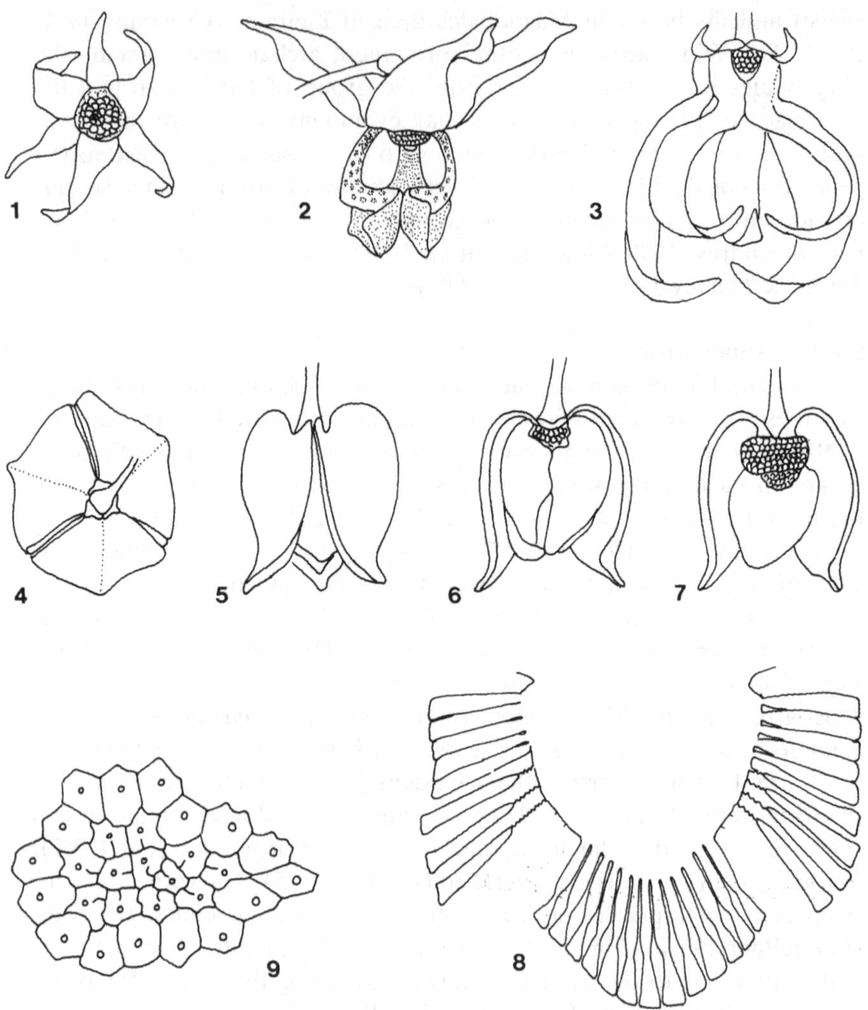

Figure 8.1. Flowers of Annonaceae. 1. *Polyalthia longifolia*, widely open flower (× 1.5). 2. *Monodora junodii*, the three inner tepals with brown spots and distally coherent, forming three lateral entrances, probably fly-pollinated (× 1). 3. *Artabotrys hexapetalus*, a tepal of the second whorl removed to show the narrow basal entrance formed by the inner tepals (× 1). 4–7. *Annona muricata* (× 0.5). 4. Flower from the base showing the three reduced tepals of the outer whorl and the three valvate tepals of the second whorl. 5. Open flower from the side. 6. A tepal of the second whorl removed to show the tepals of the inner whorl. 7. Tepals of one side removed to show pollination

carpels in a whorl are also common in these taxa (in *Monanthotaxis whytei* three whorls with nine carpels in each). Unicarpellate species occur in many different genera; this is apparently a frequent parallel evolutionary trend (Endress 1990b).

In *Monodora* and *Isolona* the gynoecia are syncarpous with parietal placentae (e.g. Guédès & Le Thomas 1981, Deroin 1985). However, development begins as in a single carpel (Leins & Erbar 1980, 1982). Most likely a single carpel is transformed during development into a syncarpous structure by late secondary subdivision of one into several structural elements (Endress 1990b) (see section 2.4.6).

The flowers are commonly self-compatible but pronouncedly protogynous (often with a neutral phase between the female and male phase), even synchronous protogyny has been reported (see below). The stigma is heavily secretory. In many genera the stigma abscises at the end of the female phase. The tightly spaced stamens often abscise in the male phase but remain loosely attached to the floral base by extended spiral thickenings of tracheids of their vascular bundle (Endress 1985) (Fig. 8.1.8). This allows better pollen exposition. The extrorse anthers open by valves (Endress & Hufford 1989). The inner 'petals' often contain diffuse food tissue, well-defined food bodies or nectaries (Norman & Clayton 1986, Schatz 1987, Morawetz 1988). In two tropical species of *Asimina* the female phase is 2–6 days, the male phase 1 day, with a neutral phase of 1 day in between (self-compatible) (Norman & Clayton 1986). In *Cananga odorata* the flowers start emitting scent a day before the stigma becomes secretory. The female phase begins at 1800; the next day at 1800 the floral organs drop (Deroin 1988b). In *Monodora myristica* the female phase is 7–8 days, the male phase 4–5 days (presumably self-incompatible) (Lamoureux 1975). In *Polyalthia littoralis* anthesis is about 2–3 days (self-compatible) (Okada 1990).

Synchronous dichogamy was reported from *Rollinia jimenezii* in Mexico (Murray & Johnson 1987). Flowers opened at 1500 on 10 May. Stigmas became receptive (secretory) before 1900, accompanied by a yeasty scent.

Fig. 8.1 (*cont.*)

organs. 8. *Annona muricata*; longitudinal section of floral centre in male stage; innermost stamens fallen, middle stamens abscised but still retained by extended spiral thickenings of vascular bundle (× 3). 9. *Cananga odorata*; transverse section of gynoecium and innermost stamens of old bud to show tight packing and organ outlines influenced by neighbouring organs (× 17).

At 0500 on 11 May the stigmas were still shiny but the scent was reduced. By 1400 the stigmatic caps had abscised. Anthers dehisced betwen 1500 and 1600. Corollas dropped from the flowers between 1715 and 1900. On this day no asynchronous flower opening occurred on the tree. Synchronous dichogamy was also indicated for *Annona senegalensis*. Flowers are in the female stage from 0600 till 0900 and in the male stage from 1600 till 1700. There are no flowers in anthesis on the next day (Deroin 1989).

Pollinators are small or large beetles (Gottsberger 1970, 1986, 1988, 1989a,b) that are attracted by the fruity scent of the flowers. Thrips use flowers as a brood site and may also pollinate them; beetles also lay eggs in flowers, but it is not known whether they might complete their life cycle in them (Gottsberger 1970). Observations on pollinators include the following taxa: *Guatteria* (Nitidulidae, Chrysomelidae, Curculionidae, thrips), *Duguetia* (Nitidulidae, Curculionidae), *Xylopia* (Nitidulidae, thrips) (Gottsberger 1970); *Cananga odorata* (Chrysomelidae and Curculionidae) (Deroin 1988b), *Annona senegalensis* (Curculionidae) (Deroin 1989), *Annona* species (Nitidulidae) (Podoler *et al.* 1984), *Rollinia jimenezii* (Nitidulidae) (Murray & Johnson 1987). *Cyclocephala* (Dynastinae–Scarabaeidae) is the pollinator of probably over 100 Neotropical (large-flowered) Annonaceae (Schatz 1987; see also Gottsberger 1989b). Other Scarabaeidae and Cerambycidae pollinate *Asimina* species (Norman & Clayton 1986). Some flowers smell like carrion and are probably fly-pollinated (*Sapranthus*, dark brown flowers, up to 15 cm long, Janzen 1983a; Olesen (1992) found beetles (Tenebrionidae, Nitidulidae, Scarabaeidae) and *Trigona* bees as potential pollinators).

Large-flowered *Annona* species have wide floral chambers and very thick petaloid organs and are pollinated by large scarab beetles, as studied by Gottsberger (1989a). They are strongly thermogenic. Different species studied showed anthesis of one to three days with functional and thermogenic phases at night, when the pollinators are active. A larger number of beetles may congregate in flowers and mate in this chamber. They also fly to other flowers during the active nocturnal phase. In *A. coriacea* and *A. montana* anthesis is two nights. The peak of the male phase (of second-night flowers) is about an hour earlier than the peak of the female phase (of first-night flowers), which may enhance pollination. *A. corniflora* has a diversity of anthesis rhythms from one to three nights. The male phase is always short during the last night of flowering. *Annona crassiflora* flowers for only one night. At the end of the night the coherent petals drop with scattered stamens inside. The beetles remain in this 'chamber' during the day.

Many other Annonaceae have flowers that allow access to the pollination chamber for small beetles only. In these flowers the valvate petals remain tightly closed except for a small basal part, where small beetles are able to enter. A similar construction but with larger lateral openings may be used for fly-pollination, which may be inferred for the lantern flowers of *Monodora* species.

The morphological and biological context make it probable that relatively small flowers with few organs in regular positions, with stamens that lack an elaborate sterile apex, and without a specialized cantharophily syndrome, are primitive for the family. They occur e.g. in the Madagascan *Ambavia* (Deroin & Le Thomas 1989). However, such small flowers may also have arisen by reduction in some genera, as indicated by the presence of elaborate corollas and stamens in some of them. Possibly *Anaxagorea*, distinctive by enlarged inner staminodes, represents a basal group (J.A. Doyle, personal communication).

8.1.2 Eupomatiaceae

The Eupomatiaceae consists of two species in the genus *Eupomatia*, which occur in Eastern Australia and in New Guinea. Fully open flowers are 2–5 cm in diameter. Floral phyllotaxis is spiral (Endress 1977b). The number of floral organs is quite variable. In bud the flower is completely enclosed by one (or two?) sheathing bract, which is congenitally closed except for a minute (postgenitally closed) apical pore (Endress 1977b). A perianth is lacking, if the floral envelope is considered to be a bract. The *ca.* 20–100 stamens have a broad base with three main vascular bundles (Fig. 8.3.6). The anthers are introrse. The 'sunken' thecae open with valves (Endress & Hufford 1989). The inner staminodes, *ca.* 40–80, are large and attractive (Fig. 8.2). Stamens and inner staminodes are congenitally united at the base to form a synandrium. The gynoecium is a flat disc, consisting of *ca.* 13–70 congenitally fused carpels. They are largely plicate, but since they are spirally arranged, a compitum is not formed, except, sometimes, for the very innermost carpels (Endress 1977b). The carpels are stout, tapering towards the base and lacking a style. Each carpel contains 2–11 ovules, which are arranged in two rows along the ovary.

The flowers are protogynous. The flowering process lasts one day in *Eupomatia laurina*, and two days in *E. bennettii* (Endress 1984a). In *E. laurina* the flowers open at 0500. The female phase with the unicellular-papillate and slightly wet stigmas exposed is from *ca.* 0530 till 1100. Then the inner staminodes close over the gynoecium. The male phase begins at 1700 and extends through the night. Stamens and inner staminodes are now tightly bent over the floral centre. The next morning the synandrium has abscised and lies on the ground. In *E. bennettii* the flowers open at 0800. The female phase with the gynoecium exposed is from *ca.* 0930 till in the evening. The next day the flower is in the male phase. The synandrium abscises in the second night.

The flowers are white in *E. laurina*, yellow and red in *E. bennettii*. They have a strong fruity smell (overpowering in *E. laurina*), which is produced by the inner staminodes. These staminodes have an oily covering containing the volatiles (Endress 1984a, Bergström *et al.* 1991). They also have

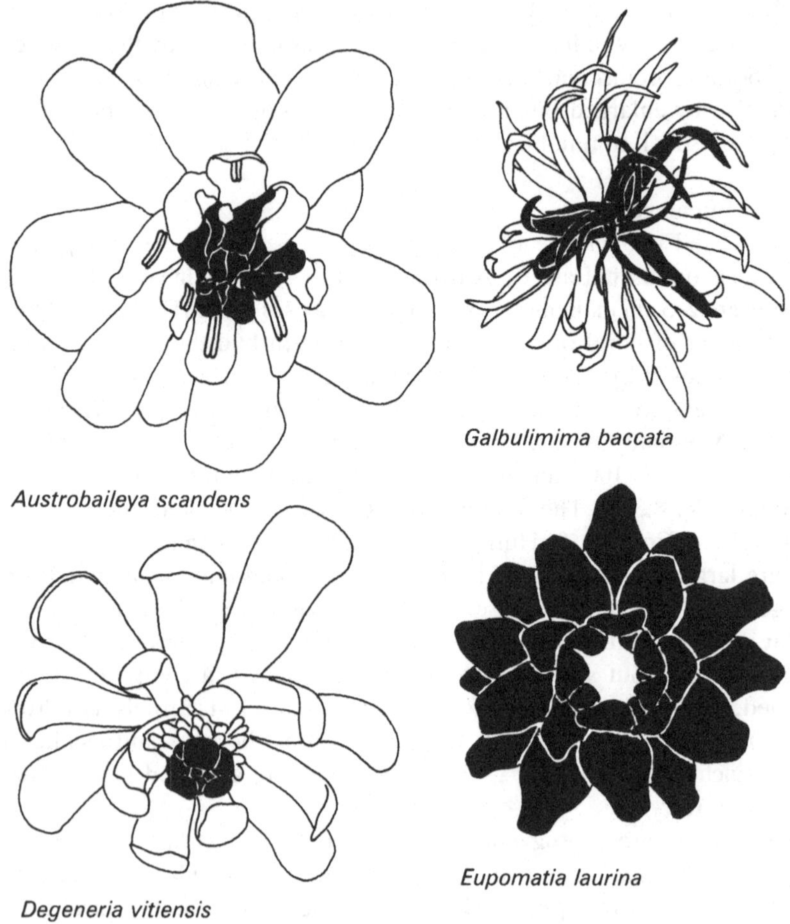

Figure 8.2. Flowers of some small, relic families of Magnoliales with conspicuous inner staminodes (black) and laminar stamens: Austrobaileyaceae, Degeneriaceae, Eupomatiaceae, Himantandraceae (from Endress 1984b).

distinct mats of long, unicellular hairs, which secrete a sticky substance (described as 'food bodies' in the earlier literature) (Diels 1916, Endress 1984a). Pollination in both species is exclusively by curculionid beetles of the genus *Elleschodes* (Hamilton 1898, Hotchkiss 1959, Armstrong & Irvine 1990). They feed on the nutritive tissue of the inner staminodes, which also provide them a sheltered chamber. Armstrong & Irvine (1990) have shown for *E. laurina* that the weevils lay their eggs in the flowers. The larvae develop in the abscised synandrium on the ground and pupate in the soil. These beetles are possibly monotropic.

Both species are self-compatible. Several levels of anthesis synchronization, however, favour outcrossing. In *E. laurina* all flowers at anthesis are synchronous and dichogamous with a pronounced neutral phase between female and male stage. Further, they tend to flower in flushes with at least one flowerless day in between (Endress 1984a). In individuals of *E. bennettii* often only a single flower is produced per year.

8.1.3 Degeneriaceae

The family contains the single genus *Degeneria* with two species restricted to Fiji (Bailey & Smith 1942, Miller 1988). The following description holds for *D. vitiensis*. The flowers are 4–8 cm in diameter and pendent (Fig. 8.2). Floral phyllotaxis is spiral (personal observation). Aestivation is pronouncedly imbricate (Swamy 1949). The number of floral organs varies considerably (Smith 1981 and personal observations). The outermost three sepaloid organs are congenitally fused at the base (bracts or tepals? see also Winteraceae, below). They are followed by 12–22 petaloid organs. The 20–43 stamens are laminar, the thecae pronouncedly extrorse, 'sunken' and exhibiting valvate dehiscence (Endress & Hufford 1989) (Fig. 8.3.5). The 5–15 inner staminodes are larger than the stamens. The gynoecium has a single carpel and contains 22–30 ovules. The carpel is slightly ascidiate. It is tightly closed on a broad front along the ventral slit by postgenital fusion (not open, as often stated in the literature!). The stigma is formed by two convoluted crests along the upper half of the ventral slit. The ovules are arranged in two rows along the ventral slit in the ovary. The carpel much resembles that in *Tasmannia* (Winteraceae, see below) (Figs 2.14, 8.4).

The flowers are protogynous (Thien 1980). The flowering process lasts 21–23 hours (Miller 1989). The flowers open early in the evening. At 2100 they are completely open and the gynoecium exposed. The stigma is unicellular-papillate. At 2200 the inner petaloid organs cover the floral centre. Thus the flower is in a neutral phase, which lasts until just before dawn. At that time the male phase begins. The stamens have dehisced and are exposed but the gynoecium is covered by the inner staminodes. The petaloid organs and stamens, however, are reflexed. The flowers drop between 1700 and 1800. (In *D. roseiflora* the flowering process is shorter.) The flowers are white to light beige (pinkish to magenta in *D. roseiflora*, Miller 1988). The inner staminodes are distally brown and coated with bright yellow secretion (purple secretion in *D. roseiflora*), which may help in making the pollinator's body sticky. The flowers are pollinated by nitidulid beetles (*Haptoncus*) (Takhtajan 1973, Thien 1980), and perhaps also by other beetles, Diptera and Microlepidoptera (Miller 1989). They have a foul (Thien 1980) or fruity smell (Smith 1981), which is produced by the inner staminodes (Miller 1989).

8.1.4 Austrobaileyaceae

This is an extremely isolated monotypic family in a small area of northeastern Australia (Endress 1980c). The pendent flowers of *Austrobaileya scandens* are 5–6 cm in diameter (Fig. 8.2). Floral phyllotaxis is spiral (Endress 1980c, 1983). The imbricate perianth with *ca.* 19–24 organs exhibits a series of morphs transitional from small, green bracts to large, yellowish petaloid organs without a delimitation between bracts, sepaloid and petaloid organs. The 7–11 broad and flat stamens also show a gradual change towards narrower organs with shorter pollen sacs. The two thecae are laminar on the inner surface and strongly protruding (not 'sunken'), but they are marginal in early ontogeny (Endress 1983) (Fig. 8.3.4). They open with simple longitudinal slits (Endress & Hufford 1989). The stamens are followed by 9–16 inner staminodes with a ventral furrow and irregular ridges on the dorsal side. The 10–13 carpels are free, stipitate, completely ascidiate, with transversally bilobate stigmatic tips. The stylar canal is not postgenitally fused but filled with secretion. The 4–13 ovules are arranged in two longitudinal rows on the ventral side of the ovary. They are perichalazal as in many Annonaceae.

The flowers are protogynous. Duration of anthesis of a flower is unknown. In the female phase the bright yellow gynoecium is exposed. In the male phase it is covered by the inner staminodes. The unicellular-papillate stigmas are heavily secretory and have a common cover of exudate at anthesis (Endress 1980c). Pollen tube transmitting tracts in styles and ovaries are also heavily secretory. The petaloid organs and stamens are greenish to yellowish. Except for the outer petaloid organs they are speckled with dark brown spots. The thecae are bright yellow. The inner staminodes are covered with large, dark brown spots and have a pronouncedly papillate epidermis. The sapromyophilous flowers emit a fetid smell of rotten fish, which is produced by the inner staminodes. They attract carrion flies (Endress 1980c). Drosophilid eggs were found in the ventral furrows of the inner staminodes and in the floral base (Endress 1984b). The plants are self-incompatible, with arrest of pollen tubes in the mucilaginous cap of the stigma (Prakash & Alexander 1984).

8.1.5 Winteraceae

Winteraceae, with 9 genera (4 according to Vink 1988) and *ca.* 65 or more species, has a circumtropical but scattered, relictual distribution in the Neotropics, Madagascar, Malesia, Australia, New Caledonia, and New Zealand. The flowers show a broad spectrum of size, organ number and organ arrangement (Fig. 8.3.1–3). There may be enormous variation in floral organ number within a single species. For instance, *Tasmannia piperita* (in the broad circumscription of Vink, 1970) has 0–15 petaloid perianth organs, 7–109 stamens, 1–15 carpels, and 2–46 ovules per carpel. The flowers are bisexual (unisexual in *Tasmannia*). Irregular floral phyllotaxis patterns predominate, but spiral ones also occur (Hiepko 1966, Vink 1970, Erbar & Leins 1983, Endress 1986a). In *Takhtajania* trimerous whorls seem to be present (Vink 1978). The perianth contains 2 or 3 outer parts, which are usually congenitally united and rupture or abscise as a calyptra, as the flower opens. The question is still unanswered, whether they should be regarded as bracts or tepals (see section 9.3). The inner, petaloid perianth parts, 0–25 in number (with this entire range in *Tasmannia*; Vink 1970), tend to decrease considerably in size from the outer to the inner ones. The phyllotaxis often starts with two in a whorl and continues with doubling or even repeated doubling of organ number in inner whorls; irregularities are common (Vink 1970, 1977, 1988; personal observation). Stamens, in contrast to the petaloid perianth members, tend to increase in size towards the floral centre and the inner ones often open earlier than the outer ones at anthesis, although they are initiated centripetally. Stamen number counts vary between 4 (lowest number in *Pseudowintera*, Sampson 1980) and 371 (highest number in *Zygogynum*, Sampson 1983). In *Zygogynum* alone stamen number varies between 20 and 371. Stamen arrangement is irregular in most cases, even in flowers with few organs (Vink 1988, personal observation). In *Drimys* it is relatively regular (Hiepko 1966, Erbar & Leins 1983). The anthers are terminal on flattened or rotate filaments; in *Belliolum* they are more lateral and have an apical appendage. The thecae open by simple longitudinal dehiscence. The stamens have one to three main vascular bundles (Sampson 1987). The 1 to *ca.* 50 carpels are free or, in *Zygogynum*, congenitally united into a globular structure without a compitum, since each carpel has a separate inner morphological surface (see section 2.4.1). They are either largely plicate (e.g. *Tasmannia*) or largely ascidiate (e.g. *Drimys*, *Zygogynum*) (Tucker 1959, Leinfellner 1969a). A style is lacking and a stigmatic crest flanks the orifice of the carpel (Figs 2.14, 8.4). The carpel is tightly closed at anthesis by postgenital fusion of its flanks (Fig. 2.15). The single carpel in the flowers of *Pseudowintera traversii* is terminal (Sampson & Kaplan 1970). Whether the poorly known Madagascan genus *Takhtajania* has a unicarpellate or a syncarpous, bicarpellate gynoecium with parietal placentation, is still disputed (Vink 1978, Tucker *et al.* 1979, Endress 1990b) (see section

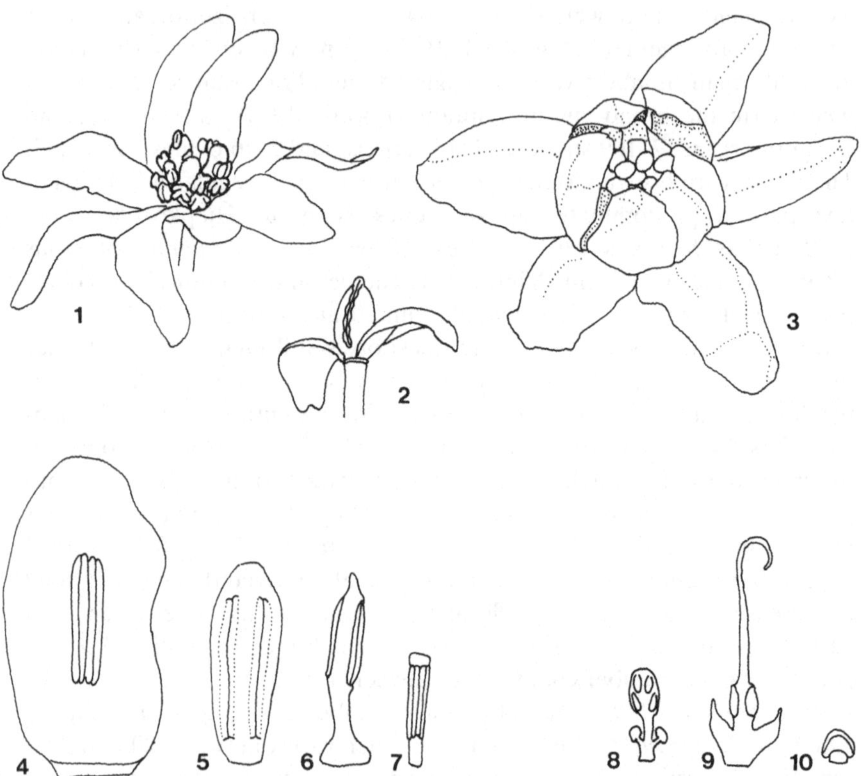

Figure 8.3. 1–3. Flowers of Winteraceae. 1. *Drimys winteri*; bisexual flower (× 2). 2. *Tasmannia membranea*; female flower with two tepals and a carpel (and scar of outer tepal) (× 2). 3. *Zygogynum baillonii*; bisexual flower in female stage (× 2). 4–7. Stamens of Magnoliales (× 3). 4. *Austrobaileya scandens*. 5. *Degeneria vitiensis*. 6. *Eupomatia laurina*. 7. *Annona muricata*. 8–10. Stamens of Laurales (× 3). 8. *Litsea glaucescens* (Lauraceae). 9. *Doryphora sassafras* (Monimiaceae). 10. *Steganthera ilicifolia* (Monimiaceae).

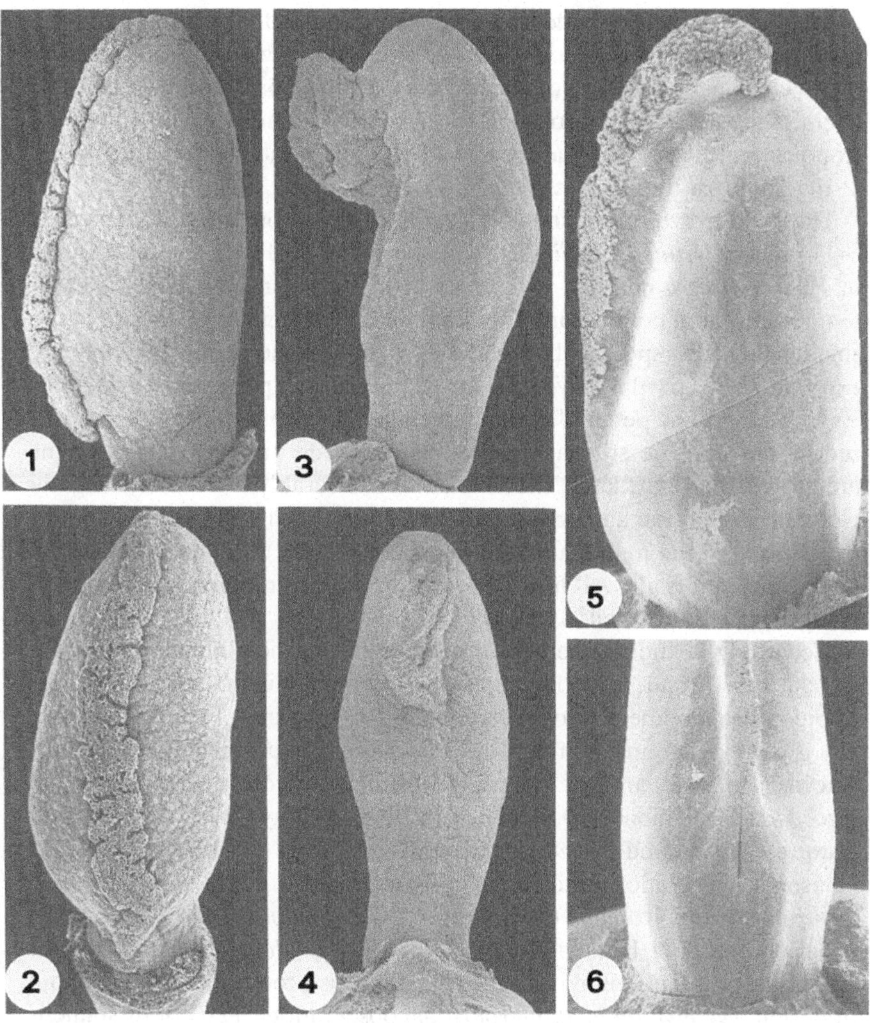

Figure 8.4. Carpels of Winteraceae and Degeneriaceae. 1–2. *Tasmannia membranea* (Winteraceae) (× 18); 1. From lateral side. 2. From ventral side. 3–4. *Drimys winteri* (Winteraceae) (× 20). 3. From lateral side. 4. From ventral side. 5–6. *Degeneria vitiensis* (× 14). 5. From lateral side. 6. From ventral side (lower end of ventral slit).

2.4.6). The 2 to *ca.* 85 ovules per carpel are arranged in two vertical rows, flanking the inner orifice of the carpel; the two rows may merge at the base (U-shaped placenta) or, in addition, at the top (O-shaped placenta) (Leinfellner 1969a, Sampson & Tucker 1978). The micropyles of the anatropous ovules are tightly appressed to the pollen tube transmitting tissue at the flanks of the placenta (Fig. 2.15.3).

The flowers are protogynous (Endress 1990b). Longevity of a flower is 6–11 days in *Drimys brasiliensis* (Gottsberger *et al.* 1980), and 11–15 days in *Pseudowintera colorata* (Wells & Lloyd 1991), in both depending on temperature and pollination. The flowers are open and bowl-shaped in the smaller-flowered species. In some larger flowers the centre is not freely exposed but partly covered by stamens or perianth (*Belliolum*, *Zygogynum*). The petaloid perianth members are white, yellow, brownish, red or purple. The stigma is unicellular-papillate and wet. The stamens are crowded at first; later, the filaments considerably thicken and slightly elongate so that the anthers are separated from each other and pollen can be released in the male phase (Carlquist 1981, 1982). Pollen is predominantly shed in tetrads. All four pollen grains of a tetrad may germinate on a stigma (Lloyd & Wells 1992) (Fig. 2.28.3).

According to the simple gestalt of the flowers, the pollinator spectrum is relatively broad, with flies and beetles predominant but also thrips, Lepidoptera, and bees (Gottsberger *et al.* 1980, Thien *et al.* 1990). Specificity is more pronounced in the larger-flowered groups: thrips in *Belliolum*, weevils (*Palontus*) and micropterigid moths (*Sabatinca*) in *Zygogynum* and *Exospermum* (Thien 1980, Thien *et al.* 1985, Pellmyr *et al.* 1990). Small-flowered *Bubbia* and *Pseudowintera* tend to have a weak musky floral scent (personal observation), indicative of fly-pollination, while larger-flowered groups are more aromatic (like orange peel in *Belliolum crassifolium*, personal observation; banana-like in *Zygogynum baillonii*, Thien 1980) (for scent analyses see Pellmyr *et al.* 1990). Scent is produced by the petaloid organs in *Zygogynum baillonii* (personal observation), and not by stamens or carpels. Food bodies that are eaten by insects occur in the petaloid organs and stamens of *Belliolum* (Carlquist 1983, Endress 1986a), *Zygogynum*, *Exospermum*, and *Bubbia* (Thien *et al.* 1990). Stigmatic secretions are also eaten by pollinators (Lloyd & Wells 1992). Self-incompatibility occurs in *Pseudowintera colorata*, with pollen tube inhibition in the ovules (Godley & Smith 1981, Lloyd & Wells 1992). Self-compatibility was recorded in *Drimys brasiliensis* (Gottsberger *et al.* 1980).

8.2 Laurales (Magnoliidae)

The Laurales are characterized by carpels with a single median ovule. Furthermore, the gynoecium is predominantly unicarpellate (except for Monimiaceae and the doubtfully placed Calycanthaceae). In Monimiaceae p.p., Lauraceae, Hernandiaceae, and Gomortegaceae the stamens tend to have paired appendages that function as nectaries. The same four families tend to have anthers in which the two pollen sacs of a theca do not have a common stomium but each open separately with a valve. This is often associated with a reduction to one pollen sac for each theca (Endress & Hufford 1989, Endress & Stumpf 1990). The flowers are commonly small, and there is a tendency toward unisexual flowers. Of the two families selected here the Chloranthaceae is especially archaic, and the Monimiaceae exhibits particularly unusual trends of floral evolution.

8.2.1 Chloranthaceae

The Chloranthaceae (4 genera, *ca.* 75 species, Todzia 1988) have a scattered distribution in Eastern Asia, Malesia, some Western Pacific Islands, Madagascar, and the Neotropics.

The record of Cretaceous Chloranthaceae-like pollen, flowers and vegetative parts is particularly extensive (Crane *et al.* 1989). The flowers are commonly in groups of spikes or in solitary spikes. They are small and extremely simple: without a perianth (except for the female flowers of *Hedyosmum*), with 1–5 stamens, and with a single carpel (Fig. 8.5). In

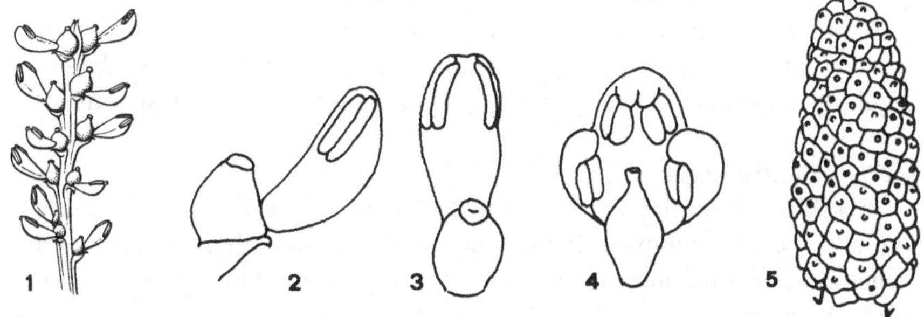

Figure 8.5. Inflorescences and flowers of Chloranthaceae. 1–3. *Sarcandra chloranthoides*. 1. Inflorescence (× 1). 2. Flower, consisting of a stamen and a carpel, subtended by a bract, from the side (× 4). 3. Same flower from adaxial side (× 4). 4. *Chloranthus spicatus*; flower, consisting of a three-parted stamen and a carpel, from adaxial side (× 8). 5. *Hedyosmum mexicanum*; young male inflorescence, consisting of unistaminate flowers (× 2).

Sarcandra and *Chloranthus* they are bisexual; in *Ascarina* and *Hedyosmum* unisexual. In *Sarcandra* a flower consists of a stamen and a carpel; in *Ascarina* the female flowers consist solely of a carpel, the male flowers in most species of a single stamen. In *Hedyosmum* the male flowers are also unistaminate and are devoid of a subtending bract. Thus the inflorescence looks like a multistaminate flower. However, its interpretation as a flower (Leroy 1983) cannot be supported, if it is viewed in the entire systematic context (Endress 1987b). In floral buds of *Sarcandra* and *Chloranthus* the broad stamen covers the gynoecium like a scale. In *Chloranthus* the stamen is especially broad and three-lobed; each theca is divided into an upper part on the middle lobe and a lower part on the outside of the respective side lobe; in some species the middle lobe is sterile (Endress 1987b). In *Sarcandra* and perhaps *Chloranthus* each theca opens with two narrow valves, in *Ascarina* and *Hedyosmum* with a simple longitudinal slit (Endress 1987b) (Fig. 2.7.1). The carpel is completely ascidiate up to the stigma in all representatives. The canal leading to the ovule is not postgenitally fused, but since the inner surface is heavily secretory, it is filled with secretion. The single pendent ovule is orthotropous, bitegmic and crassinucellar (Fig. 2.15.4).

Sarcandra and *Chloranthus* are most probably insect-pollinated. They have conspicuous anthers with broad yellow or white connectives that produce scent. They contain little pollen. The stigma is wet. *Hedyosmum* and *Ascarina* have anemophily syndromes with copious dry pollen in the inconspicuous anthers, and large dry stigmas. This is extreme in some *Hedyosmum* species (Endress 1971, 1987b, Todzia 1988).

In *Sarcandra chloranthoides* the flowers are weakly protogynous; other species have not been studied. Greenhouse plants of *S. chloranthoides* regularly set fruit, indicating either self-compatibility or agamospermy.

8.2.2 Monimiaceae

The Monimiaceae is a pantropical family with about 33 genera and 320 species (Philipson 1986). They are evolutionarily plastic in floral structure and some are highly peculiar in their floral biology. The infra-familial classification is still problematical.

A few genera (e.g. *Hortonia*, Hortonioideae; *Daphnandra*, Atherospermatoideae) have protogynous, bisexual flowers and the spirally arranged floral organs well exposed on a shallow floral cup (Endress 1980a,b, 1992a). Floral organ number, also in the gynoecium, is variable. In *Dryadodaphne*, which is similar, the organs are in tetramerous whorls. In all of these genera the stamens have two lateral outgrowths that function as nectaries. Inner staminodes also occur. The thecae have a single pollen sac that opens with a valve (in *Hortonia* two pollen sacs that open in common with simple longitudinal slit).

In the largest subgroup, Monimioideae s.l., there are unique evolutionary trends (Endress 1980a, Endress & Lorence 1983) (Fig. 8.6). The flowers are unisexual. The floral organs are commonly in dimerous or trimerous whorls, often with double positions; the phyllotaxis may also be irregular (Endress 1980a). The perianth organs become reduced in size but the floral cup becomes concomitantly enlarged and tends to include the inner floral organs. In male flowers the anthers are often sessile and open with a single slit over the apex. Pollen can be obtained through a more or less widely open pore formed by the upper rim of the floral cup. In female flowers the ovaries become included in the floral cup. The stigmas may be at the end of styles protruding out of the floral pore (e.g. *Palmeria*) or at level of the floral pore (e.g. *Steganthera*). In the extreme case the stigmas are also completely included in the floral cup, which only leaves a narrow pore in the centre. The upper rim of the cup or the entire pore becomes secretory, forming a 'hyperstigma', where pollen grains ger-

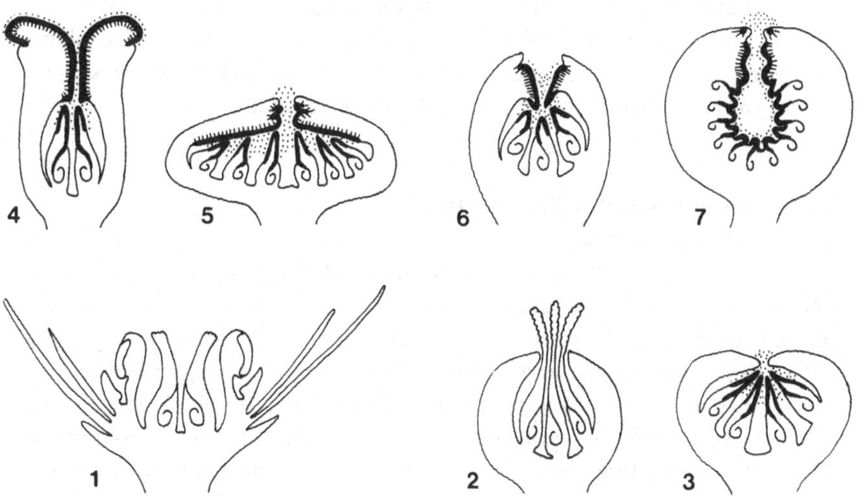

Figure 8.6. Flowers of Monimiaceae, longitudinal sections (all flowers female, except for *Hortonia*). Black, secretory areas; stippling, secreted material (modified after Endress 1980a). 1–3. Flowers without a hyperstigma. 1. *Hortonia angustifolia*. 2. *Palmeria gracilis*. 3. *Steganthera ilicifolia*. 4–7. Flowers with a hyperstigma (hatched black areas). 4. *Hennecartia omphalandra*. 5. *Wilkiea huegeliana*. 6. *Kibara coriacea*. 7. *Tambourissa purpurea*.

minate and pollen tubes grow down the pore towards the carpellary stigmas. Hyperstigmas have probably evolved in at least five genera (*Tambourissa*, *Wilkiea*, *Kibara*, *Hennecartia* and *Faika* (Endress 1980a, Philipson 1986). In addition, in *Tambourissa* the carpels are united and the ovaries inferior. There is, however, no internal compitum, but the hyperstigma may act as an 'external' compitum. Hyperstigmas may also be seen as a defence of female flowers against insect infestations, and they may have evolved from secretory structures on the perianth organs that play a role in bud protection. Flowers of various Monimiaceae are often infested by insects (Gottsberger 1977, Endress 1980a). In *Siparuna* gall midges are pollinators and at the same time fulfil their life cycle in the flowers (Feil 1992). Another trend is seen in the Madagascan genera, especially *Tambourissa*. While a few species have developed closed flowers with hyperstigmas, other species have flowers that secondarily split open at anthesis and exhibit the carpels and stamens on flat disks (Endress & Lorence 1983, Lorence 1985). In this group large flowers with exceedingly many organs have developed: in *T. ficus* the expanded flowers are up to 7.5 cm in diameter and have up to almost 2000 stamens or carpels. They are pollinated by flies, beetles (curculionids) and bees (Lorence 1985). At least in some species the stigmas produce copious secretion. The plants are monoecious or dioecious. In five monoecious *Tambourissa* species tested, only one was self-compatible (Lorence 1985).

8.3 Aristolochiales (Magnoliidae)

8.3.1 Aristolochiaceae

The Aristolochiaceae, pantropical and also temperate, with some 400 species, have predominantly trimerous flowers and other patterns derived therefrom. They are herbs, vines or shrubs. There are many similarities with Annonaceae and a close relationship with Magnoliales is surmised. The monotypic *Saruma* from China with a double perianth and regularly trimerous and polysymmetric small, simple flowers is perhaps the most conservative genus. It is the only genus with nearly free carpels and more or less superior ovaries. *Thottea* is a tropical genus with small polysymmetric flowers. They have three sepals and six or more stamens with double positions in the same sectors. Organs in the petal positions are highly delayed in development and reduced. The gynoecium is tetramerous (Leins *et al.* 1988); the styles are also increased in number by irregular lateral doubling (A. Weber, personal communication; personal observation). All species of Aristolochiaceae observed to date have protogynous flowers (Endress 1990b). The flowers are commonly brown or yellowish. There is perhaps no other family of comparable size that is so exclusively pollinated by Diptera. Sapromyophily is widespread (*Aristolochia*). Pollination by fungus gnats is also prominent (*Asarum*, *Heterotropa*) (Vogel 1978b and others). *Aristolochia grandiflora* has the longest flowers in the world, more than 1 m in length; in *Pararistolochia goldieana* they reach 60 cm (Poncy & Lobreau-Callen 1978).

Aristolochia

Aristolochia, with about 300 predominantly tropical species, especially in America, is by far the most diverse genus in the family. One may suspect that it is a polyphyletic group (Huber 1985). Diversity is mainly in the flowers. However, the floral groundplan is relatively stable with regard to other genera of the family. The perianth is monosymmetric and has only three sepals. They are congenitally united into a long tube forming a trap and an often very conspicuous expanded upper part. Further, the flowers have mostly six stamens (in double position) and six carpels. Stamens and carpels are congenitally fused into a gynostemium. The perianth is most plastic in shape. All three sepals may be well developed and each form a tail (Fig. 8.8.4–6). The perianth may also have only one tail-like appendage or only a short lobe. It has been argued that in these cases only one sepal forms the perianth (e.g. Guédès 1968). However, early ontogeny suggests the presence of three primordia (Leins & Erbar 1985). If there is only one tail it may be adaxial (*A. macroura*) (Fig. 8.8.1–3) or abaxial (*A. grandiflora*) (Fig. 8.7.1–3). If the position of the three sepals is constant, one adaxial and two lateral, the tail in *A. grandiflora* may be formed by two united sepals. This is an open question. In a flower of *A. grandiflora* studied by me the tail was 60 cm long, the length of the entire flower more than 1 m. The expanded part of the perianth may be flat and shield-like (*A. gigantea*, *A. elegans*) (Fig. 8.7); it may also be prominently two-lipped (*A. brasiliensis*) or one-lipped (*A. arcuata*). The tubular part of the perianth is either straight (*A. lindneri*) or strongly curved (*A. grandiflora*, *A. gigantea*, *A. brasiliensis*, *A. macroura*) (Figs 8.7, 8.8). The inner surface of the tube, which acts as a pitcher trap, is elaborated. In the tube there are commonly narrower parts, often in the form of a diaphragm, that restrict access to smaller flies (*A. grandiflora*, *A. tricaudata*, *A. lindneri*). Often there are stiff hairs around the narrow parts that prevent an early exit of flies (*A. grandiflora*).

Floral colours vary between light greenish and almost black. Often there are patterns with dark spots and stripes on a lighter ground colour. A window-pane at the base of the floral tube is common. In *A. lindneri* the pollination chamber is light and the tube dark on the first day, and almost reverse on the second day after pollination (Lindner 1928).

Floral scent is varied, often like carrion (*A. cathcartii*, *A. grandiflora*)

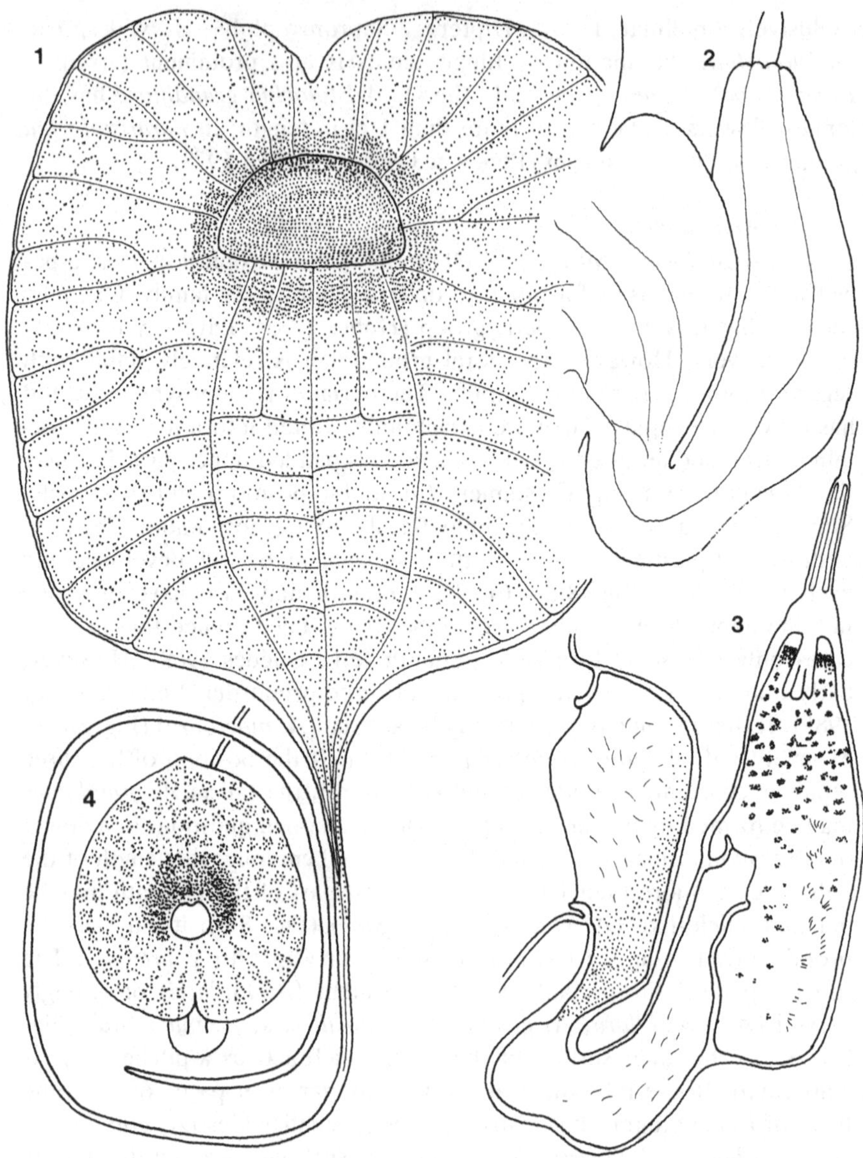

Figure 8.7. *Aristolochia* flowers. 1–3. *A. grandiflora*. 1. Frontal view, with dark floral entrance, appendix below, and incision above (× 0.3). 2. Side view of curved trap (× 0.4). 3. Longitudinal section of trap with internal narrow parts, hairs and window-pane at the base, and of the small pollination organs in the base of the trap (× 0.5). 4. *A. elegans*; frontal view, with light (yellow) floral entrance, surrounded by darker area, incision below (× 0.5).

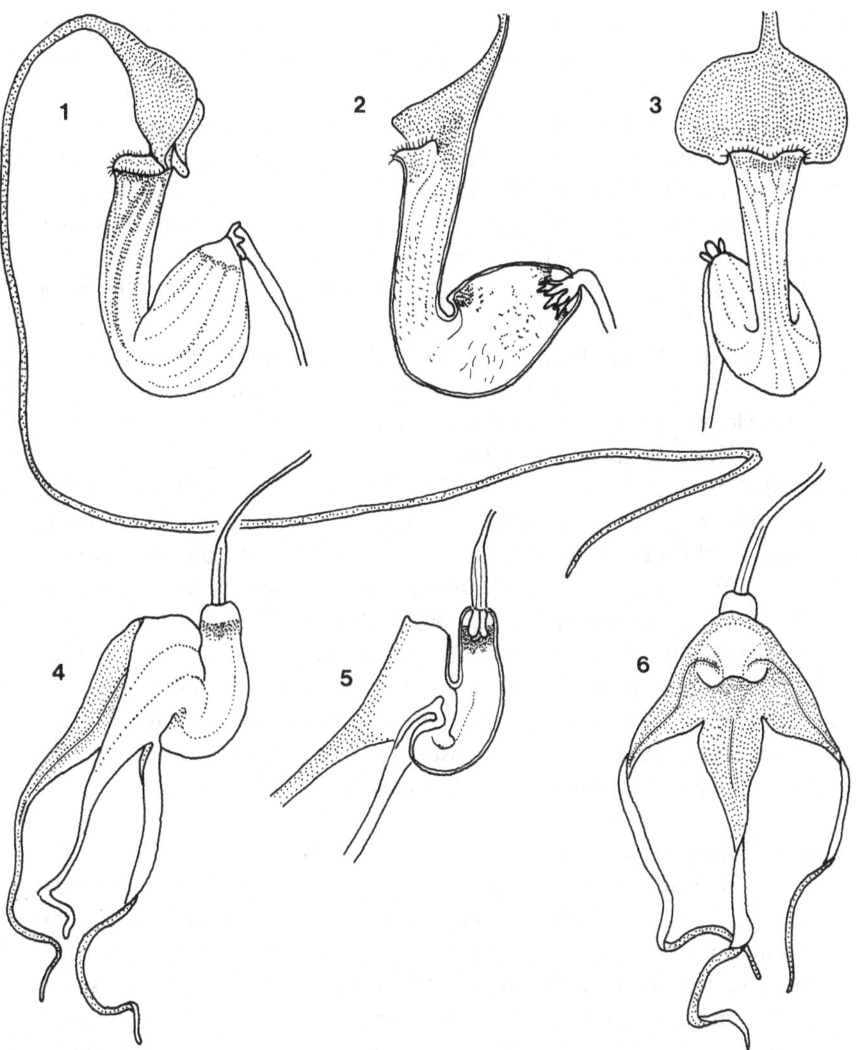

Figure 8.8. *Aristolochia* flowers (× 0.5). 1–3. *A. macroura*. 1. Side view, appendix above, trap with brown stripes and basal window-pane. 2. Longitudinal section of trap and pollination organs. 3. Frontal view. 4–6. *A. tricaudata*. 4. Side view, three appendices, one below and two lateral ones, trap with basal window-pane. 5. Longitudinal section of trap, with very narrow inner part after the entrance, and pollination organs. 6. Frontal view, with light entrance.

or like faeces (*A. macroura*, *A. lindneri*); it may also be fruit-like (*A. gigantea*) or inconspicuous (*A. elegans*, *A. tricaudata*). In *A. grandiflora* the expanded limb of the perianth emits a carrion-like smell but the floral entrance smells rather like rotten plants (personal observation). The expanded part of the perianth around the floral entrance may have a furry or velvety indument in flowers with carrion scent. It may also have conspicuous hairs with thickened ends (*A. fimbriata*, *A. redicula*). Nectar is secreted at the inner wall of the trap (e.g. *A. grandiflora*, Cammerloher 1923; *A. macroura*, Costa & Hime 1983) and, in addition, at the floral entrance (*A. fimbriata*, Daumann 1959).

Anthesis of a flower takes two days in *A. grandiflora* (first day female, second morning male, carrion smell disappearing on the second day) (Cammerloher 1923), *A. lindneri* (Cammerloher 1933), *A. gigantea* (Costa & Hime 1982) and *A. pilosa* (Wolda & Sabrowsky 1986).

In all *Aristolochia* species studied to date, flies have been recorded as pollinators: *A. grandiflora* (mainly Anthomyiidae) (Cammerloher 1923); *A. elegans* (Phoridae) (Brues 1928); *A. lindneri* (flies of 7 different groups) (Lindner 1928); *A. fimbriata*, *A. odoratissima*, *A. triangularis*, *A. giberti*, *A. elegans* (flies not specified) (Lindner 1928); *A. brasiliensis* (Syrphidae) (Noronha 1949); *A. arcuata* (Chloropidae and Milichiidae) (Brantjes 1980); *A. melanostoma* (Lauxanidae and Phoridae) (Brantjes 1980); *A. gigantea* (Phoridae) (Costa & Hime 1982); *A. macroura* (Sarcophagidae and flies of five other families) (Costa & Hime 1983); *A. pilosa* (females of Milichiidae and Chloropidae) (Wolda & Sabrowski 1986).

8.3.2 Rafflesiaceae

The Rafflesiaceae s.l. is a mainly tropical family with 8 genera and *ca.* 50 species. Whether it belongs in or near the Aristolochiales, as commonly treated (e.g. Takhtajan 1987, Thorne 1992), or whether it is a member of the Rosidae (Cronquist 1981), has not been settled. At any rate, it is an outstanding family of parasitic plants that exhibits a number of peculiar similarities in flower structure and biology with the Aristolochiaceae. The flowers vary from small (*Pilostyles*) to enormously large (*Rafflesia*). They are unisexual and have a simple, fleshy perianth with a whorl of 3–16, often 4–5, united organs (two whorls in *Sapria*). The structure of the stamens and the gynoecium is various and highly peculiar in the most specialized groups. The stamens are united into a tube or fused with the (nonfunctional) gynoecium into a gynostemium. The ovules are very small and produced in large numbers in the commonly inferior ovary with mostly parietal or highly modified placentae. The flowers are predominantly sapromyophilous.

Rafflesia

Rafflesia is a genus with 14–15 species in Eastern Asia and Malesia, famous for its giant and strange flowers and for its threatened status (Meijer 1984, Salleh & Latiff 1989). Accounts with coloured photographs of the flowers of some species have been published lately (e.g. Meijer 1985, Salleh 1991). Delpino (1873) predicted pollination of *Rafflesia* by carrion flies more than a hundred years ago; only recently was this confirmed by detailed field studies for two species, *R. pricei* and *R. kerrii* (Beaman *et al.* 1988, Bänziger 1991).

The dull red flowers are 15–90 cm in diameter, the largest being *R. arnoldii* and *R. schadenbergiana* (Meijer 1984) (Figs 8.9, 8.10). They are surrounded by several series of large bracts. The perianth forms a broad tube with a diaphragm and has five broad expanded lobes. The gynostemium in the floral centre is similar in male and female flowers, but different in detail. It is expanded into a flat disk at the end, which is slightly below the diaphragm. The 20–50 sessile anthers or the stigmatic region, respectively, are situated on the undersurface of the disc. Each anther has an irregular number of microsporangia (*ca.* 60 in *R. arnoldii*, Endress & Stumpf 1990) (Figs 2.11, 8.10.4–5). They converge toward the apex, where pollen is released through a single central pore. Pollen comes out as a slimy mush (Fig. 8.10.3). The structure of the gynoecium is poorly known. The stigma is probably formed by a velvety ring (Bänziger 1991). The inferior (superior according to Ernst & Schmid 1913) ovary contains a number of irregular cleft-like locules with laminar-diffuse placentae (Ernst & Schmid 1913). It is not clear how the internal gynoecium structures develop. Carpel numbers have not been established to date. The main difficulty is that the organs are too large for microtome and SEM studies. The ovules are highly reduced, almost orthotropous, tenuinucellar and with only a rudimentary outer integument (Ernst & Schmid 1913, Bouman & Meijer 1986).

The floral organs are adorned with peculiar structures, some of which have only now been tentatively interpreted in terms of pollination biology. The perianth lobes and diaphragm have an uneven surface with warts and pustules and are speckled with lighter areas, which may simulate infected animal tissue. The undersurface of the diaphragm has large white spots that are slightly translucent (window-panes) (Beaman *et al.* 1988, Bänziger 1991). The inner wall of the perianth cup is covered with branched black outgrowths. The upper surface of the gynostemium disk is beset with *ca.* 15–50 large processes whose apices are covered with bristles. The anthers are somewhat hidden by the reflexed rim of the gynostemium disk, and the access to each anther is channelled by hairy ribs (Fig. 8.10.2–3). The

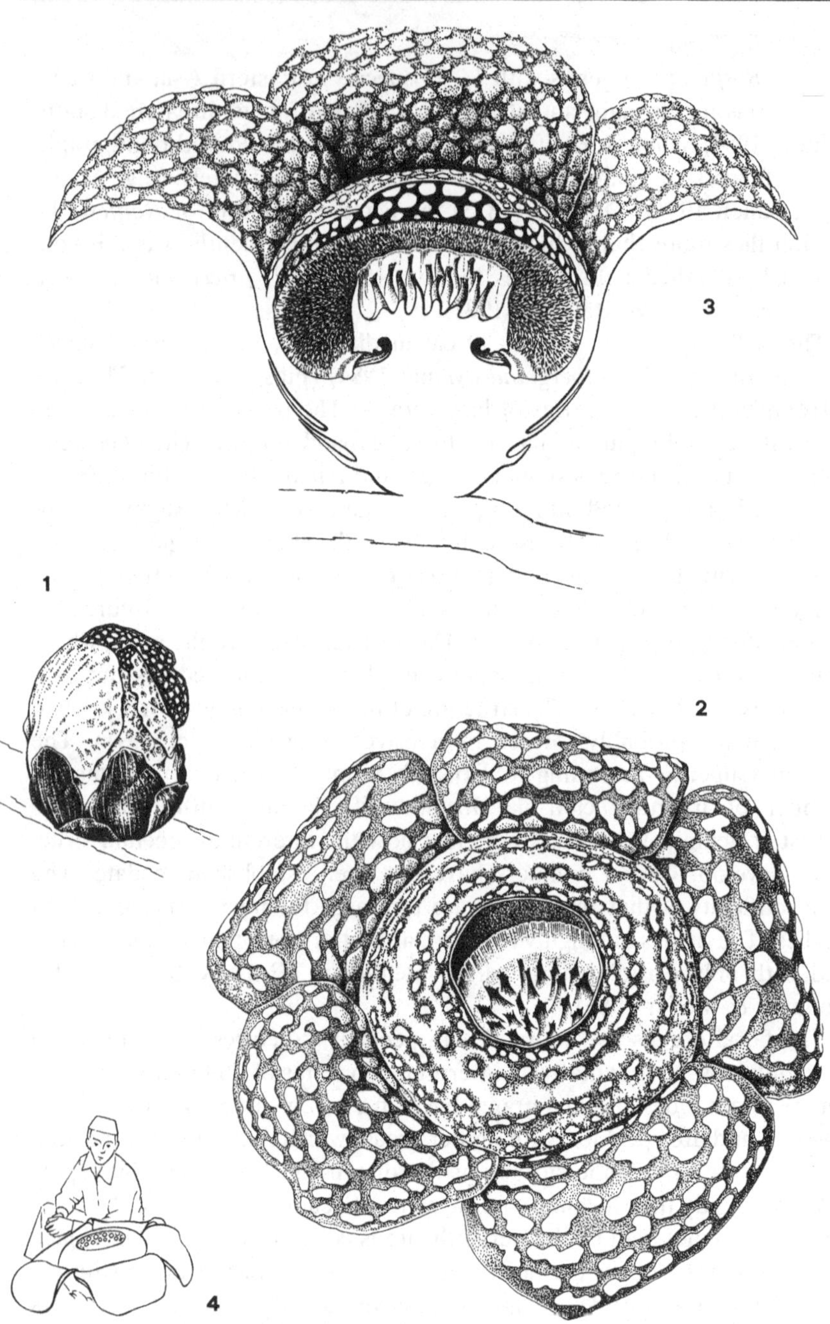
3
1
2
4

flowers emit a carrion-like smell. In *R. kerrii*, however, the smell is carrion-like only at a distance and outside the flower (produced by the perianth lobes), whereas in the chamber of the perianth cup a weaker fruity fragrance prevails, the two scents being kept separate by the diaphragm (Bänziger 1991).

Longevity of *Rafflesia* flowers has not been studied in detail but is estimated at *ca.* 5 days (Beaman *et al.* 1988, Bänziger 1991). The pollination process is similar in both species studied (Beaman *et al.* 1988, Bänziger 1991). Calliphorid flies, predominantly females, visit an open flower, first attracted by the carrion odour. A fly lands on a perianth lobe or on the diaphragm; it advances to the rim of the diaphragm and flies onto a process of the disk or onto the perianth tube wall (Fig. 8.10.1). The checker design, including the white blotches under the diaphragm, and the branched outgrowths and the processes may have a secondary, visual and tactile attractive function in conjunction with the primary olfactory cue. Whether these structures produce scent or heat or whether they may absorb rainwater in a wet environment is unknown (Beaman *et al.* 1988). The fly then crawls to the underside of the disk. The fruity smell may announce plant food and not give the stimuli for oviposition. Drops of pollen mush in the flower base and a slimy exudate from the column of the gynostemium may be taken up. Searching for more, the fly climbs up the channel until its thorax comes into contact with the pollen mush (Fig. 8.10.2). It has to retreat backwards and may search for more or fly off. In the female flower the situation is similar with the exception that there is no pollen mush and that channels near the stigma are lacking. Thus the fly can proceed in any direction and deposit pollen mush over a wide sector of the stigma. The female flower is therefore a kind of a roundabout flower (see section 3.4), in contrast to the male flower. According to Bänziger (1991) it may be emphasized that calliphorid flies may fly over distances of more than 20 km within a few days and may live for over 3 months. They may thus be able to pollinate flowers that are far apart from each other.

Are *Rafflesia* flowers deceptive? Obviously they mimic carrion. However, if the pollinators find food in the flowers and if they are not stimu-

Figure 8.9. *Rafflesia* flowers (modified after Meijer 1985 and Salleh 1991). 1. *Rafflesia keithii*; floral bud, surrounded by bracts. 2. *Rafflesia arnoldii*; open flower. 3. *Rafflesia arnoldii*; open flower in schematic longitudinal section, ovary not drawn. 4. *Rafflesia arnoldii*, showing enormous size of flower.

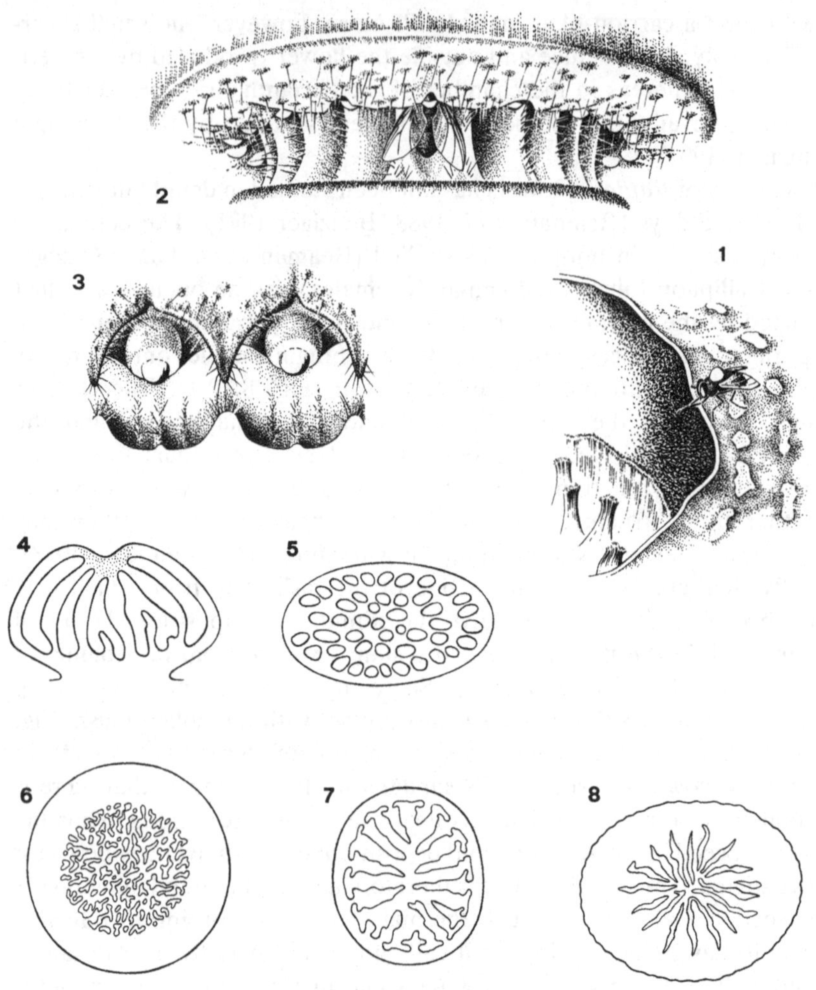

Figure 8.10. Rafflesiaceae (1–3 modified after Meijer, 1985, and Salleh, 1991). 1. *Rafflesia pricei*; carrion fly with pollen mush on its back, sitting on margin of diaphragm. 2. *Rafflesia pricei*, male flower; carrion fly below central disk in channel with anther applying a slimy pollen mush to the fly's back. 3. *Rafflesia arnoldii*, male flower; two channels with anthers presenting a slimy pollen mush. 4–5. *Rafflesia arnoldii*, stamen (× 5) (modified after Endress & Stumpf 1990). 4. Longitudinal section; with converging pollen sacs; dehiscence area stippled. 5. Transverse section. 6–8. Ovaries, transverse sections (modified after Harms 1935). 6. *Sapria himalayana*. 7. *Mitrastemon yamamotoi*. 8. *Rhizanthes zippelii*.

lated to oviposit in the flowers, they do not really parasitize the pollinators. In this sense, at least *R. kerrii* seems not to be deceptive (Bänziger 1991).

8.4 Lecythidales (Dilleniidae)

The systematic position of the Lecythidaceae is not settled. It has been removed from the Myrtales of the Rosidae and placed as a separate order Lecythidales in the Dilleniidae (Cronquist 1957, discussion by Dahlgren & Thorne 1984). Notwithstanding its separate position, there are many parallels especially in the reproductive structures, between the Lecythidaceae–Barringtonioideae and Myrtaceae or Combretaceae; a remote dissociation of these groups does not seem satisfactory.

8.4.1 Lecythidaceae

The Lecythidaceae (brazil nut family) is predominantly Neotropical and has about 280 species. The Neotropical representatives are an important woody element of the lowland rain forests; they are often large trees (Prance & Mori 1979).

The flowers are often large and conspicuous (*Gustavia* up to 20 cm diameter), solitary or often on long, pendent or arching spicate inflorescences. The most prominent feature is the androecium with a large number of basally united stamens (Fig. 8.11). Flowers of *Gustavia* have 500–1200 stamens. Apart from the androecium the flowers are commonly 4- or 6-merous (Fig. 8.12). The androecium is initiated as a flat doughnut-shaped mound around the gynoecium primordium, and the stamen primordia appear in a centrifugal direction on this mound, as studied in *Couroupita* (Hirmer 1918, Thompson 1927, Leins 1972b) (Fig. 8.13) and in other genera (Thompson 1927). The same pattern occurs in *Barringtonia* (Fig. 8.15). This centrifugal pattern was a major argument for the removal of the Lecythidaceae from the Myrtales, which have predominantly centripetal stamen initiation. The carpels are united up to the stigma and have a semi-inferior or inferior ovary with axile placentation. In some genera the calyx is congenitally united in bud and abscises as a cap or ruptures into valves as the flower opens. Petals may be present or absent. Many representatives have nectarless flowers. In Planchonioideae nectar is produced by an intrastaminal disk; in some Lecythidoideae nectar is produced by a modified part of the androecium. All these features are also common in the Myrtaceae. The flowers are white or cream to pink or purple in the Neotropical genera, white or pink in the Palaeotropical ones.

The androecium is not only conspicuous by its size and high stamen number, but it is also evolutionarily plastic; its plasticity is responsible for much of the evolutionary radiation within the family. In many cases the outer or inner series of organs are staminodes. In *Asteranthos* and *Napoleonaea* a conspicuous corona that replaces the missing petals is interpreted as the outermost series of the androecium, representing modified staminodes (Prance & Mori 1979).

In many Neotropical genera the androecium is highly monosymmetric.

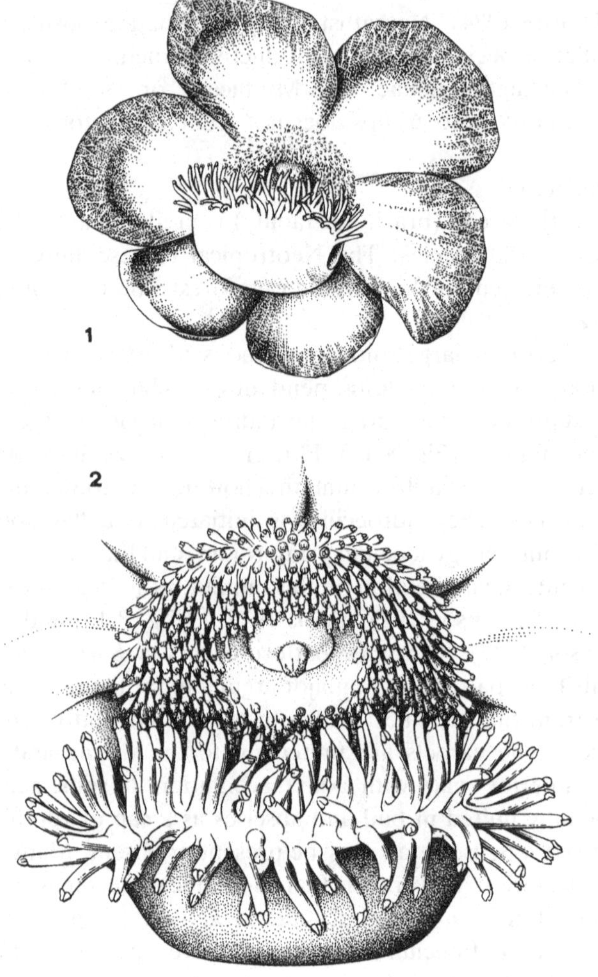

Figure 8.11. *Couroupita guianensis* (Lecythidaceae). 1. Flower in frontal view (× 0.7). 2. Androecium consisting of ring of smaller fertile stamens, surrounding the gynoecium, and basal extension of united sterile stamens (× 2).

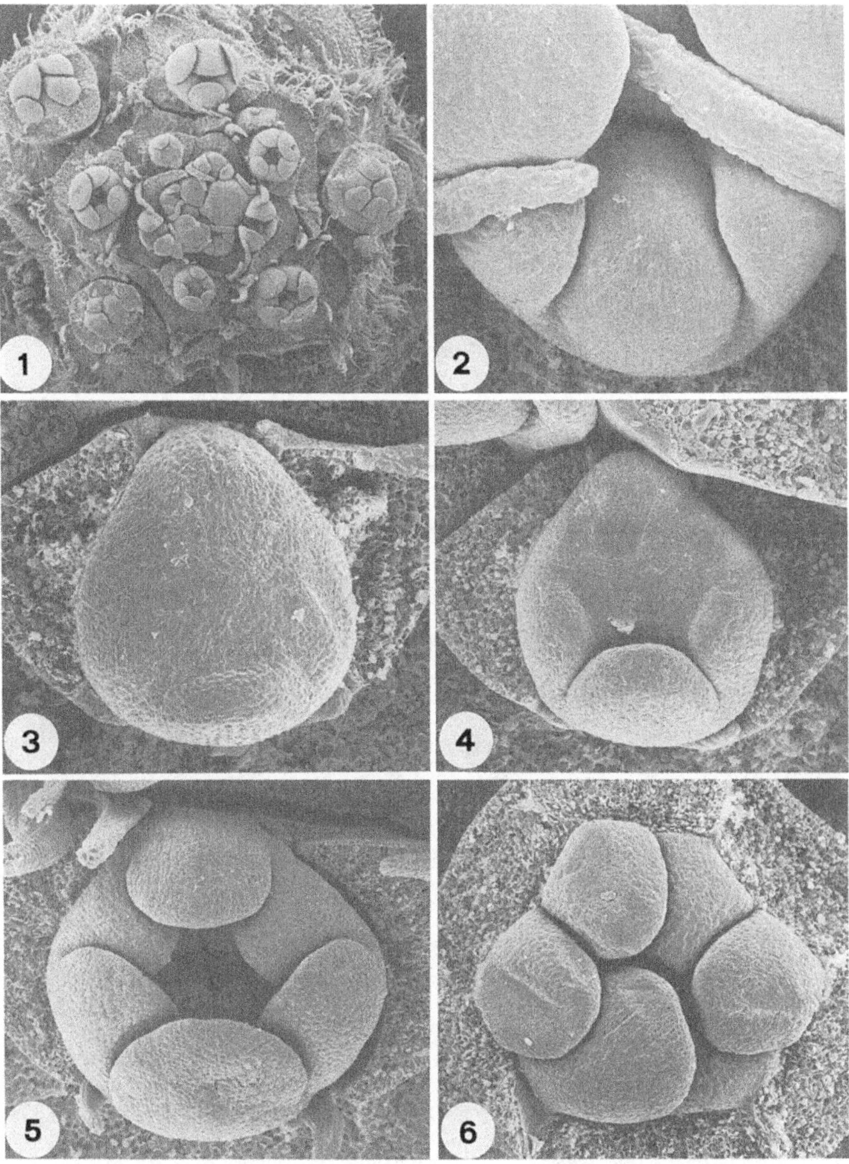

Figure 8.12. *Couroupita guianensis* (Lecythidaceae); early floral development. 1. Young inflorescence apex with spirally arranged flowers in different developmental stages (× 15). 2. Flower primordium, partly covered by two lateral bracteoles (× 150). 3. Flower primordium, abaxial sepal primordium becoming apparent (bracteoles removed) (× 150). 4. The adaxial and the two abaxial–lateral sepal primordia becoming apparent (× 110). 5. Sepal aestivation according to sequence of appearance (× 80). 6. The six petals, alternating with the sepals (sepals removed) (× 80).

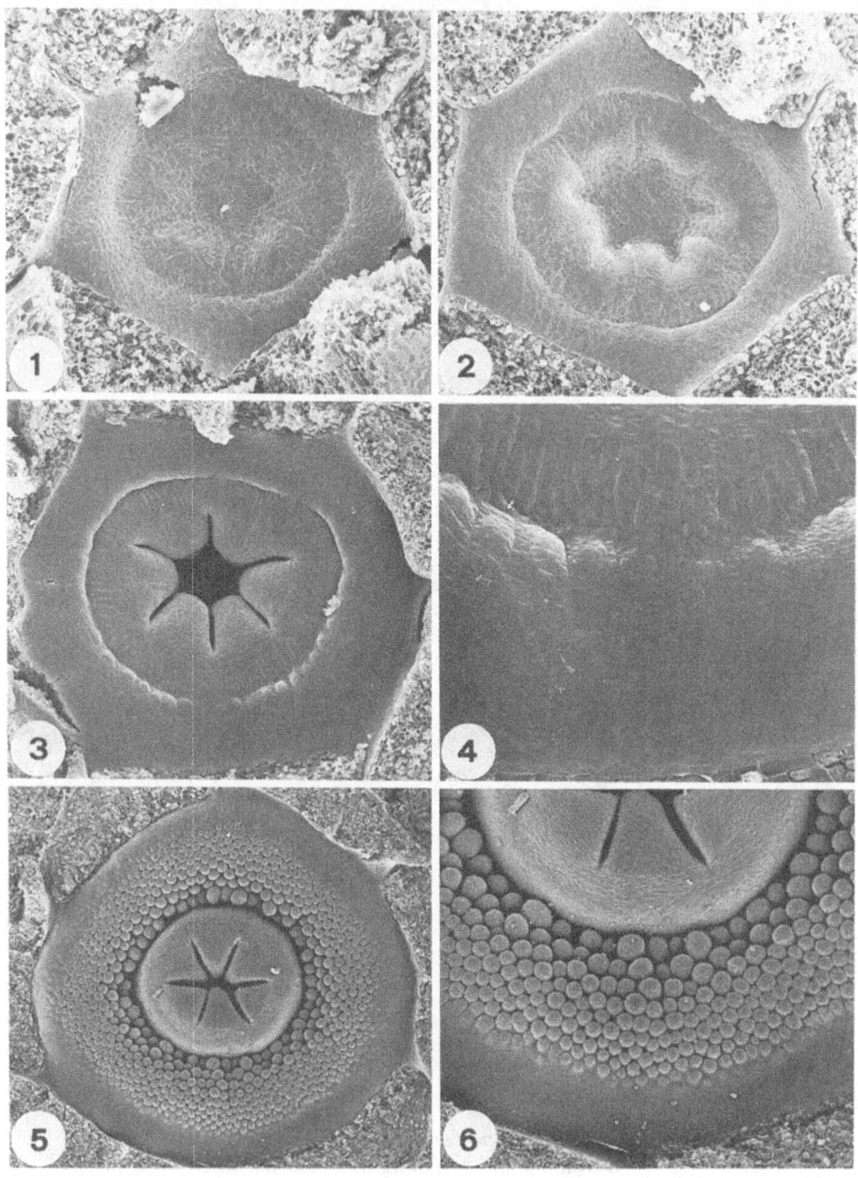

Figure 8.13. *Couroupita guianensis* (Lecythidaceae); early development of androecium and gynoecium. 1. Floral centre, androecium ring primordium surrounding the gynoecium primordium (perianth organs removed) (× 150).
2. Androecium ring primordium still undifferentiated, while individual carpels have become distinct (× 130).

In addition to the regular ring of several series of stamens around the gynoecium a median-abaxial extension of the androecium forms a large lip or hood with numerous modified fertile or sterile stamens that is often curved over the floral centre (Figs 8.11, 8.14). This hood may be quite complex. It is rolled up in *Eschweilera*, and it is twice folded in *Couratari*, and large parts are without stamens or staminodes (Prance 1976, Prance *et al.* 1983, Mori *et al.* 1978). These highly monosymmetric androecia can be derived from simple, polysymmetric forms as in *Grias*, *Gustavia* or *Allantoma* via slightly monosymmetric forms as in *Cariniana* (Prance 1976).

Important comparative studies on the floral biology of Neotropical Lecythidaceae were provided by Prance (1976), Mori *et al.* (1978), and Mori & Boeke (1987).

Large bees (*Xylocopa*, euglossines) were found to be the principal pollinators of the monosymmetric flowers (Prance 1976). The bees land on the hood and then push into the floral centre, lifting up the hinged hood and thereby brushing against the fertile stamens with the back (Prance 1976). They collect pollen from the staminodes on the hood, which are often yellow, in contrast to the white stamens. This pollen may be fertile or sterile and larger than and morphologically different from the fertile pollen, as shown for two *Couroupita* species (Mori *et al.* 1980) (see section 5.1). Some groups only produce pollen as a reward (*Grias*, *Gustavia*, *Lecythis idatimon*). In more elaborate groups nectar is produced at the base of the staminodes on the hood (e.g. *Bertholletia*, *Lecythis* p.p., *Couroupita*, *Eschweilera*).

The hood gives the flowers the architecture of lip-flowers. The hood is a landing platform. It limits entry to strong insects; it also acts as a spring which forces the bee's back against the staminal ring and ensures pollen deposition at the right place (Mori *et al.* 1978).

In *Gustavia*, which lacks a hood, the stamens on the massive synandrium are curved towards the floral centre and less force is needed to exploit the flowers; *Melipona* was observed as a main pollinator in *G. superba* (Mori & Kallunki 1976), *Bombus* in *G. hexapetala* (Prance 1976).

Couroupita is somewhat different, too. In *C. subsessilis* the hood is

Fig. 8.13 (*cont.*)

3. First stamen primordia appearing on inner rim of ring primordium, the six united carpels beginning to close (× 90). 4. Some stamen primordia from 3, magnified (× 300). 5. Appearance of stamen primordia in centrifugal sequence on ring primordium (× 35). 6. Part of 5, magnified (× 70).

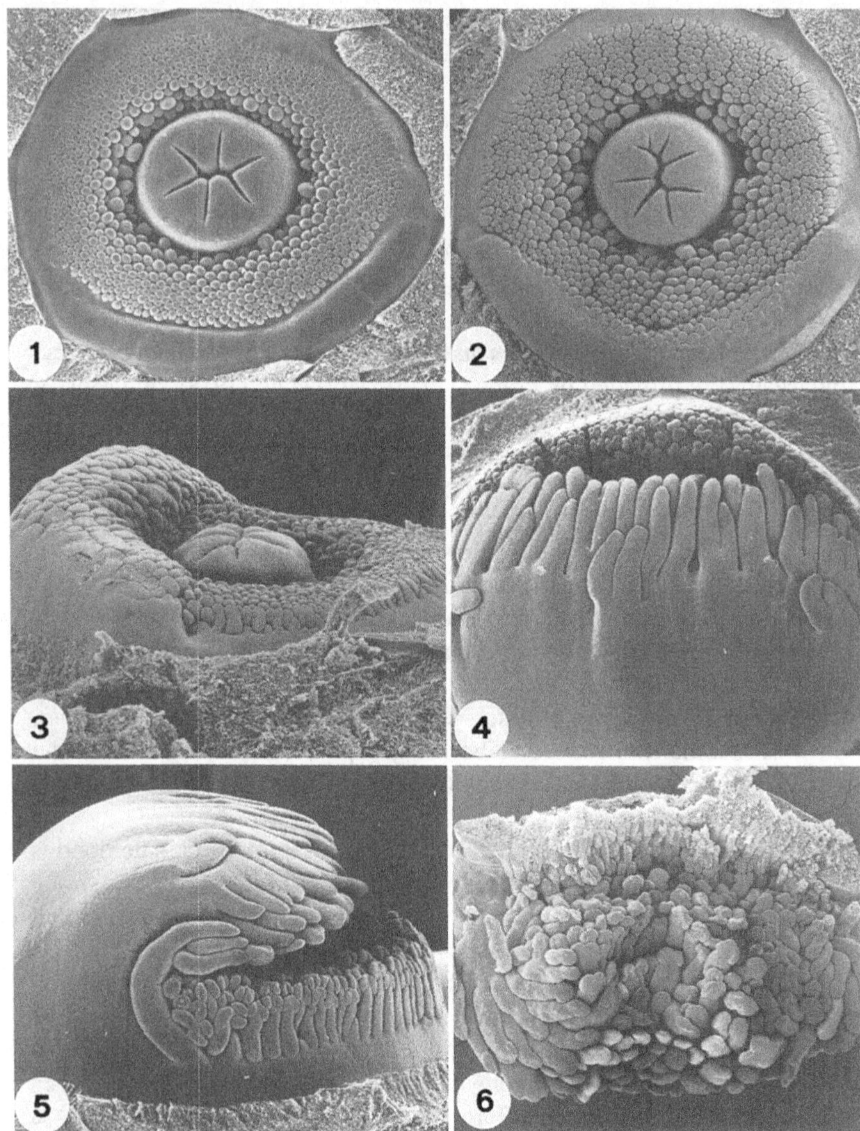

Figure 8.14. *Couroupita guianensis* (Lecythidaceae); later androecium development. 1. Androecial 'hood' becoming apparent as a crescent-shaped differentiation of the abaxial side of the ring primordium (× 30). 2. Hood with stamen primordia on inner side (× 25). 3. Hood bulging up, as seen from the side (× 20). 4–5. Later stage of hood development. 4. From above (× 15). 5. From the side (× 13). 6. Isolated hood of the same flower from below with numerous stamens (× 13).

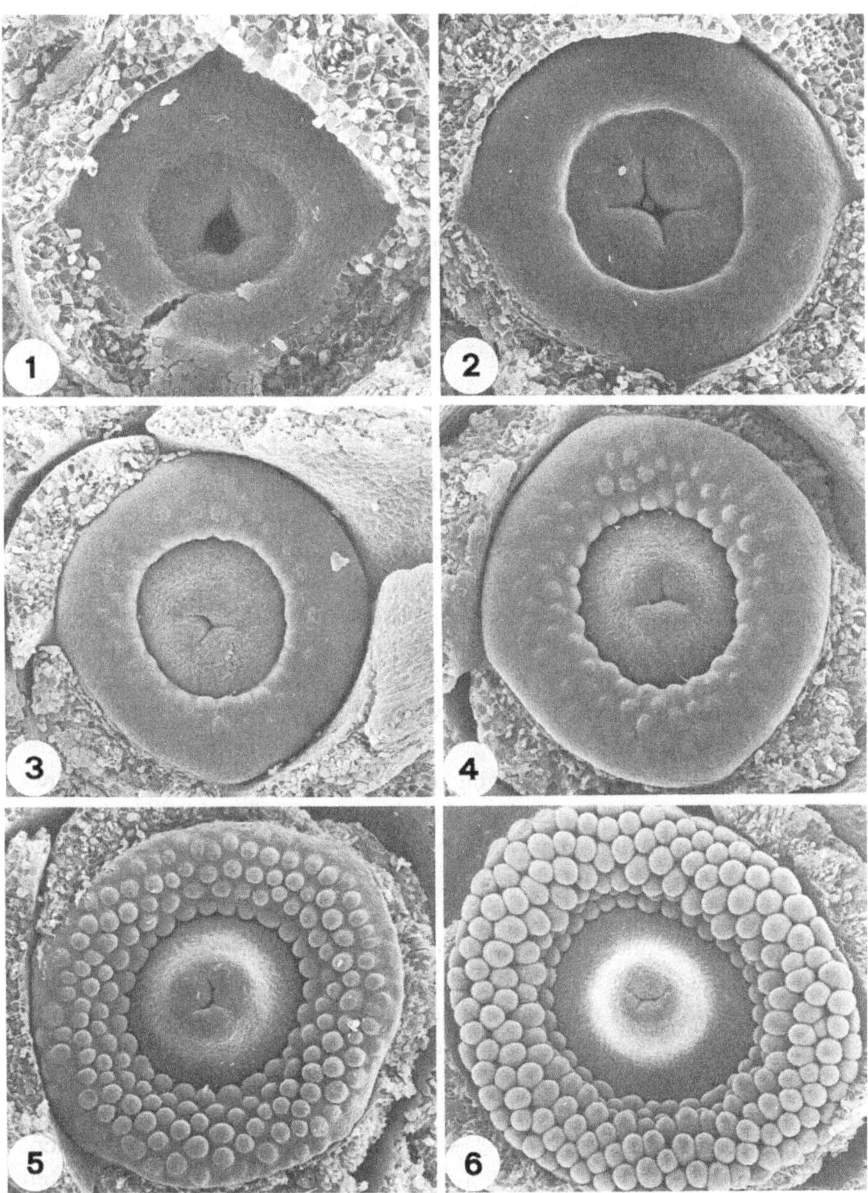

Figure 8.15. *Barringtonia* cf. *samoensis* (Lecythidaceae); androecium development. 1. Androecial ring primordium surrounding the gynoecium (× 180). 2. Androecial ring primordium still undifferentiated, while the carpels are almost closed (× 130). 3. Stamen primordia on the inner side of the ring primordium (× 100). 4. Stamen primordia appearing in centrifugal sequence on ring primordium (× 110). 5. Outermost stamens are initiated, a hood is not differentiated (× 100). 6. All stamens present (× 80).

V-shaped at the end and the floral centre is therefore not covered. Accordingly, smaller wasps and bees act as pollinators (Prance 1976). Also in *C. guianensis* the hood does not cover the floral centre but arches only loosely over it. Accordingly, it is visited by a large array of insects, although *Xylocopa* seems to be the only efficient pollinator (Ormond *et al.* 1981). The strong sweet scent is produced by the petals and by the thick apical parts of the staminodial filaments on the hood.

In *Lecythis poiteaui* the petals are green and reflexed, the androecium is white, the flowers produce abundant nectar and emit a musty-fruity odour. They are bat-pollinated, and also visited by birds, opossums and monkeys (Mori *et al.* 1978).

In Neotropical Lecythidaceae an inflorescence commonly produces many flowers, but only one or at the most two open each day on each inflorescence, thus prolonging the the flowering period over several weeks or months (Prance 1976). At the level of entire plants three strategies are present in the different groups: 'steady-state' flowering, 'big bang' flowering and 'cornucopia' flowering (see section 7.4) (Mori & Prance 1990). The flowers of most species last only one day (Prance 1976). Most species have diurnal flowers that open at dawn: e.g. the bee-pollinated *Bertholletia excelsa* opens between 0430 and 0500 and petals and stamens drop in the afternoon of the same day. Some species are nocturnal: *Couratari atrovinosa*, bee-pollinated (opening at 1800–1900, dropping off the following morning after 1100; most bee visits between 0545 and 0600); *Lecythis poiteaui*, bat-pollinated (most visits between 1900 and 2100); also *Gustavia augusta*, *Eschweilera pedicellata* and *E. grandiflora* open at night (Mori & Boeke 1987). Most species studied showed self-incompatibility (Mori & Kallunki 1976, Prance 1976), with the exception of *Couroupita guianensis* (Ormond *et al.* 1981) and *Bertholletia excelsa* (Mori & Boeke 1987).

It has been argued that among the Neotropical representatives pollen-flowers are the most primitive condition. A further step was the emergence of the hood and heteranthery with fertile stamens and fodder stamens. The advent of nectar production in this type of androecium, such as in *Lecythis corrugata*, led, then, to loss of fodder stamens and more efficient nectar production in the most advanced groups, which are pollinated mainly by euglossine bees (Mori *et al.* 1978). Evolution in another direction led to bat-flowers.

It would be interesting to follow up the question of a relationship between euglossine and other large bees and monosymmetric flowers in the family (Prance & Mori 1987). Brazil nut (*Bertholletia excelsa*) plantations have not been successful, perhaps because the euglossine and other large bees that are pollinators of the species in the wild are lacking (Nelson *et al.* 1985, Prance & Mori 1987, Mori & Prance 1990).

The Old World genera *Barringtonia* and *Planchonia* have more generalized brush-flowers. They are nocturnal, produce abundant nectar and have a sweetish-musty smell (personal observation). In *Barringtonia calyptrata* the flowers have about 170 white stamens. They are visited by a wide array of animals: bats at night, lorikeets and bees in the morning (see section 3.4). In contrast to the Neotropical groups, in *B. calyptrata* and *B. racemosa* several flowers may be open on one inflorescence at the same time (see also Weber 1986). In *Planchonia careya* the flowers open synchronously at dusk. They are white with red stamen bases. There are about 250 stamens and 50 inner staminodes. The next morning at dawn most flowers have fallen.

8.5 Violales (Dilleniidae)

8.5.1 Passifloraceae

Passiflora

Passiflora is a large genus with about 400 species of predominantly Neotropical vines. Its curious flowers have long fascinated morphologists and naturalists (e.g. Masters 1871) (Figs 8.16, 8.17). They are basically pentamerous with two perianth whorls and a stamen whorl; the gynoecium is trimerous and the ovary has three parietal placentae as in many other Violales (Fig. 2.33.1). Stamens and ovary are raised on an androgynophore. The three stylar branches bear three capitate stigmas with dry, unicellular papillae (Fig. 2.27.1). The most peculiar organ is the corona between perianth and stamens, which consists of several series of numerous narrow parts, often thread-like or forming folded rings (frills). Puri (1948), from a comparative anatomical point of view, and de Wilde (1974), from a systematical point of view, regard the outer parts of the corona as emergences of the perianth, the inner parts as staminodial. However, an evolutionary interpretation may be that all corona parts are multiplied staminodes of an originally polyandrous androecium with centrifugal stamen inception. The centrifugal androecium pattern is common in the Dilleniidae, and with this pattern there is always a tendency for the last-formed organs to become staminodes instead of stamens (see also section 8.4.1). A feature that may have encumbered such an interpretation is the separation of the stamens from the corona by an androgynophore (Fig. 8.18). However, in more simple-flowered relatives of the genus *Passiflora*, such as *Adenia*, *Basananthe* and *Deidamia*, an androgynophore is not present (de Wilde 1974). The much-needed comparative developmental study of the flowers is still wanting for the family.

Lindman (1906) discussed diversity and functional aspects of the different parts of the corona. The two inner series of the corona ('limen', innermost) and ('operculum', second inner) form two frills that encompass and cover the nectar disk and the nectar room from inside and outside.

Figure 8.16. *Passiflora* flowers; bee-pollinated species (× 1). 1. *P. quadrangularis*. 2. *P. rubra*. 3. *P. holosericea*. 4. *P. suberosa*.

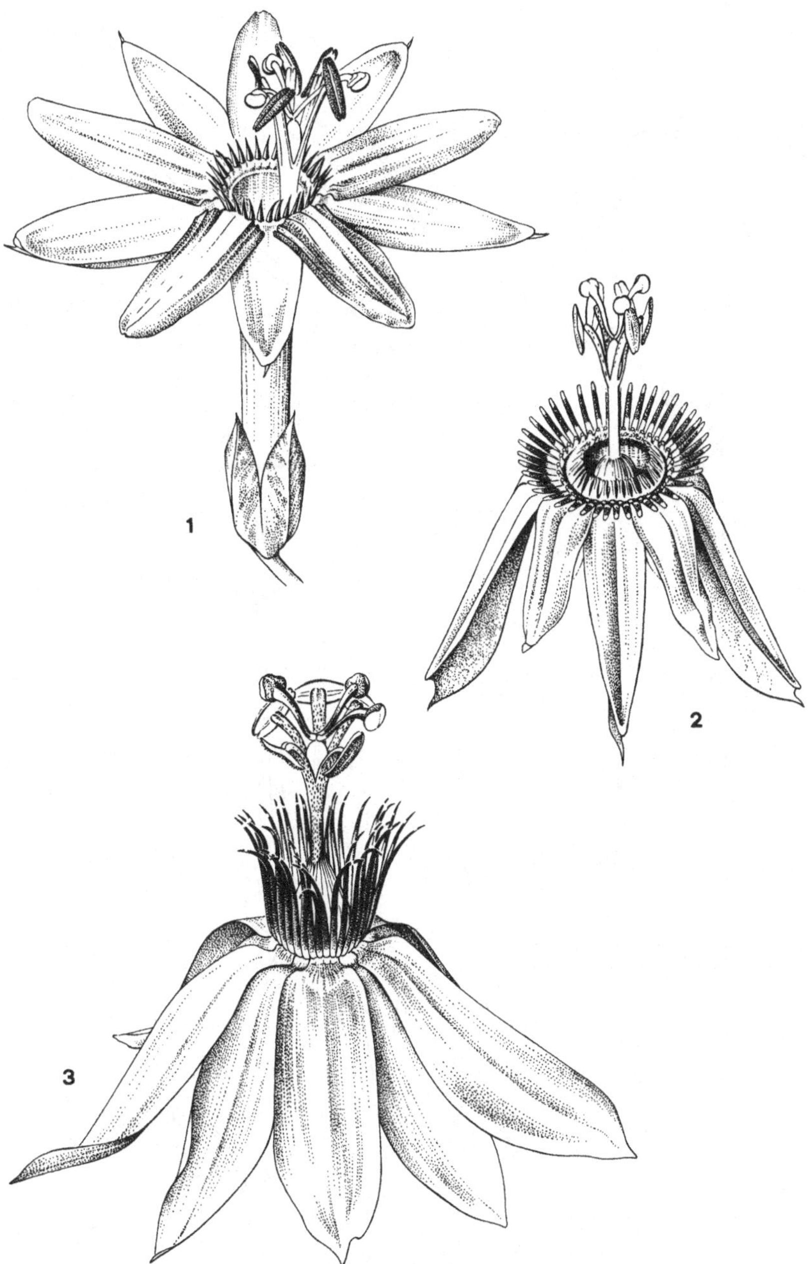

Figure 8.17. *Passiflora* flowers; bird-pollinated species (× 1). 1. *P*. sp. (subg. *Tacsonia*). 2. *P. racemosa*. 3. *P. coccinea*.

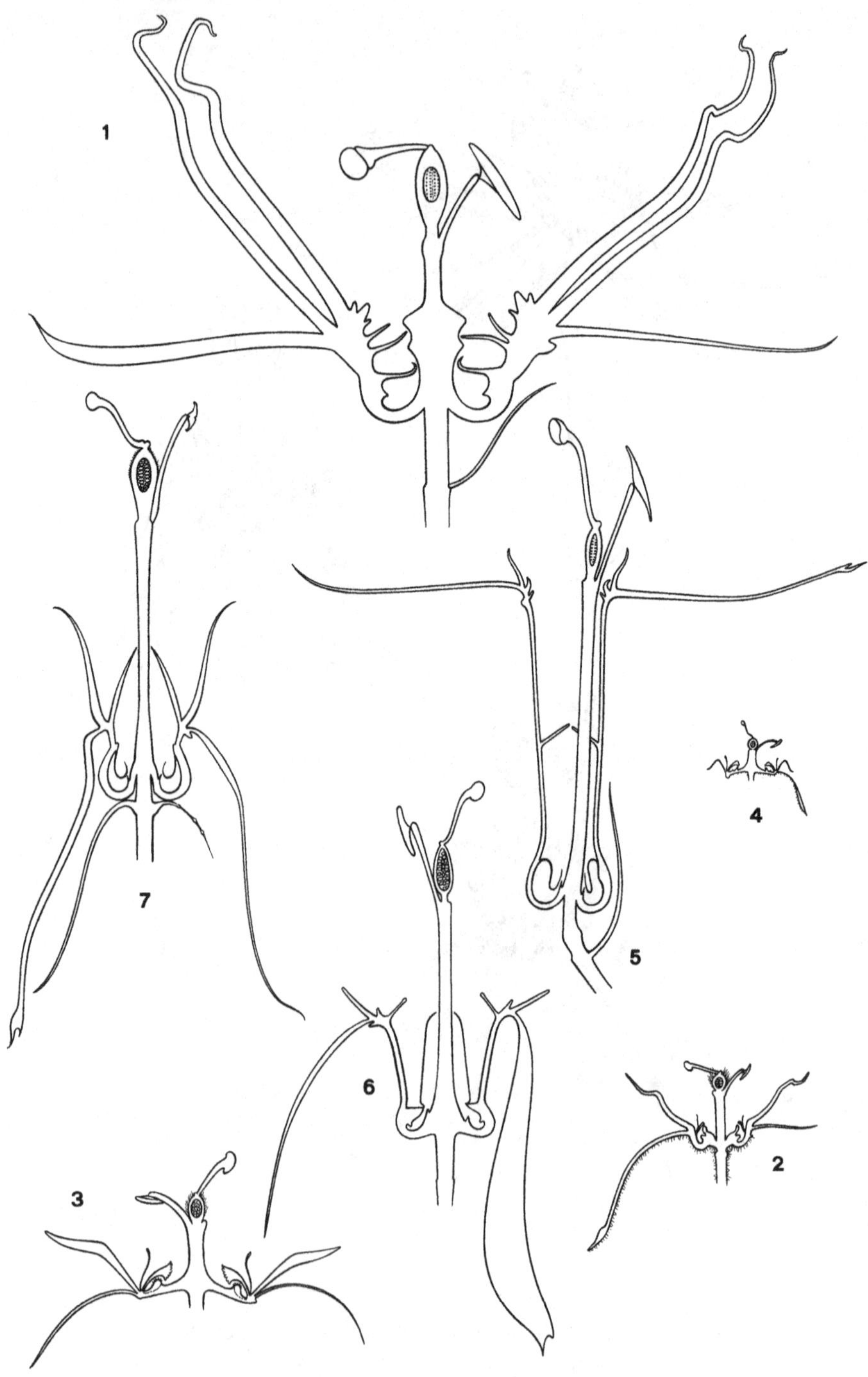
1
7
5
4
6
2
3

The third inner series may also be involved in nectar protection. Thus, in many species nectar is available only to large bees; other pollinators are excluded. The floral cup may form a conspicuous trough in the nectariferous region. One or more outer series of the corona are on top of the floral cup. They form rings of fringes that are often attractive by contrasting colours. In some bee-pollinated species the outermost series may become very long, even longer than the petals (e.g. *P. quadrangularis*). On the other hand, in some bird-pollinated species they are much reduced (e.g. subg. *Tacsonia*).

The floral groundplan is the same in both large and small flowers. The diameter of flowers may be smaller than 1 cm in *P. suberosa* and may reach 17 cm in *P. vitifolia* (Holm-Nielsen *et al.* 1988). However, the construction varies from shallow, disk-shaped to salverform or funnel-shaped flowers with long floral tubes, the former in both large and small flowers, the latter only in large flowers. The funnel-shaped flowers of *P. harlingii* may attain 18 cm in length; in the salverform flowers of *P. mixta* the floral cup reaches 15 cm length and, in addition, the free parts of the petals are 5 cm long (Holm-Nielsen *et al.* 1988). The differentiation of the corona is diverse in flowers of different shapes and sizes. Further, in the salverform flowers with long tubes of subg. *Tacsonia* (Figs 8.17.1, 8.18.5), and the funnel-shaped flowers of subg. *Rathea*, the three bracts (bracteoles) below the flower have come so close to the floral base that they simulate a calyx. The sepals of most *Passiflora* species are about the same size and colour as the petals. However, they commonly have a prominent dorsal tip, which does not occur in the petals (see section 2.2.3). In some small-flowered species the petals are wanting (e.g. *Passiflora suberosa*).

The floral mode also varies (Vogel 1990c, Fig. 10). Probably most species, at least those with blue and white flowers, are pollinated by large bees (Croat 1978). *Xylocopa* plays an outstanding role (Frankie & Vinson 1977, Corbet & Willmer 1980, I. & M. Sazima 1989). Hermit hummingbirds are pollinators of very different floral forms within the genus. *Passiflora mixta* (subg. *Tacsonia*), with floral tubes up to 15 cm long, is pollinated by the Andean swordbill (*Ensifera*), which has a very long bill (D. & B. Snow 1980). The scarlet flowers of *P. coccinea* have only short floral tubes. However, the corona collar is stout and upright; in addition to excluding nectar thieves, it controls the posture of the nectar-feeding

Figure 8.18. *Passiflora* flowers; schematic longitudinal sections (× 1). 1. *P. quadrangularis*. 2. *P. rubra*. 3. *P. holosericea*. 4. *P. suberosa*. 5. *P.* sp. (subg. *Tacsonia*). 6. *P. racemosa*. 7. *P. coccinea*.

hermits (*Phaethornis*) in order to hold the head in the right position to brush against the pollination organs (Skutch 1980). The flowers may face upwards or downwards. *Passiflora vitifolia* is also pollinated by *Phaethornis* (Gill *et al.* 1982). It is possible that butterfly pollination also occurs, as observed for *P. kermesina* and *P. coccinea* (Benson *et al.* 1976). Especially interesting are species where both bees and hummingbirds may be effective pollinators, such as *P. edulis* (Corbet & Willmer 1980). Another extreme is the bat-pollinated *P. mucronata* (M. & I. Sazima 1978). The whitish flowers are about 8 cm wide. The pollination apparatus is monosymmetric.

The following case studies may give an indication of the variations within the basic behaviour of *Passiflora* flowers, before some general conclusions are drawn.

Passiflora foetida

Janzen (1968) gives a detailed account of the flowering behaviour of this species in Mexico and other parts of Central America. It is a small vine that flowers near the ground. On each major branch of the vine a new flower opened at about 0530 within 10–60 seconds. Flower opening was highly synchronized on a single plant and among the plants in a small area. The flowers were visited and usually pollinated by at least three species of large bees of the genus *Ptiloglossa* (Colletidae). These bees were hovering and flying over the plants before the flowers opened. Within 1–15 seconds of opening, virtually every flower either had a bee on it or had been visited once. The flight activity period lasted 45 minutes after the flowers opened. Each flower was visited 10–18 times during this period. The bees landed on the horizontal platform of petals and corona, paused for 1–10 seconds, then placed the mouth parts in the doughnut-shaped trough around the base of the ovary and moved around the flower. The flowers are protandrous. At first the styles are upright and bees do not come in contact with them. Only the dehisced anthers are bent downwards, and the bee's back brushes pollen out of the anthers, as it moves around the flower. Only 12 minutes after flower opening the styles began to bend downward, and after 4–10 minutes were in the position to receive pollen. As early as 45 minutes after flower opening all bee activity had ceased. The petals were wilted by 1100. A female *Ptiloglossa* visited at least 14 flowers before contacting one with receptive stigmas. Even if the plants are self-compatible, the probability of a flower being fertilized by its own pollen is relatively small. Thus, there is an efficient outcrossing mechanism.

This species was also studied by Gottsberger *et al.* (1988) in Brazil. Most flowers opened between 0415 and 0445 and began to wilt at 0730. At this time visits by pollinators ceased. The main flower visitors were a *Xylocopa* and two *Centris* species. The corona produces floral scent and is also a kind of a lever to open the access to the nectar. When a larger bee pushes toward the stiff stamens, the corona is bent downwards and so opens the nectar trough. One third of the flowers were functionally male in being smaller and lacking the stylar movements. Self-incompatibility was found to occur. Raju (1954) observed autogamy of this species in a Botanical Garden in India.

Passiflora edulis

This species, a crop plant, was studied in a plantation in the West Indies by Corbet & Willmer (1980). The flowers begin to open shortly after noon and close during the night or early the following morning. Flower opening is rapid and synchronized. The anthers dehisce before the perianth segments separate. Within a few minutes the anthers have reached their final position. Only after an hour do the stigmas curve downward in a position to receive pollen. Later the styles turn again toward an erect position. A *Xylocopa* species and three species of hummingbird were observed as potential pollinators. The plants studied were self-incompatible. Another study carried out in Brazil showed the flower to open at 1300–1400 and to close at *ca.* 2200 (I. & M. Sazima 1989). Two species of *Xylocopa* were effective pollinators, while smaller bees (*Trigona*) acted as pollen and nectar thieves. Here the plants were also found to be self-incompatible.

Passiflora vitifolia

This hummingbird-pollinated species was studied in Costa Rica by Janzen (1968). It is a large vine with flowers as high as 30 m up in the canopy. The flowers are exposed from above. Compared with *P. foetida*, anthesis is longer and the timing less precise. Flowers open between 0550 and 0730 and wilt between 1400 and 1800. The bending downward of the stigma begins 30–180 minutes after flower opening, and the movement requires up to 120 minutes. Nectar content of a flower is about 100 times more than in *P. foetida*. Hummingbirds of at least four species were observed visiting flowers. Plants are self-incompatible (East 1940).

Passiflora coccinea (Figs. 8.17.3, 8.18.7)

Another hummingbird-pollinated species, observed by Skutch (1980) in Costa Rica. It is a vine that grows 10–15 m high. It is more abundant in secondary growth than in primary forest. The flowers start opening at 0300. After 0500 the flowers are more or less open and the anthers have dehisced. They are visited by hummingbirds throughout the day. Anthesis ends in the evening or at night. *Phaethornis* was observed as a floral visitor.

Passiflora mucronata

Observations of this bat-pollinated species were made in Brazil by M. & I. Sazima (1978). It is a vine that covers low vegetation of a few metres in height. The flowers open between 0100 and 0200 and wilt by 0700 to 1000. The opening is explosive and takes only 15 seconds. Opening is synchronous in an individual within about 15 minutes. The anthers are already open as the flower opens. At first they curve radially outwards and then they all move towards one side, away from the foliage. Thus the pollination apparatus becomes monosymmetric and is directed towards a potential bat visitor. In addition, the androgynophore is slightly arching. Pollen application is on the head. During the sideways movement of the anthers the styles begin to curve downward to position the stigmas between the anthers. About 15 minutes after flower opening two species of bat arrived. The flower observed to be visited most often received five visits. Nocturnal visits of sphingids and wasps and diurnal visits of bees and birds were also observed. The species seems to be self-incompatible.

However large the diversity between small- and large-flowered species and even among the large-flowered species may be, there are some consistent features in the behaviour of the flowers, as shown by these case studies and already discussed by Janzen (1968). There is rhythmicity in flowering. A small number of flowers is produced regularly each day on an individual for periods of two (*P. vitifolia*) to nine months (*P. foetida*) (see also Skutch 1980 for *P. coccinea*). Anthesis of a flower is less than a day. Timing of opening can be precise (*P. foetida*) (Janzen 1968). The species are not rare but widely scattered. They are pollinated by animals that fly large distances on regular rounds. The concept of traplining (see section 7.5) was developed with the observations on *Passiflora* and solitary bees (Janzen 1968, 1971, 1974b). Other *Passiflora* pollinators, hermit hummingbirds and bats, are also typical trapliners.

The flowers show specific movements of stigmas and anthers, which favour allogamy. The sequence of these movements may be different between species. Protandry is common (detected by Sprengel, 1793, in *P. coerulea*) but perhaps also protogyny may occur, as reported for *P. quadrangularis* (Moncur 1988). Large-flowered *Passiflora* species are self-incompatible. The reduced, small-flowered, inconspicuous *P. rutilans* (and other species) is self-compatible (Lewis 1966). In the widely cultivated *P. edulis* with large flowers, self-incompatible and self-compatible races occur (Moncur 1988), genotypes with different degrees of style curvature were also observed (Ruggiero & Andrade 1989).

The anthers are introrse, but by a backward curvature of the filament and backward tilting of the anthers they are directed to the floral periphery and downwards at anthesis. Furthermore, in bee-pollinated species, they become rotated through 90° between filament and connective and thus change from a radial to a tangential direction. This anther torsion comes about by asymmetric differential growth in the transition region between filament and connective (Troll 1922, Moncur 1988). Interestingly, in hummingbird-pollinated species, such as *P. coccinea* and *P. mixta*, and in the bat-pollinated *P. mucronata*, the anthers remain radially directed in the open flower. This difference corresponds with the different approach directions of the smaller bees and the larger hummingbirds and bats. Bees alight on the flower and move tangentially around the flower; they may make at least one complete circuit ('roundabout flowers'). Hummingbirds and bats, in contrast, approach the flower from a radial direction and pollinate it while hovering.

The same ordered sequence of movements and the short anthesis that occurs in all species studied seems to constitute a programmed set of fixed events in the genus *Passiflora*. What changes is the timing of anthesis onset and the duration of anthesis (within the limits of a day), and in coordination, the speed of the floral movements. Most interesting are Janzen's (1968) observations of *P. foetida* in Mexico, where the entire programme is compressed into a very short time, and the onset of anthesis and pollination is extremely precise and synchronous. Janzen (1968) argues that its symbiosis with *Ptiloglossa* may form a coevolved system with the *Ptiloglossa* flying earlier and thereby reducing competition with other bees, and *P. foetida* flowering earlier and more precisely and increasing the pollinating efficiency of the bees. This is most interesting in view of the somewhat longer anthesis duration of this species as observed in Brazil by Gottsberger *et al.* (1988), where it is pollinated by other bees.

8.6 Fabales (Rosidae)

The Fabales are an extremely successful group. Each of its three families contains several thousand species, the Fabaceae, with about 10 000 species, being the third largest family of the angiosperms. Part of its success is due to its very elaborate floral constructions. The Caesalpiniaceae and Mimosaceae are almost confined to the tropics, but the worldwide family Fabaceae also has its greatest diversity in the tropics. Their fascinating biology and the occurrence of a large number of crop plants in the group have attracted many biologists. No other big group of the angiosperms has been given such balanced treatments of systematics and evolutionary aspects as the Fabales. This is especially well demonstrated by the compendia edited by Polhill & Raven (1981), Stirton (1987), Stirton & Zarucchi (1989), Herendeen & Dilcher (1992) and Ferguson & Tucker (1994). Further, in no other big group has there been such a broad comparative study of floral development (Tucker 1987, 1989, Tucker & Douglas 1994, and other works of Tucker's school). On the other hand the immense gaps in our knowledge about the whole biology of the flowers also become apparent in this group (e.g. Schrire 1989). It therefore seems sensible to treat the Fabales here in a more detailed manner than some of the other groups.

Although the extreme floral forms are superficially very different, the three families share many features in floral structure (Fig. 8.19). A unicarpellate gynoecium in otherwise pentamerous flowers with a double perianth and two whorls of stamens are common. In Caesalpiniaceae and Fabaceae monosymmetric flowers with a characteristic imbricate petal aestivation (ascending in Caesalpiniaceae, descending in Fabaceae) abound. In both families the uppermost of the (usually) five petals is the median one. In contrast, in Mimosaceae the petals are mostly valvate, and the lowermost petal is the median one (Tucker 1989). The order of initiation of sepals and petals in the three families is not uniform. In each family a

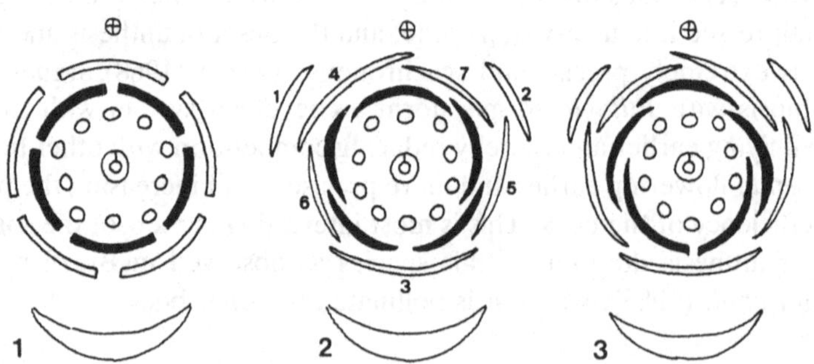

Figure 8.19. Floral diagrams of Fabales. Black, petals; crossed circle, inflorescence axis; lowermost organ in diagrams, subtending bract. In 2 and 3 two lateral bracteoles outside the flower are present; in 2 the initiation sequence of the two bracteoles and five sepals is indicated (modified after Tucker 1989). 1. Mimosaceae. 2. Caesalpiniaceae. 3. Fabaceae.

smaller or larger piece of the potential spectrum is realized. Fabaceae are more uniform than the other two families in this respect. Sepal initiation is spiral (helical) or unidirectional (from abaxial to adaxial) in Caesalpiniaceae, simultaneous or spiral in Mimosaceae, but uniformly unidirectional in Fabaceae. Petal initiation is spiral or unidirectional in Caesalpiniaceae, simultaneous in Mimosaceae and unidirectional in Fabaceae (Tucker 1989). Thus, floral monosymmetry is established at an early stage. Stamens are often more or less united. The anthers are pronouncedly introrse; in Fabaceae this is combined with secondary pollen presentation (Endress & Stumpf 1991). The carpel is often long, slender and stipitate. It is predominantly epeltate (Leinfellner 1969b, 1970). The ovules alternate in two rows along the two carpel margins, as is familiar from bean pods. They are variously curved and become pronouncedly campylotropous in some Fabaceae. Nectar is commonly produced by a disk or collar around the gynoecium base (Waddle & Lersten 1974).

Mimosaceae and Fabaceae are the most divergent families within the order. In Mimosaceae the stamens (filaments) are strongly developed at the expense of the perianth and become almost the only optically attractive organs (brush-flowers). In contrast, in most Fabaceae, with their 'papilionate' flowers, the stamens do not take part at all in the optical apparatus in that they are included (together with the gynoecium) in the two lowermost petals, which form the 'keel' ('carina'); often the other pair of petals, the 'wings', are also mechanically linked with the keel so that a complicated mechanical apparatus is formed by gynoecium, androecium and four of the five petals (see section 3.6).

Because of the descending corolla aestivation in Fabaceae, the upper (median) petal is the outermost one. It is mostly also the largest one and is frontally expanded; it is the 'standard' or 'flag' petal (or vexillum). In bee-pollinated species it often has contrasting colours (yellow markings or convergent lines at the base of the exposed part) so that it is the most attractive of the five petals with a mainly advertising function. Because of the ascending corolla aestivation in Caesalpiniaceae, in contrast, the upper petal (standard) is the innermost one. Although it is mostly not larger (often smaller!) than the outer ones, it may also have contrasting colours, and it shows other manifold elaborations connected with its innermost position, which will be explained later.

Let us focus now on the innermost petals of Caesalpiniaceae and Fabaceae. In both families they are especially flexible allowing them to become involved in special elaborations, notably synorganizations with the androecium and gynoecium. In Fabaceae they form the keel, as we have seen, which encloses the pollination organs and coacts in various ways with them (see section 8.6.3 with the classical four cases of Delpino, 1867) (Fig. 3.7). In Caesalpiniaceae, as already mentioned, the innermost petal is the standard petal. Its various elaborations and possible synorganizations with the androecium are described in the next section.

Unexpectedly, the difference in position of the standard petal and keel petals in Caesalpiniaceae and Fabaceae is not due to a different organogenetic sequence. In both families the keel petals are the first initiated and the standard petal the last initiated. The difference arises only during later development, and this is possible because of the narrow young petals whose margins cover each other only in relatively late developmental stages (Tucker 1987).

In almost all Fabales nectar is concealed (if produced at all), except for the widely open flowers of *Ceratonia* and *Dialium* (Caesalpiniaceae). In addition, pollen is concealed in the Fabaceae, but in some way also in *Cassia* s.l., where it is contained in more or less tubular anthers (in *Cassia* s.l. no nectar is produced). In Caesalpiniaceae and Mimosaceae there is a tendency to hold pollen grains together, in some Caesalpiniaceae by viscin threads, in Mimosaceae by the formation of polyads.

Self-incompatibility occurs in representatives of all three families, although in different proportions. It is more widespread in Caesalpiniaceae and Mimosaceae than in Fabaceae. Mainly in Caesalpiniaceae and Mimosaceae there is high level of fruit abortion (Bawa & Buckley 1989).

It is assumed that bee-pollination was important in the divergence of papilionoid and mimosoid legumes from caesalpinioid ancestors. The presence of fossil flowers of both mimosoids and papilionids at the Palaeocene–Eocene boundary suggests that diversification of Fabales began in the Late Cretaceous or earliest Tertiary (Crepet & Taylor 1985). Among extant Fabales the least specialized flowers may be those of *Ceratonia* or *Gleditsia* (Polhill *et al.* 1981). In *Ceratonia* they are unisexual and widely open with a large nectary disk (Tucker 1992a), and have a superficial resemblance to those of some Sapindaceae. In *Gleditsia* they are also unisexual and have variable numbers and chaotic position of organs (Tucker 1991). Similarities of *Ouacapoua* (Caesalpinieae) with Connaraceae have also been mentioned (Dickison 1981).

8.6.1 Caesalpiniaceae

Corner (1964), in *The life of plants* (p. 203), writes: 'Among dicotyledons, the most wonderful array [of zygomorphic flowers] occurs in the caesalpinioid trees and climbers of the bean family (Leguminosae). Here belong *Cassia*, *Bauhinia*, *Amherstia*, and a hundred more genera of the tropical forests, displaying the zygomorphic specialization along several lines to end in three, two or one stamens.' The most spectacular series of floral elaboration and reduction are in the tribes Detarieae and Amherstieae (Cowan & Polhill 1981).

Indeed, the flowers of some Caesalpiniaceae are extremely attractive not only by their size and coloration but also by their elegant shapes (Fig. 8.20). This is mainly due to the fact that the stamens and gynoecium are not secluded but take part in the outer shape of the flower. In genera like *Bauhinia*, *Delonix* and *Amherstia*, the petals are also pronouncedly

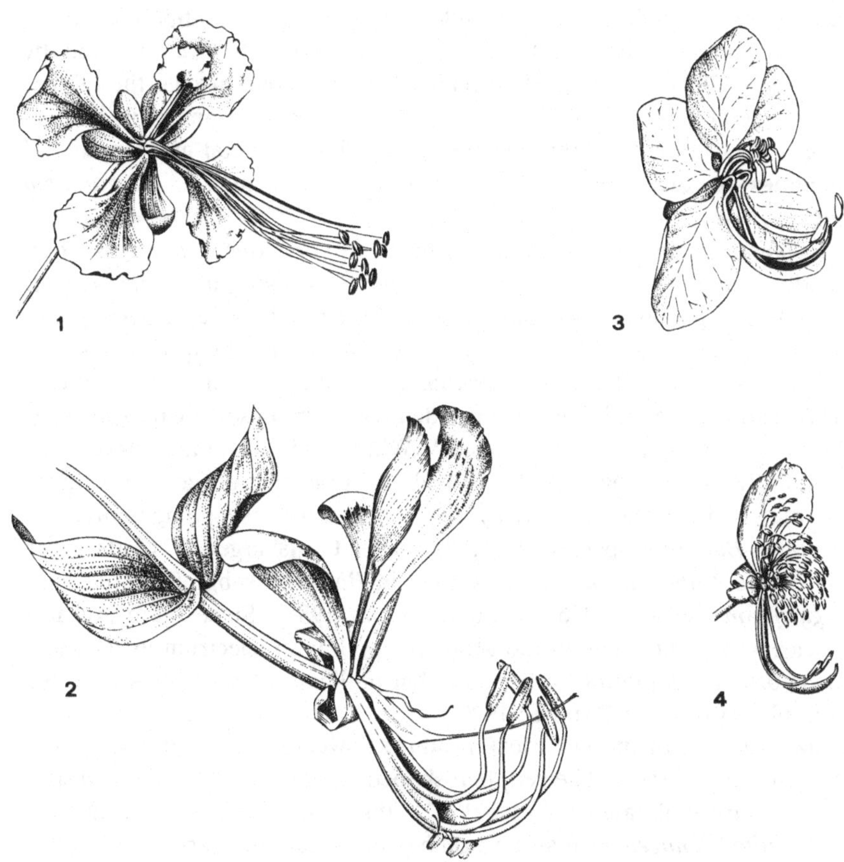

Figure 8.20. Flowers of Caesalpiniaceae (1–3) and of *Swartzia* (Fabaceae) (4) (× 1). 1 and 2 are butterfly-pollinated. 3 and 4 are buzz-pollinated by large bees; they have three or two large lower stamens, respectively, associated with the style and several to numerous smaller upper stamens. 1. *Caesalpinia pulcherrima*. 2. *Amherstia nobilis*. 3. *Cassia fistula*. 4. *Swartzia pinnata*.

clawed, so that the construction appears light and elegant. It is, however, not true that these flowers are 'widely open' in every respect, as it may appear at first sight, but nectar is hidden in all these genera, in intricate ways (as we have already seen in *Delonix* and will now see in other cases). Some of these monosymmetric flowers are very large, e.g. *Delonix* and *Amherstia*, and especially *Baikiaea*, where stamens and petals are 15 cm long. Flowers of *Camoensia maxima* may reach more than 20 cm in length, and their tube is 7 cm long (Thompson 1931) (*Camoensia* is, however, classified with Fabaceae–Sophoreae by Polhill, in Polhill & Raven 1981). In many taxa the floral centre is hidden by the closely contiguous stamen bases, which are frequently hairy; this had already been noticed by Sprengel (1793)! Access to the nectar is only possible through the tubular banner petal and the one lateral opening between two stamens in *Delonix regia* (see section 1.3), through the tubular banner petal and two lateral openings at the side of the median adaxial stamen in *Caesalpinia pulcherrima*.

In some groups the floral cup is fairly to very long; since it is often slender it may be superficially mistaken for the upper part of the pedicel. In some groups the ovary sits at the base of the floral cup (e.g. *Caesalpinia*, *Delonix*), in others higher up at one side of the floral cup (inner spur), either on the abaxial side (e.g. *Bauhinia*, Fig. 8.21.4,6, or extremely so in *Griffonia*, thus resembling the one-sided condition in some Chrysobalanaceae, e.g. *Dactyladenia*, see Prance & White 1988) or on the adaxial side (e.g. in *Amherstia*, Fig. 8.21.1). The nectar at the base of the narrow floral cup may then be reached only by long-tongued insects (sphingophily, e.g. in some *Bauhinia* species, see Silberbauer-Gottsberger & Gottsberger 1975); the floral cup is extremely long in *Bauhinia* subg. *Bauhinia* sect. *Gigasiphon* (de Wit 1956, Wunderlin *et al.* 1987). Since there are also species with a short cup in the genus the pollinator spectrum in *Bauhinia* s.l. is broad; adaptation to bee- and bat-pollination also occurs (see, for example, Hokche & Ramirez 1990).

Some genera of the Detarieae produce flowers in dense, pendent clusters with large bracts. The red, bell-shaped flowers of most *Brownea* species are bird-pollinated (Arroyo 1981); the white, fruity-scented flowers of *Maniltoa lenticellata* may be pollinated by bats or marsupials. Other Detarieae that are chiropterophilous include *Hymenaea*, *Eperua*, and *Elisabetha*. *Jacqueshuberia* (Caesalpinieae) and *Dicymbe* (Amherstieae) are also chiropterophilous (Vogel 1968). Other bird-pollinated species may be found among *Chidlowia*, *Acrocarpus* (both Caesalpinieae) and *Phyllocarpus* (Detarieae) (Arroyo 1981).

In Caesalpiniaceae the innermost position of the standard petal is used

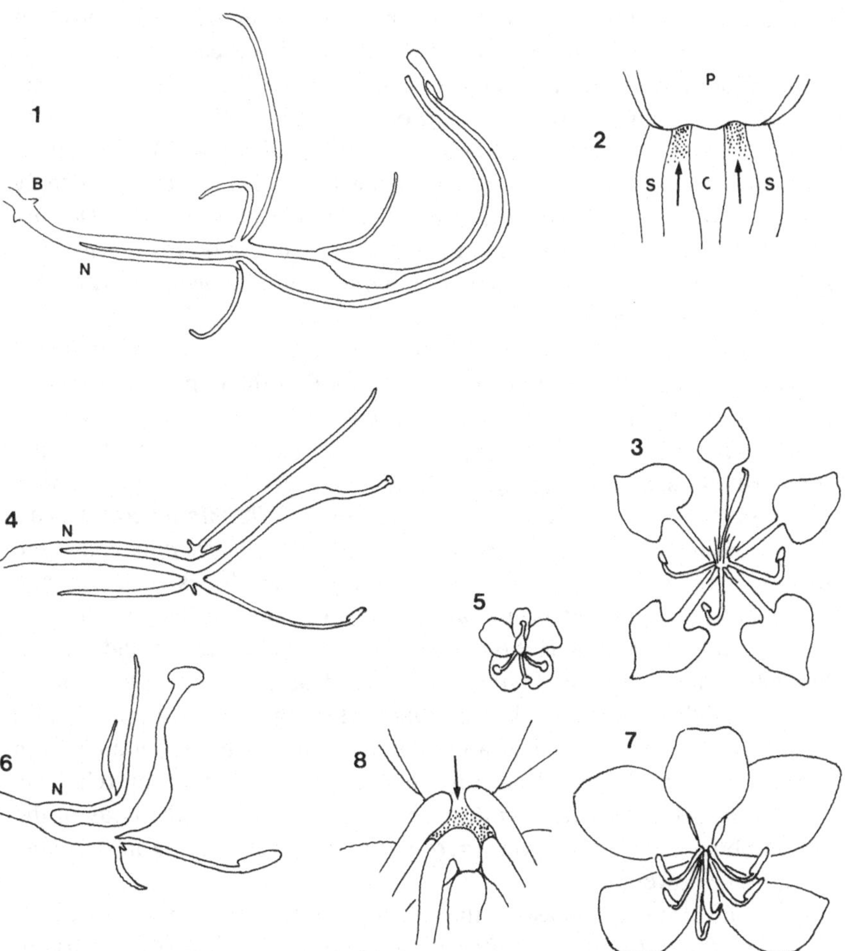

Figure 8.21. Flowers of Caesalpiniaceae. 1–2. *Amherstia nobilis*. 1. Median longitudinal section; B, attachment region of bracteoles; N, lower end of inner spur with nectary (× 1). 2. Floral centre from above; arrows indicate the two entrances into the spur between the carpel (C), the flanks of the nine united lower stamens (S), and the upper petal (P) (× 3). 3–4. *Bauhinia galpinii*. 3. Flower in frontal view (× 0.5). 4. Median longitudinal section; N, lower end of inner spur with nectary (× 1). 5–6. *Bauhinia integrifolia*. 5. Flower in frontal view (× 1). 6. Median longitudinal section: N, lower end of inner spur with nectary (× 3). 7–8. *Bauhinia variegata*. 7. Flower in frontal view (× 0.5). 8. Floral centre; arrow indicates the entrance into the spur between the two upper stamens and the carpel (× 3).

in at least three different ways in elaborations including synorganization with the androecium. In contrast to the keel in Fabaceae these different elaborations have so far not been comparatively studied. In each form the potential of folding or rolling inward of the flag is used.

(1) The claw of the standard may be tubular (rolled inward). The upper end of the tube is the only access to the nectar, which is therefore deeply buried in the otherwise open flowers. This has been shown for *Delonix regia* (Lindman 1902) (see section 1.3, Fig. 1.1); it also occurs in *Caesalpinia* (Troll 1951). However, a similar apparatus has also been described for the fabaceous *Psoralea pinnata* (Lindman 1902).

(2) In the second phase of anthesis the plate (blade) of the standard petal may roll in or fold down in such a way that the upper surface is no longer visible to insect visitors. This may change the appearance of flowers even more drastically than simple colour change (which occurs in the standard in many Caesalpiniaceae, also in the ultraviolet range) (see Jones & Buchmann 1974 and Gori 1983 for *Parkinsonia aculeata*). This disappearance of the inner standard surface occurs in *Delonix regia* (see section 1.3, Fig. 1.1.3–4), *Caesalpinia eriostachys* (Jones & Buchmann 1974) and *Cercidium* (*Parkinsonia*) *floridum* (personal observation).

(3) A case of synorganization of the standard petal with the androecium has been suggested for *Chamaecrista* (Gottsberger *et al.* 1988, Gottsberger & Silberbauer-Gottsberger 1988). Here the standard petal is also rolled in and forms a kind of a tubular extension of the small anthers in the floral centre. These (small) flowers are buzz-pollinated by relatively large bees; pollen is canalized through the banner tube to the body of the bee in about the same region where it may touch the stigma in the same or in other flowers.

It should also be emphasized that the units of attraction in many Caesalpiniaceae are not only single flowers but entire inflorescences or parts of them, even in species with large and widely spaced flowers. Often flowers or conspicuous floral parts remain on the inflorescence after anthesis and add to the flag function (e.g. *Amherstia*, *Caesalpinia*, *Delonix, Senna*) (see below).

Enantiostyly is well developed (e.g. *Cassia*, *Copaifera*). However, opposite flowers on inflorescences may be mirror images of each other, so that a twin group may be monosymmetric again (Arroyo 1981) (see section 6.3).

Pollen is transported in aggregates held together by viscin threads in the butterfly-pollinated *Delonix* and *Caesalpinia pulcherrima*, and in the bat-pollinated *Bauhinia pauletia* (Cruden & Jensen 1979).

There is some diversity in the structure of the unicarpellate gynoecium

(Owens 1989). Two very different kinds of stigma differentiation occur in the family. Stigmas are expanded, capitate (e.g. *Bauhinia*) or concave (with two subtypes: 'crateriform' without dilated chamber and 'chambered' with small entrance and dilated chamber, e.g. Caesalpinieae, Cassieae) (Owens & Lewis 1989, Owens 1989, 1990) (Fig. 8.22.1–3). In the cases studied they are wet and mostly non-papillate (Owens & Stirton 1989). The concave stigmas are often surrounded by a fringe (collar) of stiff, non-receptive hairs. The hairs may be short or long. If they are long, the stigmatic crater may be shallow (*Delonix*). The function of these hairs is unknown (Owens 1990). It is possible that the hairs scratch like a comb over the body of the large pollinators (butterflies, birds) and that the loosened pollen is attracted from the pollinator's body to the stigma by electrostatic forces. Styles are hollow from the beginning (*Baikiaea*) or lysigenously hollow (*Caesalpinia*) or solid (*Gleditsia*). These features are diverse even within tribes.

The presence of gametophytic self-incompatibility in some Caesalpiniaceae has been suggested based on indirect evidence (Arroyo 1981). This is difficult to assess practically, since there is normally a high degree of fruit abortion in many representatives.

Caesalpinia (Caesalpinieae)

Caesalpinia is a pantropical genus of about 100 species. *Caesalpinia pulcherrima* is one of the best studied species of the Caesalpiniaceae in terms of floral biology. The flowers are adapted to butterfly pollination (Vogel 1954, Cruden 1976a, Cruden & Hermann-Parker 1979). They are in upright racemes and are directed toward the side. There are bisexual and male flowers. In male flowers the gynoecium is reduced and they produce only about half as much nectar as bisexual flowers. Sepal aestivation in bud is quincuncial-imbricate, a result of spiral sepal initiation. However, the adaxial sepal is higly precocious and covers all other sepals like a hood (Tucker *et al.* 1985). The open flowers are weakly scented. Sepals, central parts of the petals, stamen filaments and gynoecium are red, while the peripheral parts of the petals are yellow at the beginning of anthesis and then turn red, too. The lateral petals are expanded, while the standard petal is more narrow, involute for most of its length, and forms a long tube (Fig. 8.20.1). Stamens and gynoecium are long and slender and arch forward. The anthers are highly versatile (seesaw-type). They do not open widely but release pollen clumps (pollen held together by viscin threads) through narrow slits. The anthers are not caducous but after anthesis the entire stamens are shed. The stigma is crateriform and surrounded by stiff hairs. The crater is filled by a droplet of clear stigmatic

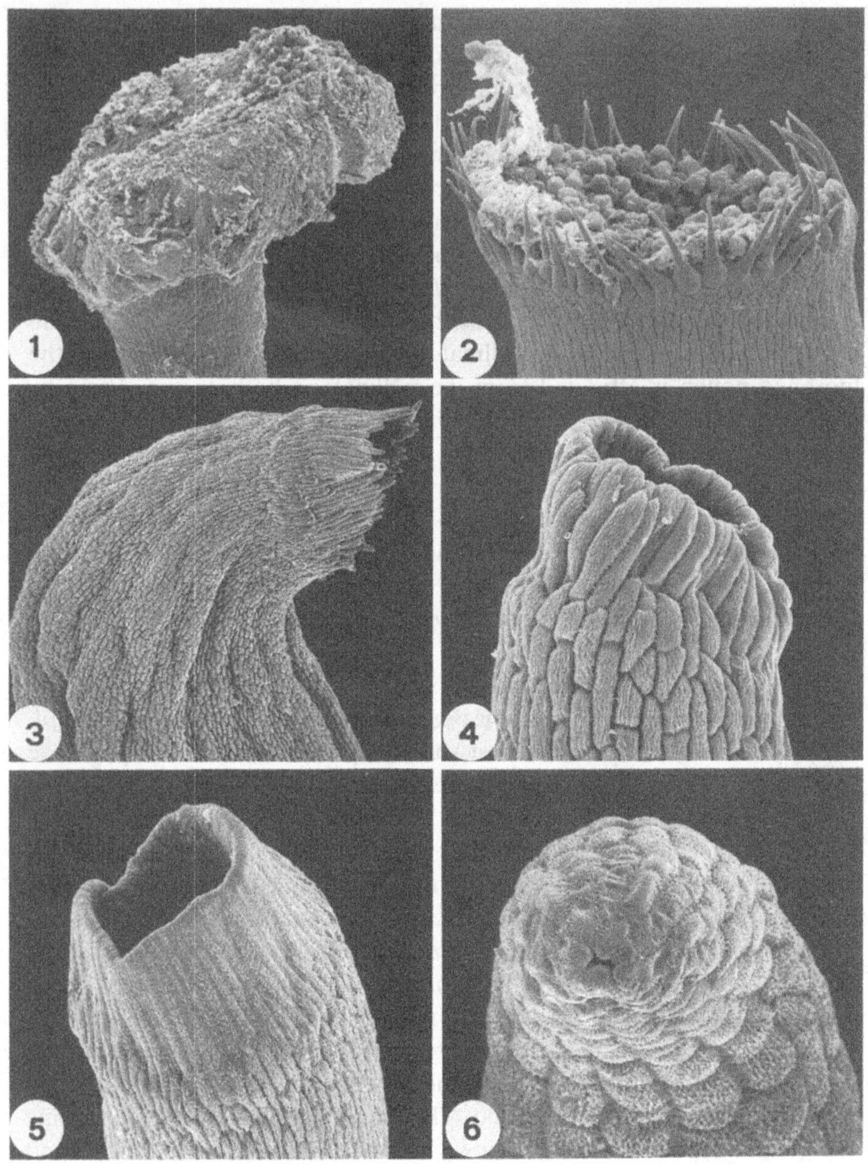

Figure 8.22. Stigma forms of Caesalpiniaceae (1–3) and Mimosaceae (4–6). 1. *Peltophorum pterocarpum* (× 35). 2. *Chamaecrista* aff. *mimosoides* (× 110). 3. *Cassia fistula* (× 80). 4. *Adenanthera pavonina* (× 300). 5. *Prosopis pubescens* (× 180). 6. *Mimosa spegazzini* (× 800).

fluid (Owens 1990). Nectar is secreted by a conical nectary at the gynoecium base and is only accessible through the narrow tube of the standard petal. This tube leads to two narrow gates on both sides of the adaxial median stamen (of the inner stamen whorl). This is similar to the situation in Fabaceae (see section 8.6.3). However, in *Caesalpinia* the other nine stamens are not fused. Their bases are just so close together that they form a tight wall around the nectar. In addition, hairs on the filaments further prevent access into the floral centre. The hairs are absent only at the two nectar gates. This hairless region involves all three stamens that form the nectar gates: the median and the two adjoining uppermost lateral ones (Fig. 8.23.1–2). The two upper lateral petals also have a short tubular part at the base but without allowing access to the nectar. The two lower lateral petals lack a tubular part.

Anthesis of a flower takes one day. The flowers open and the anthers dehisce before 0800. Two-day-old flowers remain on the plant. Nectar secretion begins before the flowers open and persists until mid-afternoon. Nectar is sucrose-dominated and has a high amino acid concentration, as in other butterfly-pollinated plants (Cruden & Hermann-Parker 1979).

Swallowtails (Papilionidae) are the predominant pollinators. They approach the flowers from above or in front and flutter their wings continuously while visiting a flower (Cruden & Hermann-Parker 1979, DeVries 1983a,b). Pollen clusters are transferred to the stigma by the wings. The greater amount of nectar in the bisexual flowers keeps the butterflies at the flowers for a longer time, thus increasing the likelihood of pollination (Cruden *et al.* 1983). The butterflies usually do not visit older flowers that have turned red and no longer offer nectar (Cruden & Hermann-Parker 1979). Plants seem to be self-compatible. However, the behaviour of the butterflies promotes outcrossing. They usually visit four to six flowers and then fly to another tree (Cruden & Hermann-Parker 1979). Arroyo (1981) mentions hummingbirds as pollinators in Costa Rica, del Coro-Arizmendi & Ornelas (1990) in Mexico.

Caesalpinia gilliesii has flowers of similar proportions, but they are pale yellow and flower at night (opening at 1930); they are pollinated by hawkmoths (Cocucci *et al.* 1992). The standard petal does not have a tubular base and is not much different from the other petals either in shape or in colour (including UV). Each flower lasts only one night, while an inflorescence lasts about a month (Cocucci *et al.* 1992).

The flowers of *C. eriostachys* are bee-pollinated (Janzen 1967b, Jones & Buchmann 1974). The pollination organs and the tubular base of the standard petal are much shorter than in *C. pulcherrima*. The flowers look uniformly yellow. However, the standard petal is UV-absorbent, while the

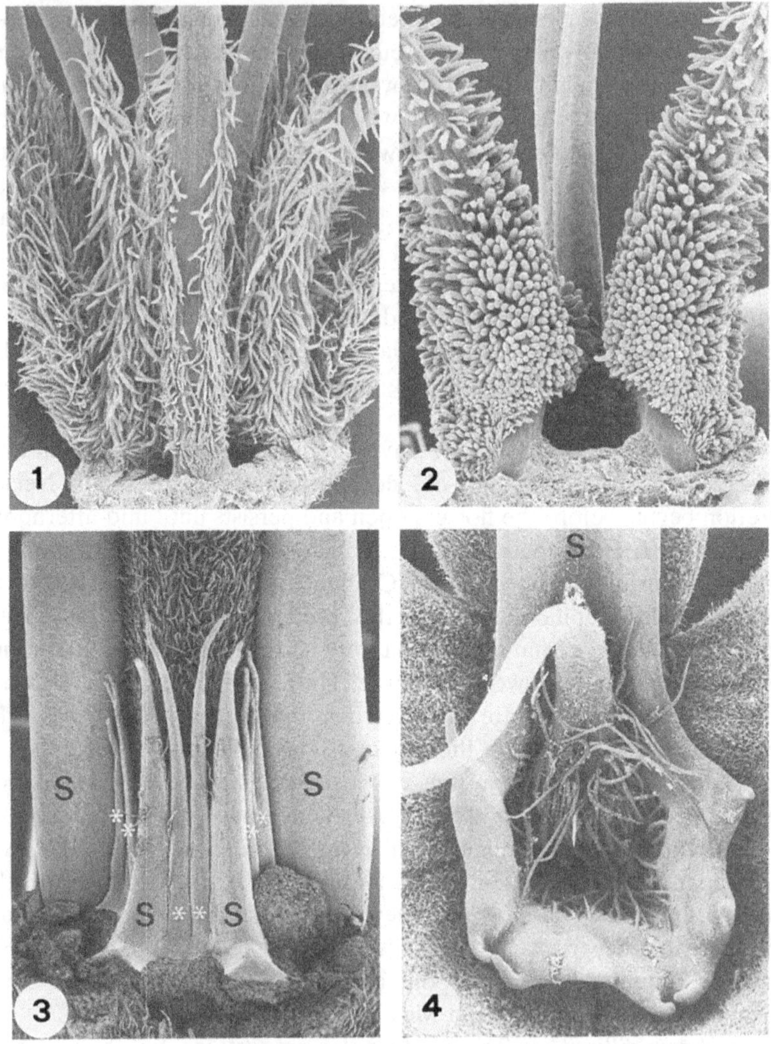

Figure 8.23. Androecium base in flowers of Caesalpiniaceae (1–3) and Chrysobalanaceae (4), all flowers shown from adaxial side. 1. *Caesalpinia sepiaria*; two entrances towards the nectar besides the base of the adaxial stamen (× 15). 2. *Caesalpinia* sp.; entrance towards the nectar, adaxial stamen removed (× 15). 3. *Bauhinia galpinii*; inner staminodes in pairs (marked with asterisks) between the outer stamens/staminodes (marked with S) (× 20). 4. *Dactyladenia barteri*; entrance into inner spur inside the united stamen bases; adaxial stamens reduced, abaxial stamens united into a band (S) (× 18).

other petals are UV-reflective. In old flowers the standard petal folds downwards over the pollination organs so that its lower side is now exposed, which is UV-reflective. Anthesis takes one day. Flowers open synchronously before 0415 (Frankie *et al.* 1983). They have a strong fragrance. The species is self-incompatible (Bawa 1974).

Caesalpinia coriaria has a similar flower form and is probably also bee-pollinated. *Caesalpinia nuga* is pollinated mainly by carpenter bees and a flower lasts for two days (Aluri 1990). *Caesalpinia conzattii* and *C. exostema* are hummingbird-pollinated (Opler 1983, Vogel 1990c).

Delonix (Caesalpinieae) (part II)

Delonix contains several species in tropical Africa and Madagascar. In the introductory chapter (section 1.3) our fragmentary knowledge of *Delonix regia*, this most familiar ornamental tree throughout the tropics, was pointed out. Some new observations made in 1981–82 on cultivated specimens in Brisbane (subtropical Australia) and 1990 in Atherton (tropical Australia) are presented here.

The flowers of *Delonix* are similar in many respects to those of *Caesalpinia*. They differ in a conspicuous feature: the thick, valvate calyx in contrast to the thinner, quincuncial calyx of *Caesalpinia* and related genera (e.g. Polhill & Vidal 1981). In addition, the sepals are postgenitally united. However, the valvate calyx of *Delonix* is evidently derived from a quincuncial-imbricate pattern (Fig. 8.24). It starts early development with a quincuncial-imbricate pattern (Fig. 8.25.4), but early on this pattern is superimposed upon by excessive thickening growth so that a valvate pattern results. In microtome sections of floral buds the early quincuncial-imbricate pattern can be clearly seen by quincuncial indentations of the innermost part of the postgenitally fused flanks of the sepals (Fig. 8.24). *Delonix* is an instructive example to show the evolutionary derivation of valvate aestivation from the less elaborate quincuncial imbricate pattern.

The flower primordia are subtended by bracts, which become hairy in early stages (Fig. 8.25.1). Sepal primordia show irregular furrows, imprints of the neighbouring bracts (Fig. 8.25.2). The sepals are initiated in a spiral, the abaxial one being the first (Fig. 8.25.3). Soon, however, a gradient is superimposed on the development. The sepals on the abaxial side grow more rapidly than those of the adaxial side (Fig. 8.25.4). This makes *Delonix* an interesting intermediate between the spiral and the unidirectional initiation patterns described by Tucker (1989) for Caesalpiniaceae (see also section 9.2.1). By the early thickening of the sepals the floral apex becomes imprinted by the inner sepal surfaces so that it appears as a shallow 5-sided pyramid (Fig. 8.25.5). At the five angles the petal prim-

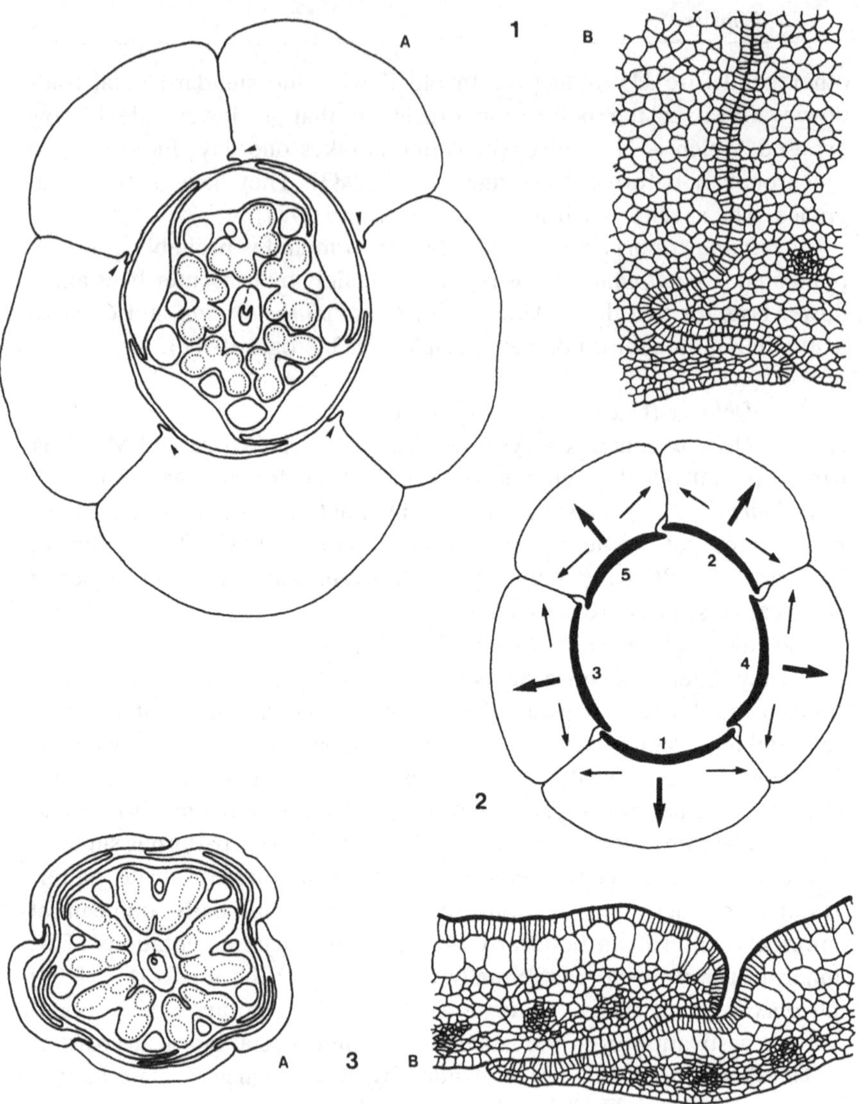

Figure 8.24. Derivation of the valvate calyx of *Delonix regia* from the imbricate-quincuncial calyx of related groups. 1–2. *Delonix regia*. 1 A. Transverse section of floral bud; arrows mark the sinuses derived from outer sepal flanks in early ontogeny (× 11). 1 B. Sinus between the two upper sepals (× 110). 2. Diagram showing the development of the valvate pattern by extreme thickening combined with broadening of the sepals (arrows), ontogenetic sequence of sepals indicated. 3. *Parkinsonia aculeata*. 3 A. Transverse section of floral bud, calyx imbricate-quincuncial (× 11). 3 B. Contiguous flanks of the two upper sepals, corresponding to the sinus in 1 B (× 110).

ordia soon appear (Fig. 8.25.6). Still later the two stamen whorls and the adaxially directed carpel appear (Fig. 8.26.1). Later the petals become broader and cover the stamens, and the ventral slit of the carpel is formed (Fig. 8.26.2). The stamen filaments show differential elongation allowing dense anther packing: the anthers of the outer stamen whorl are above those of the inner whorl (Fig. 8.26.3). As in *Caesalpinia pulcherrima* the mature stigma is crateriform (Owens 1990) and surrounded by non-receptive, stiff, unicellular hairs of unequal length (Fig. 8.26.4). The mature anthers are highly versatile and of the seesaw-type (Figs 8.27.1, 8.26.5). They release clumps of pollen held together by viscin threads (Fig. 8.26.6).

At anthesis the mass of flowers may make the trees completely red at times. The flowers are produced in more or less horizontally arching racemes. In trees with densely arranged shoots the racemes may be quite crowded. Flowers open in acropetal sequence. The time interval between opening of flowers is one or more days, so that no two flowers of the same stage are present in a raceme. The flowers open early in the morning. Shortly after flower opening the anthers open. The stigma appears dry at first, but then a droplet becomes visible in the surrounding basket of hairs. The time at which the stigma becomes receptive has not been determined. However, the stigmatic droplet seems to appear only after anther dehiscence. The dehisced thecae do not gape widely. Clumps of pollen may remain in the anther until the following day. During rainy weather the open anthers close again. Pollen is yellow on the first day; later it becomes grey. The asymmetry of the floral base caused by the one-sided nectar gate has been mentioned in section 1.3 (Fig. 8.28). Lindman (1902), who detected this asymmetry, found the nectar gate always on the right side of the flower. However, according to my own observations the nectar gate may occur on either side.

On the first day of anthesis the standard petal is widely expanded and frontally exposes its upper surface, which is white or light bluish and contains red stripes, contrasting with the uniformly red lateral petals and the red upper surface of the sepals and the red stamen filaments (Fig. 1.1.2). On the second day the standard petal is folded together or downwards (Fig. 1.1.3). It is not known whether the stigma is still receptive. Nectar is most abundant on the first day, slightly less on the second, and almost absent on the following days. It may take about five days until all floral organs have dropped in irregular sequence. Anthers may drop before the filaments or they may drop together. Usually, organ dropping starts on the third day. However, it may be earlier after rains or storms. At the end the remnant flower drops at an abscission zone about 5 mm below its

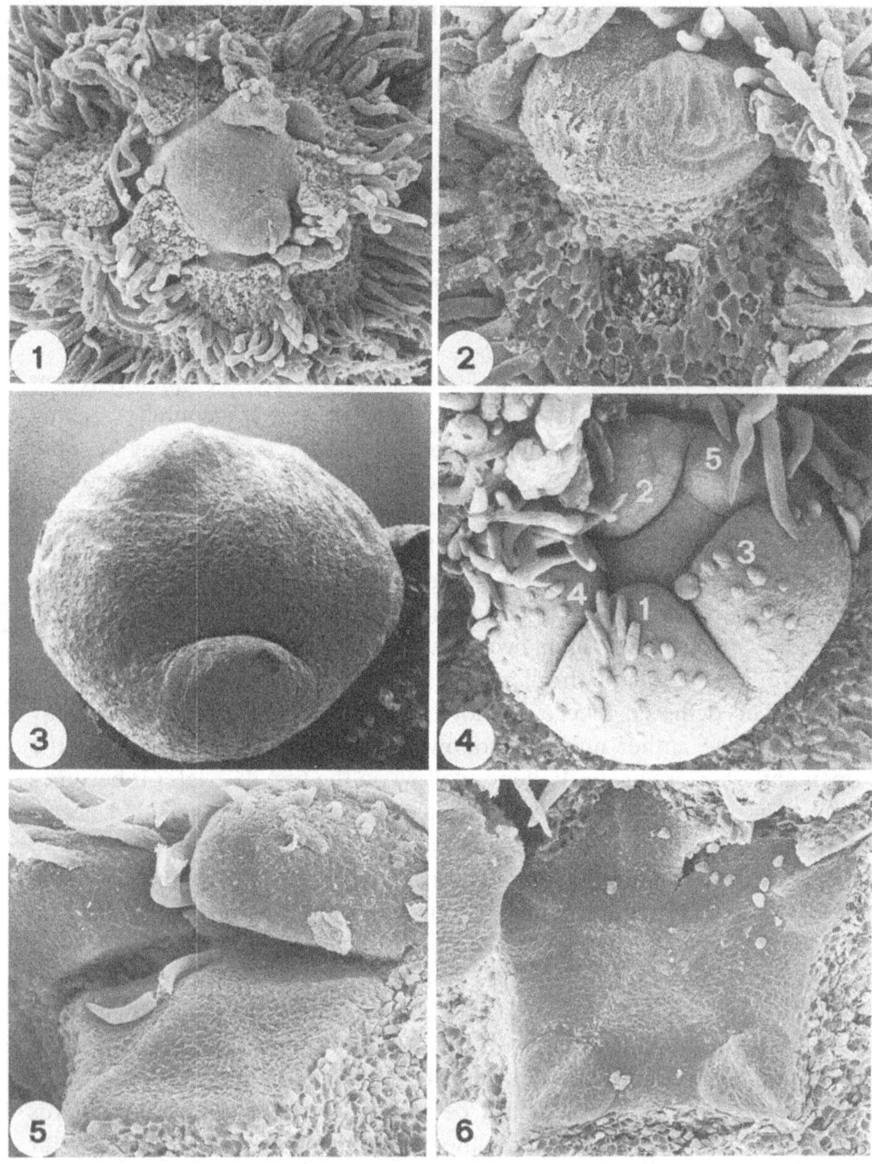

Figure 8.25. Early floral development of *Delonix regia*. 1. Inflorescence apex with spirally arranged floral primordia, subtending bracts of the lower primordia removed (× 100). 2. Flower primordium with scar of removed subtending bract, abaxial sepal initiated, with surface markings of hairs and remains of secreted material (× 200). 3. Isolated floral primordium

base. Thus, several flowers may remain on an inflorescence and form a flag, although only one first-day flower with an exposed standard petal is among them. The presence of red flags by the cooperation of functional and non-functional (old) flowers, as in *Caesalpinia pulcherrima*, also occurs in some other butterfly-pollinated plants (see section 4.1.4). As mentioned earlier, in both *Delonix regia* and *Caesalpinia pulcherrima* Papilionidae have been observed as pollinators (Owen 1971).

A second species, *D. elata* from Tanzania, has white flowers that turn cream in late anthesis. The standard petal has a bright yellow guide mark.

Amherstia (Amherstieae)

Like *Delonix regia*, *Amherstia nobilis* ('Queen of the Flowering Trees') is one of the most beautiful tropical trees but its flower structure and biology are poorly known (Corner 1964) (Figs 8.20.2, 8.21.1–2). *Amherstia*, a monotypic genus, is native to Burma. The flowers have not been studied in the natural habitat. The inflorescence is pendulous, the flowers therefore resupinate. The pink flowers are about 10 cm long. They seem to be widely open, since all floral organs are freely exposed. However, as in most Caesalpiniaceae, nectar is hidden, but, in contrast to *Caesalpinia* and *Delonix*, it is in a narrow floral cup (or inner spur) about 3.5 cm in length. Two large, spreading prophylls (bracteoles) are at the base of the floral cup. Of the five recurved sepals the upper two are fused. Only the three upper petals are developed; the two lower ones are present as two rudimentary teeth. The three petals have a bright yellow tip, which is UV-absorbing, in contrast to the UV-reflecting lower parts of the petals (Kay 1987). This distribution is an unusual pattern. The banner petal, in addition, has pink stripes and a white centre. The yellow marks are exposed both laterally and frontally. Of the ten stamens only five are fertile (those in the same radii as the sepals). The nine lower stamens are united, while the median upper one is free (or more or less fused with the gynoecium) or replaced by a second carpel (personal observation) or lacking (Knuth *et al.* 1904). The gynoecium is on the adaxial side of the floral cup. The nectar gate towards the inner spur is on both sides of the gynoe-

Fig. 8.25 (*cont.*)
of about the same stage, sepals initiated, abaxial sepal most advanced (× 250). 4. Sepals laterally contiguous, ontogenetic sequence indicated (× 180). 5. Slightly older, three sepals removed, floral apex five-angled because of the five appressed sepals (× 250). 6. Slightly older, sepals removed, petal primordia apparent in the five angles of the floral primordium (× 180).

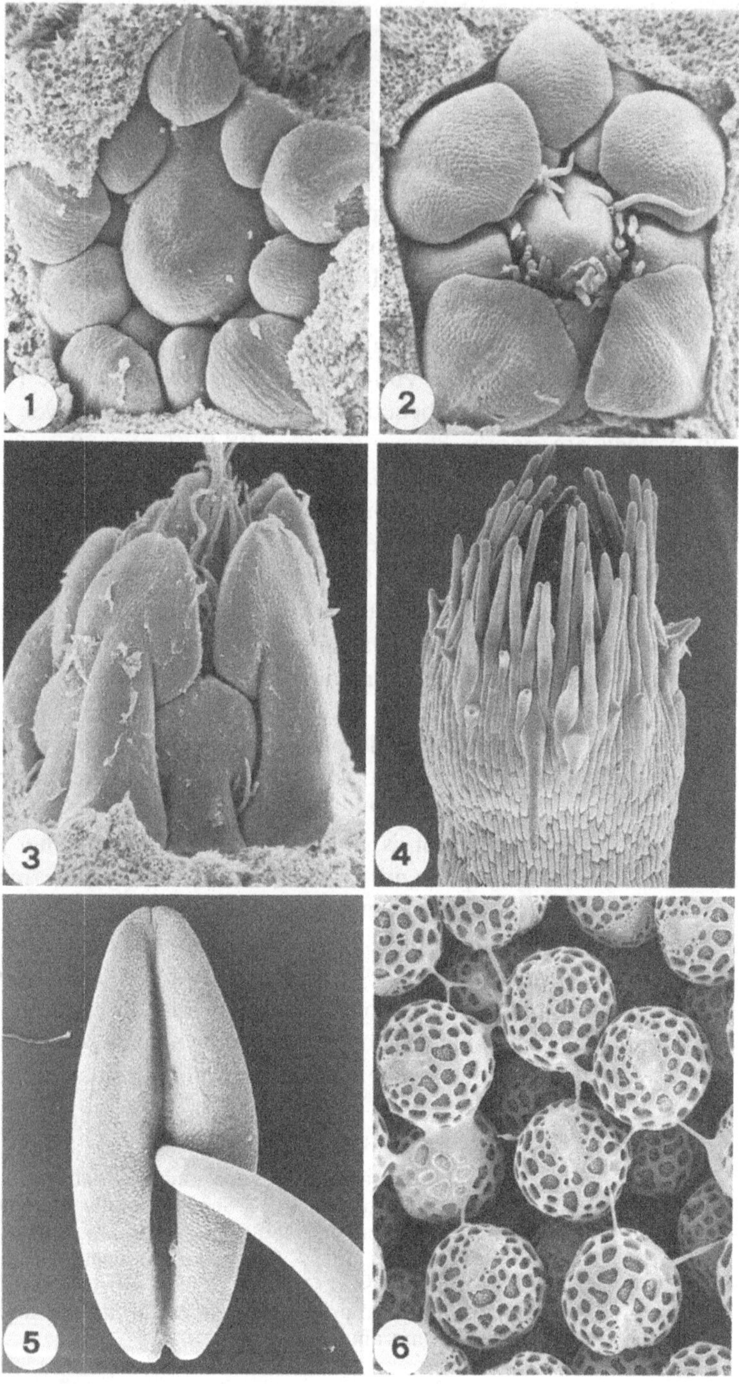
1
2
3
4
5
6

cium attachment, delimited towards the sides by the gutter-shaped synandrium. Stamens and gynoecium are pronouncedly curved upwards. The five fertile anthers are all frontally presented towards an approaching pollinator. They open with only narrow slits and are extremely versatile (seesaw-mechanism) (Figs 8.20.2, 8.27.2). Since they are dorsifixed, the frontal exposition results by a backward tilting. Pollen hangs together in large clumps; however, in contrast to *Caesalpinia* and *Delonix*, clumping is achieved not by viscin threads but by abundant pollenkitt. The floral features suggest lepidopteran pollination, although bird-visits have been reported (Werth 1915, van der Pijl 1938). Knuth (in Knuth *et al.* 1904) observed honeyeaters (Meliphagidae), butterflies (Pieridae) and *Xylocopa* bees as pollinators in Java.

Trees observed in the Philippines showed continuous flowering over several months (Garcia 1975). Of the *ca.* 12 flowers of an inflorescence, two to three are in anthesis at the same time (Knuth *et al.* 1904). Flowers open in the morning. They are slightly protandrous. Anthers open at 0600. The stigma becomes receptive at 1000 and remains receptive till noon of the next day. One or two days after anthesis the floral organs drop, but the showy bracteoles remain and keep adding to the showiness of the inflorescence. Self-incompatibility seems to occur (Garcia 1975).

Bauhinia (Cercideae)

Bauhinia, pantropical and with about 300 species, is one of the largest and florally most diverse genera within the Caesalpiniaceae (de Wit 1956, Wunderlin *et al.* 1987). Floral size, colour and display are highly diverse. Many species are cultivated as ornamentals because of their large, elegant flowers or their rich display of smaller flowers (Ledin & Menninger 1956). In some species the flowers appear widely open because of the narrow claws of the expanded petals. However, as in *Amherstia*, they have a narrow floral cup (or inner spur), of varying length, that conceals the

Figure 8.26. Later floral development of *Delonix regia*. 1. Sepals removed, gynoecium primordium present but still undifferentiated (× 140). 2. Petals covering stamens, upper petal beginning to be covered by its neighbours (ascending aestivation), carpel folded (× 90). 3. From the side, petals removed, showing the two-storied packing of anthers (× 70). 4. Stigma immediately before anthesis (× 90). 5. Dorsal side of anther, immediately before anthesis (× 13). 6. Pollen grains connected by viscin threads (× 400).

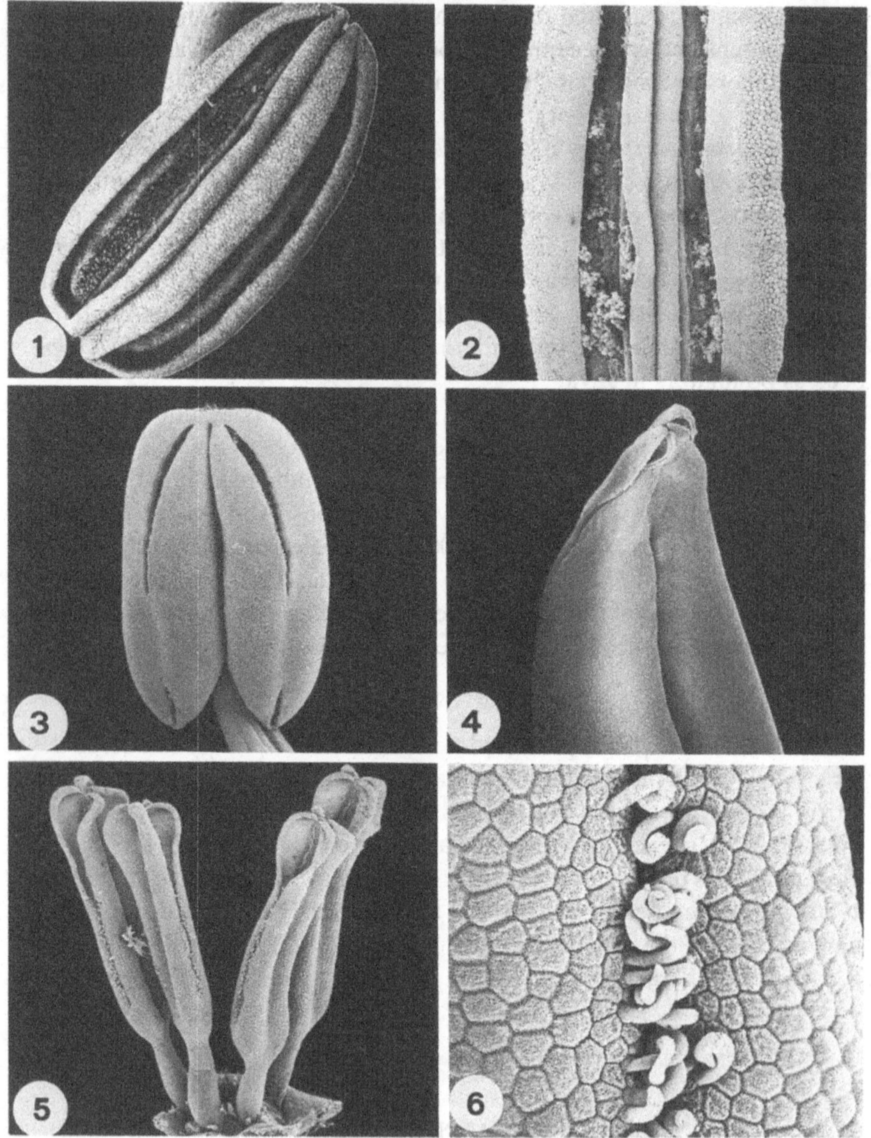

Figure 8.27. Anther forms of Caesalpiniaceae. 1–2. From butterfly-pollinated flowers. 3–6. From buzz-pollinated flowers. 1. *Delonix regia* (× 13). 2. *Amherstia nobilis* (× 13). 3. *Cassia fistula* (× 13). 4. *Senna didymobotrya* (× 15). 5–6. *Chamaecrista* aff. *mimosoides*. 5. The four stamens of a flower (× 15). 6. Interlocking hairs flanking the thecal furrow in lower half of the anther (× 200).

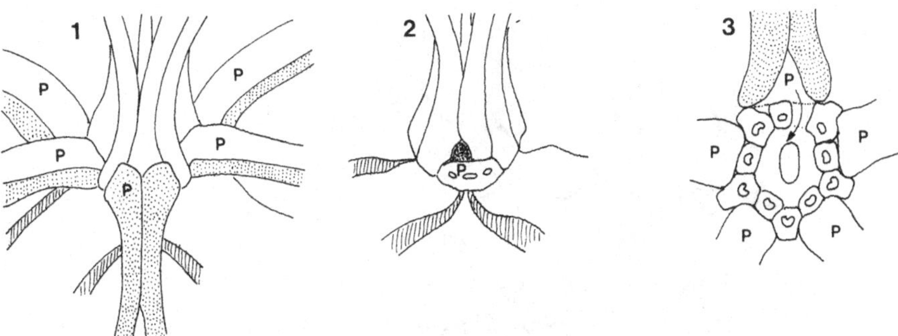

Figure 8.28. Floral centre of *Delonix regia* (× 3). 1. From adaxial side; hatching, flanks of the valvate sepals; stippling, undersurface of petals (P); base of the adaxial petal folded to form a tube towards the nectar. 2. Same, but all petals removed to show the asymmetric access to nectar; P, scar of the adaxial petal. 3. From above, adaxial side up, stamens and gynoecium removed; arrow marks asymmetric access to nectar.

nectar. In contrast to *Amherstia*, the gynoecium is on the upper abaxial side of the cup.

In subg. *Phanera*, sect. *Phanera* (classification after Wunderlin *et al.* 1987), small-flowered species (such as the Palaeotropical *B. integrifolia* with flowers of 1 cm diameter) may display their bi-coloured flowers, with colour change from orange to red in large numbers in flat, umbel-like inflorescences (Figs 8.21.5–6, 8.29). They have a sweet scent reminiscent of *Phlox* flowers. They are probably pollinated by butterflies and a number of other smaller pollinators. This and other related species produce an oily exudate of unstudied function (probably secondary pollenkitt) along the stomium of the mature anthers (as described earlier for *B. scarlatina* by Cammerloher 1929b). The anthers are caducous. Nectar is available in a one-sided floral cup (inner spur) only 2–3 millimetres in length (personal observation). The calyx ruptures into 2–3 irregular parts as the flower opens.

Neotropical representatives of subg. *Phanera*, sect. *Caulotretus*, such as *B. glabra*, *B. guianensis* and *B. rutilans*, have small diurnal flowers (opening in the morning and lasting 9–24 hours) that are visited by a variety of bees, wasps, butterflies and hummingbirds (Hokche & Ramirez 1990).

The large, white flowers of some species of subg. *Bauhinia*, sect. *Pauletia*, are mainly nocturnal (they open at dusk and last one night). Some

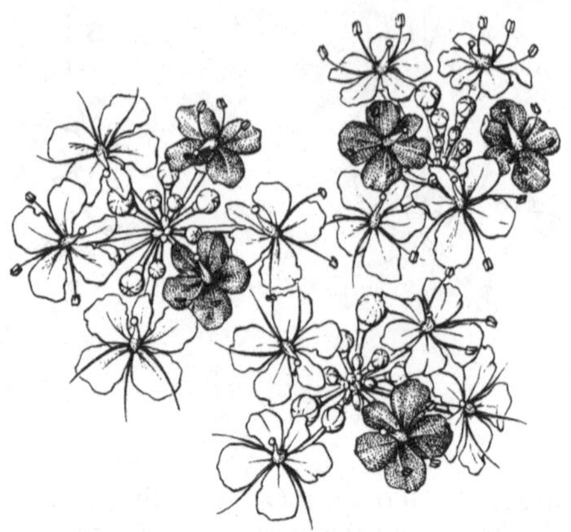

Figure 8.29. Flowers of *Bauhinia integrifolia* (Caesalpiniaceae), in umbel-like aggregations with colour change from yellow (shaded) to red (unshaded) (× 1).

species are bat-pollinated (*B. pauletia*, Heithaus *et al.* 1974; *B. ungulata*, Ramirez *et al.* 1984; *B. aculeata*, *B. multinervia*, Hokche & Ramirez 1990). Dobat & Peikert-Holle (1985) also list a number of other Neotropical bat-pollinated species; especially noteworthy among them is *B. megalandra* with a single large anther. Other species are hawkmoth-pollinated (probably the African *B. macrantha*, Vogel 1954; and the South American *B. forficata*, *B. candicans*, *B. platypetala*, Lindman 1902, and *B. cupulata* and *B. curvula*, Silberbauer-Gottsberger & Gottsberger 1975). They all have ten stamens. *Bauhinia platypetala* is protandrous. As in *Delonix* the nectar gate is on one side of the adaxial stamen, and therefore the floral base is slightly asymmetric (Lindman 1902).

Some commonly cultivated ornamentals also belong to subg. *Bauhinia*. The African *B. galpinii* (sect. *Afrobauhinia)* with relatively large (6 cm diameter), scentless, orange flowers is butterfly-pollinated (Vogel 1954) (Fig. 8.21.3–4). The petals have long, narrow claws and relatively small blades. The inner spur leading to the nectar is narrow and about 2 cm long. The three stamens arch slightly upwards, the gynoecium slightly downwards. The five sepals are not united. As the bud opens they reflex to the abaxial side. The Asian *B. purpurea* (sect. *Purpureae*) with large, red flowers is pollinated by large bees (*Xylocopa*) (van der Pijl 1954). This is probably also true for the similar *B. variegata* (Fig. 8.21.7–8). The flowers are about 7 cm in diameter. The nectar gate is between the upper two of the five stamens present and the gynoecium. The inner spur is 1.7 cm long. Stamens and gynoecium arch upwards. The sepals are united.

As the flowers open they rupture at one radius to form a one-sided sheath-like calyx at the abaxial side. In the presumably hawkmoth-pollinated species of sect. *Gigasiphon* (subg. *Bauhinia*) the narrow inner spur may reach 10 cm length (de Wit 1956). Almost equally long tubes have evolved independently in sect. *Pseudophanera* and *Afrobauhinia* (both subg. *Bauhinia*) and in sect. *Meganthera* (subg. *Phanera*) (Wunderlin *et al.* 1987).

A striking trend within the genus is a reduction in the number of fertile stamens from ten (as in some taxa of all four subgenera) to five (as in subg. *Bauhinia*, sect. *Purpureae*), three (as in certain taxa of subg. *Bauhinia*, and predominantly in subg. *Phanera*) or even a single one (in 17 species of subg. *Bauhinia*: convergent in the Neotropical sect. *Bauhinia*, the African and Madagascan sect. *Afrobauhinia*, and the Southern Asian sect. *Telestria*) (Wunderlin 1983, Wunderlin *et al.* 1987). More rarely, functionally unisexual flowers have evolved (e.g. *B. malabarica* of subg. *Elayuna*, and *B. divaricata* of subg. *Bauhinia*; Tucker 1988b). The reduced stamens are often still present as staminodes. These reduced stamens may be increased in number, which may be due to the more narrow circumference of the primordia (e.g. *B. galpinii*, see Fig. 8.23.3, and *B. malabarica*, Tucker 1988b) (see also section 2.5.3).

Another trend of reduction is in petal number. Most species have five petals. In five of the Neotropical species with a single stamen the petals are reduced to between four and one (Wunderlin 1983).

Both self-compatibility and self-incompatibility occur within the genus (Heithaus *et al.* 1974). Andromonoecy, gynomonoecy and dioecy are known in various species (Wunderlin *et al.* 1981).

Cassia sensu lato (Cassieae)

The former genus *Cassia* has been subdivided into the genera *Cassia* (30 species), *Senna* (260 species) and *Chamaecrista* (265 species), based largely on differences in floral structure (Irwin & Barneby 1981, 1982). All three segregated genera are circumtropical. They have in common nectarless flowers that provide only pollen to pollinators (Fig. 8.20.3). The filaments are firm and elastic. The anthers are not versatile and are relatively more basifixed than in the other genera discussed. They usually open by pores. The dry pollen is collected by bees, which vibrate it out of the 'tubular' anthers (see section 5.1). A genus related to *Cassia* s.l. that is probably also buzz-pollinated is *Duparquetia*; here the four or five stamens with short filaments have their large anthers postgenitally fused into a firm structure. In the large-flowered *Cassia* and *Senna*, *Xylocopa* bees are important pollinators (Dulberger 1981).

Interestingly, in the three genera *Cassia*, *Senna*, and *Chamaecrista* the anther pores are formed by different methods. In all of them a normal longitudinal stomial furrow appears at first. However, later development diverges. In *Cassia*, commonly the middle part of the stomium does not dehisce, but only the apical and the basal part, resulting in two short slits ('pores') per theca: a larger upper one and a smaller lower one (Fig. 8.27.3). In *Chamaecrista* aff. *mimosoides* the stomium dehisces along its entire length, but in earlier stages in the lower half of the anther hooked hairs develop along the two sides of the stomium, which interlock and so prevent an opening. In this way a terminal pore arises (Fig. 8.27.5–6) (see also Tucker 1992b). In *Senna* the stomium begins dehiscence along its entire length but the theca fails to open completely in its lower part resulting in a small apical pore (Fig. 8.27.4).

The flowers of *Cassia* and *Senna* are also characterized by their heteranthery, another feature often seen in pollen-flowers (Vogel 1978a) (see section 5.1); here, 3–4 stamen forms are often present in one flower, the upper (adaxial) ones often much smaller than the lower (abaxial) ones (Venkatesh 1956, Lasseigne 1979, Irwin & Barneby 1981). In *Cassia* the three abaxial stamens of the outer whorl are enlarged and their firm, elastic filaments sigmoidally curved, while all other stamens are much smaller. In *Senna*, in contrast, the two abaxial stamens of the inner whorl are enlarged. In *Chamaecrista* the androecium is often more or less polysymmetric and all stamens are of about the same size (Irwin & Barneby 1981). However, in *C.* aff. *mimosoides* only the four upper stamens of the outer whorl are normally developed (Fig. 8.27.5). Flowers are most often yellow in all three genera but there are some *Cassia* species with pink petals, which contrast with the large yellow stamens. As in other similar groups, dummies have developed to reinforce the optical effect of yellow anthers. Here, in these pink-petalled species, the filaments of the largest stamens are partly inflated (*Cassia javanica* and related species).

The highly specialized chambered stigmas with only a small pore as entrance to the receptive chamber occur in all three genera (Owens & Lewis 1989). They seem to be especially pronounced in the large-flowered *Cassia* species. They may be seen as a further adaptation to buzz-pollination, where electrostatic forces may be involved in pollen reception (see section 5.1), or if a secretion droplet appears at the pore during buzzing it may afterwards suck in deposited pollen even through a narrow pore. This awaits investigation.

Some species have very long fruits. Ovule number may vary between 5 and 266 in *Senna* (Irwin & Barneby 1982).

8.6.2 Mimosaceae

In Mimosaceae, polysymmetric flowers with small petals and stamens with long and showy filaments but small anthers prevail (Fig. 8.30). The flowers produce nectar in most groups. Usually the flowers are arranged in dense heads and form classical brush-blossoms. The valvate petals, however inconspicuous, are protecting organs for floral buds and may become relatively thick (e.g. *Wallaceodendron*, Ramirez-Domenéch & Tucker 1990). Many Mimosaceae produce polyads of constant number per anther and with a constant number of microspores (Fig. 8.31.4). At the most primitive evolutionary level are perhaps flowers with 10 stamens, these not beeing much larger than the petals. Such flowers occur, for example, in *Adenanthera* (Adenanthereae or Mimoseae) and are similar in appearance to some unspecialized Caesalpiniaceae (e.g. *Haematoxylon*: Caesalpinieae). Elias (1981) regards *Pentaclethra* (Parkieae) with imbricate petals, five stamens (and 5–15 staminodes) as the most primitive genus of the family because the flowers are so similar to those of *Dimorphandra* (Caesalpiniaceae–Caesalpinieae). Tucker (1994) also regards *Dimorphandra* as the sister group of the mimosoids.

From here two main evolutionary trends can be seen:

(1) In the Acacieae and Ingeae an increase in stamen number per flower and often synandry takes place; an increase in carpel number also occurs

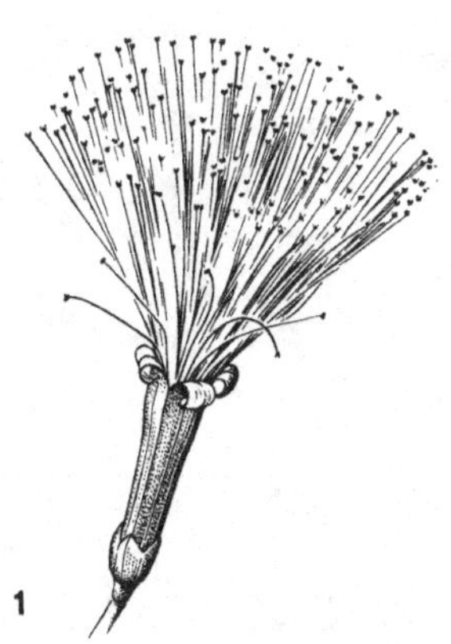

Figure 8.30. Flowers of Mimosaceae. 1. *Archidendron vaillantii*; large individual brush-flower (× 1). 2. *Parkia javanica*; brush-blossom composed of numerous densely arranged, small, heteromorphic flowers; the constricted zone contains sterile flowers that are specialized to produce nectar (× 0.6).

in some genera (*Affonsea*, *Archidendron*). In *Acacia myrtifolia* up to 537 stamens per flower were recorded by Bernhardt & Walker (1984). On the other hand, the anthers are exceedingly small. The number of flowers per head (or smallness of flowers) and number of stamens per flower show a reciprocal relationship (Kenrick & Knox 1989). The increased number of stamens is initiated on a meristematic ring around the gynoecium primordium in a centripetal sequence (Gemmeke 1982) and partly also in a centrifugal sequence (Derstine & Tucker 1991), often starting from five distinct sectors. The stamen filaments are already long in bud and often highly distorted (Fig. 3.3.6). In some genera, such as *Archidendron* and *Inga*,

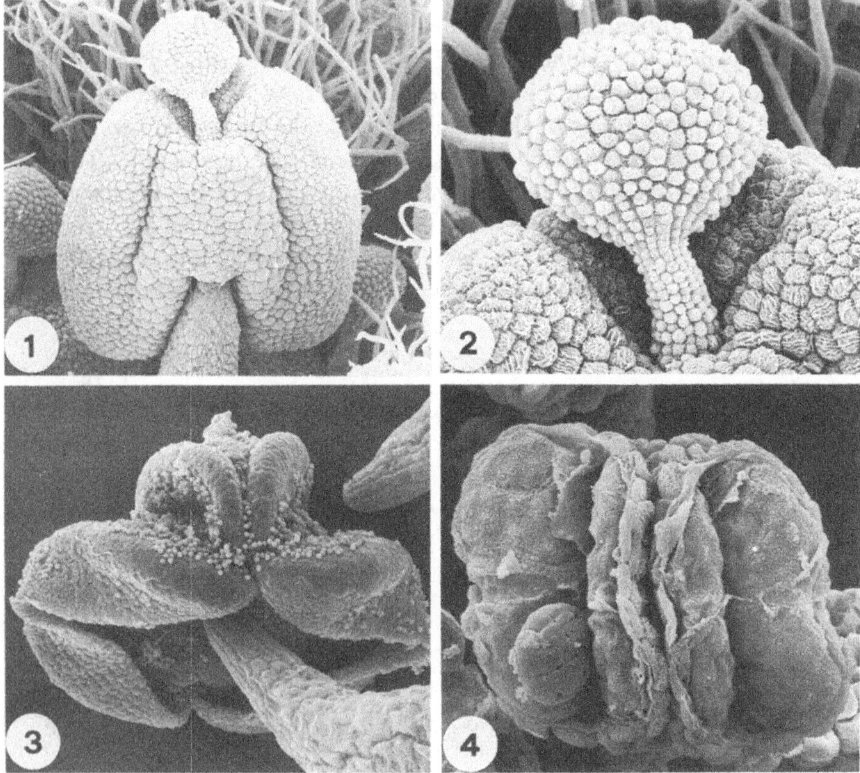

Figure 8.31. Anthers of Mimosaceae. 1–2. *Prosopis pubescens*. 1. Anther with apical protrusion, before anthesis (× 70). 2. Anther protrusion (× 180). 3. *Mimosa spegazzini*; open anther with numerous minute pollen tetrads (× 130). 4. *Acacia sphaerocephala*; open anther with one of the few produced polyads left (× 600).

individual flowers become quite large. Pollen is shed in polyads; stigmas are cup-like.

(2) In the Mimoseae, in contrast, stamen number per flower is often decreased (to 4–5). Individual flowers are diminished in size but form aggregates of head-like or spicate inflorescences, which are so dense that superficially the flowers lose their individuality. The consequence is that in several genera the inflorescence becomes a kind of pseudanthium with heteromorphic flowers. This occurs in *Dichrostachys*, *Neptunia* (Mimoseae) and *Parkia* (Parkieae) (Fig. 8.30.2). Here, differentiation results in lower, larger, showy or scent-producing, sterile flowers, intermediate, smaller, nectariferous, sterile flowers and upper, fertile flowers (for American *Parkia* see Vogel 1968, 1969a,b, for *Neptunia* see Tucker 1988c, 1989). *Neptunia* (Tucker 1988c, 1989) and *Parkia* (Elias 1981) are, in addition, exceptional among Mimosaceae in having monosymmetric flowers. In some Ingeae flowers are also heteromorphic, in that the central flowers of a head often have thicker staminal tubes (*Abarema*, *Albizzia*, *Calliandra*, *Klugiodendron*, *Punjuba*, see Nielsen 1981, Classen-Bockhoff 1990) and they may be the only flowers in the entire head to have nectaries (Maheshwari 1931). Here also floral monosymmetry may occur (*Calliandra*, Arroyo 1981).

The large and richly nectariferous heads of most *Parkia* species are bat-pollinated (Baker & Harris 1957, Hopkins 1984) (see below). This is also true for the large single flowers of some *Inga* species (Vogel 1968) (see below). Ornithophilous taxa may be found especially in *Calliandra*. In some other, less specialized genera, bird-pollination may occur alongside other modes (see *Inga*, below).

Many Mimosaceae have anthers with a caducous apical protrusion that may secrete a sticky substance (Fig. 8.31.1–2). Polyads may also have a sticky appendage by which they are attached to the body of floral visitors (pollinators) (F. Müller in Ludwig 1897). It is also mentioned for *Calliandra* by Faegri & van der Pijl (1979). More recent studies on anther glands seem to be lacking (Kenrick & Knox 1989).

The structure of the unicarpellate gynoecium is less diverse than in Caesalpiniaceae, probably owing to its commonly smaller size. As in many Caesalpiniaceae the stigma is concave. However, it is smaller, devoid of a fringe of bristles and looks rather like the top of a chimney (see Fig. 8.22.4–6). It is generally wet and non-papillate (Kenrick & Knox 1989, Owens & Stirton 1989) and concave to various degrees (Kenrick & Knox 1989). Self-incompatibility has been reported in many Mimosaceae by indirect evidence (Arroyo 1981, Kenrick & Knox 1989). In addition to the difficulties mentioned for Caesalpiniaceae, the small size and exceedingly large number of anthers make experimental manipulations almost impossible (Kenrick & Knox 1989).

Inga (Ingeae)

Inga is a Neotropical genus of about 400 species. Flowers are in spikes or heads in relatively low number. The perianth is rather small but the numerous conspicuous white stamens and the style are long (stamens up to 8 cm long in the chiropterophilous *Inga sessilis*, Vogel 1968). The flowers have a sweet odour. Relatively dilute nectar, which is produced copiously inside the floral cup (formed by stamens and corolla), is accessible to a wide range of pollinators. Koptur (1983) found bats, hummingbirds, sphingids and other moths, and various bees to be efficient pollinators within a single species, *I. brenesii*. The importance of hummingbird-pollination for some *Inga* species was also emphasized by Feinsinger (1976), and of bat-pollination by Vogel (1968) and by Salas (in Croat 1978).

Blooming strategies are various, as observed in Costa Rica (Koptur 1983). Some show mass-flowering toward the end of the wet season (e.g. *I. brenesii*); *I. punctata* has an extended blooming period of up to six months. In some species the flowers open at only one time of the day (*I. oerstediana* in the late afternoon), while others open continuously at all times of day and night (*I. brenesii*); in *I. densiflora* most flowers open in the early morning, a few in the late afternoon (Koptur 1983). Individual flowers are functional for only 6–10 hours, but two days in *I. mortoniana* (Koptur 1983). The flowers are slightly protandrous in most species. Although the stigma cup is small, more than one polyad will fit in. The *Inga* species studied by Koptur (1983) are all self-incompatible. Species that bloom simultaneously are not cross-compatible (Koptur 1984).

Acacia (Acacieae)

Acacia, with about 1200 species, is one of the largest genera of the angiosperms. It has minute yellow or white flowers in large numbers densely packed in spherical or spicate inflorescences. Despite the diminished size of the individual flowers, stamen number may be very high (see above). There is a tendency towards mass-flowering with synchronized inflorescences, andromonoecy, protogynous flowers and a high degree of self-incompatibility, although some species are considered partially self-compatible (Bernhardt 1989, Kenrick & Knox 1989). Knox & Kenrick (1983) proposed gametophytic control of self-incompatibility for *A. retinodes*. A large number of mainly short-tongued bees foraging for polyads may be found as pollinators on a species (Bernhardt 1989). In Australian *Acacia* species, which are devoid of floral nectar (in contrast to African and American species!), alongside pollen, extrafloral nectaries on the leaf petioles serve as food sources for pollinating bees and birds (Vanstone & Paton 1988, Bernhardt 1989, Thorp & Sugden 1990). These are, then, extrafloral but nuptial nectaries in the sense of Vogel (1977). Du Toit (1990) asks whether giraffes pollinate certain African *Acacia* species.

The cup-like stigma of *Acacia* is minute. In many species only a single polyad can be received by a stigma. The results of Kenrick & Knox (1982) show that the number of ovules per gynoecium is correlated with the number of pollens per polyad in that the ovule number is never larger than that of the pollen per polyad (with only one exception out of the 46 species with a constant number of pollen per polyad investigated). In 40 species with 16 pollens per polyad the ovule number was 5–16 (16 in only one species); in 3 species with 8 pollens per polyad the ovule number was 2–10; in 1 species with 32 pollen per polyad the ovule number was 24; in 1 species with 4 pollens per polyad (with tetrads) the ovule number was 1–3.

Mimosa (Mimoseae)

Mimosa contains *ca.* 480 species of predominantly Neotropical trees, vines, shrubs or herbs; some are notorious weeds (Barneby 1991). The delicate flowers are in heads that last only half a day in *M. pudica* (Percival 1974). Mass-flowering may occur (Barneby 1991). The flowers are commonly bisexual but the lower ones in a head are often male. They are (3-)4–5(-6)-merous. The fused calyx is commonly minute and sometimes irregularly lobed (5–7-lobed in the 4-merous flowers of *M. albida*, Ramirez-Domenéch & Tucker 1990), giving the impression of sepal multiplication by reduction, or it may be extremely reduced to a rim (Barneby 1991). The commonly four petals are united. The androecium is haplo- or diplostemonous. Reduction in stamen number is probably a multiple parallel evolutionary trend within the genus (Barneby 1991). The stamens are either free or basally united into a tube or basally fused with the petals (Barneby 1991). The anthers are crescent-shaped and the thecal walls widely open by extreme reflexion (Fig. 8.31.3). Connective protrusions are lacking. The ovary commonly contains 4–20 ovules (Barneby 1991). The stigma is concave but minute, pore-like; the tetrad size is the smallest recorded in the angiosperms, 6.5 μm in diameter (Guinet 1981) (Fig. 8.22.6).

Parkia (Parkieae)

The genus *Parkia*, with about 30 species, occurs in tropical America, Africa, Madagascar and Asia. The inflorescences are ball-like or oblong heads of up to more than three thousand small red or yellow flowers (Hopkins 1984) (Fig. 8.30.2). *Parkia gigantocarpa* has heads of 23 cm length and 10 cm breadth (Vogel 1968). In contrast to most other Mimosaceae the calyx is imbricate. A few species have only bisexual flowers. However, andromonoecy is predominant in the genus. Some species even have three floral morphs in an inflorescence: bisexual, male and sterile flowers that form an elaborate functional unit. The lowermost, sterile flowers are the largest ones because of their enlarged staminodes, which act as osmophores ('scent-flowers') (Vogel 1968). Then follow male flowers with copious nectar production by large nectaries ('nectar-flowers') (Vogel 1968). Since they do not entirely expand at anthesis, they are the smallest ones in the inflorescence and, therefore, in some species they form a circular groove around the inflorescence where nectar accumulates. The bulk of the flowers are bisexual, although only a few fruits are produced per inflorescence. The inflorescences are either pendent or upright and extend out of the foliage of the trees (Hopkins 1984). In *P. pendula* the thin, pendulous peduncles reach 1 m length.

Parkia is predominantly bat-pollinated. The first detailed studies on bat-pollination in the natural habitat were on the African *P. clappertoniana* by Baker & Harris (1957), following several earlier reports on species in Java (e.g. Docters van Leeuwen 1938). Other African and probably a Madagascan species are also bat-pollinated (Hopkins 1983). De Carvalho (1960) and Vogel (1968) found bat-pollination in Neotropical species. Of the eleven Neotropical species, nine (in two sections) are bat-pollinated; a third section is entomophilous and visited mainly by trigonid bees (Hopkins 1984). In Eastern Asia chiropterophily has also been demonstrated for at least three species (Start & Marshall 1976). Inflorescence structure and biology are not much different in Palaeo- and Neotropical species. In both regions medium-sized bats are the main pollinators. However, in the Neotropics the bats land head downwards, and in the Palaeotropics head upwards (Hopkins 1984).

In bat-pollinated species an inflorescence is open for only one night (Hopkins 1984). However, individual trees of *P. bicolor* flower for 6–8 weeks, and trees within a population are more or less synchronized (Hopkins 1983). In the Neotropics Hopkins (1984) found, in addition to the 'steady-state type', also the 'cornucopia type' and 'mass-flowering' (see section 7.4). The flowers open in late afternoon and start abundant nectar production at dusk or later. At the onset of nectar production the first bat visits occur (Vogel 1968). Several species appear to be protandrous (Vogel 1968, Hopkins 1984). The flowers have a smell of particular over-ripe fruit. In *P. clappertoniana* Baker & Harris (1957) found up to 5 ml of nectar produced in one night in a single inflorescence, Grünmeier (1990) up to

12 ml in *P. bicolor*. In bee-pollinated species nectar was not found to occur (Hopkins 1984). *Parkia nitida* has a high pollen:ovule ratio, which suggests outcrossing (Hopkins 1984). In the African *Parkia bicolor*, in addition to fruit bats, small primates and a rodent were observed feeding on the nectar of the inflorescences (Grünmeier 1990).

8.6.3 Fabaceae

This largest family of the Fabales has the least variation in the floral groundplan. However, in other respects the evolutionary plasticity of flowers is extensive. 'Most of the 31 or 32 tribes have a recognizable papilionoid corolla, although they show innumerable variations on the theme in such features as connation, twisting, elongation, claw formation, beak, pits, projections, sculpturing, resupination, and colour' (Tucker 1987). The well known flag-flowers (papilionate, papilionoid flowers) (see section 3.6, Fig. 3.7.1) in Fabaceae have two lips: the upper lip, the 'standard' formed by the adaxial petal, and the lower lip formed by the other four petals, which also enclose androecium and gynoecium (Fig. 8.19.3, 8.32). Androecium and gynoecium are usually hidden in the 'keel', formed by the two abaxial petals, while the two lateral petals, the 'wings', are often in some way laterally connected with the keel. In the simplest forms they are free from each other, but in more elaborate forms they become more intimately connected by means of mutual outgrowths and pockets, which allow mechanical cooperation. The standard is often the main optical display organ and is thus frontally exposed and often contains a nectar guide of colours contrasting with the more peripheral region. Nectar guides may be even more conspicuous and diverse in the ultraviolet range of the spectrum, with UV-absorbing marks mostly in the floral centre; wing petals may also be included (Kay 1984, 1987). However, nectar guides have not been found in bird-pollinated taxa (Kay 1987). The keel, together with the wings, forms a landing platform for pollinating insects, which are predominantly Hymenoptera. Sculpturing on the outer surface of the wing petals is common. It may be only epidermal and provide footholds for insect pollinators, but it may also be at the morphological level, when the surface forms little furrows that serve as thrusting pads for forcible entry by bees (Stirton 1981). These furrow patterns may be different in each species and also play a role as a prepollination isolating mechanism between species. The sepals are united into a tubular structure, which is an additional means to provide firmness to the complicated pollination apparatus of petals, stamens and carpel. Thus, the papilionate flower as a whole is an extremely elaborate construction by

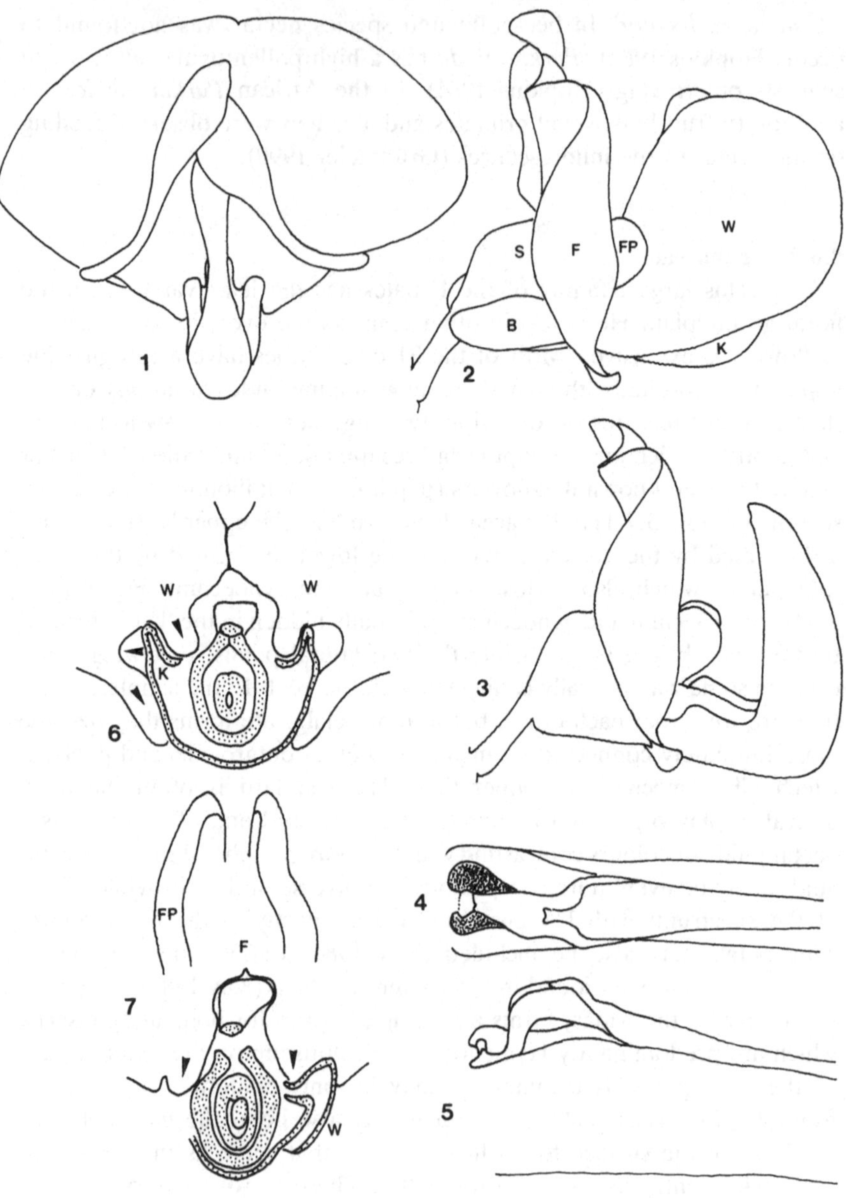

Figure 8.32. Flower structure of *Lablab purpureus* (Fabaceae). 1. Flower in frontal view (× 3). 2. Flower from the side; B, bracteole; S, united sepals; F, flag; FP, ventral protrusion of flag, clasping the wings (W); K, keel (× 3). 3. Flower from the side, bracteoles and wing petals

the synorganization of all its parts. Sprengel (1793) and Delpino (1867, 1868–74) were the earliest to appreciate this phenomenon.

Concomitant with this elaborate construction, postgenital fusions occur on various floral organs. The two flanks of the standard petal are sometimes fused in bud and later open again. The two keel petals fuse in early ontogeny at the lower side and remain fused. Partial fusion at the upper side may also occur in later ontogeny, especially in types 3 and 4 of pollen presentation (see below). Wing and keel petals may fuse, as mentioned above. The free upper stamen may fuse with the nine congenitally united lower stamens (see below).

Some large-flowered Fabaceae are visited almost exclusively by large polytropic bees (*Xylocopa*, *Apis*, *Centris*, *Bombus*); tropical *Centrosema*, *Canavalia*, *Vigna* and *Harpalyce* species are pollinated principally by *Xylocopa* (van der Pijl 1954, Arroyo 1981, Gottsberger *et al.* 1988). Many smaller-flowered Fabaceae in tropical and temperate regions are pollinated by a wide array of bee species (Batra 1967). Interestingly, in *Erythrina* and *Mucuna* (both Phaseoleae), bird- and bat-pollinated flowers have evolved from elaborate bee-pollinated ancestors. The bell-shaped flowers of *Alexa* (Sophoreae) with freely exposed pollination organs are also bat-pollinated (Vogel 1969a). *Camptosema* (Phaseoleae) and some other small genera are bird-pollinated (Arroyo 1981). *Camoensia* (Sophoreae) is hawkmoth-pollinated (Vogel 1969a).

Also, surprisingly, a number of large-flowered tropical Fabaceae have resupinate flowers, either in pendent inflorescences or by curvature of the pedicel. Such upside-down flowers occur in some groups that are pollinated by large bees (e.g. *Centrosema*, *Clitoria*, *Canavalia*, see Vogel 1954, Stirton 1977b) as well as bird-pollinated species (*Erythrina crista-galli*) (Fig. 8.33). Stirton (1977b) found that in *Canavalia virosa*, *Xylocopa* bees operate the flower from the standard petal and *Megachile* bees from the wing-keel complex. Standard petals in *Centrosema* and *Clitoria* have radial furrows, which may act as footholds for bees. In *Centrosema* the standard is reinforced by a dorsal spur near the floral entrance (Fig 8.33.2). For the

Fig. 8.32 (*cont.*)

removed (× 3). 4. Base of androecium from above, the two accesses to the nectar between the free upper stamen and the nine united lower stamens shaded (× 6). 5. Same as seen from the side (× 6). 6–7. Transverse sections of pollination organs with adjacent petals at two different levels; stippling, cut planes (× 6). 6. Interdentations between wings (W) and keel (K) marked with arrows. 7. Section slightly more proximal; interdentations between flag (F) and wings (W) marked with arrows; FP, ventral protrusion of flag.

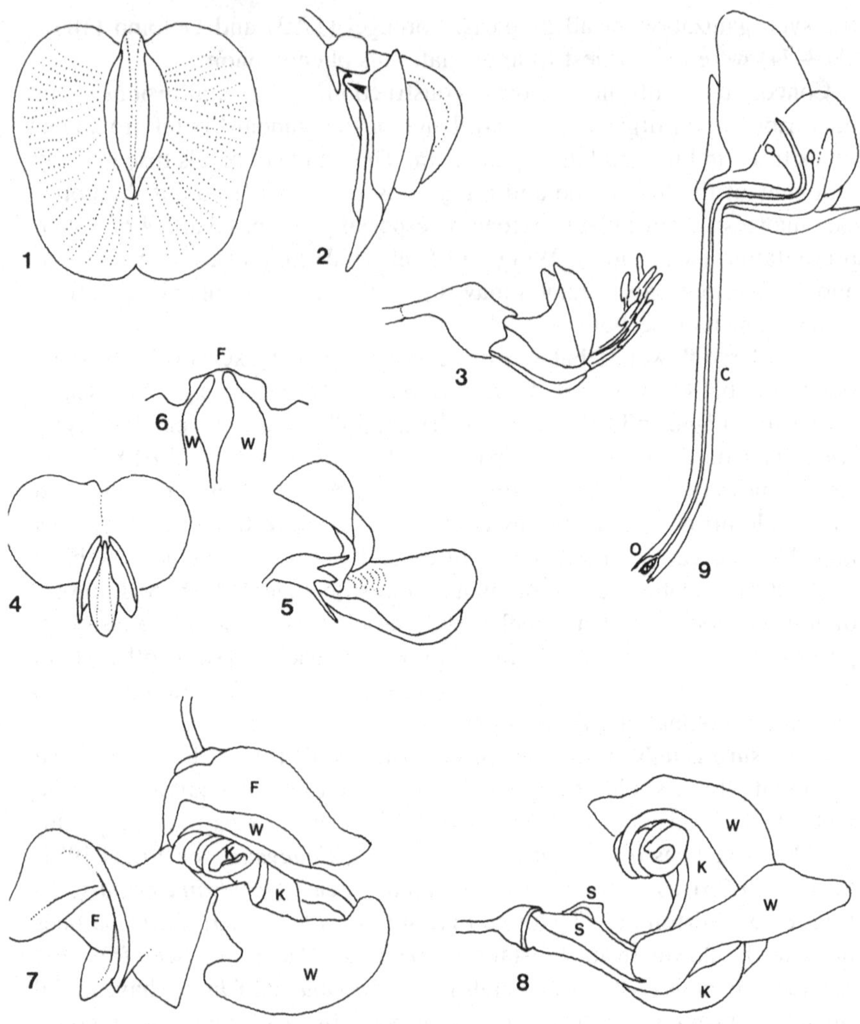

Figure 8.33. Flower forms of Fabaceae. 1–2. *Centrosema pubescens*, flower pendent (× 1.5). 1. Frontal view. 2. From the side; arrow points to dorsal spur on flag. 3. *Castanospermum australe*, pollination organs exposed (× 1). 4–6. *Vigna unguiculata*. 4. Frontal view (× 1). 5. From the side (× 1). 6. Interdentation of flag (F) and wings (W) (× 6). 7–8. *Vigna caracalla*, highly asymmetric flower; F, flag; K, keel; S, stamens; W, wings (× 1) (modified after Lindman 1902). 7. Frontal view. 8. From the side, flag removed. 9. *Arachis hypogaea*; longitudinal section of flower; C, floral cup; O, ovary (× 2.5) (modified after Smith 1950).

somewhat less stout flowers of *Clitoria*, van der Pijl (1954) found *Apis* and *Anthophora* as pollinators.

The pollination organs are also synorganized in that the stamens, which are mostly fused and form a tube, tightly surround the rigid gynoecium. They therefore move together if a pollinator alights on the keel. In most Fabaceae the flowers produce nectar below the gynoecium. Then, the adaxial stamen is not incorporated in the staminal tube, which allows access to the nectar by two openings ('diadelphy', 9+1-pattern) (Figs 8.19, 8.32, 8.34). However, the free stamen commonly becomes postgenitally fused with the two flanks of the congenitally united lower nine stamens, except for the very base ('pseudomonadelphy', Tucker 1989). By this elaboration the complete staminal tube is restored and yet the two nectar gates are present. Rarely, the stamens are free (e.g. Sophoreae, Swartzieae), or, in contrast, all ten stamens are congenitally fused into a tube ('monadelphy') (e.g. in Genisteae, which provide only pollen to their pollinators).

When a flower is visited by a pollinator, the stamens and the gynoecium are somehow exposed. Usually the anthers open before anthesis and pollen is deposited within the tip of the keel around the upper part of the style. Already Delpino (1867, 1873) had distinguished four different methods by which pollen is presented at pollination within the family: (1) the keel is moved downwards, so that the stigma and the anthers are exposed at the tip of the keel, the keel moving back to the initial position after the pollinator has left the flower (ordinary type; occurring in the majority of the genera); (2) pollen is brushed out of the keel tip by the hairy tip of the style (style brush type; occurring in Vicieae and Phaseoleae); (3) pollen is pressed out (together with the stigma) in macaroni-like portions at the tip of the keel; this is possible because the keel petals are postgenitally fused at most of their length, except for their tip and base (macaroni pump type; occurring in, for example, *Lupinus* and *Lotus*); (4) gynoecium and stamens forcibly spring out of the keel when the flower is visited, and cannot move back afterwards; pollination therefore takes place by a single event (explosion type; occurring in, for example, *Mucuna*, *Desmodium*, *Indigofera* and the (nectarless) Genisteae).

The occurrence of the style brush type was analysed in detail by Lavin & Delgado (1990). Many Fabaceae have hairs of some sort at the style. However, only those with upwardly directed hairs that can brush pollen upward are considered here to be of the style brush type. These hairs may be simple or branched. Lavin & Delgado (1990) estimate that this brush type has arisen independently in eight taxa among Fabaceae. It often occurs consistently in a larger group irrespective of the pollination system (e.g. *Phaseolus* with bee-pollinated, bird-pollinated and cleistogamous flowers) (Lavin & Delgado 1990).

It seems that groups with a style brush were particularly suited to developing large flowers. The largest flowers among Fabaceae with the pol-

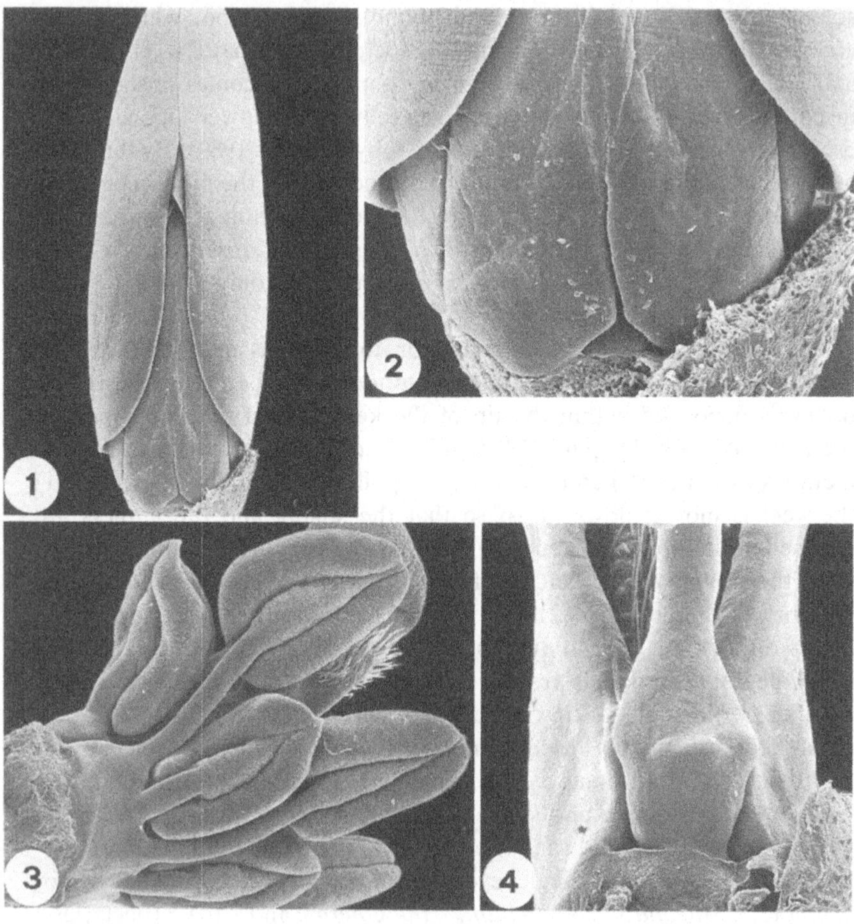

Figure 8.34. Floral bud structure of *Lablab purpureus* (Fabaceae). 1. Bud from abaxial side, sepals removed, flag petal surrounding the bud, its flanks postgenitally united in upper part, the two keel petals free at the very base, but postgenitally united higher up (× 13). 2. Free base of the two keel petals (× 40). 3. Androecium and gynoecium from the side, the lower 9 stamens united, the upper stamen free (× 25). 4. Androecium base from adaxial side, showing the free upper stamen (× 35).

lination organs enclosed in a keel occur predominantly in taxa that typically have a style brush (e.g. *Clianthus*, *Clitoria*, *Mucuna*, *Phaseolus*, *Vigna*). This is in parallel with the Campanulales where the large-flowered Campanulaceae have an elaborate style brush, while in the smaller-flowered Asteraceae it is often more reduced.

A further elaboration of the style brush type is exhibited by the strongly asymmetric flowers in some genera of the Phaseoleae and Vicieae, often involving lateral (pleurotribic) pollen deposition on the pollinator (Hedström & Thulin 1986, Rao *et al.* 1986, Teppner 1988, Lavin & Delgado 1990). The large, bizarre flowers of *Vigna caracalla* and *V. appendiculata* have attracted special attention (Delpino 1868, Lindman 1902, van der Pijl 1954; see below).

The inclusion of the pollination organs in the keel has led to a particular differentiation of the stigma. Although many Fabaceae are self-compatible, pollen which is normally deposited around and on the stigma before anthesis cannot germinate on the stigma unless 'tripping' at anthesis is caused by a flower visitor. This was first discussed by Darwin (1858). The stigma is covered by a membrane which prevents pollen germination (Fig. 8.35). Only by the tripping process is the membrane disrupted, allowing the pollen, now in closer contact with the mostly wet, papillate stigma surface, to germinate (Jost 1907, Pazy 1984, Alon *et al.* 1987). In only a few groups is a stigmatic membrane absent and the stigma dry (e.g. *Arachis*) or wet (e.g. *Lupinus nanus*) (Juncosa & Webster 1989). Often, as in many Caesalpiniaceae, the stigma is surrounded by hairs (independent of a style brush) which may also promote outcrossing (Juncosa & Webster 1989, Lavin & Delgado 1990) (Fig. 8.35.1). It may

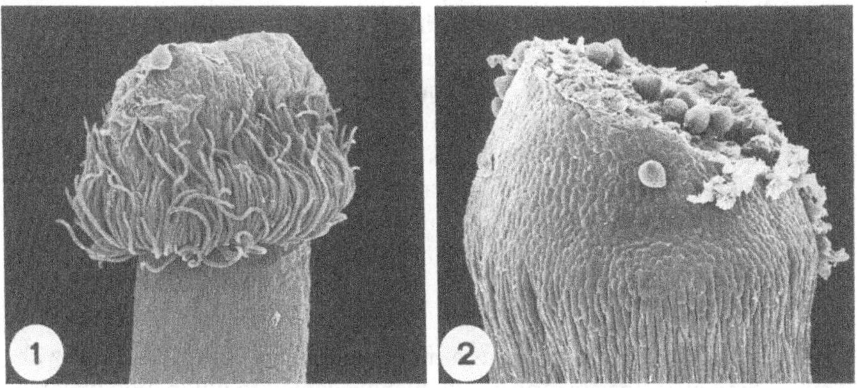

Figure 8.35. Stigma forms of Fabaceae. 1. *Mucuna gigantea* (× 60). 2. *Erythrina crista-galli* (× 150).

be absent in closely related autogamous species (see *Arachis*). In contrast to Caesalpiniaceae and Mimosaceae the stigma is not concave but convex. In the Sophoreae the stigma is the least differentiated.

The style is hollow, at least the upper part of the canal being derived lysigenously (Shivanna & Owens 1989). However, it remains 'solid' immediately below the stigma. In members of at least six tribes the style is thickened towards the end to form an 'entasis' (Shivanna & Owens 1989). The lower end of the entasis has been interpreted as the site of the incompatibility reaction (J. & Y. Heslop-Harrison 1982c) (see sections 2.4.1, 6.5). Self-incompatibility was found in about 20% of the species tested (Arroyo 1981). It is gametophytically controlled. Only a few temperate representatives have been investigated in detail, however (Arroyo 1981). It seems that tropical herbaceous Fabaceae are less commonly self-incompatible than temperate ones (Arroyo 1981). Nevertheless, many Fabaceae have high levels of outcrossing because of their tripping device. Many of the crop plants in the Fabaceae are self-compatible or even autogamous. Almost all Phaseolinae studied to date are self-compatible (Arroyo 1981).

There is good reason to suppose that the evolution of the papilionate flower in Fabaceae has been largely influenced by pollinating Hymenoptera (Leppik 1966, Arroyo 1981). Many extant papilionate flowers are highly adapted to hymenopteran pollination. There is, however, also diversification beyond the limits of melittophily. The array of pollinators is broad. Some successful genera, such as *Erythrina* or *Mucuna*, rely heavily on other animals as pollinators.

On the other hand, there is scarcely any diversification beyond the limits of papilionate flowers. The very few exceptions include *Petalostigma*, *Amorpha*, and *Ateleia*. *Ateleia* is perhaps the most extreme case. It has highly reduced unisexual flowers with but a single remaining petal (Tucker 1990) and is said to be the only wind-pollinated legume (Janzen 1989). However, the systematic position of *Ateleia* (Caesalpiniaceae or Fabaceae?) has not been settled (Tucker 1990).

Thus, again, the floral bauplan is highly conservative but, in contrast, the floral biological mode is very plastic. Details of the floral architecture are also plastic, as is shown by the multiple parallel evolution of the four types of pollen presentation in papilionate flowers (see below).

Lablab (Phaseoleae)

Lablab purpureus forms a monotypic genus of African bean with monosymmetric flowers, apparently highly adapted to pollination by large bees (Figs 8.32, 8.34). The flowers (standard petal) are 1.5 cm broad.

Standard and wing petals are faintly purple at first and then change to lavender. In the centre of the standard-blade there are three converging white guide marks. Each wing petal has a tooth on the adaxial margin, which embraces the somewhat involute flank of the keel (Fig. 8.32.6). The standard petal has two pronounced, firm lateral ridges in the lower half of its extended blade which embrace the two wing petals and hold them in place (Fig. 8.32.7). Lower down the median part of the standard petal is elaborated into a firm groove, which tightly holds the uppermost (free) stamen. This connection is so intimate that the standard petal cannot be removed from the flower without breaking the clasped stamen. To reach the nectar a bee has to force its head into the groove of the standard above the uppermost stamen filament. At this site the filament itself has a shallow groove on its dorsal side, which leads the mouthparts of the pollinator to the nectar space in the floral centre (Fig. 8.32.4–5). Copious nectar is secreted by the nectary, which consists of finger-like projections surrounding the ovary base. The keel with the pollination apparatus is pronouncedly curved. The style has a brush on its ventral side. Van der Pijl (1954) found *Lablab* to be pollinated by *Xylocopa* in Java.

Vigna (Phaseoleae)

The genus *Vigna*, with about 150 species, is the largest genus of the beans, and shows high diversity in floral symmetry. There are species with more simple, monosymmetric flowers (e.g. *V. unguiculata*) (Fig. 8.33.4–6) in addition to those with highly asymmetric ones.

Some species have flowers so extremely distorted that their architecture is difficult to visualize. One of the most extreme forms is the South American *V. caracalla* (Fig. 8.33.7–8). Its structure has been studied by Lindman (1902) (as *Phaseolus caracalla*) (see also Delpino 1868, p. 265; Troll 1951). The white and faintly pink and purple flowers are about 5 cm broad. Only the calyx is monosymmetric. The other floral parts are asymmetric. The floral bud becomes coiled (almost two full coils) like a snail shell. The mature style together with the distal part of the keel is tightly coiled (about four coils). The style, if artificially expanded, is 11 cm long. In the open flower the standard petal remains coiled and is curved backwards. It has a yellow guide mark near the nectar gate. The two very unequal wings are horizontally exposed, one upon the other. The lower wing forms a holding device for pollinators, the upper covers the coiled keel. Large bees (Lindman mentions *Bombus* and *Xylocopa*) press their head towards the base of the standard petal in a furrow between two auricles to reach the nectar. As in *Lablab*, the base of the upper free stamen is clasped in this furrow. The keel is pressed downwards. The coils become tighter. The stylar end with the stigma comes out of the keel. Pollen reaches the back of the visitor (nototribic or pleurotribic). Since the flower offers a large amount of nectar a visitor may remain for 30 seconds.

Vigna vexillata was studied in Costa Rica by Hedström & Thulin (1986) (earlier in South Africa by Scott-Elliot, 1891). The flowers are also asymmetric. The one-day flowers open before sunrise. They are pollinated by *Xylocopa* in the early morning. During a visit the style (provided with a brush) and the free parts of the stamens slip out of the rigid keel-beak. They slide over the dorsal part of the head and thorax of the bee and deposit and take up pollen from there. A visit lasts 7–8 seconds. Afterwards the pollination organs slip back into the keel. Hedström & Thulin (1986) note that the asymmetric, nototribic or pleurotribic flower form has evolved independently in several groups of the genus *Vigna* and in other genera of the Phaseolinae (probably in *Physostigma* and *Wajira*).

Erythrina (Phaseoleae)

Erythrina, with about 110 species, occurs throughout the tropics. It is a distinct and striking genus and many species are cultivated as ornamentals for their attractive flowers and seeds. Studies on pollination biology are relatively extensive. Probably all species are bird-pollinated (Raven 1979), perhaps except for *E. fusca*, which may be bat-pollinated (Paulus 1978). Almost all species have brilliant red or orange flowers. Exceptional species have yellow or multicoloured flowers with green parts (Raven 1974, 1977). The flowers may attain several centimetres in length. The petals are thick and fleshy. Probably all species are self-compatible (Neill 1987). Flowering is staggered. Anthesis of a flower is two days. An inflorescence flowers for two to three weeks. An individual may bloom for two or three months, and a population for four to five months or more (Neill 1987).

Two main pollination syndromes have evolved, one with perching birds and the other with hovering birds (Cruden & Toledo 1977). All of the *ca.* 40 Old World species and 15 of the 70 New World species are pollinated by perching birds of several families of the Passeriformes, while 55 of the New World species are pollinated by hummingbirds (Neill 1987).

Species adapted to perching bird pollination in general have horizontal inflorescences with reflexed flowers so that nectar can be reached from the peduncle. The flowers are widely open. The standard is expanded. The pollination organs are exposed. Pollen is often covered by abundant pollenkitt and is sticky (Hemsley & Ferguson 1985). Nectar is copious but low in sugar concentration; hexose is the dominant sugar (I. & H. Baker 1982). There are two subgroups: one pollinated by larger passerine birds, the other (African) by sunbirds (Cruden & Toledo 1977). The first subgroup has somewhat more open flowers and produces more nectar, which is rich in amino acids or proteins. Flowers are mostly homogamous (Neill 1987).

In contrast, species adapted to hummingbird pollination mostly have upright inflorescences and horizontally directed flowers, so that they can be easily reached by hovering. The peduncles are often spiny (Raven 1982). The flowers appear more tubular. The keel tends to be reduced; the standard is folded and conceals the other petals including the pollination organs (Neill 1987). Pollen is usually powdery and shows little pollenkitt (Hemsley & Ferguson 1985). Nectar content is smaller than in the other group but sucrose concentration is higher (Cruden & Toledo 1977). The flowers are mostly protandrous: male on the first day and female on the second day of anthesis (Neill 1987). The hummingbirds touch the reproductive parts with the throat, chest or base of the bill (Neill 1987).

In both syndromes traplining occurs: by orioles (Cruden & Toledo 1977) and by hummingbirds (Neill 1987).

The hummingbird-pollinated taxa are more uniform than the other group and are probably evolutionarily younger (Hemsley & Ferguson 1985, Neill 1987). However, they are included in six different sections and are probably derived from passerine-pollinated groups in several independent lineages (Neill 1987). In this context species that do not fit the two groups in every respect are of special evolutionary interest (Toledo & Hernandez 1979).

Mucuna (Phaseoleae)

The genus *Mucuna*, with about 100 species, occurs in the New and Old World tropics. Some species have spectacular inflorescences with large, brilliantly coloured flowers. In *M. longipedunculata* the keel of the flowers may reach 8 cm in length (van der Pijl 1941b). Some have greenish or whitish flowers. They are often in pendent racemes, in some species dangling on peduncles up to 3 m long (Vogel 1969a). The flowers have the normal orientation, however, with the keel turned downwards by twisting of the pedicels (Baker 1970). As in *Erythrina* the floral organs of *Mucuna* are thick and fleshy. Abundant nectar secretion has its structural basis in a highly lobed nectary collar around the ovary and massive secretory tissue.

In Papuasia, where 20 species are known (Verdcourt 1979), an impressive radiation involving different pollinator groups has taken place. Insects, lorikeets, small bats and nocturnal possums have been recognized as pollinators of different species (Hopkins 1992). In the Neotropics, too, bird-

and bat-pollinated species occur (Baker 1970). It would be interesting to study whether this diversification occurred independently in the Palaeotropics and in the Neotropics.

In bat-pollinated species the flowers open in the evening (e.g. *M. pruriens*, *M. reticulata*, *M. junghuhniana*, *M. gigantea*). The exposure of the pollinating organs is explosive. The tip of the keel is hard and probably acts as a support to bring about the explosion by a floral visit. Triggering of the explosion mechanism requires forceful action by the bats (van der Pijl 1941b, Vogel 1957). The explosion may be so intense that pollen is thrown out of the flowers (Vogel 1969a). Once opened the flowers have a strong smell, which disappears in the morning. Bird-pollinated species do not seem to be explosive (Vogel 1969a).

Arachis (Aeschynomeneae)

Arachis contains about 70 species; the principal cultivated species is the annual *A. hypogaea* (peanut or groundnut). Peanut flowers are strange in several ways (Fig. 8.33.9). They appear near the base of a shoot in the axil of foliage leaves or basal bracts; the axillary products of these leaves are very short shoots (partial inflorescences) containing a single lateral flower. The flowers have only 8 functional stamens, since two (the median adaxial and one of its neighbours) are highly reduced and form small, thread-like staminodes. All ten organs are united into a tube. The most unusual feature is the exceedingly long (*ca.* 2–3 cm) and narrow floral cup, which has the appearance of a pedicel. However, the flowers are sessile. The ovary with 2–5 ovules is sessile in the base of the floral cup and the long, thin style passes through the floral cup into the keel (Smith 1950). The 8 stamens are in two whorls; the outer has long anthers and the inner has smaller, reniform anthers. In an old bud the style surpasses the anthers. The long anthers release their pollen first. Then the lower, reniform anthers open and act like a piston by filament elongation: they push the released pollen upwards to the level of the stigma (Periasamy & Sampoornam 1984b). The flowers usually self-pollinate within the closed keel. They open early in the morning and are open for a few hours. In contrast to other Fabales the stigma is of the dry type. In *A. hypogaea* it is capitate; in perennial species it is surrounded by hairs (Lu *et al.* 1990). Possibly the capitate stigma without surrounding hairs is an adaptation to self-pollination. However, the breeding systems of the perennial species are poorly known. After anthesis the solid basal part of the ovary greatly elongates by intercalary growth. The resulting peg with the ovary on top curves towards the ground (positively geotropic) and grows into the soil (Periasamy & Sampoornam 1984a). The ovary remains covered by the remains of the outer floral parts, which form a protecting cap (Smith 1950). Fruit growth usually takes place underground. In wild species of *Arachis* the peg may grow horizontally for as much as one metre; the intercalary meristem also extends to the ovary, so the fruit may become long and break readily into segments (Pickersgill 1983).

The evolutionary history of this peculiar behaviour has not been analysed. In several Fabaceae amphicarpic behaviour with chasmogamous flowers that develop into normal fruits and cleistogamous flowers that develop into subterranean fruits is known (Cheplik 1987). The ancestors of *Arachis* may have been amphicarpic, and only cleistogamous flowers and subterranean fruits may have remained in *Arachis*.

In the closely related genus *Stylosanthes* with similar flowers (although with 10 well-developed stamens) *S. guianensis* was observed to be pollinated be a large range of different bees; *S. gracilis* was found to be partly autogamous (Pereira-Noronha *et al.* 1982).

8.6.4 Parallel evolutionary trends

The Fabales are a fabulous group to show parallel evolutionary trends in flowers (e.g. Polhill *et al.* 1981, Arroyo 1981). An often cited aspect is the parallel occurrence of 'papilionate flowers' in Caesalpiniaceae (*Cercis*, *Caesalpinia* p.p.) and Fabaceae (most genera).

Increase in stamen number

Another most obvious feature is the parallel increase in stamen number (from the probably basically ten) occurring in parts of all three families, and, at least in the Caesalpiniaceae, even in several subgroups (see Polhill & Raven 1981):

Caesalpiniaceae: Caesalpinieae (*Campsiandra*, 15–20; *Orphanodendron*, 16–17, see Barneby & Grimes 1990); Cassieae (*Mendoravia*, 11–12; *Storckiella*, 4–12); Detarieae (*Maniltoa*, up to 100; *Colophospermum*, 20–25; *Brownea*, 10–15); Amherstieae (*Polystemonanthus*, more than 25);

Mimosaceae: Ingeae (all groups, especially *Archidendron*, more than 100); Acacieae (all groups, *Acacia myrtifolia*, with up to 537, see Bernhardt & Walker 1984);

Fabaceae: Swartzieae (30 or more, esp. *Aldina*, *Swartzia*, *Cordyla*); Sophoreae (*Holocalyx*, 10–12, *Cyathostegia*, 20–30).

Increase in carpel number

A few of the genera with pronounced increase in stamen number in Caesalpiniaceae and Fabaceae also show a moderate increase in carpel number (e.g. *Archidendron*, *Affonsea*, *Swartzia*) (see Tucker 1987); more than one carpel also occasionally occurs in other Fabales. Multicarpely has sometimes been believed to be a primitive condition for Fabales, but it seems more plausible as a secondary phenomenon linked to polyandry, which is likewise an evolutionary secondary phenomenon in these highly specialized flowers.

Reduction in stamen number

The reverse, reduction in stamen number, also occurs in all families, in Mimosaceae e.g. in Mimoseae and Parkieae (4–5, rarely 3), in Fabaceae (rare), and most pronouncedly in the open monosymmetric flowers of Caesalpiniaceae, where reduction series occur in several groups. In *Bauhinia* there are many species with 2–3 fertile stamens, and reduction to a single (median?) stamen with a big anther has evolved in parallel in 3 sections of subg. *Bauhinia* (Wunderlin *et al.* 1987). Multiple parallel reduction to 3 or 2 (fertile) stamens is also present in three other tribes: Cassieae (*Apuleia*, *Dialium*, *Cicorynia*, *Labichea*, *Chamaecrista*), Detarieae (*Heterostemon*, *Elizabetha*, *Leucostegane*, *Endertia*), and Amherstieae (*Cryptosepalum*, *Gilbertiodendron*, *Pellegriniodendron*).

Synandry

Synandry has also evolved in all three families, although the stamens are always free at inception. Synandry is more prominent in monosymmetric flowers. Only rarely, however, is synandry equal between all stamens in monosymmetric flowers. Here, synandry is mostly one-sided in that the uppermost stamen is free. The '9+1-pattern' of most Fabaceae (see above) also occurs in many Caesalpiniaceae (esp. Detarieae and Amherstieae) and has a similar function in providing nectar gates.

Taxa with synandrous androecia:

Caesalpiniaceae: Caesalpinieae (*Jacqueshuberia*, *Sympetalandra*); Cercideae; Detarieae (*Brachycylix*; *Heterostemon*, 9+1?; *Baikiaea*; *Sindora*, 9+1; *Sindoropsis*, 9+1); Amherstieae (*Amherstia*, 9+1; *Berlinia*, 9+1);
Mimosaceae: Ingeae; Mimoseae;
Fabaceae: most tribes, 9+1.

Heteranthery

Heteranthery, where the functional stamens have different shapes within a flower, also occur in some pollen-flowers of Caesalpiniaceae (Cassieae) and Fabaceae (Swartzieae).

Corolla reduction

Reduction of the corolla to a single petal occurs in some Caesalpiniaceae and Fabaceae (*Swartzia*, *Ateleia*). Rarely there is no petal at all. In *Saraca* (Caesalpiniaceae) tepals have become petaloid and petal primordia have been 'transformed' into stamens (Corner 1958: 'transference of function'; Tucker 1989: ontogenetic 'transformation' of petals to stamens). In *Petalostemon* (Fabaceae–Psoraleae) five stamens have become petaloid to replace the lost petals; the papilionoid flower form has dissolved because the individual flowers are congested in dense spikes or heads (Faegri & van der Pijl 1979). However, Wemple & Lersten (1966) gave another morphological interpretation of these flowers: five stamens are reduced and four petals (with the exception of the standard) have become incorporated in the staminal tube in the place of the reduced stamens. This example shows the difficulty in applying the homoeosis concept (e.g. Sattler 1988) to flowers (see section 9.2.1).

Asymmetric flowers

In some Caesalpiniaceae and Fabaceae flowers are not only monosymmetric but tend to be asymmetric. In Caesalpiniaceae this is the case in species of *Cassia* s.l. and *Labichea* (Holm 1988) with enantiostyly, in *Labichea*, in addition, with two anthers of different size (Holm 1988); and in *Delonix* with the asymmetric nectar gate (also *Bauhinia*, Lindman 1902). In *Chamaecrista* the corolla is slightly asymmetric in that the standard petal has one outer (!) flank in bud (Okpon 1969) and also the standard petal tube is asymmetric (see Gottsberger *et al.* 1988, see above). In Fabaceae pronouncedly asymmetric flowers are known from *Vigna* and *Phaseolus* species (Lindman 1902, Hedström & Thulin 1986). *Vigna caracalla* and *V. appendiculata* are extreme cases, where only the very small calyx is symmetrical (Lindman 1902, Troll 1951). Slight asymmetry also occurs in *Lathyrus* (Teppner 1988) and in some other genera in conjunction with a style brush and pleurotribic pollination (Lavin & Delgado 1990).

On the other hand, some flowers are almost polysymmetric, often in connection with reductions of the petals. This occurs, for example, in *Ceratonia* and *Gleditsia* among Caesalpiniaceae, and in *Cadia* and *Amorpha* among Fabaceae (Tucker & Douglas 1994). In these genera the expansion of the petals occurs late, if at all, and there are also irregularities in the aestivation pattern.

Among Fabaceae the brush-type mechanism has arisen independently in eight larger groups (Lavin & Delgado 1990). The pump mechanism occurs in Liparieae, some Crotalarieae and Genisteae (Polhill 1976). The explosive mechanism occurs in Brongniartieae, and in some Genisteae, Indigofereae and Desmodieae, which are not all closely related within the family (Arroyo 1981).

Colour change

Pronounced colour change in flowers during anthesis occurs in numerous Caesalpiniaceae (*Delonix*, *Caesalpinia*, *Parkinsonia*) and Fabaceae (*Lablab*, *Castanospermum*).

Bird- and bat-pollination

Bird- and bat-pollination have evolved in all three families. Especially impressive is the frequent occurrence of bat-pollination.

8.7 Gentianales (Asteridae)

8.7.1 Asclepiadaceae

The Asclepiadaceae have the most elaborate, complicated flowers of all the dicots. These flowers are so unusual that it seems hard to imagine how they could have evolved. However, it is fascinating that they do not stand isolated. The neighbouring family, the Apocynaceae, helps in understanding the main probable evolutionary steps that have led to the extreme in the Asclepiadaceae.

On the one hand, the flowers show all the usual floral organs: sepals, petals, stamens, carpels; and their number is extremely stable: 5,5,5,2; among the *ca.* 2900 species of the family, not a single one is known with another regular organ number. Very rare deviations are six instead of five floral sectors, which have been observed in some species (but always as exceptions). Such an uniformity is probably not known from any other dicot family of comparable size. On the other hand there is unusual synorganization between parts, also between organs of different categories (Fig. 8.36). This has led to the evolutionary origin of new organs that are not present in other angiosperm groups. Some of these new organs are highly plastic in contrast to the relative uniformity of the conventional organs and may also show intraspecific variability. Synorganization of corolla and androecium has led to the origin of the 'corona' and the complicated canal system for nectar deposition. Synorganization of androecium and gynoecium has led to the formation of the gynostegium and the pollinaria. Synorganization of neighbouring stamens has led to the formation of 'guide rails (slits)' to attach the pollinaria to parts of the pollinator's body, and again,

Figure 8.36. Floral diagram of *Asclepias*; congenital and post-genital fusions between organs marked with thick lines; pollinaria black (modified after Endress 1990a).

to take up the pollinia sticking to the insect's body into the stigmatic chambers.

Not only are sympetaly and (in Asclepiadoideae) synandry consistent but also postgenital fusion of the apex of the totally apocarpous (!) gynoecium into a 'style head' and postgenital fusion of the anthers with the style head.

The mutual position of the inner organs (parts of the corona, stamens, carpels) are strongly fixed and precise.

The size of the flowers is very diverse. It fluctuates between a few millimetres (*Secamone*, *Echidnopsis*) and 40 cm (*Stapelia gigantea*) (Reese 1973). However, it is only the corolla that is so diverse in size, while the reproductive parts are always relatively small, owing to functional constraints.

The flower organs are thick and firm: a precondition for the high synorganization and precise fitting of the parts of the highly symmetrical flower.

With regard to its architecture, the flower is an extremely elaborate 'revolver flower' with compartmentalization into five sectors (sometimes secondarily more or less into ten). The flower has complicated internal chambers, and the fertile parts are hidden in these chambers (Kunze 1982b speaks of 'internation' of the fertile parts).

In early floral development the conventional floral organs are initiated and develop normally. Postgenital fusion of the style head occurs early. Formation of the new organs, corona and translators, takes place only later. The corona is initiated after corolla and androecium have reached a certain size (Hofmann & Specht 1986, Kunze 1990). It develops at the sympetalous and synandrous base of these primary organs. In this intercalary zone new morphogenetic activity arises and forms the novel organs. This late origin, after a certain consolidation of the primary organs, allows an extended development of the corona. Therefore, it now has the potential to develop into three-dimensionally complicated architectures.

Perhaps the most amazing organ of the asclepiadaceous flowers is the translator of the pollinarium. It consists solely of secreted material. The entire pollinarium, i.e. translator plus 2 pollinia, is morphologically heterogeneous: partly gynoecial and partly androecial. The pollinia, of course, are formed in the stamens, while the translator is secreted at the surface of the style head. (This secretion is evolutionarily derived from stigmatic secretion; see section 8.7.2.) The firm connection between anthers and style head by postgenital fusion is necessary for formation of the pollinarium. This complex organ, anthers plus style head, is called the gynostegium.

Pollinators are largely bees (Asclepiadeae) and flies (Stapelieae). *Sarcostemma viminale* is predominantly pollinated by Hymenoptera; anthesis of a flower is 4–5 days (Liede & Whitehead 1991). The highly elaborate pitcher flowers of *Ceropegia* show extreme adaptation for pollination by small flies (see below) (Fig. 8.37). But also the open flowers of other Stapelieae are highly elaborate fly-flowers, with carrion scent, complicated petal surface patterns (Ehler 1975), vibratile hairs or (in *Tavaresia*) peculiar vibratile blobs on the corona (Fig. 8.37) or (in *Stapelia longii*) bizarre black glistening bodies on the corona (Rauh 1979). Other extreme forms are *Calotropis* with large, massive flowers and highly protected nectar that are pollinated by *Xylocopa* (Cammerloher 1931, van der Pijl 1954, Wanntorp 1974, Ramakrishna & Arekal 1979, Eisikowitch 1986, Ali & Ali 1989) and *Stephanotis* with long, tubular, nocturnally scented flowers (Matile & Altenburger 1988). A flower of *Stephanotis* may be in anthesis for seven days. In *Hoya carnosa*, also with nocturnal scent, anthesis of a flower is four days (Matile & Altenburger 1988). Both are probably hawkmoth- or moth-pollinated (Vogel 1954). Forster (1992) found skippers (Hesperiidae) as pollinators in *Hoya australis*. In some species of several genera autogamy was observed (Kunze 1991).

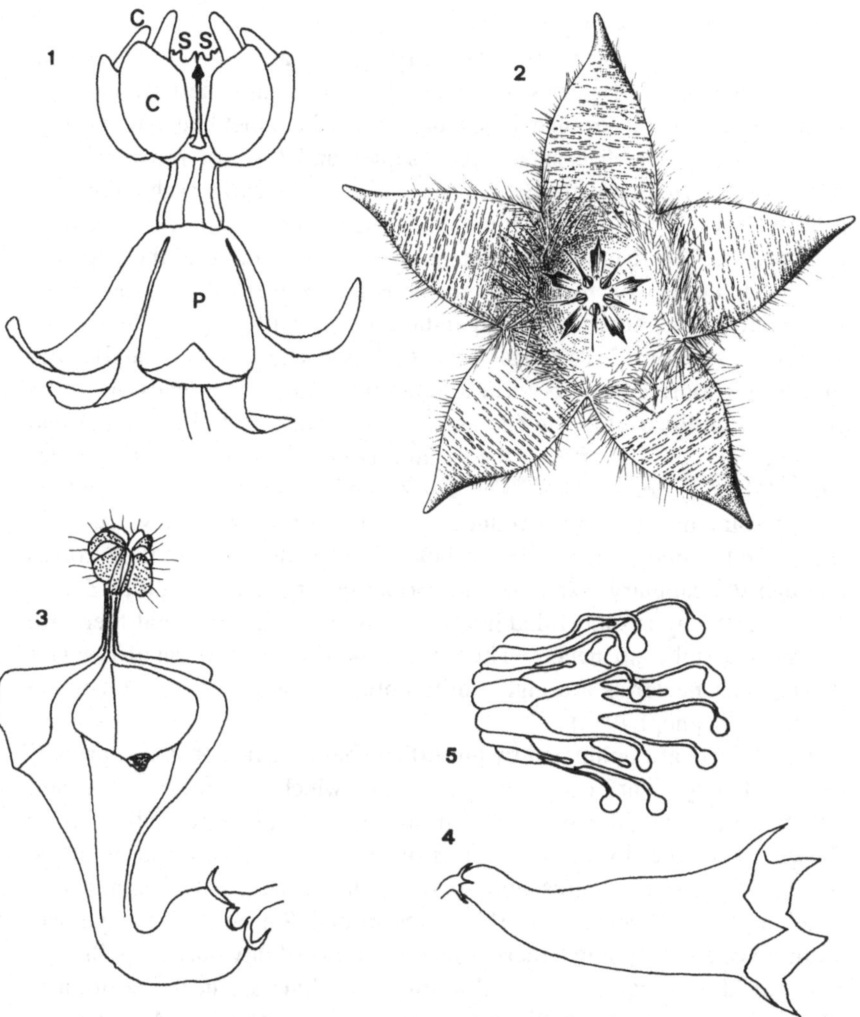

Figure 8.37. Flower forms of Asclepiadaceae. 1. *Asclepias curassavica*, from the side; C, parts of corona; P, petal; S, stamens, between them the black clip of a pollinarium visible, butterfly-pollinated (× 4). 2–4. Fly-pollinated flowers. 2. *Stapelia hirsuta*, from above, petals large, corona and pollination organs small (× 1). 3. *Ceropegia distincta* ssp. *haygarthii*, from the side, trap-flower with highly differentiated petals, forming five lateral entrances and a distal flag with vibratile hairs and osmophores, pollination organs and corona enclosed in the base (× 2). 4–5. *Tavaresia barclyi*. 4. Flower with funnel-shaped corolla from the side (× 0.3). 5. Corona with 15 parts, 10 of them with apical dangling blobs (× 2).

Asclepias (Asclepiadeae)

Among the predominantly North American genus *Asclepias* with more than a hundred species *A. curassavica* is a common herbaceous weed in the wet tropics. Its orange-yellow flowers are in umbel-like clusters (Figs 7.2.1, 8.37.1). The five small green sepals and the five red petals are reflexed. Alternating with the petals are five yellow cups, each containing a horn. These cups contain abundant nectar. Cups and horns constitute the corona. The five yellow stamens on the same radii as the cups are somewhat hidden by them. The stamens are congenitally united in the lower part. The anthers are postgenitally united with the style head. The anther flanks protrude as wings beside the cups. The firm (sclerified) wings of the neigbouring anthers form a long guide rail, which leads up to the clip of the pollinarium. The clip appears as a black spot at the upper end of the guide rail. The arms of the pollinarium are hidden below the anther wings (Galil & Zeroni 1969). The gynoecium is almost completely hidden by the stamens. Nectar is produced in the five narrow furrows between the united stamens, thus exactly below the guide rails. It then comes through the capillary system of the corona into the cups (Galil & Zeroni 1965). Each cup is subdivided into two parts by its horn so that there are ten sites to take up nectar. Thus, the flower is an elaborated revolver flower. The nectaries are highly differentiated epithelia without stomata (Christ & Schnepf 1988).

The flowers are pollinated by butterflies that take nectar. Legs, proboscis or other parts are caught by guide rails, which lead them to the clip. The butterfly's organ gets wedged in the clip and the entire pollinarium is drawn out of the flower. As it dries out in the air, the two arms of the pollinarium curve inwards and so the pollinia are easily inserted into another guide rail, as the butterfly moves around. Since this bending movement takes a few minutes, there is increased probability that the pollinator has moved to another individual when the pollinia are in the position to be inserted. Thus cross-pollination becomes more probable. Another feature to enhance cross-pollination is that insertion of pollinia is impeded if the pollinaria have been removed from a flower (Wyatt 1978; also mentioned for *Calotropis* by Ali & Ali 1989). There are five stigmatic regions on the undersurface of the style head, which are exactly located at the inner end of the guide rail. Thus a pollinium inserted in a guide rail will automatically come into contact with the receptive region, will stick to it and be detached from the translator, as the butterfly moves further.

The development of the flowers is shown in Figs 8.38, 8.39 and 8.40. The five sepals are initiated in a spiral and attain an imbricate-quincuncial aestivation (Fig. 8.38.3). In contrast, the five petals are initiated simultan-

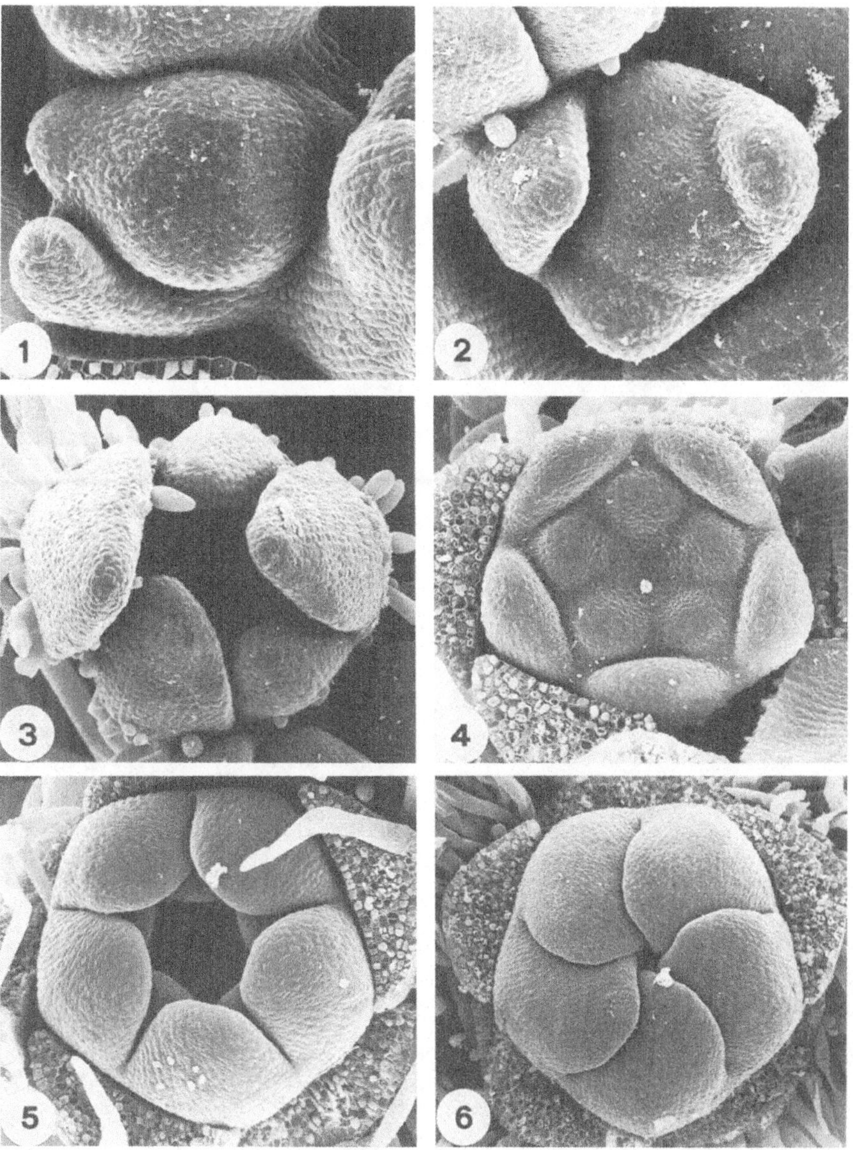

Figure 8.38. Early floral development of *Asclepias curassavica*. 1. Flower primordium with two lateral sepals initiated (× 250). 2. Sepals appear in spiral sequence (× 250). 3. Sepals with quincuncial-imbricate aestivation (× 180). 4. Petals and stamens appearing simultaneously in two whorls, sepals removed (× 200). 5. Beginning of contort (dextrorse) aestivation of petals (× 150). 6. Floral centre closed by contort petals (× 110).

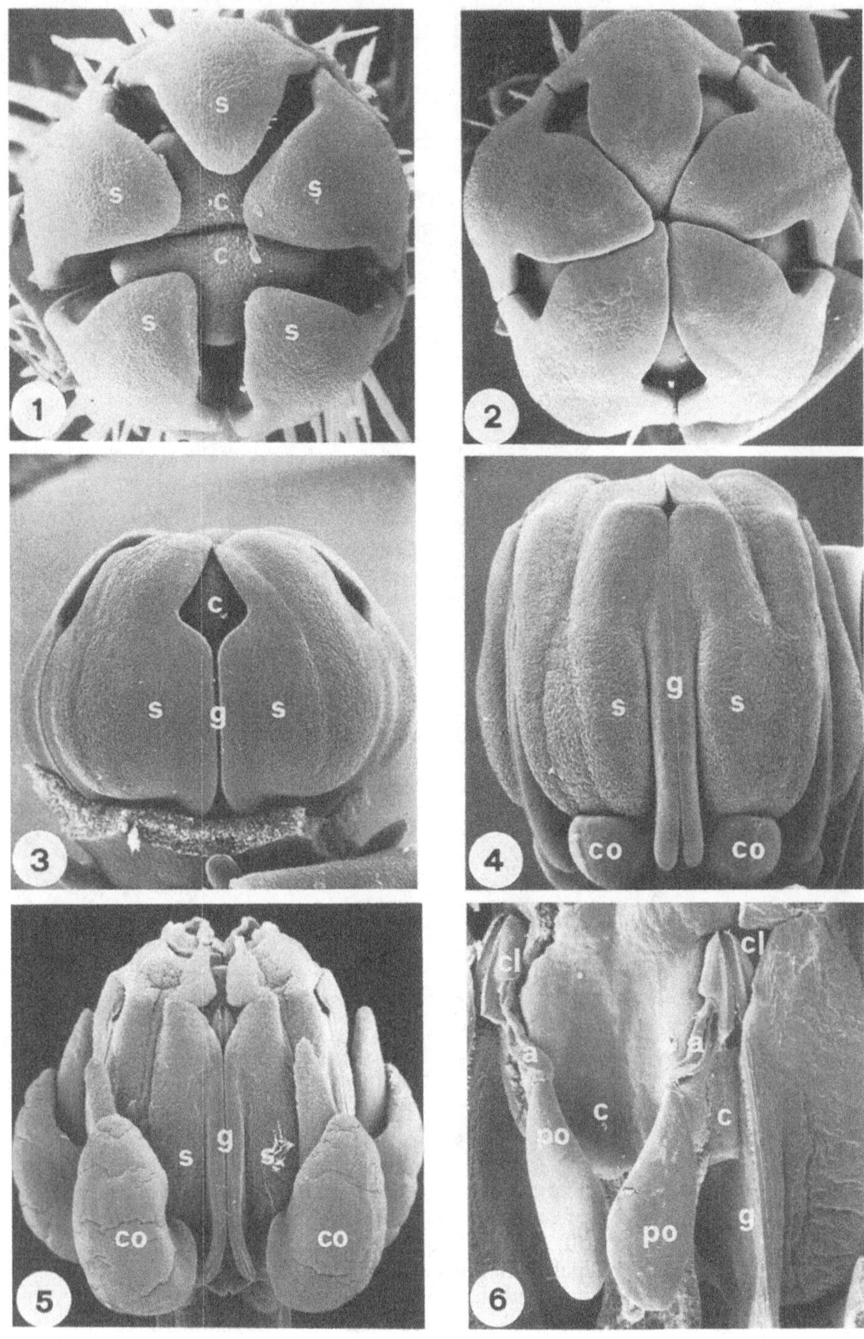
s
s
c
s
c
s
s
1
2
c
s
g
s
3
g
s
s
co
co
4
g
s
s
co
co
5
cl
cl
a
a
c
c
po
po
g
6

eously and attain a contort aestivation (dextrorse) (Fig. 8.38.4–5). The five stamens are also initiated simultaneously. Their flat apical appendages soon bend over the two young carpels in the centre (Fig. 8.39.1–2) and then completely cover the gynoecium apex. The two free carpels unite very early at the apex where later the style head arises. The style head attains five vertical edges because of the surrounding five anthers, which act like a mould (Fig. 8.40). The anther flanks that later form the guide rails are differentiated early, as is the narrower part, where later the clip of the translator is secreted on the style head (Fig. 8.39.2–3). Only at this stage do the young corona cups and horns become visible at the basal dorsal surface of each stamen (Fig. 8.39.4). The translators are secreted only shortly before anthesis (Fig. 8.39.5–6). Secretion begins in the centre of each of the five vertical edges of the style head between the stamens. This gives rise to the clip. Since each vertical edge is slightly concave, this concavity serves as a mould for the curved clip (Fig. 8.41). In detail, the two lateral parts of the clip are formed first, and then the middle part connecting them (Demeter 1922, Schnepf *et al.* 1979). Later the translator arms are secreted centrifugally, and finally the arms are attached to the neighbouring pollinia in the now opened thecae.

There are usually about seven to ten open flowers in a flower cluster (Willson & Melampy 1983). A flower stays open for several days. Nectar sugar concentration varies between 8 and 24%.

Asclepias curassavica is thought to form a guild together with *Lantana camara* and *Epidendrum radicans* (or *E. ibaguense*) (see section 7.6).

There is a comparatively large literature on the reproductive biology of various *Asclepias* species, particularly, however, on temperate species, such as *A. syriaca* and *A. tuberosa*. In particular, the balance between nectar and seed production, the effect of nectar robbing on pollination, and the success of pollinium insertion have been studied at the population

Figure 8.39. Later floral development of *Asclepias curassavica*; a, arm of pollinarium; c, carpel; cl, clip of pollinarium; co, part of corona; g, guide rail; po, pollinium; s, stamen (modified after Endress 1990a). 1. Anther apices beginning to cover the style head (× 80). 2. Gynoecium completely covered, except at the sites where later the clips of the pollinaria are formed (× 60). 3. Same from the side, with the first stage of the guide rail (× 60). 4. Later stage, corona elements arising (× 35). 5. Old flower bud, corona and guide rails differentiated, clips of pollinaria formed (× 19). 6. Flower at anthesis, one stamen removed to show the two pollinaria associated with it (× 35).

and species level (e.g. Morse & Fritz 1985, Willson & Melampy 1983). Anthesis of a flower takes three to six days. The plants are predominatly self-incompatible. Pollination is generally by bees and moths in *A. syriaca*. Sympatric species may share pollinators (Kephart 1983). It would be interesting to know whether pollinium placement or the specific site of pollen

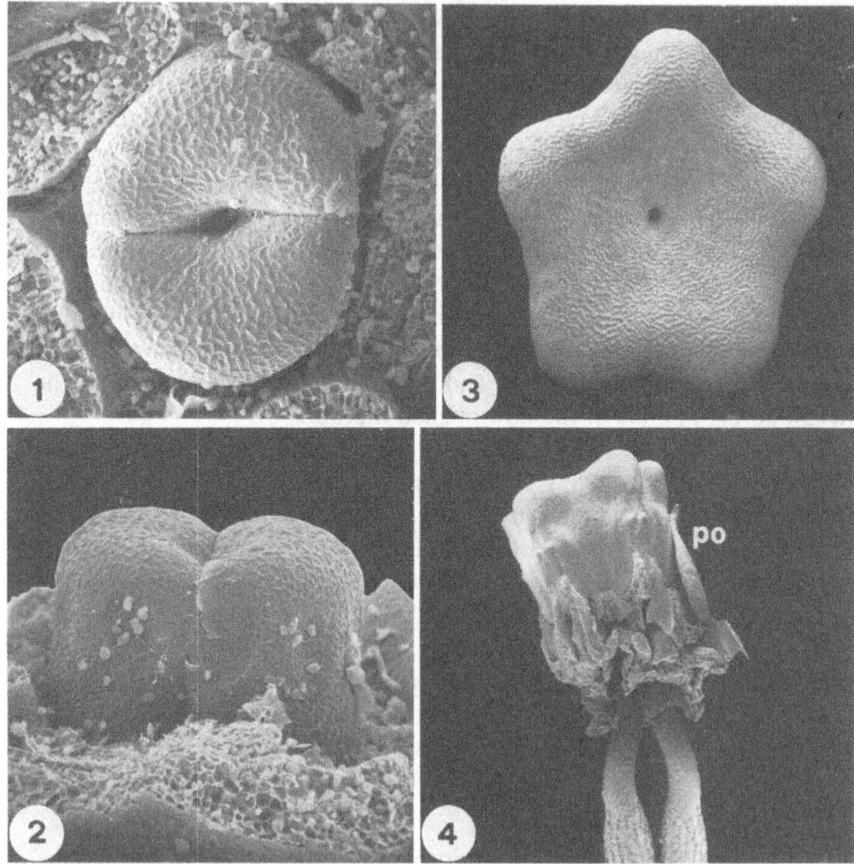

Figure 8.40. Gynoecium development of *Asclepias physocarpa* (modified after Endress *et al.* 1983). 1–2. Young carpels beginning to unite postgenitally. 1. From above (× 210). 2. From the side (× 220). 3–4. At anthesis, style head differentiated by complete postgenital union of carpel apices. 3. From above, five-angled shape of style head attained by the influence of the five stamens (× 25). 4. From the side, stamens and pollinaria removed, except for pollinarium at right (po), the two carpels free at the base (× 11).

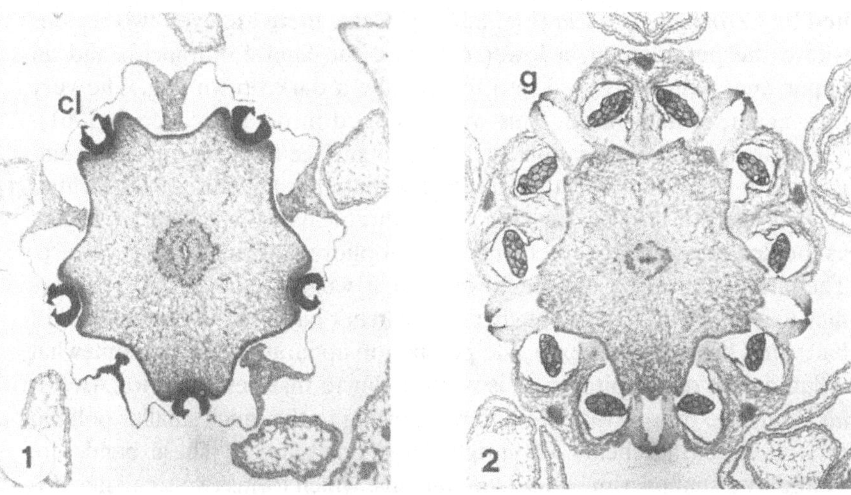

Figure 8.41. Style head of *Asclepias curassavica* at early anthesis, transverse sections (× 23). 1. At the level of the clips (cl) of the pollinaria, which have been secreted by the style head. 2. At the level of the guide rails (g) formed by the flanks of the anthers, pollinia visible in the anthers.

tube emergence from pollinia affects the extent of reproductive isolation between species (Macior 1965, Schill & Jäkel 1978, Rao & Kumari 1979). Pollinia may remain viable for as long as two weeks (Morse 1985). Commonly not more than one fruit per flower cluster and only one of the two carpels of the gynoecium develops (Morse 1985).

Ceropegia (Stapelieae)

The Stapelieae are principally adapted to fly-pollination and have reached a great diversity of flower architectures (Leach 1978, Rauh 1979, Dyer 1983, Bruyns & Forster 1991). The most complicated flowers are found in the genus *Ceropegia*, which, in itself, is a wealth of wonderful radiation (Huber 1957, Vogel 1961). It contains about 150 species in dry parts of Southern Africa and tropical Asia. In addition to the complicated pollination apparatus, as described above, the corolla is highly elaborated into a pitcher trap construction. The corolla forms a tube by congenital union of the five petals. The free parts of the petals are postgenitally united in bud. Either they open and spread at anthesis, or, in most species, they open but remain attached to each other at the very tip, thus forming five lateral entrances. One of the most complicated architectures is exhibited by *Ceropegia distincta* (Fig. 8.37.3). Here, there are even two regions where the petals open, a lower one with the lateral entrances, and an upper one, which is raised into the air like a dark brown flag. The very tips, again, remain fused. This was described in detail by Vogel (1961). There are several elaborations on the inner surface of the corona, concomitant with fly-pollination, that are differentiated in the closed bud (Müller 1926, Vogel 1961). The flag has a set of vibratile hairs, which are presented as the flower opens, together with an osmophore covering the flag surface. The lateral entrances into the flower have a wax-secreting surface, causing flies to slide down into the pitcher. The pitcher has a narrow part with stiff hairs that prevent their exit. The pollination apparatus is at the somewhat inflated base of the pitcher. It is very similar to that described for *Asclepias*. Only the proportions are different, owing to the much smaller pollinators. Each anther bears an upright band on its back. These bands are united by a shallow rim behind the stamens, which forms two small nectariferous bowls between the stamens, thus, again ten in the entire flower (Fig. 8.44). The corolla is dark in colour inside the pitcher, except for a window-pane at the very base around the pollination apparatus. In contrast, the corona is vividly coloured with brown, yellow and red and may therefore have optical guiding functions.

In contrast to *Asclepias* the anthers do not form deep guide rails but the five vertical edges of the style head protrude outwards between the anthers. Thus the clip of the pollinarium is exposed at the surface of the pollination apparatus. In addition, the pollinia are at a higher level than the clip. This may help in positioning them in the right place at pollination. In contrast to *Asclepias*, the guide rails are so narrow that they cannot take up an entire pollinium but only a narrow part at its inner edge. Insertion is facilitated by a hook-like projection of the pollinium edge. This edge contains the only site where pollen tubes can emerge from the pollinium (Volk 1950, 1951).

Pollinators are small flies (Milichiidae, Chloropidae, Ceratopogonidae), groups that are usually attracted by the smells of various animals and live as nest commensals or parasites (Vogel 1961). They grasp the corona bands behind the anthers and take nectar from the nectar bowls.

Anthesis of a flower takes two days (it may be four days in other *Ceropegia* species) (Vogel 1961). At the end of anthesis the flower is tilted in a horizontal position, the hairs in the narrow part of the pitcher wilt, and the flies can escape.

The elaboration of the corolla has its foundation in early development. In contrast to *Asclepias*, the petals do not become imbricate-contort but valvate (Fig. 8.42). This valvate configuration enables the postgenital fusion of their margins and formation of the complicated apparatus with the lateral entrances and the apical flag (Fig. 8.43). Further, the pitcher tube is formed by early elongation of the congenitally united petal bases (Fig. 8.42).

The entire group of the Stapelieae (Bruyns & Forster 1991) has valvate petal aestivation, also those with widely open corollas (e.g. *Stapelia*). The valvate aestivation of the relatively thick organs enables a tight bud closure. This may also be seen as a preadaptation to the successful radiation of the group in dry regions of southern Africa. Dannenbaum & Schill (1991) also point to the compact pollinia with thick exines that protect against water loss, which could be seen as an additional preadaptation for life in arid habitats.

8.7.2 Evolutionary steps towards the asclepiad flowers

The evolution of the highly complex flowers of Asclepiadaceae–Asclepiadoideae can be traced from more simple flower forms in Apocynaceae–Plumerioideae via successively more complicated forms through Apocynaceae–Apocynoideae and Asclepiadaceae–Periplocoideae (Schick 1980, 1982a, Fallen 1986, Kunze 1982a,b, 1990, 1991).

(1) In the Apocynaceae–Plumerioideae the anthers are still free from the style head. The anthers have four pollen sacs in two thecae. In the great majority of genera pollen is released as single grains. In some genera the entire style head constitutes the stigma, which is covered by a sticky secretion. Lobes in the petal sinuses in some taxa are precursors of a corona. The stamens are congenitally fused with the petals. Nectar is secreted on a disk around the ovary or on the ovary itself or on the petals (Schick & Remus 1984). Five nectar gates between the five stamens represent the only sectorial differentiation of these revolver-flowers.

(2,3) In the Apocynaceae–Apocynoideae the style head becomes involved in the revolver construction. The anthers become postgenitally fused with the style head in various ways. On the style head there are five zones between the stamens with sticky secretion functioning as a glue for pollen transport. In *Apocynum*, in addition to the sticky secretion, a discrete platelet for pollen reception (translator) is secreted that rolls around a probing proboscis and takes up pollen (tetrads) with its sticky upper surface. The stigma becomes restricted to the lower edge of the style head. Pollen is dispersed either as single grains or as tetrads.

(4) In the Asclepiadaceae–Periplocoideae the secreted platelet (translator) is differentiated into a spoon, which receives pollen, and a stalk. Sticky secretion sticks pollen to the spoon and the stalk to the pollinator (Schick 1982b). Pollen is released in tetrads. The corona has become

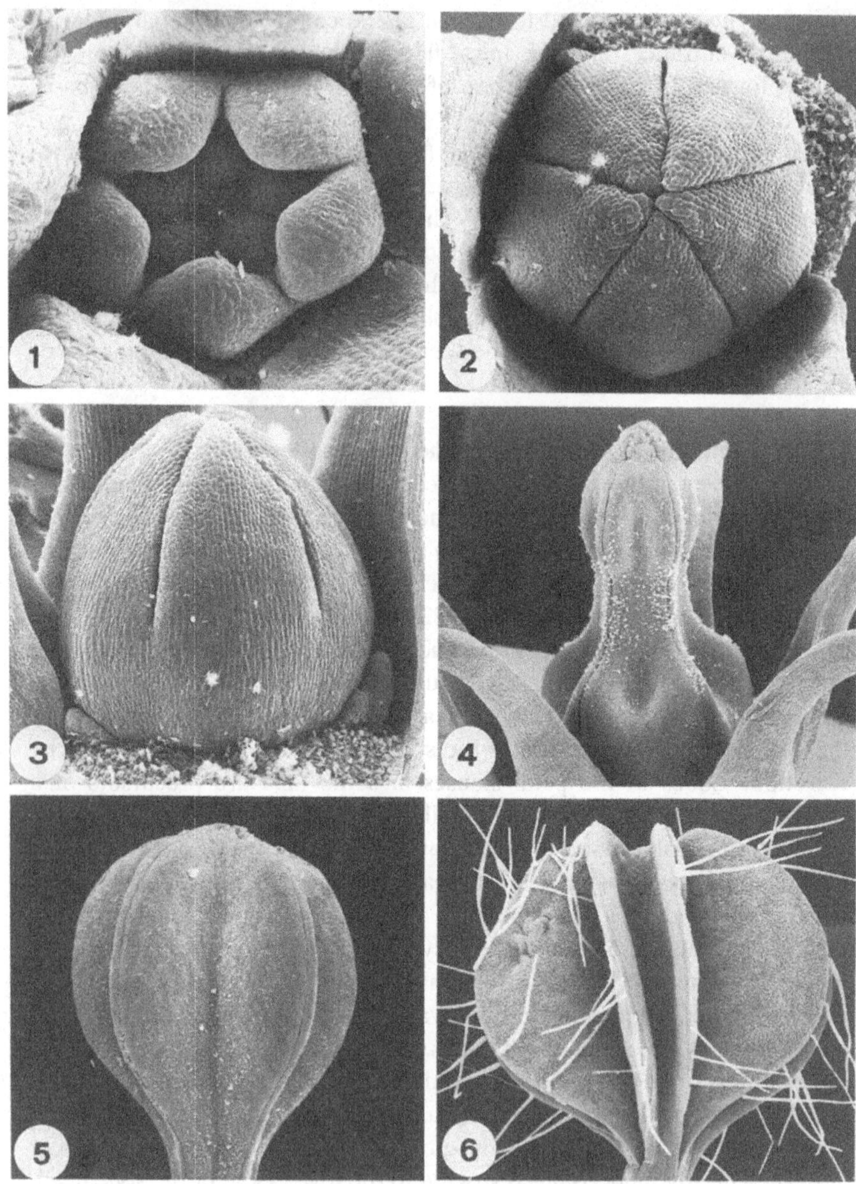

Figure 8.42. Floral development of *Ceropegia distincta* ssp. *haygarthii*. 1. The five young petals forming a whorl, in the centre the five stamen primordia visible (× 180). 2. The five petals in valvate aestivation (× 90). 3. Same from the side to show the congenitally united petal base (× 90). 4. Upper part of

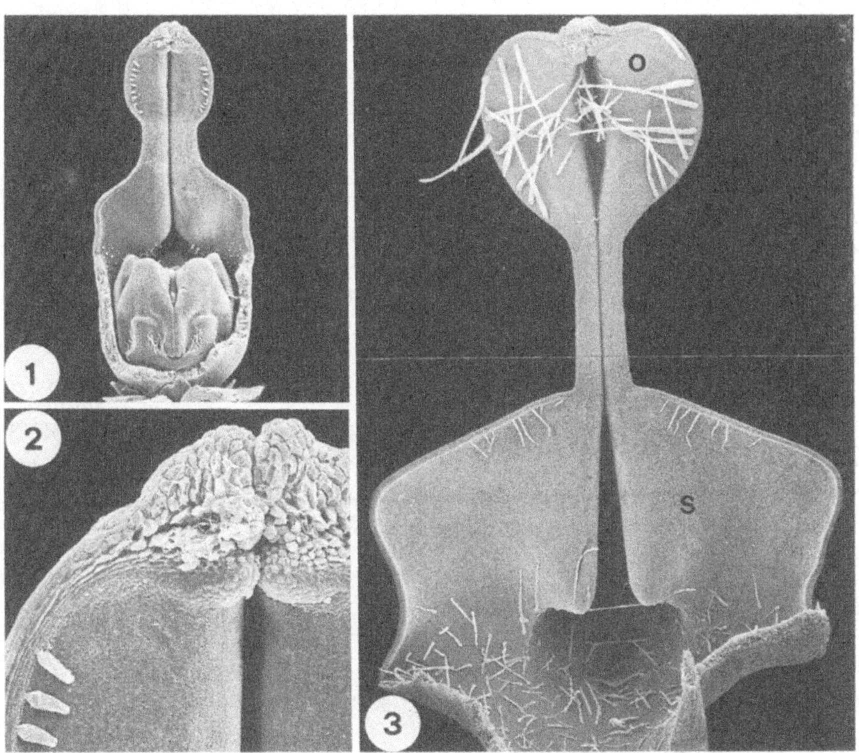

Figure 8.43. Differentiation of inner side of corolla in *Ceropegia distincta* ssp. *haygarthii*. 1. Flower from the side with three petals removed (× 15). 2. Uppermost part of the same flower, showing postgenital union and young vibratile hairs (× 90). 3. Upper part of corolla of flower at anthesis with three petals removed; o, osmophore; s, slide zone of the pitcher (× 13).

closely associated with the pollination apparatus, thus more closely with the androecium than with the corolla. It is optically attractive and serves as a holding device for pollinators. As in Apocynaceae the stamens are congenitally fused with the corolla, forming a common tube with it. Nectar is secreted on the inner wall of this tube in the petal sectors between the

Fig. 8.42 (*cont.*)
corolla differentiated into constricted and expanded zone (× 25). 5. Apical expanded zone of corolla in older bud (× 20). 6. Apical expanded zone of corolla at anthesis, with postgenital sutures opened and osmophores and vibratile hairs exposed (× 13).

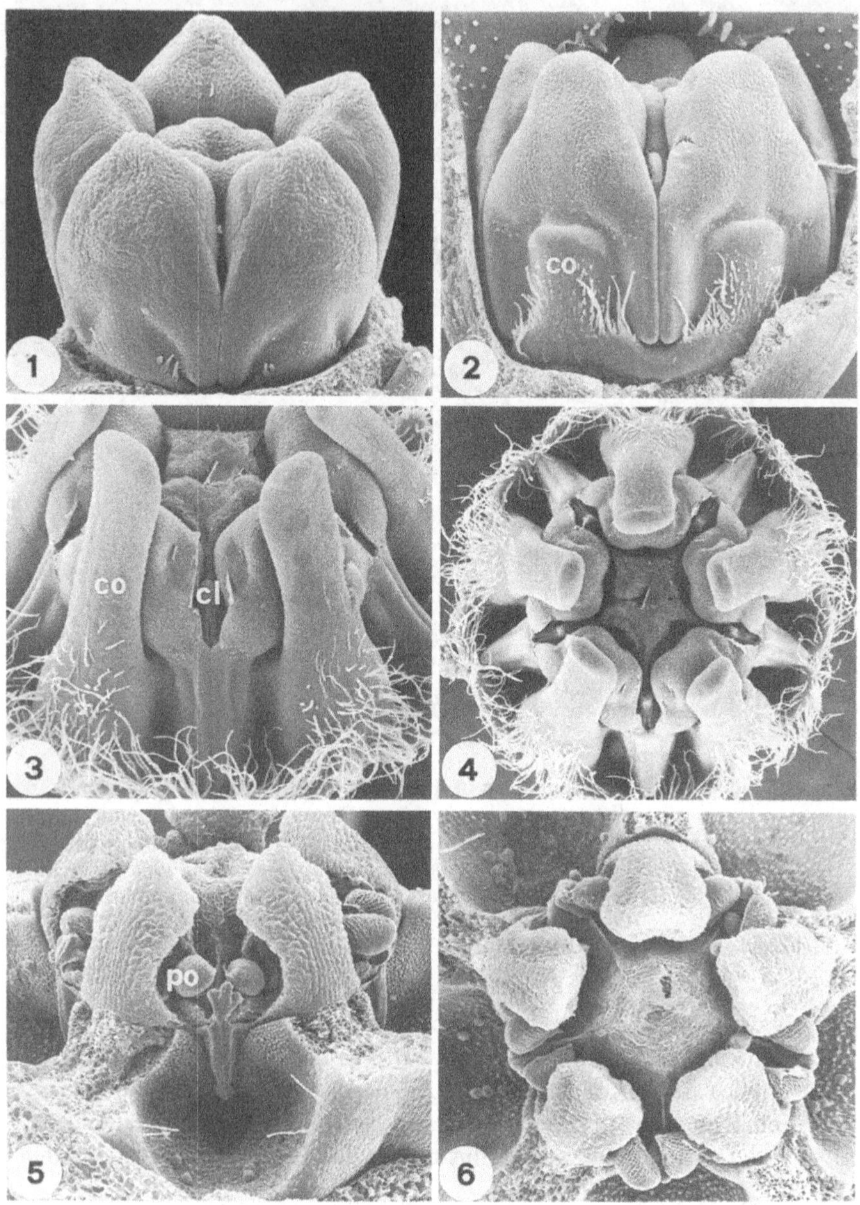

Figure 8.44. Development of pollination apparatus in *Ceropegia*. 1–4. *C. distincta* ssp. *haygarthii*. 5–6. *C. woodii*. 1. Stamens before corona initiation (× 70). 2. Corona elements (co) initiated on dorsal side of stamens (× 45). 3. Older bud, corona elements (co) surpassing the anthers,

stamens. In some genera with open, disk-like flowers that are fly-pollinated (e.g. *Periploca*) the nectaries are partly exposed, flaring out onto the petals.

(5) In the Asclepiadaceae–Asclepiadoideae the stalk of the translator is transformed into a clip (Schick 1982b). Adhesive attachment has been replaced by the clamp mechanism of the clip. Instead of a stalk two lateral arms are differentiated as new formations. The spoon has been lost. Pollen is shed as pollinia. Only the inner one of the two pollen sacs of a theca is fertile (Dannenbaum & Schill 1991), while the outer anther flank is differentiated into one half of a guide rail. As an interesting exception, *Secamone*, with extremely small flowers, still has tetrads and two pollen sacs per theca (reminiscent of Periplocoideae) (Safwat 1962). The corona is still more differentiated in forming a complicated ring system around the pollination organs; it serves, in addition, as a nectar holder. The stamens are congenitally united at the base, which forms a massive ring around the gynoecium. Nectar is secreted on the outer side of this ring in deep sinuses between the stamen bases immediately below the five stigmatic sites of the style head, and, at least in some groups, apparently plays a role in pollen germination, in that germination occurs without any contact with the stigma (Kunze 1991).

The evolution of these complicated flowers shows successive synorganization of previously more independent organs and, at the same time, successive steps of differentiation by separation and transference of functions to newly emergent organs or organ parts (Tables 8.1, 8.2). In detail, there is again considerable diversity on each of these five evolutionary levels (Fallen 1986).

Other evolutionary trends involve the unusual step from syncarpy (Apocynaceae–Plumerioideae–Carisseae) to apocarpy (other groups) and change in the function of the compitum within Asclepiadaceae. In the apocarpous groups the carpel apices are postgenitally fused into a style head thus forming a compitum, which allows regular distribution and centralized selection of male gametophytes (Endress *et al.* 1983). In the Asclepiadaceae–Asclepiadoideae, however, the change from free tetrads to pollinia and the frequent specialization in forming fruits of only one of the two carpels per flower makes a compitum not only superfluous but even disadvantageous. The lack of a compitum in some Asclepiadoideae

Fig. 8.44 (*cont.*)

clip (cl) of pollinarium secreted between the anther flanks (× 35). 4. Same flower from above (× 25). 5. At anthesis, the two released pollinia (po) flanking the visible clip of a pollinarium (× 50). 6. Same flower from above (× 45).

Table 8.1. *Initiation and synorganization of structural elements during development in asclepiad flowers (modified after Endress 1990a)*

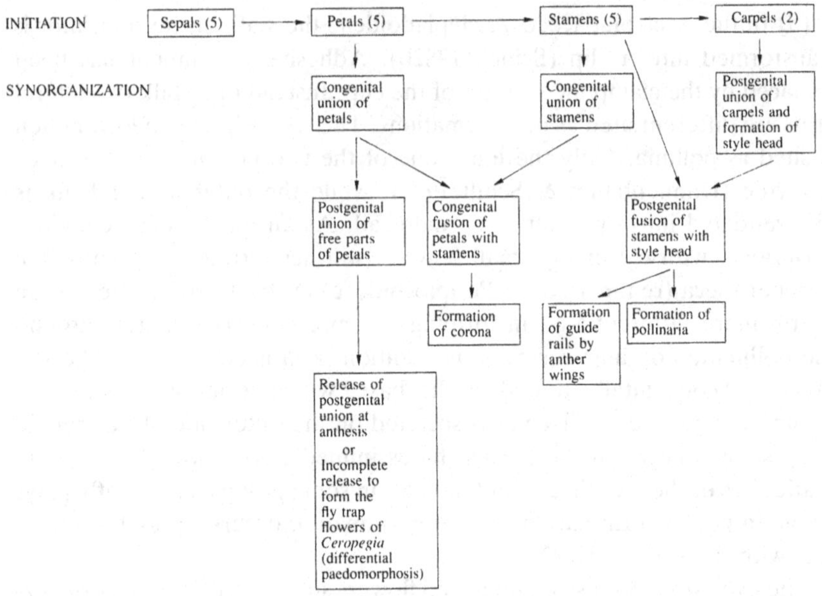

may be interpreted in this light. The lack of a compitum in at least some Asclepiadoideae had been surmised by Endress *et al.* (1983) and was established by Sage *et al.* (1990) for *Asclepias* and by Kunze (1991) for some Ceropegieae and Asclepiadeae, while a compitum is present in Secamoneae and Marsdenieae. Evidence for a functional compitum was also given by Walker (1978) for *Catharanthus* (Apocynaceae–Plumerioideae–Plumerieae). In these most advanced taxa the postgenitally united style head has concentrated on the function of translator formation.

A peculiarity is the change in the petal aestivation pattern. In Apocynaceae–Plumerioideae it is mostly contort to the left. In Apocynaceae–Apocynoideae, Asclepiadaceae–Periplocoideae and the lower Asclepiadaceae–Asclepiadoideae it is quite constantly contort to the right. Only in the Asclepiadoideae–Stapelieae does it become valvate, concomitant with the more elaborate corolla (see above).

In the Apocynaceae salverform flowers are predominant. They also occur in various groups of the Asclepiadaceae, e.g. *Cryptostegia*, *Ectadium* and *Gymnanthera* (Periplocoideae) and *Stephanotis* (Asclepiadoideae). However, in Asclepiadaceae the outstanding pollination apparatus has acquired other means of canalizing the array of pollinators. Therefore, flowers with a reduced floral tube occur more often.

Table 8.2. *Evolution of the asclepiad flower by progressive elaborations and separation of functions*

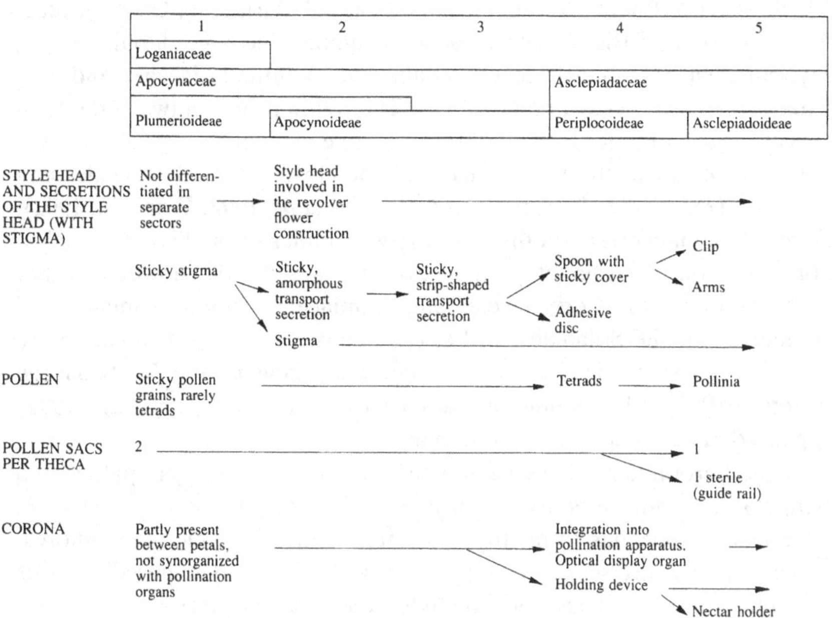

As to the floral mode in Apocynaceae, emphasis is on lepidopteran and hymenopteran pollination, while in Asclepiadaceae hymenopteran and dipteran pollination are predominant. However, it is interesting to see how, in the course of evolution, as outlined above, similar radiations of floral biological modes have arisen. Fly-pollination with open flowers occurs in *Apocynum* (Apocynaceae–Apocynoideae) (Schick 1982a), and in many Asclepiadaceae–Periplocoideae, and is a main component of Stapelieae (Asclepiadaceae–Asclepiadoideae). Pollination by *Xylocopa* occurs in *Cryptostegia* (Asclepiadaceae–Periplocoideae) (Schick 1982b) as well as *Araujia* (Scotti 1911), *Pachycarpus* (Vogel 1954), and *Calotropis* (Asclepiadaceae–Asclepiadoideae) (see above). A sphingophily syndrome occurs, for example, in *Carissa* (Apocynaceae–Plumerioideae), *Beaumontia*, *Macrosiphonia* and *Nerium* (Apocynaceae–Apocynoideae) (Church 1908, Vogel 1954, Schick 1982a, Herrera 1991), *Stephanotis* and *Hoya* (Asclepiadaceae–Asclepioideae) (see above). In the few Apocynaceae studied to date, self-incompatibility seems to be common (Rowley 1980), as also mentioned above for *Asclepias* (Asclepiadaceae). Pollination by birds and mammals seems to be largely absent in Apocynaceae and Asclepiadaceae because of the complicated pollination apparatus that requires precise action by smaller animals.

8.8 Scrophulariales (Asteridae)

The Scrophulariales contain about ten families and more than 10 000 species, the largest families being Scrophulariaceae, Gesneriaceae, Acanthaceae and Bignoniaceae. Smaller families include Globulariaceae, Myoporaceae, Orobanchaceae, Pedaliaceae, Lentibulariaceae, and perhaps also Buddlejaceae and Oleaceae. The relationships between the families are very intimate. However, their relative phylogenetic positions have not been resolved; the same is true for the closely related Solanales and Lamiales (Downie & Palmer 1992, Olmstead *et al.* 1992, Wagenitz 1992). Here I will concentrate on the four largest families of the Scrophulariales. For the demonstration of parallel evolutionary trends, however, other families, including Verbenaceae and Lamiaceae of the Lamiales and Solanaceae of the Solanales, will also be considered. The Lamiaceae are probably polyphyletic from more than one clade of the Verbenaceae (Cantino 1992). The volume on Lamiaceae by Harley & Reynolds (1992) appeared after completion of the manuscript.

In the Scrophulariales and Lamiales monosymmetric, synsepalous and sympetalous lip-flowers with a prominent corolla are very common. In terrestrial herbaceous forms they are often arranged in elongate inflorescences (spikes, thyrses) where they can be horizontally exposed towards the side. In trees, vines and epiphytes the inflorescences are different. As in Gentianales the pentamerous floral groundplan with a dimerous gynoecium is very stable. Concomitant with monosymmetry is the almost universal loss or at least strong reduction of the median stamen (which is always the adaxial one). In the corolla there are two upper and three lower petals. They mostly form an upper and a lower lip. The lower lip is usually differentiated as a landing platform and is provided with nectar guides. In the most elaborate flowers the pollination organs are more or less enclosed in the upper lip. These flowers are then comparable with reversed flag-flowers (see section 3.6) (Fig. 3.7). As in flag-flowers the main pollinator group in these lip-flowers are Hymenoptera, although there is a wide spectrum of pollinators: many tropical groups are bird-pollinated (the Neotropical Gesneriaceae predominantly so), some are bat-pollinated. The stamens are always congenitally fused with the corolla to some extent (see 'stapet', section 2.2.3). Because of this and the complicated, firm structure of the corolla, the corolla is not retarded in development and it has a protective function for the inner floral organs for a long time during floral development.

A feature that occurs in very few other angiosperms is a 'pollen sac placentoid', a protrusion of parenchymatic tissue into each pollen sac from the connective side. It was found to be characteristic for most families of the Scrophulariales and, in addition for Verbenaceae, Lamiaceae, Solanaceae, and some Convolvulaceae (Hartl 1963) (Fig. 8.53.2). It is of special phylogenetic interest because it also occurs in a few families of the Rosidae that have been considered as potential 'ancestors' of Asteridae (Endress & Stumpf 1991).

The gynoecium is syncarpous, and the two carpels are congenitally united up to the stigmatic region. As is common in Asteridae, the ovules are tenuinucellar, unitegmic and predominantly anatropous.

Parallel evolutionary trends abound in these lip-flowers (Vogel 1954, 1963b). They will be discussed below (section 8.8.5).

8.8.1 Scrophulariaceae

The cosmopolitan Scrophulariaceae, with *ca.* 4500 predominantly herbaceous species, are outstanding among the Scrophulariales and Lamiales by their range of diversity. The internal systematics are far from settled. Two subfamilies, Scrophularioideae (=Antirrhinoideae) and Rhinanthoideae, are recognized (Thieret 1967). Tribal subdivision, however, is unclear to date, perhaps largely owing to floral plasticity and parallel evolution.

Owing to the great plasticity there are few distinctive characteristics in floral structure. The anthers are crescent-shaped in many groups and the thecae tend to be confluent over the apex or to dehisce over the apex (Trapp 1956, Endress & Stumpf 1990). The bilocular ovary tends to contain a large number of ovules on axile, protruding-diffuse placentas, although there are reductions in ovule number in several taxa (Hartl 1956a). In contrast to Bignoniaceae with likewise bilocular ovaries, the synascidiate portion tends to be more developed and, concomitantly, each ovary locule has a single placental protrusion (not two separate ones) and the ovary wall contains only two (dorsal) major vascular bundles (not four, i.e. two dorsal and two lateral ones) (van Steenis 1927, 1949, Armstrong 1985).

There are few studies on the anthecology and breeding systems of tropical representatives of the Scrophulariaceae. The bee- and hummingbird-pollinated Maurandyinae (Antirrhineae) have nectar-flowers, open or personate, with a diversity of breeding systems, including homogamy, dichogamy (protandry and protogyny) and dichogamy combined with herkogamy, obligate outcrossing or autogamy (Elisens 1985). Other Antirrhineae are predominantly self-incompatible (Sutton 1988). The large genus *Calceolaria*, together with *Angelonia* and *Monttea*, all in South America and partly tropical, largely have oil-flowers pollinated by anthophorid bees (Vogel 1974, Vogel & Machado 1991, Molau 1988, Simpson *et al.* 1990).

In *Calceolaria* outcrossing is predominant, although many species are self-compatible (Molau 1988). Many *Calceolaria* species have protogynous flowers with the stigma receptive 2–5 days before the stamens open; sect. *Calceolaria* with predominantly annual herbs is autogamous (Molau 1988). About a quarter of the *Calceolaria* species are pollen-flowers, lacking an elaiophore. However, the occasional occurrence of elaiophores in them suggests that they have evolved from oil-flowers (Molau 1988). *Angelonia* has protandrous flowers; the perennial species are self-incompatible, but the annual *A. pubescens* is self-compatible and may be autogamous (Vogel & Machado 1991). *Monttea* has one-day flowers; it is weakly self-compatible and minimally autogamous (Simpson *et al.* 1990).

All other families of the Scrophulariales–Lamiales complex seem to be more narrow in their circumscription. Among the larger families the closest relationships are probably with Acanthaceae, Bignoniaceae and Gesneriaceae. There are a number of genera of doubtful position at the familial borderlines. Almost all of them have been placed in the Scrophulariaceae after closer examination mainly of the floral structure. This may reflect the greater diversity of this family, which may accept problematical groups more easily than the more narrowly defined neighbouring families.

At the borderline between Scrophulariaceae and Gesneriaceae, the genera *Rehmannia* and *Brookea* (Burtt 1963), *Cyrtandromoea* (Burtt 1965), and *Charadrophila* (Weber 1989b) find a better place in the Scrophulariaceae.

At the borderline between Scrophulariaceae and Bignoniaceae, the genera *Paulownia* (Campbell 1930, Armstrong 1985), *Wightia* (van Steenis 1949, 1977), *Schlegelia* and *Gibsoniothamnus* (Williams 1970, Leinfellner 1973a, Armstrong 1985) are better placed in the Scrophulariaceae. However, Cronquist (1981) prefers to place Schlegelieae and *Paulownia* in the Bignoniaceae.

At the borderline between Scrophulariaceae and Acanthaceae, Nelsonioideae are placed in the Scrophulariaceae (Bremekamp 1953). However, Johri (1963) and Cronquist (1981) prefer a position in the Acanthaceae (also of the Thunbergioideae).

At the borderline between Scrophulariaceae and Myoporaceae, *Oftia* fits better with Scrophulariaceae (Dahlgren & Rao 1971).

8.8.2 Gesneriaceae

In contrast to Scrophulariaceae, the family Gesneriaceae is almost exclusively tropical. It contains *ca.* 3000 species (Wiehler 1983). Both of the major subfamilies, the New World Gesnerioideae and the Old World Cyrtandroideae, have *ca.* 1500 species. Herbaceous and shrubby species are predominant; a relatively large number are epiphytes. Among the tropical Scrophulariales, the Gesneriaceae is the most versatile family with respect to floral structure and biology.

A prominent feature in the family is that the four anthers tend to be postgenitally united, either pairwise or all four together; if four together, they may be arranged in a parallel or diagonal pattern. The postgenital union occurs only shortly before anthesis (Trapp 1956, Lamond & Vieth 1972) (Figs 8.45, 8.46). These patterns may be constant within genera. However, in *Monophyllaea*, for example, three different arrangement patterns occur, which sheds light on their potential for parallel evolution (see Burtt 1978). Synorganization is especially elaborate in *Drymonia* (Wiehler 1983). The four anthers become connate by the length of the thecal mar-

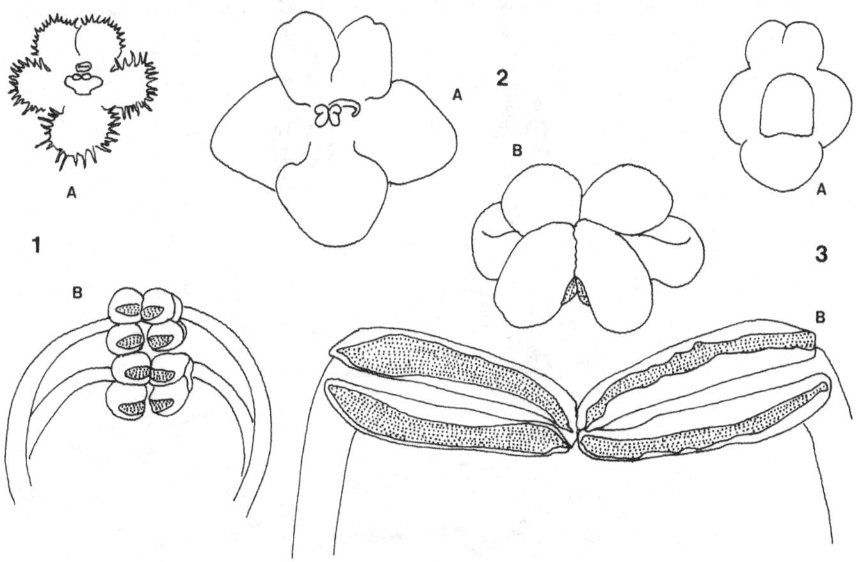

Figure 8.45. Flower forms, anther synorganization and anther opening in Gesneriaceae. 1. *Episcia dianthiflora*; pollinated by female euglossine bees. A. Frontal view of flower (× 1). B. Synorganization of the four stamens (× 6). 2. *Saintpaulia ionantha*; buzz-pollinated by bees. A. Frontal view of flower, style asymmetric (enantiostyly) (× 1). B. Synorganization of the two stamens, anthers with pore-like openings (× 6). 3. *Aeschynanthus speciosus*; pollinated by birds. A. Frontal view of flower with wide mouth, pollination organs not drawn (see also Fig. 8.46) (× 1). B. Synorganization of two of the four stamens with large anthers (× 6).

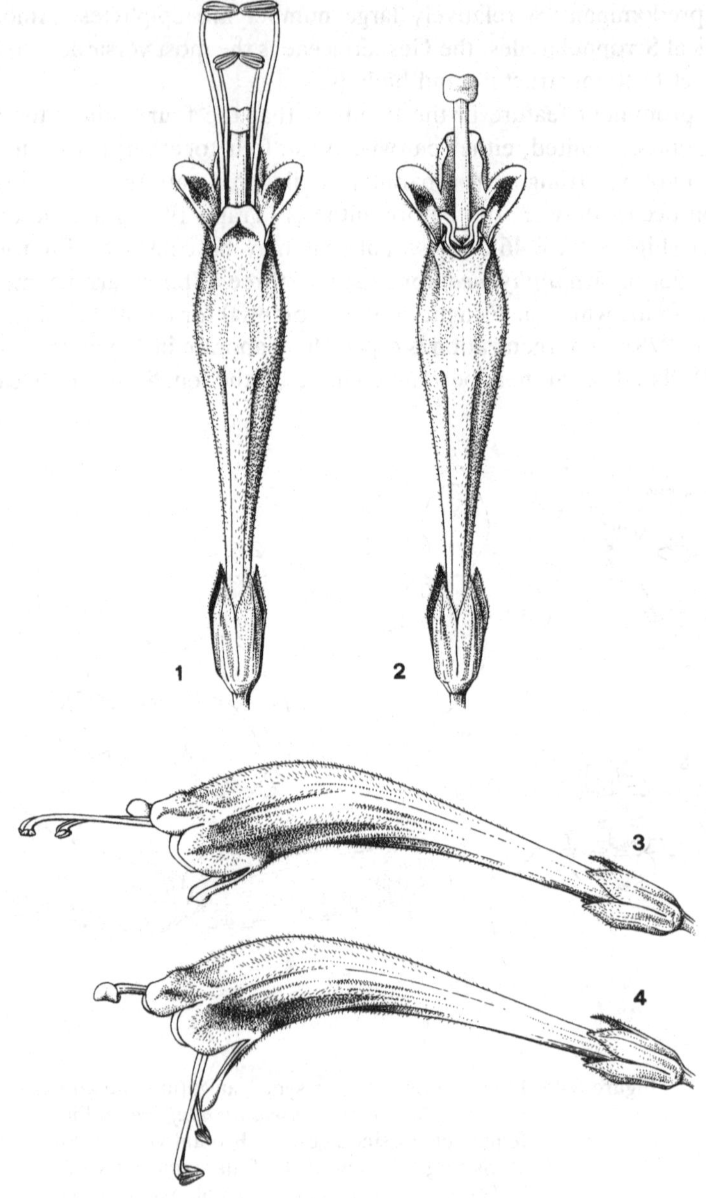

Figure 8.46. Flowers of *Aeschynanthus speciosus* (Gesneriaceae) showing marked protandry combined with herkogamy (× 1). 1–2. Flowers from abaxial side. 1. Male stage. 2. Female stage. 3–4. Flowers from the side. 3. Male stage. 4. Female stage.

gins and by their upper end, thus forming a column with the connectives outside. The thecae have basal pores. Shortly before anthesis the anther unit flips upside down, and the anther pores are now on the upper side. If the anther unit in the open flower is touched by a pollinator, it tips over briefly like a salt shaker and dry pollen falls on the visitor. The anthers are emptied after five to eight visits.

In contrast to the neighbouring Scrophulariaceae the ovary is more often unilocular with parietal placentae than bilocular (Weber 1971). The synascidiate portion tends to be short. There is a trend to produce a very high number of ovules, and there are commonly four protruding-diffuse placentae. Gesnerioideae is the only large taxon in the Scrophulariales–Lamiales complex with a strong tendency toward inferior ovaries. In the Gesnerieae and Gloxinieae there is a range from semi-inferior to inferior ovaries.

The Gesneriaceae are extremely versatile in their floral modes, which may show exceptionally broad ranges or frequent switches within genera; speciation and evolution are most intimately linked with changes in the mode of pollination (Wiehler 1983). Therein, the gesneriads may be comparable with the orchids. About 60% of the Neotropical Gesneriaceae are hummingbird-pollinated, *ca.* 30% are pollinated by female and male euglossine bees in search of nectar, and *ca.* 10% are pollinated by bats, butterflies, hawkmoths, moths, flies or male euglossine bees in search of perfumes (Wiehler 1983).

Wiehler (1983) distinguishes four different kinds of corollas in hummingbird-pollinated gesneriads: (1) narrow tube with small limb; (2) narrow but expanding tube with large, hooded upper lip of 4 petals and a lower lip of one petal (columneoid); (3) urceolate corolla with ventrally inflated tube, constricted throat and narrow limb (hypocyrtoid); (4) moderately wide tube and wide limb (converted 'euglossine corolla'). Each of these four types occurs in 4–15 genera (often in different tribes), either in the majority of species or in single species, suggesting multiple parallel evolution (Wiehler 1983). The genus *Drymonia* is especially plastic in this respect: Wiehler (1983) found six species flowering sympatrically within a range of 50 metres. According to their corollas two were pollinated by euglossine bees, two by hummingbirds, one by bats, and one by moths. Concomitant with this is a high degree of interspecific fertility in their artificially produced hybrids (Wiehler 1983). In *Gesneria* the transition from ornithophily to chiropterophily has occurred in at least two independent groups of species (Skog 1976).

There is a correlation between hummingbird-pollination and the occurrence of a long adaxial nectary (in contrast to the otherwise more common ring-like disk). The additional occurrence of adaxial nectaries in some euglossine-pollinated groups suggests an evolutionary shift from hummingbird- to euglossine-pollination in these groups (Wiehler 1983). Concomitant with this adaxial nectary in many Episcieae the lower portion of the four filaments is sheath-like and united into an adaxially open tube, which channels the tongues of the pollinators to the nectar (Wiehler 1983).

In flowers visited for nectar by euglossine bees there is a peculiar trend to have fimbriate petals on salverform corollas. This occurs in species of *Alsobia*, *Episcia*, *Nautilocalyx*, *Paradrymonia*, *Drymonia* and *Neomortonia* (Wiehler 1978).

In Gesneriaceae the evolution of polysymmetric or nearly polysymmetric flowers from monosymmetric ancestors occurred several times (Burtt 1970), as did the evolution of nectarless pollen-flowers (see above) (e.g. Vogel 1978a, Weber 1989a, Vokou *et al.* 1990). Some nectarless flowers even seem to be deceptive (*Ornithoboea arachnoidea*, *Didymocarpus geitleri*, Weber 1979, 1989a).

Observations of the longevity of individual flowers are scarce. Wiehler (1983) mentions four days for *Drymonia* species. In many other groups flower longevity seems to be even greater. In many-flowered inflorescences, single flowers may open in sequence over a longer period (*Epithema saxatile*) or all may open more or less at once (*Agalmyla tuberculata*) (Weber 1976b, 1982). The function of the often two-flowered cymes in gesneriad inflorescences has not been evaluated with respect to the flowering strategy (Weber 1982). Studies on breeding systems are also scarce.

8.8.3 Bignoniaceae

Bignoniaceae, a pantropical family of more than 800 species, with its main centre of diversity in America, is subdivided into seven tribes (Gentry 1988). Vines and trees are predominant. Flowers are, in general, very showy and tend to be larger than in Gesneriaceae.

In many Bignoniaceae the anthers are extremely sagittate with the two thecae diverging at right angles, and the connective has a protrusion (Figs 8.47, 8.52). The sepal lobes are often highly reduced so that the calyx tube appears either truncate or it is almost closed at the end and ruptures with one or two slits at maturity. Calyces that are closed in this way are often differentiated as 'water calyces' (see section 2.2.4) (Fig. 8.53). Nectaries are often present on the outer calyx surface. In contrast to the neighbouring Scrophulariaceae, the corolla tends to be appressed to the calyx and not narrowed at the base (van Steenis 1949) and the ovary has four placental ridges (not just two) and contains four (not just two) main vascu-

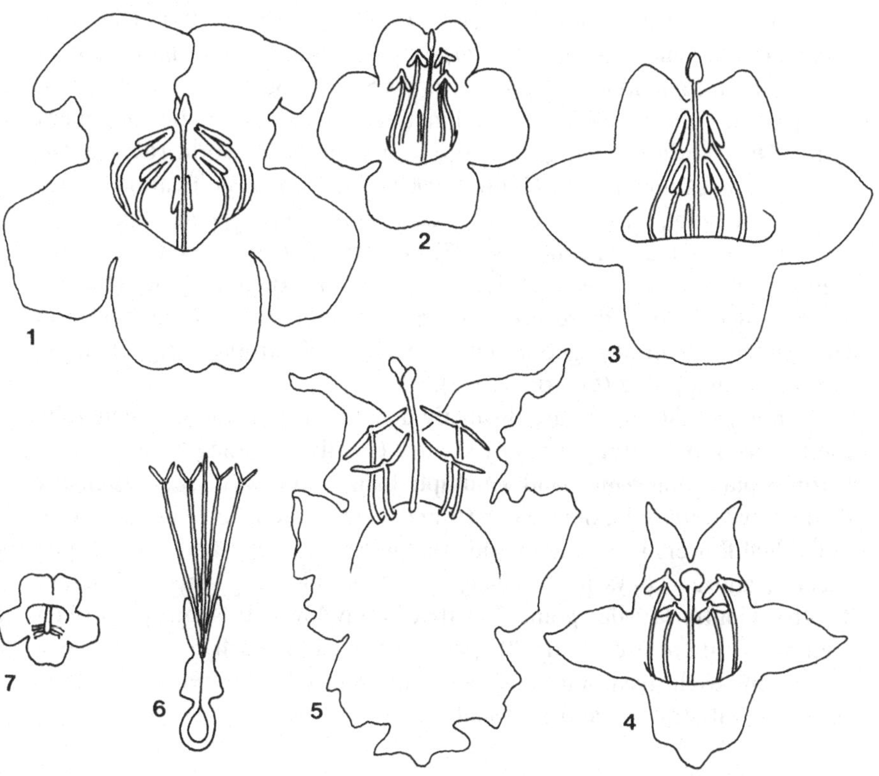

Figure 8.47. Flower forms of Bignoniaceae (× 0.5). 1–4. Bat-pollinated flowers. 1. *Oroxylum indicum*; five stamens. 2. *Parmentiera aculeata*; four stamens and a staminode. 3. *Kigelia pinnata*; four stamens and a staminode. 4. *Crescentia cujete*; four stamens, staminode lacking. 5–6. Bird-pollinated flowers. 5. *Spathodea campanulata*; four stamens, staminode minute or lacking. 6. *Deplanchea tetraphylla*; four stamens, staminode lacking (modified after Weber & Vogel 1984). 7. *Jacaranda mimosifolia*; bee-pollinated, four stamens enclosed in floral tube, staminode large and protruding out of the floral tube, with glandular hairs (see also Fig. 8.52).

lar bundles (two dorsal and two synlateral ones) (Leinfellner 1973b, Armstrong 1985). The presence of four main vascular bundles in the ovary is functionally correlated with the tendency to produce siliqua-like fruits with a replum. The symplicate zone of the ovary is relatively long but only exceptionally (*Crescentia*) is the ovary unilocular (Leinfellner 1973b).

Extensive comparative studies on the reproductive biology of Bignoniaceae have been carried out by Gentry (1974a,b, 1976, 1990a). The main pollinators are medium-sized to large bees. Birds are also important in many genera (*Spathodea*, *Deplanchea*, Weber & Vogel 1984; 20 Neotropical genera, Gentry 1990a). Bat-pollination occurs in a number of genera in the Old World (e.g. *Kigelia*, *Oroxylum*) and New World (*Crescentia*, *Parmentiera*, *Amphitecna*) (Gentry 1990a). Hawkmoth-pollination is present in, for example, *Leucocalanthe*, *Tanaecium*, *Sphingiphila*, *Nyctocalos* and *Radermachera* (van Steenis 1977, Gentry 1990b). A comparison of Neotropical Gesneriaceae (Wiehler 1983) and Bignoniaceae (Gentry 1990a) suggests that there are more genera with mixed floral modes in Gesneriaceae than in Bignoniaceae. Nearly all Neotropical Bignoniaceae are self-incompatible (Gentry 1990a).

Phenological strategies are distinctly diverse. In fact, the now generally used concept of strategy types by Gentry (1974b) including 'steady-state', 'cornucopia', 'big-bang' and 'multiple-bang' flowerers, was originally devised with the Bignoniaceae in mind (see section 7.4). Longevity of individual flowers is commonly one day in 'steady-state' and 'cornucopia' species, and 2–3 days in 'big-bang' and 'multiple-bang' species (Gentry 1974b). Gentry (1990a) points out that in any lowland Neotropical plant community there are about 20 species of bignons, each of which has a unique and exclusive pollination niche determined by different pollinators, phenological strategy and seasonality.

8.8.4 Acanthaceae

The Acanthaceae is a pantropical family, containing more than 4000 species. The major two subfamilies are Acanthoideae and Ruellioideae, while Thunbergioideae and Mendoncioideae are much smaller (Bremekamp 1965). The family is more similar in habit of inflorescences and flowers to the Lamiaceae than the preceding three families. The most distinctive familial trend, however, is in the fruit and not in the floral structure. The gynoecium contains only 2–4 ovules in most groups; the ripe fruits open explosively and the often disk-shaped seeds are forcibly expelled. Only in few groups (e.g. Mendoncioideae) the mature fruits do not open.

Another tendency, although less prominent, is that the lower lip is much larger than the upper lip or (rarely) that even all five petals are incorporated in the lower lip.

Justicia (sect. *Harniera* and *Ansellia*) is more or less self-compatible, but most species are allogamous because of protandry; the major pollinators are large bees (Hedrén 1989, Kelbessa 1990). Other species of *Justicia* are hummingbird-pollinated (McDade & Kinsman 1980). In ten species of

Carlowrightia and in eight species of *Tetramerium* studied, the flowers opened in the early morning and lasted not more than five hours; they were homogamous, but to some extent herkogamous, and self-compatible (Daniel 1983, 1986). Halictid bees and bombyliid flies, and to a lesser extent also butterflies, are the main pollinators in both genera. The same pollinator groups were found in *Henrya* (Daniel 1990). The hummingbird-pollinated *Aphelandra golfodulcensis* has inflorescences with 12–48 flowers. Each inflorescence produces one or two open flowers per day. They open in the early morning and fall in the late afternoon of the same day (McDade & Kinsman 1980). Similar patterns occur in other species of the *Aphelandra pulcherrima* complex (McDade 1984). In this group of about 40 species, all are hummingbird-pollinated. It is suggested that pollination by hermit hummingbirds is primitive for the group, whereas pollination by trochiline hummingbirds, concomitant with shortening of the corolla tube, has evolved twice in distinct lineages, each with one or two species, respectively (McDade 1992).

Thunbergia

Thunbergia, a prominent and somewhat isolated genus of the Acanthaceae, with a hundred species in the Old World tropics, has large showy flowers with a highly elaborate pollination apparatus (Bremekamp 1955) (Figs 8.48, 8.49). It shows a spectrum of bee-, hawkmoth- and bird-pollinated flowers (Vogel 1954). Since several species are often cultivated as ornamentals or occur as weeds in many parts of the tropics, they can easily be observed (Bor & Raizada 1990). The most elaborate flowers are among those that are pollinated by bees.

Thunbergia grandiflora is an Indian vine with mostly pendent inflorescences and then resupinate flowers (Fig. 8.48.1). The flowers have often been studied and figured (e.g. Burck 1891, Burkill 1906, Backer 1916, Cammerloher 1931, van der Pijl 1941a, 1954, Faegri & van der Pijl 1979, Bor & Raizada 1990, Krebs 1990) but the elaborate lever mechanism of the anthers has never been accurately described. Van der Pijl (1954) discusses these flowers as the most perfectly adapted to pollination by large *Xylocopa* species (esp. *X. latipes*); they cannot be pollinated by other animals and are therefore good examples of monophilic flowers. Flowers are produced over a long period by an individual and only a small number is open on a given day. They are homogamous one-day flowers, opening early in the morning when it is still dark and falling in the afternoon or evening of the same day (Burkill 1906). Crepuscular species of carpenter bees are already at work on these flowers in the dark early morning and other species can be observed until into the afternoon (Burkill 1906).

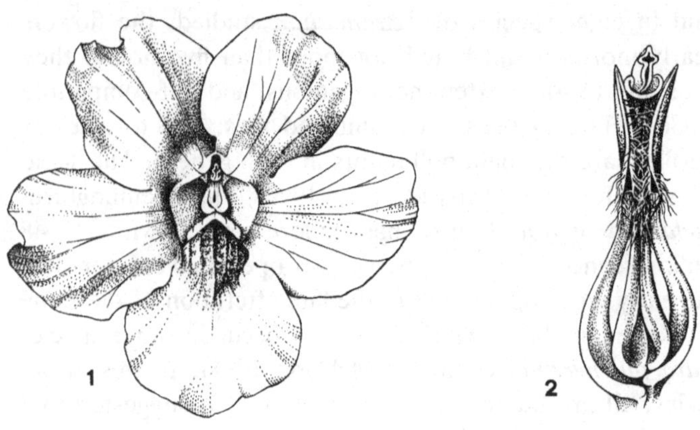

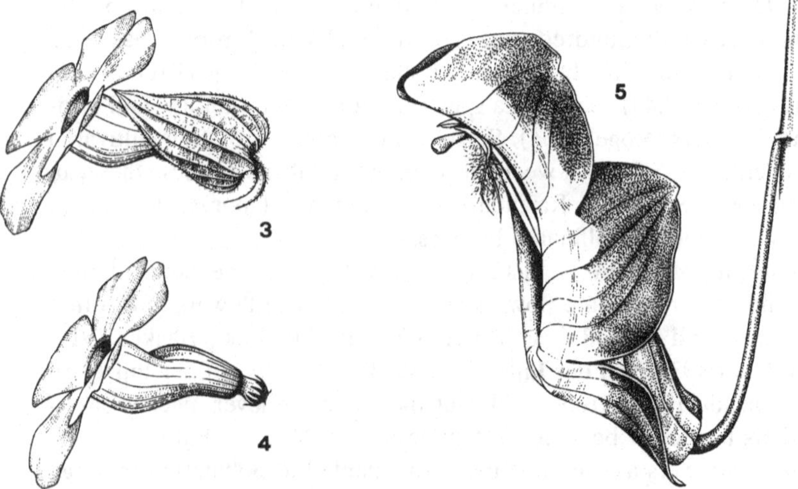

Figure 8.48. Flower forms of *Thunbergia* (Acanthaceae).
1–2. *T. grandiflora*; pollinated by *Xylocopa* bees.
1. Frontal view of flower (× 0.7). 2. Pollination apparatus with four stamens and convoluted stigma (× 2).
3–4. *T. alata*; flower with bee-pollination syndrome, from the side (× 1). 3. Flower with large bracteoles.
4. Bracteoles removed to show the reduced calyx.
5. *T. mysorensis*; flower with bird-pollination syndrome, from the side (× 1).

The light blue and almost 10 cm wide flowers have a relatively wide, light yellow throat. The lower lip is marked with purple longitudinal lines leading into the throat. It has a firm architecture: it arches upwards and has longitudinal and transverse ribs (like a washboard) (Fig. 8.51.4), which provide holding devices for the entering carpenter bees (van der Pijl 1954). The firmness of the lower lip has been described as based on internal lignification of the ribs (van der Pijl 1954). My own observations, however, showed that there is no lignification at all. In contrast, the tissue even contains large intercellular spaces. The firmness is achieved at the morphological level alone: by arching upwards like an arched bridge and by the ribs. The pollination organs are adpressed to the upper lip and covered by it; the stigma, with two differently convoluted lips forming a funnel-like structure, and the style are situated in a groove in the upper lip, immediately above the four anthers. The ovary and the yellow nectariferous disk are hidden in a lower chamber, which is separated from the upper chamber by a constriction in the corolla. Only a small passage, further narrowed by a ring of hairs, is left between upper and lower chamber. This passage is, in addition, occluded by the broad and firm bases of the filaments of the upper stamen pair, which cling tightly together by transverse ridges and grooves on their contiguous surfaces, when the flower is in a relaxed state (Figs 8.48.2, 8.49.8).

When a carpenter bee visits the flower, it presses the filaments apart with its head and can so reach the nectar (van der Pijl 1954). It also presses the anthers upwards, which results in a shower of pollen grains onto its back (Burkill 1906). The upper anther on each side lies behind the lower anther. The dehisced anthers in the open flowers remain more or less closed and conceal their pollen content when in a relaxed position. At the base of each theca there is a somewhat excentric firm horn. These horns are levers; when they are pressed upwards by a carpenter bee the four anthers are aligned more into a plane, since each anther is slightly twisted outwards, and the thecal slits slightly open by this twist, so that pollen falls out of them. All openings of the strongly introrse anthers are directed towards the back of the visitors. The entire construction is firm and elastic so that the thecal slits close again after the visit. The thecal slits are bordered by long uniseriate (moniliform) hairs, whose function in pollen transfer has not been studied.

According to van der Pijl (1954) probably only the lower of the two stigmatic lobes is receptive. Since in *Thunbergia erecta*, which has a similar pollination apparatus, the upper lobe has been reported to be receptive, while the lower was interpreted as a protective device for the upper lobe against self-pollination (see Knuth *et al.* 1905), this needs critical study

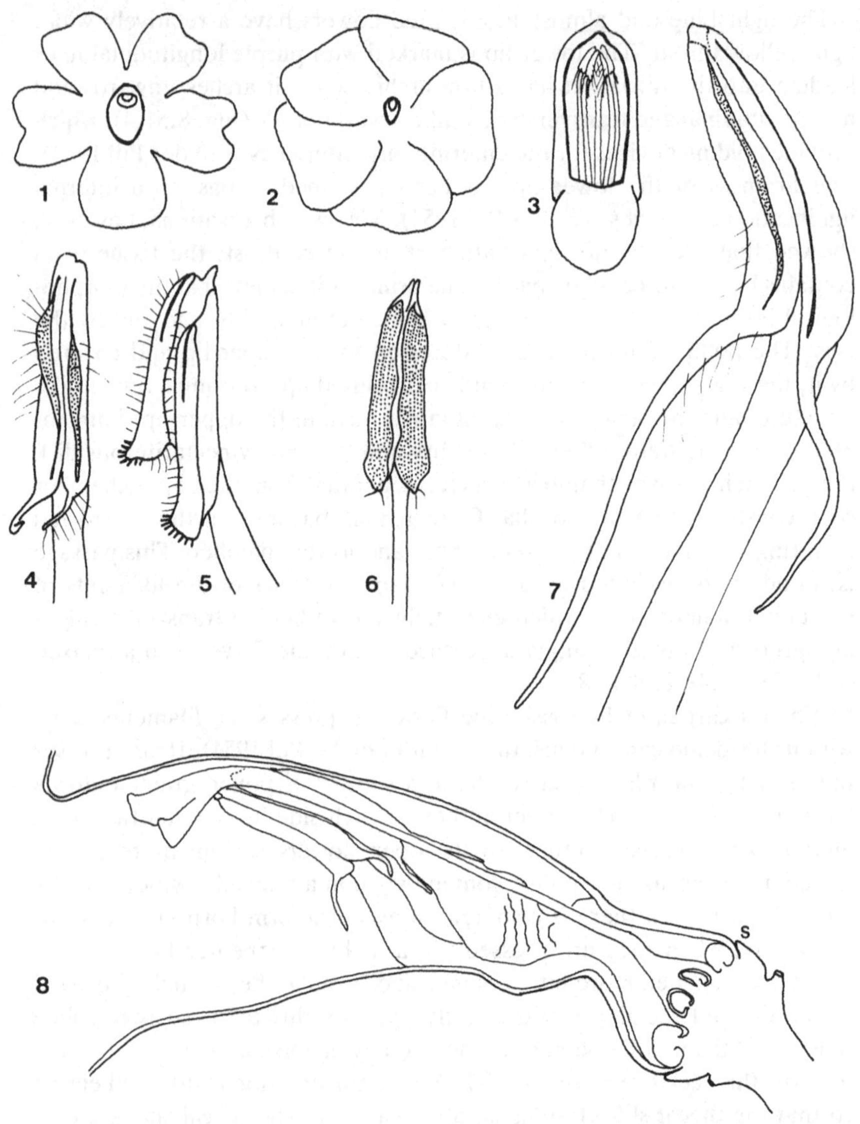

Figure 8.49. Flower forms of *Thunbergia* (Acanthaceae). 1–7. Flowers with different width of floral tube and different anther forms according to different pollination mechanisms.
1–3. Flowers in frontal view (× 0.5). 1. *T. alata*; bee-pollination syndrome. 2. *T. fragrans*; hawkmoth-pollination syndrome. 3. *T. mysorensis*; bird-pollination syndrome. 4–7. Anthers (× 6). 4. *T. alata*; basal horns to open the closed thecae. 5. *T. erecta*; basal plates to open the closed thecae. 6. *T. fragrans*; anthers enclosed in floral tube, thecae

(Fig. 8.51.3). Preliminary observations with the SEM and microtome sections do not show any difference between the two stigmatic lobes.

The young flower is enclosed by the two enlarged and postgenitally fused bracteoles (Fig. 8.50.1–2). In contrast, the calyx is reduced to a small rim around the floral base and individual sepals cannot be recognized (Fig. 8.50.5). Notwithstanding reduction, the surface of this rim is densely covered by a large number of cup-like nectaries. They attract ants, which play an important role in floral defence against nectar thieves and other parasites, as first found by Burck (1891) and confirmed by Krebs (1990). Thus, the flowers have two kinds of floral nectaries: nuptial around the ovary and extranuptial on the reduced calyx (Fig. 8.51.5–6). The wall of the lower floral chamber containing the nectar for the pollinators is, in addition, especially thick and hard and cannot easily be penetrated from the outside by potential nectar robbers.

The bee-pollinated species (e.g. *Thunbergia grandiflora*, *T. erecta*, *T. alata*, *T. gregorii*) have blue, purple or orange flowers, all with a prominent nectar guide in the relatively wide throat. They all have a similar pollination apparatus, although *T. grandiflora* has the largest and most robust flowers. Only the proportions of the anther parts show significant variation. In *T. erecta* (formerly the genus *Meyenia*) the thecal slits are restricted to the upper third of the anther and, instead of a horn, there is a broad thickening with a bristly fringe at the base of each theca.

The probably bird-pollinated species (e.g. *Thunbergia coccinea*, *T. mysorensis*) have scarlet flowers with widely open yellow throats. Those of *T. mysorensis* are pendent and resupinate. The thecal horns of *T. mysorensis* are longer than the anthers themselves but function as in the bee-pollinated species.

The presumably hawkmoth-pollinated species (e.g. *T. guerkeana*, *T. gigantea*, *T. fragrans*) have white flowers, a very narrow throat and a narrow tube, which attains 15 cm length in *T. gigantea* (see Vogel 1954). In *T. fragrans* the anthers are hidden deep within the floral tube. They are smaller than in the other species described and simplified: they lack horns and a lever mechanism and are, instead, widely open after dehiscence. In addition, the two stigmatic lobes are simple in shape.

Fig. 8.49 (*cont.*)
opening automatically, opening devices for pollinators lacking. 7. *T. mysorensis*; long basal horns to open the thecae. 8. *T. grandiflora*; longitudinal section of flower to show the constricted zone of the lower floral tube, the large nectaries in the base of the floral tube, the broad stamen filaments, and the reduced sepals (S) (× 2.4).

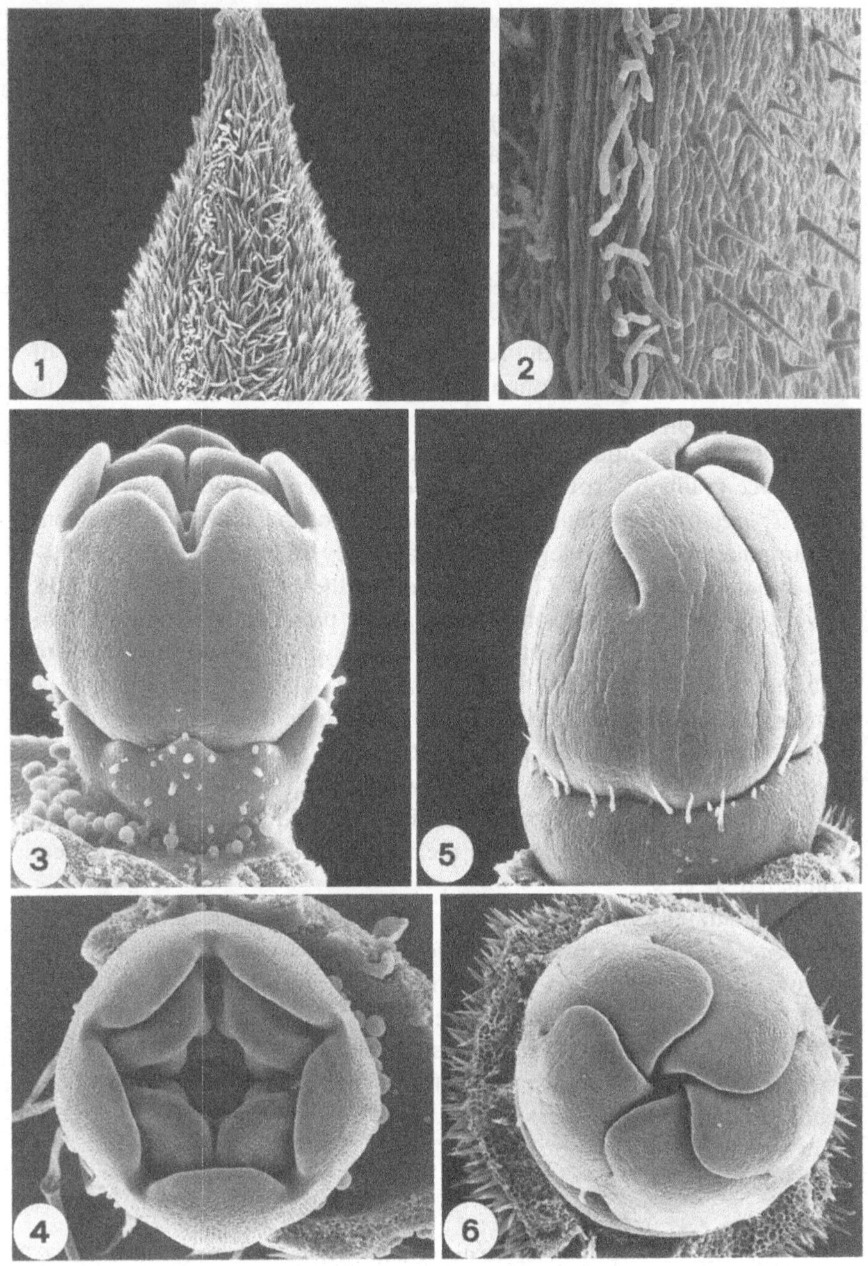

Figure 8.50. Structure of floral buds of *Thunbergia* (Acanthaceae). 1–2. *T. grandiflora*. 1. Floral bud, enclosed in two hairy, postgenitally united bracteoles (× 30). 2. Postgenital

In contrast to *T. grandiflora*, in most other species of the genus the sepals have not completely disappeared but are still present as tiny teeth of irregular size and number (*ca.* 10–23). Complete protection of the floral buds by the two bracteoles is consistent in the genus as well as in the neighbouring genus *Mendoncia*.

8.8.5 Evolutionary plasticity and parallel evolutionary trends involving more than one family of the Scrophulariales (including Verbenaceae, Lamiaceae and Solanaceae)

Evolutionary plasticity and parallel evolutionary trends are particularly impressive in the Scrophulariales, too. The features discussed here are tentatively arranged in a sequence according to organization, construction, mode, and breeding systems, although some aspects do not fit strictly into one of these categories. A large suite of trends concomitant with the elaboration of lip-flowers involves different modes of enclosure of pollen and the evolution of different modes of pollen dispensing.

Petal aestivation

The petals mostly have an imbricate aestivation with four relatively constant major patterns, which are correlated with floral monosymmetry. They are ascending ('rhinanthoid'), descending ('scrophularioid'), quincuncial, and contort (Fig. 2.4). The contort pattern tends to be associated with flowers where monosymmetry is less pronounced. Most families exhibit a certain range of these patterns: a prominent example of parallel evolutionary trends. The pattern may be constant in larger groups (e.g. at the family level in Lamiaceae) or it may be labile even within species (e.g. *Collinsia*, *Lathraea*, Hartl 1965–74).

In Scrophulariaceae petal aestivation is descending cochlear in most Antirrhineae, Scrophularieae and Cheloneae, and ascending in most Digitaleae and Pedicularieae (Hartl 1965–1974, Armstrong & Douglas 1989). In Lentibulariaceae it is (usually) descending in *Pinguicula*, and ascending in *Utricularia* (Eichler 1875). In Bignoniaceae it is descending (ascending in Crescentieae). In Acanthaceae it is ascending (*Stenandrium*, *Aphel-*

Fig. 8.50 (*cont.*)
fusion of bracteoles released, showing the ridge of interdented epidermis cells (on left side) (× 100). 3–4. *T. alata*. 3. Floral bud with reduced (but increased number of) sepals (× 70). 4. Floral bud from above with five petals and four stamens (the adaxial stamen suppressed) (× 70). 5–6. *T. grandiflora*. 5. Floral bud from the side, sepals more reduced than in 3 (× 45). 6. Same bud from above, with contort petals (× 45).

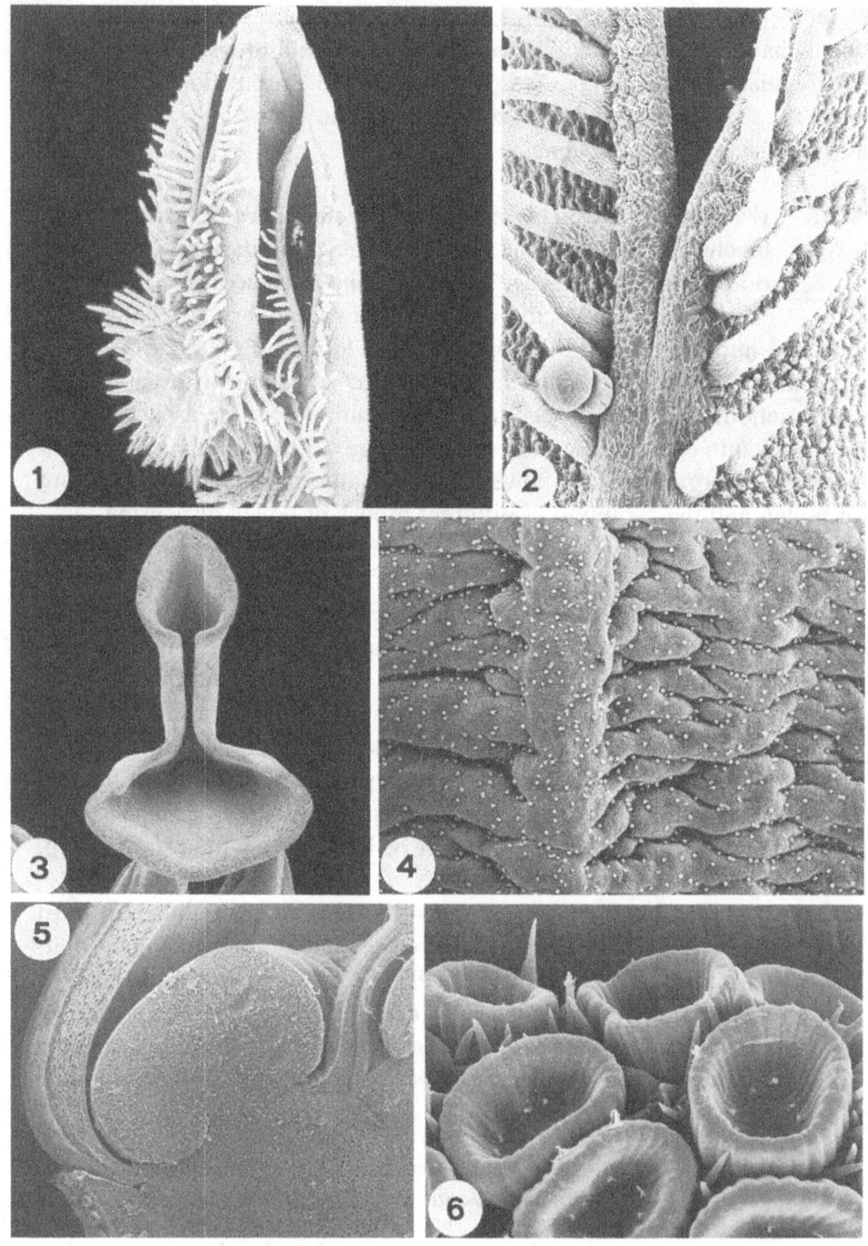

Figure 8.51. Floral parts of *Thunbergia* (Acanthaceae). 1–2. *T. erecta*. 1. Anther with basal opening plates (see also Fig. 8.49) (× 18). 2. Lower end of opening of a theca (× 110). 3. *T. alata*; stigma (× 15). 4–6. *T. grandiflora*.

andra), quincuncial (*Adhatoda*) or contort (to the left) (*Eranthemum*, *Whitefieldia*, *Thunbergia*) (Bremekamp 1965, Scotland *et al.* 1994). In Gesneriaceae it is usually ascending (*Rhynchoglossum*, *Epithema*, *Loxonia*, *Stauranthera*, Weber 1978) but descending in *Monophyllaea* and *Whytockia* (Burtt 1978). In Orobanchaceae it is descending or quincuncial (Eichler 1875). In Martyniaceae it is quincuncial (*Martynia*). In Pedaliaceae it is valvate (*Sesamum*). In Globulariaceae it is apert. In Myoporaceae it is quincuncial or cochlear. In Verbenaceae it is mostly descending, in some genera quincuncial (*Tectona*, *Callicarpa*, *Duranta*, *Lantana*, *Cornutia*) (Bocquillon 1863, Eichler 1875). In Lamiaceae it is always descending.

In Solanaceae polysymmetric flowers are contort (*Datura*, *Fabiana*) or valvate (*Solanum*, *Salpichroa*, *Jaborosa*, *Trechonaetes*, *Cestrum*, *Vestia*) (Huber 1980); lip-flowers are descending (*Hyoscyamus*, *Salpiglossis*) (Eichler 1875, Robyns 1931) or ascending (*Petunia*, *Brunfelsia*, *Schizanthus*, Robyns 1931). The notion that the monosymmetric flowers in Solanaceae have an oblique plane has been criticized by Huber (1980) and Ampornpan & Armstrong (1991): owing to the peculiar monochasial inflorescence ramification, the median plane is not easy to establish.

Petal number

There are very few deviations from the pattern with five petals (Endress 1992b). *Sibthorpia* (Scrophulariaceae) has almost polysymmetric flowers where the organ number can be increased to up to 8 per whorl (Hedberg 1955). The number may rarely decrease to four as in some Scrophulariaceae (Hufford 1992), some Gesneriaceae (Burtt 1970) and *Sphingiphila* (Bignoniaceae) (Gentry 1990b), or even to zero, as in a species of *Besseya* (Scrophulariaceae) (Hufford 1992). However, in part of these genera it is unstable and can revert to five within some species (Lehmann 1919).

Reductions in stamen number and reversals to the full complement of stamen number

Very rarely are all five stamens present (Fig. 8.47.1), mainly in groups where monosymmetry is only weakly expressed; in Scrophularia-

Fig. 8.51 (*cont.*)
4. Washboard-like surface of lower lip (× 13).
5. Longitudinal section of floral base showing nectary disk on left side of ovary and extranuptial nectaries on outside of reduced sepals (lower left) (× 13). 6. Extranuptial nectaries on outside of reduced sepals (× 150).

ceae (*Verbascum*, *Sibthorpia*, *Capraria*), in Bignoniaceae (*Nyctocalos*, *Oroxylum*, *Rhigozum*, *Catophractes*, *Radermachera*, (*Rhodocolea*, see Gentry 1976), in Gesneriaceae (Cyrtandroideae: *Conandron*, *Tengia*, *Ramonda*, *Protocyrtandra*; Gesnerioideae: *Bellonia*, *Napeanthus*; Coronantheroideae: *Depanthus*), in Acanthaceae (*Pentstemonacanthus*) (review in Endress 1992b, additional information in Wiehler 1983), in Lamiaceae in *Bystropogon spicatus*. Furthermore, abnormal peloric flowers with the odd stamen fertile have been found in several species among most of the families discussed here. Presumably the genetic or epigenetic mechanism responsible for its suppression in normal flowers is relatively simple (Heslop-Harrison 1952). The systematic distribution of 5-staminate flowers in Scrophulariales suggests that this condition is apomorphic within the order, and therefore a reversal, because the 5-staminate condition is probably plesiomorphic at the more general level of the Asteridae.

In most cases there are four lateral stamens in the androecium, which form two symmetrical pairs; the median organ may be present in the form of a smaller or larger staminode (Figs. 2.20, 8.47, 8.50). In the Scrophulariaceae, the odd staminode is present in about 30% of 145 genera studied (Polak 1900). Presence and absence occur in both subfamilies: in Scrophularioideae it is more often present and in Rhinanthoideae it is mostly absent. It is mostly a small rudiment but in *Penstemon* and *Scrophularia* it is even larger than the stamens and has some function in pollination biology (Straw 1956). In almost all Bignoniaceae the odd staminode is present (if not differentiated as a fertile stamen) (Rohrhofer 1931), and in *Jacaranda* and *Digomphia* it is larger than the stamens (Morawetz 1982) with a function in pollination biology similar to that in *Penstemon* (Fig. 8.52).

In a number of groups in different families only two stamens (one pair) are fertile: another good example of parallel evolutionary trends. Either the lower or the upper pair may be left. In Scrophulariaceae the upper pair is left in *Veronica*, *Paederota*, *Wulfenia*, *Ilysanthes* (Rhinanthoideae), *Calceolaria*, *Gratiola*, *Dopatrium*, *Curanga* (Scrophularioideae) etc.; the lower pair is left in *Ixianthes*, *Anticharis*, *Micranthemum* (Scrophularioideae) etc. In Gesneriaceae the upper pair is left in *Sarmienta* (Coronantheroideae), *Epithema* and *Opithandra* (Cyrtandroideae) (Burtt 1958, Weber 1976b), while the lower pair is left in *Boea*, *Chirita*, *Cyrtandra*, *Didymocarpus*, *Ornithoboea* and *Streptocarpus* (all Cyrtandroideae) (Burtt 1958, Weber 1979). In Bignoniaceae the lower pair is left in *Catalpa*. In Acanthaceae the upper pair is left in (among many other genera) *Ebermayera*, *Beloperone*, *Rhytiglossa* and *Dicliptera*; the lower pair is left in *Eranthemum*, *Thyrsacanthus* and *Justicia*. In Lentibulariaceae

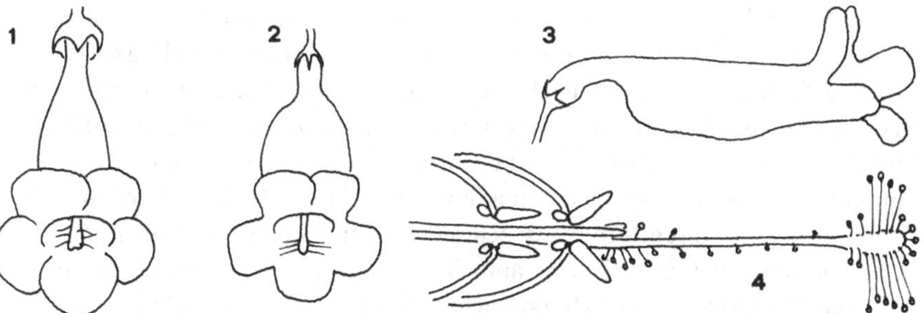

Figure 8.52. Convergence of bee-pollinated flowers with enlarged odd staminode protruding out of the floral tube. 1. *Penstemon hartwegii* (Scrophulariaceae); flower from adaxial side (× 0.7). 2–4. *Jacaranda mimosifolia* (Bignoniaceae). 2. Flower from adaxial side (× 0.7). 3. Flower from the side (× 0.7). 4. Pollination apparatus from abaxial side to show the odd staminode with glandular hairs, surpassing the four stamens, anthers of the 4 stamens with a fertile and a sterile theca (× 2.7).

always the lower pair is left. In Martyniaceae the lower pair is left in *Martynia diandra*. In Verbenaceae the upper pair is left in *Stachytarpheta*; the lower pair is left in *Oxera*. In Lamiaceae the lower pair is left in *Amethystea*, *Rosmarinus*, *Salvia* and *Monarda*. In Solanaceae the lower pair is left in *Schizanthus*; this is, however, in this case, the pair adjoining the odd stamen position (Robyns 1931). The genus *Opithandra* (Gesneriaceae), based on the presence of only two upper stamens, may be 3- or 4-fold polyphyletic (Burtt 1958). This may also be the case for other genera. The evolution of only two stamens is therefore an exceedingly multiple parallel trend in the Scrophulariales–Lamiales complex. The almost polysymmetric flowers of *Porodittia* (Scrophulariaceae) have three stamens (Molau 1988).

In many groups the anthers have only one theca, either by fusion of the two thecae over the anther apex (synthecal) or by loss (and transformation) of one side of the anther (monothecal). Both evolutionary pathways probably have occurred in various families, even within one family (see examples and discussion mainly in Trapp 1956, Endress & Stumpf 1990).

Ovary structure

Although each family has its special trends in ovary structure, as described above, there is a broad spectrum in each family with considerable overlap between families (Weber 1971). Two extreme forms are (1) bilocular ovaries with a long synascidiate portion and a single axile protruding-diffuse placenta in each locule, and (2) unilocular ovaries with a reduced synascidiate portion and parietal placentae with two protruding-diffuse branches on each side. The first is common in Scrophulariaceae and Solanaceae but also occurs in a few Gesneriaceae (Hartl 1956a, Weber 1971). The second tends to be associated with an increase in ovule number; it is common in Gesneriaceae and Orobanchaceae but also occurs in a few Scrophulariaceae and Bignoniaceae (Weber 1971, Leinfellner 1973b); among the Bignoniaceae it has arisen in parallel in some species of several genera of advanced Crescentieae (Madagascan *Colea* and Neotropical *Crescentia* and *Amphitecna*) (Gentry 1976). A third extreme form is a free central placenta that is evolutionarily derived from the first form. It occurs (still with reduced septa) in *Limosella* (Scrophulariaceae) and is common (with loss of the septa) in the Lentibulariaceae (Hartl 1956b). Apical septa are common in the few-seeded Verbenaceae and Lamiaceae (see below) but they also occur in Antirrhineae (Scrophulariaceae) (Hartl 1962) and a few Gesneriaceae (Weber 1971).

Reduction in ovule number and repercussions on inflorescence structure, breeding systems, and fruit differentiation

In Lamiaceae and most Verbenaceae the bicarpellate gynoecium contains only four ovules, each of which is in a separate niche of the ovarial cavity owing to the formation of 'false septa' (not always in Verbenaceae) (Junell 1934). Reduction to 4 (or fewer) ovules is paralleled in most Acanthaceae, some Myoporaceae, Globulariaceae, and Selagineae (Scrophulariaceae) (Junell 1961). In addition, in most Lamiaceae, each of the four ovarial compartments enclosing a single seed bulges out to form a separate propagule. A tendency to form propagules consisting of a seed included in part of the fruit wall or even indehiscent fruits also occurs in Myoporaceae, Globulariaceae, Acanthaceae (*Mendoncia*, *Justicia heterocarpa*, *Aphanosperma*). A compensation for a small seed number per flower would be a large number of flowers per plant. In this respect, Yeo (1972) points to the tendency to form numerous small flowers in Verbenaceae and Lamiaceae. To a certain extent, this tendency is also present in the Acanthaceae. In the Lamiales large flowers are rarer than in the Scrophulariales. In Lamiaceae long thyrses with many-flowered partial inflorescences are often formed. In Verbenaceae there is a tendency to form long spikes (e.g. *Stachytarpheta*, *Verbena* p.p.) or often large, umbel-like inflorescences (e.g. *Clerodendrum* p.p., *Premna*, *Vitex*, *Verbena* p.p., *Lantana*). These many-flowered inflorescences commonly have a long flowering period, since only a few flowers are open at a given time (Fig. 7.1.1). Acanthaceae, Verbenaceae (Symphoremoideae, Classen 1986) and Lamiaceae (e.g. *Salvia*) also have a tendency to produce large, showy bracts in the inflorescences.

In Lamiaceae and Acanthaceae, perhaps also Verbenaceae, self-compatibility seems to be widespread (for Lamiaceae see van der Pijl 1972). Although protandry is common, autogamy and cleistogamy are also known from various representatives. On the other hand, a trend towards the gynodioecious condition occurs in both Verbenaceae and Lamiaceae (although most representatives still have bisexual flowers) (van der Pijl 1972). In Verbenaceae a number of genera have species with unisexual flowers and even dioecy.

United and spathaceous calyces

In several groups, the sepals are congenitally united leaving only a small pore at the end and thus providing tight protection for the inner floral organs; at anthesis the calyx often splits at one or two sides ('spathaceous' calyx). This occurs in some Bignoniaceae (*Spathodea*), Solanaceae (*Brugmansia*) (Fig. 3.2) and Verbenaceae (*Faradaya*).

Water calyces and calyx nectaries

Notable in several families of the group is the scattered occurrence of water calyces (Fig. 8.53.1) (see section 2.2.4) (esp. Bignoniaceae, Gesneriaceae, Verbenaceae, Solanaceae) and likewise of nectaries in the form of multicellular protuberances on the outside of the calyx (floral but extranuptial) (esp. Bignoniaceae, Verbenaceae). In both cases, water calyces and extranuptial calyx nectaries, a more or less large number of low glands cover the calyx surface, the inner side in the former, the outer one in the latter. Since water calyces and extrafloral nectaries preferentially occur in the same families, evolutionary relationships between the two differentiations are suggested. In many cases water calyces are also associated with congenital union of the sepals (see above).

Flowers with narrow upper lip

In many lip-flowers the upper lip is relatively long and narrow and effectively encloses the anthers and stigma, which can be reached only by a narrow slit from below. Examples are Lamiaceae: *Leonotis* (bird-flowers; Gill & Conway 1979), *Salvia* (Fig. 8.54.1); Acanthaceae: *Aphelandra* species (McDade 1984). In many *Salvia* species the anthers are exposed by a lever mechanism (see later) or, in *Aphelandra aurantiaca* and *A. tetragona*, by a special joint in the upper lip (Troll 1951).

Flowers with narrow lower lip

Some groups of the orders Scrophulariales, Lamiales and Solanales have 'reversed' lip-flowers: the pollination organs are included in the lower lip. They thus resemble the flag-flowers of the Fabaceae in their architecture (see section 3.6). Among Scrophulariaceae this occurs in *Collinsia*, among Lamiaceae in Ocimoideae (e.g. *Plectranthus*, *Coleus*,

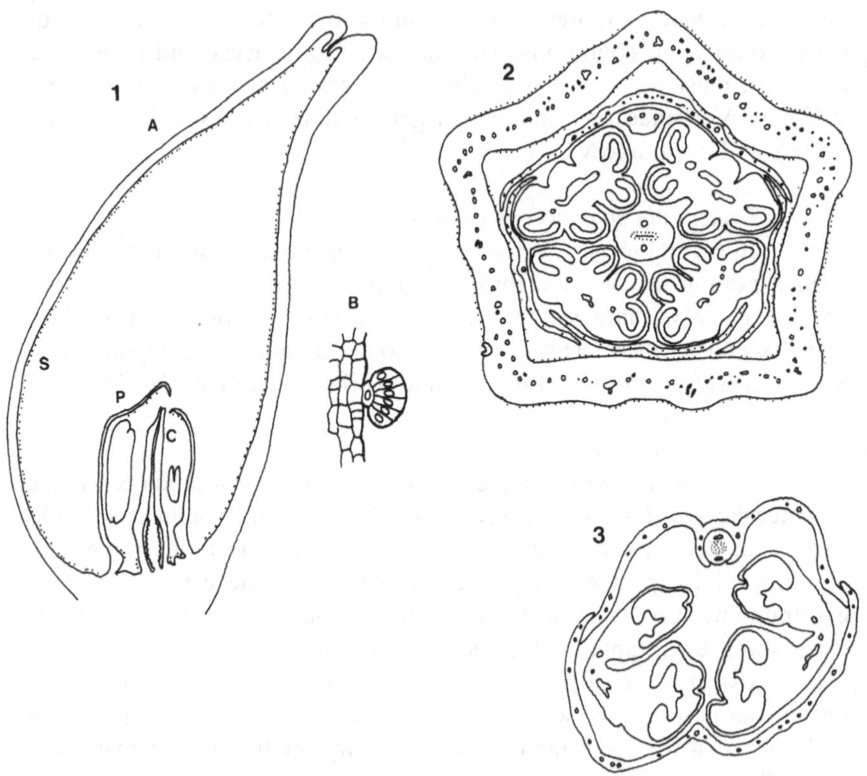

Figure 8.53. Floral bud architecture. 1. Water calyx in *Spathodea campanulata* (Bignoniaceae). A. Longitudinal section of floral bud, with congenitally united sepals (S) forming a large sac lined with glandular hairs and enclosing the retarded inner organs (P: petals; C: carpels) (× 3.5). B. Glandular hair from inside of united sepals (× 160). 2. *Kigelia pinnata* (Bignoniaceae); transverse section of bud, four anthers with pollen sac placentoids, and an odd staminode (× 10). 3. *Jacobinia pauciflora* (Acanthaceae); transverse section of bud, two anthers with pollen sac placentoids, the slender style held by two rims at the inner surface of the upper lip (× 16).

see Figs 8.54.2, 8.54.4), among Solanaceae in *Schizanthus* (van der Pijl 1972, Cocucci 1989). Another peculiar parallel feature of *Collinsia* (Kampny & Canne-Hilliker 1988) and *Plectranthus* (Burtt 1964) is a conspicuous dorsal hump in the corolla tube and, in *Coleus*, an arch-like curvature of the corolla tube (Fig. 8.54.2–4). They may be architectural necessities for this flower form (Burtt 1964).

An interesting parallel feature with the flag-flowers of Fabales (section 8.6.3) and Polygalaceae (Bamert 1990) is that in Ocimoideae congenitally united stamen filaments are present (Robyns & Lebrun 1929).

Personate flowers

In many horizontally directed flowers with the stamens included in a corolla of fused petals the anthers and pollen are protected against rain, drought, pollen thieves and other adverse influences, as Sprengel (1793) mentioned.

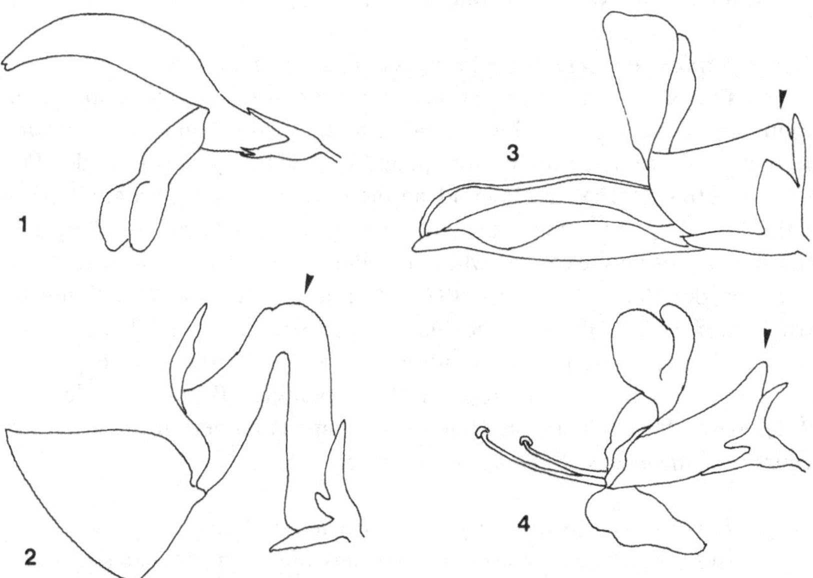

Figure 8.54. Architecture and convergence of flag-flowers compared with lip-flowers in Lamiaceae and Scrophulariaceae. 1. Lip-flower of *Salvia patens* (Lamiaceae) (× 1.5). 2–4. Flag-flowers. 2. *Coleus shirensis* (Lamiaceae) (modified after Burtt 1964). 3. *Collinsia bicolor* (Scrophulariaceae) (× 3.5). 4. *Plectranthus fruticosus* (Lamiaceae) (modified after Burtt 1964). Arrows point to the convergent spur-like dorsal protrusion in 3 and 4, and to the curvature of the floral tube in 2, which seem to be an architectural necessity (see text).

A radical mode of pollen enclosure is the closure of the corolla by upward arching of the lower lip. This has evolved in some groups of several families. It occurs in Antirrhineae (Sutton 1988) and some *Gratiola* species (Scrophulariaceae), in *Utricularia* (Lentibulariaceae), in *Linariantha* (Acanthaceae; Burtt & Smith 1964), in *Rhynchoglossum* (Weber 1978), *Didymocarpus* (Weber & Burtt 1983) (Gesneriaceae), in *Amphilophium* (Gentry 1974a) and *Stereospermum personatum* (Corner 1988) (Bignoniaceae).

Most of these flowers are pollinated by strong bees, which are able to open the flowers forcefully. For an efficient holdfast the surface of the lower lip contains transverse ribs in many cases (Müller 1929). In *Utricularia* spp., which are often free-floating waterplants, the personate flower construction provides good protection for the inner floral organs against water waves and spills; if brought under water, the flowers are automatically tightly closed by the water pressure. Personate flowers may also favour cleistogamy, which occurs in some *Utricularia* species (Jérémie 1989).

Explosive mechanisms to expose the hidden anthers

In some Lamiaceae explosive presentation of the pollination organs has to be triggered by a floral visit. In *Phlomis* and *Salvia apiana* the mature flowers are forcefully opened by *Xylocopa* bees (van der Pijl 1972, Brantjes 1981b). In Ocimoideae the stamens and stigma are hidden in the lower lip; in some representatives (*Hyptis*, *Eriope*, *Aeollanthus*) they are explosively exposed when a pollinator visits the flower (Harley 1971, van der Pijl 1972, Hedge 1972, Brantjes & de Vos 1981, Keller & Armbruster 1989). Since in the Ocimoideae structural details are quite different from group to group, multiple parallel evolution of the explosive mechanism has been suggested for the subfamily (Brantjes & de Vos 1981). A similar explosive mechanism of stamens hidden in the lower lip occurs in *Schizanthus* (Solanaceae) (Cocucci 1989).

Lever mechanisms to expose the hidden anthers

In contrast to explosive mechanisms the lever mechanisms work several times in the same flower. The lever mechanism in *Salvia* (Lamiaceae) has often been described. In each of the two fertile stamens only one theca is fertile and sits on an extremely broadened arm of the connective; the sterile theca forms a lever in the floral throat. Each time a pollinator visits a flower this lever is moved towards the floral centre and at the same time the fertile anther comes down, out of the narrow upper lip where it was hidden, and touches the back of the visitor. This apparatus seems to be especially efficient in bee-pollinated species, while in bird-pollinated species, where the stamens are longer than the upper lip and are always exposed in the open flower, it is apparently reduced (Werth 1956b). A similar mechanism is present in *Calceolaria* (Scrophulariaceae) (Vogel 1974) and perhaps in *Utricularia choristotheca* (Lentibulariaceae) (Taylor 1989).

Lever mechanisms to open the exposed but closed anthers

In a few representatives the dehisced thecae are closed in a relaxed state but each can be opened by an individual lever mechanism. Each theca has a horn at its base, which is moved towards the floral centre when a pollinator enters the flower. By slight distortion of the theca the thecal slit opens and pollen is released. This mechanism has been described for *Incarvillea* (Bignoniaceae; Trapp 1954), *Thunbergia* (Acanthaceae; Burkill 1906, see also section 8.8.4) and *Torenia* (Scrophulariaceae; Armstrong 1992). The mechanism requires a rigid support of the anthers in the flower. The upper lip is a supporting surface for the anthers in all three genera, but further support is reached in different ways: (1) by firm attachment of the anthers on stout filaments (*Thunbergia*) or (2) by mutual union of anther pairs (*Torenia*) or (3) by support of anthers by the style (*Incarvillea*).

Elaborations of the stamen filaments

Filaments may form a knob- or club-shaped outgrowth, which is exposed at the floral entrance with a conspicuous granular or hairy surface mimicking an anther with pollen or being otherwise optically attractive. This occurs among Scrophulariaceae in the tribe Gratioleae (e.g. *Lindernia*, *Torenia*, *Craterostigma*, *Crepidorhopalon*; Magin *et al.* 1989, Fischer 1989) but also in the tribe Hemimerideae (*Diascia nodosa*; Steiner 1992).

Reductions and elaborations in the anthers

Synthecal and monothecal anthers occur in certain genera of the Acanthaceae, Gesneriaceae, Lamiaceae, Lentibulariaceae, Orobanchaceae, Scrophulariaceae and Solanaceae (see Endress & Stumpf 1990) and in Bignoniaceae (Gentry 1976).

In *Salvia* (Lamiaceae) bee-pollinated species have a lever mechanism to expose the upper, fertile theca of each of the two stamens. In relatively unspecialized, small-flowered species (*S. recognita*) the lower theca is still fertile. In larger-flowered species it is sterile and elaborated as a frontally exposed lever plate. The plates of both stamens may even be postgenitally united (*S. glutinosa*). In bird-pollinated species, however, the anthers protrude out of the floral tube and no lever mechanism is needed (*S. splend-

ens, *S. coccinea*). Nevertheless, only the upper theca is fertile and the lower reduced, which may be seen as a 'burden' imposed by the earlier lever mechanism (Werth 1956b).

Sensitivity and movements of the stigmatic lobes

Sensitive stigmatic lobes occur in at least five families of the Scrophulariales (or even six families, if the Martyniaceae are separated from the Pedaliaceae). They fold together after touching of the inner, receptive surface and unfold again only after a certain length of time (Hart 1990). A common feature seems to be that these stigmatic lobes are relatively conspicuous in the flower: they are relatively large and broad and are well exposed at the floral entrance, not hidden in the throat.

Two ecological functions have been ascribed to this behaviour of evolutionarily repeated occurrence. (1) It is a means of herkogamy in that it prevents deposition of self-pollen on the stigma when a pollinator leaves the flower (Webb & Lloyd 1986). (2) It announces to potential floral visitors that the flower has been visited shortly before and that nectar is not yet available (van der Pijl 1954).

Sensitive stigmas have been found in Scrophulariaceae (*Mimulus*, *Diplacus*, *Torenia*, *Rehmannia*), Bignoniaceae (*Incarvillea*, *Bignonia*, *Tecoma*, *Catalpa*), Lentibulariaceae (*Utricularia*), Acanthaceae, Martyniaceae (*Martynia*) and Pedaliaceae (*Holubia*) (Newcombe 1922, 1924, Vogel 1954). Since in each family (except Martyniaceae) it occurs only in a small fraction of the genera, it is very probable that it has arisen in parallel at least four or five times.

Intimate synorganizations

Different floral organs become intimately connected in various ways, often by mutual partial postgenital fusions or by fixation of one part in a fold provided by the other part, or just by mutual support. These synorganizations provide special precision mechanisms.

In some long and slender flowers the thin style is enclosed in a fold of the corolla in the median plane. Among Gesneriaceae this occurs in *Chirita* (Wood 1974). Among Acanthaceae it is known from some Justicieae (*Jacobinia*, *Beloperone*, *Duvernoia*, *Himantochilus*) (Troll 1951, Vogel 1954, Bremekamp 1965); in *Duvernoia* and *Himantochilus* the stamen filaments, in addition, are enclosed in separate lateral corolla folds. Among Lamiaceae it occurs in *Phlomis* (Brantjes 1981b).

Postgential fusion of the anthers into tandems is prominent in Gesneriaceae but it also occurs here and there in other families, such as Scrophulariaceae (*Torenia*; Armstrong 1992; *Diascia*; Steiner 1992), Lamiaceae

(*Horminum*, Bokhari & Hedge 1971), Bignoniaceae (*Incarvillea*, Trapp 1954). It is especially obvious in long and slender flowers with long filaments, in which the stability of the precise median position of the anthers is enhanced, but also occurs in other flower forms.

Pollen-flowers

Pollen-flowers and lack of nectar among the Scrophulariales–Lamiales complex are perhaps best known from Gesneriaceae, where they are present in several groups (Weber 1989a). A number of genera have widely open, almost polysymmetric flowers (some with five stamens) and big, yellow anthers of the *Solanum* type (see section 5.1). They occur in the Gesnerioideae (*Bellonia*, *Phinaea*, *Niphaea*, *Napeanthus*) and Cyrtandroideae (*Saintpaulia*, *Conandron*, *Ramonda*, *Jankaea*, *Depanthus*, *Tengia*, *Protocyrtandra*) (Burtt 1970, Wiehler 1983). The anthers in *Bellonia* are poricidal. Weber (1989a) also mentions single species of *Boea*, *Loxocarpus* and *Didymocarpus* as having pollen-flowers. All these pollen-flowers are most probably derived from nectar-flowers (Weber 1989a). Pollen-flowers also occur in Scrophulariaceae (*Calceolaria* p.p., *Verbascum*) (Molau 1988) and are to be expected in other families as well. Among Solanaceae they are most prominent in *Solanum* and *Lycopersicon*; *Solanum* has poricidal anthers.

Oil-flowers and perfume-flowers

Oil-flowers occur among the Scrophulariaceae in at least four, probably unrelated, groups of genera of the southern hemisphere (*Calceolaria*, Calceolarieae; most Hemimerideae; *Monttea*, *Bowkeria*, both of uncertain position) (Vogel 1974, Vogel & Machado 1991, Simpson *et al.* 1990). All of them have trichome elaiophores on the lower lip (see section 5.3). In the Hemimerideae in *Angelonia*, and still more in *Diascia*, the elaiophores are in two pouches or two spurs, which prominently shape the floral gestalt. Oil-flowers have also evolved in Solanaceae within *Nierembergia* (Simpson & Neff 1981, Cocucci 1991). Here, trichome elaiophores are developed on the corolla around the stamens and on the base of the staminal filaments. In most species the stamens form a column around the style and in some of them the stigma is much enlarged and clasps around the column (Cocucci 1991). Nectaries are absent in all these oil-flowers. Oil-flowers had also been suspected to occur in Gesneriaceae among *Drymonia*. Steiner (1985) found, however, (for *D. serrulata*) that the secreted oil is not collected but serves to attach the powdery pollen to the pollinator's body (see above).

Perfume-flowers occur in Gesneriaceae in at least two genera, *Gloxinia*

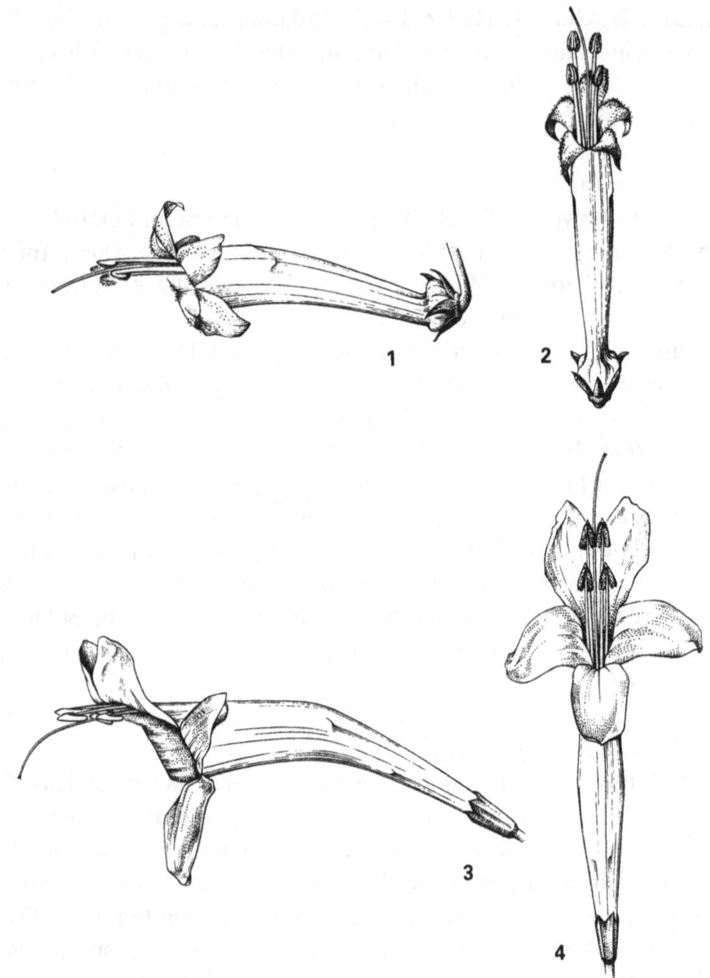

Figure 8.55. Convergence of bird-flowers ('dogfish' form) of Southern Africa (× 1). 1–2. *Phygelius capensis* (Scrophulariaceae), flower resupinate (recurved). 1. From the side. 2. From below. 3–4. *Tecomaria capensis* (Bignoniaceae). 3. From the side. 4. From below.

p.p. and *Monopyle* (Vogel 1966a, Wiehler 1983) (see section 5.5). The osmophore is inside the lower lip. Nectaries are also absent in these flowers. In Bignoniaceae, *Saritaea* has perfume-flowers (Dressler 1982). Other perfume-flowers are in Solanaceae, where they occur in species of *Cyphomandra* (M. Sazima & Vogel 1989). Here, elaiophores have evolved on the anther connectives of previously buzz-pollinated flowers.

Bird- and bat-pollination

Bird- and bat-pollination occur to a smaller or larger degree in all large families of the Scrophulariales–Lamiales complex, including Solanaceae (bat-pollination unknown in Scrophulariaceae) (Figs 8.46, 8.55 for bird-pollination). The transition from bee- to bird- and from bird- to bat-pollination has occurred many times, the reverse probably less frequently. However, in Gesneriaceae there is also a major trend from hummingbird- to euglossine bee-pollination (Wiehler 1983).

According to Vogel (1963b) the evolution of bird-pollinated flowers occurred independently in 12 subtribes among at least 26 genera of Gesneriaceae (see also Burtt 1977), in 13 subtribes among 17 genera of Scrophulariaceae, in 10 subtribes among 18 genera of Acanthaceae, in 12 subtribes among 18 taxa of Lamiaceae (Vogel 1963b). It seems, therefore, that in these 4 families alone, ornithophily originated independently at least 80 times among melittophilous taxa.

At a lower systematic level, in *Gesneria* (Gesneriaceae) bat-pollination has evolved at least twice from bird-pollination (Skog 1976), and in the *Aphelandra pulcherrima* complex (Acanthaceae) pollination by short-billed hummingbirds twice from that by hermit hummingbirds (McDade 1992).

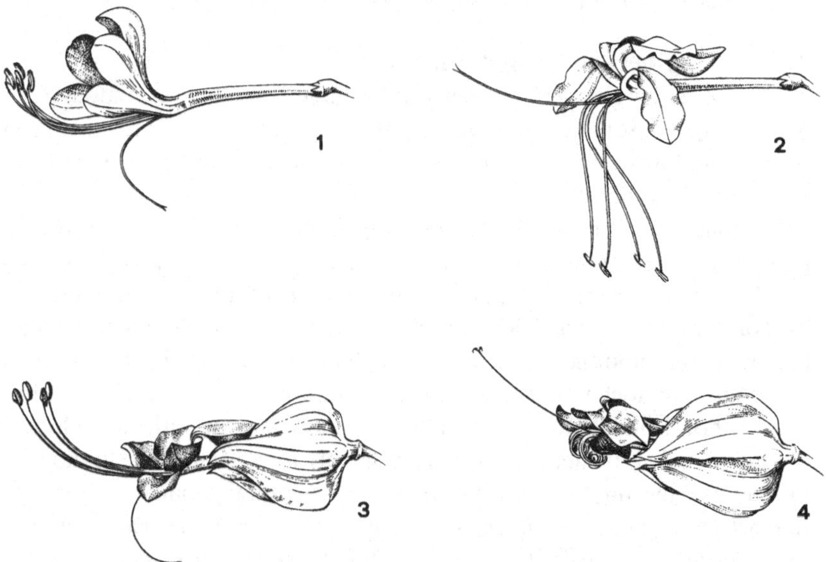

Figure 8.56. Protandry and herkogamy in *Clerodendrum* (Verbenaceae) (× 1). 1–2. *C. speciosissimum*. 1. Male stage. 2. Female stage. 3–4. *C. thomsonae*. 3. Male stage. 4. Female stage.

Forms of dichogamy and herkogamy

Protandry in conjunction with herkogamy by movement of stamens and style is perhaps the most common dichogamous trait in the Scrophulariales–Lamiales complex, although it has not been comparatively investigated for most of the families. The following pattern occurs in several different groups. At first the stamens are directed toward the incoming pollinator, the style is shorter and the immature stigma is behind the anthers. Later the stamens bend downwards and the receptive stigma comes into the former position of the anthers. This is the predominant condition in the Gesneriaceae (Ullrich 1935, Wiehler 1983) (Fig. 8.46). It is also known in Acanthaceae (*Eranthemum*, *Pseuderanthemum*, personal observation) and Verbenaceae (several species of *Clerodendrum*, e.g. Vogel 1983, Primack *et al.* 1981) (Fig. 8.56). Most Lamiaceae are protandrous (van der Pijl 1972).

Protogyny occurs in many Scrophulariaceae, although protandry is also present in the family (e.g. Stirton 1977a, Faegri & van der Pijl 1979, Elisens 1985). Among Neotropical Gesneriaceae, protogyny is known only in a few bat-pollinated species of *Gesneria* (Skog 1976, Wiehler 1983). There are as yet no indications under what circumstances which mode of dichogamy occurs within the group as a whole.

Herkogamy by stigma closure has been discussed above.

8.9 Zingiberales (Zingiberidae)

Zingiberales are predominantly large-leaved, rhizomatous forbs of wet forests. Some of them are cultivated as ornamentals for their conspicuous inflorescences and flowers. Among bananas and gingers there are important crop plants.

Comparative studies on the development of the inflorescences and flowers in Zingiberales are presented by Thompson (1933), Kunze (1985), and Kirchoff (1988b, 1991), and reviews with evolutionary considerations by Tomlinson (1962) and Kress (1990b). Generally, the inflorescences are thyrses (or compound thyrses) (Kunze 1985). The partial inflorescences, commonly monochasia, are often subtended by large, stout bracts, which protect floral buds and young fruits and may contrast with the colour of the flowers. The trimerous flowers are mostly monosymmetric lip-flowers; in Cannaceae and Marantaceae they are asymmetric but occur in pairs that are again monosymmetric. One to five of the six stamens are sterile. In many groups the sterile stamens are differentiated as conspicuous petaloid structures. In Cannaceae and Marantaceae even the fertile stamen is petaloid on one side and has only one theca; in Costaceae the fertile stamen is also petaloid but still has two thecae. The ovary is inferior and trilocular but reduced to one locule in some Marantaceae, or it is unilocular with three parietal placentae in some Zingiberaceae. Septal nectaries are common (Fig. 8.61.1), except for Zingiberaceae. Synorganizations occur between stamens and gynoecium: In Zingiberaceae and Costaceae the anther serves as a holder for the slender, lax style; in Marantaceae with secondary pollen presentation the relatively stout style serves as a pollen presenter combined with an explosive mechanism. Many representatives have large flowers that are pollinated by large animals (birds, bats, lemurs, large bees, Lepidoptera). Although most Zingiberales have many-flowered inflorescences, often only one (or very few) flowers are open at a given time. Commonly the flowers are one-day flowers. The two small families Lowiaceae and Cannaceae are not revised here, as their biology is so little known.

8.9.1 Musaceae

Musaceae is a Palaeotropical family of about 40 species in the two genera *Musa* (banana) and *Ensete*. Among the Zingiberales they are the least specialized in floral structure. The flowers have five fertile stamens, while the upper stamen is more or less reduced (Baumgartner 1913) (Fig. 8.58.1). The three sepals and the two lateral petals are united into a tubular sheath enclosing the stamens and style, while the median (upper) petal is free. The free petal forms a dome over the nectar gate (Scott-Elliot 1890). At anthesis the stamens are bent downwards and the style upwards. In an inflorescence there are usually functionally male and functionally female flowers (Baumgartner 1913).

Species with pendent inflorescences (*Ensete* and part of the genus *Musa*) have nocturnal, odoriferous flowers and are pollinated by bats (Nur 1976, Start & Marshall 1976, Gould 1978). Among the species with erect inflorescences *M. salaccensis* is pollinated by sunbirds (Nectarinidae), honeyeaters (Meliphagidae) and tree shrews (*Tupaia*); *M. velutina* with bisexual flowers is autogamous (Nur 1976). These species have diurnal, odourless flowers. In bat-pollinated species, nectar of a jelly-like consistency is produced abundantly by elaborate septal nectaries with a convoluted secretory surface. Apart from sugar it contains polysaccharide mucilage and protein (Fahn & Benouaiche 1979). The nectar of bird-pollinated species is watery (Nur 1976).

In the bird-pollinated *Musa salaccensis*, with bright pink and green bracts and flowers, flowers begin to open between 0300 and 0800 (Itino *et al.* 1991). Most male flowers dropped by the evening of the first day. Perianths of most female flowers dropped early on the third day. In the bat-pollinated *Musa acuminata*, with inconspicuous dark crimson bracts and yellowish flowers, bracts of the male inflorescence begin to open between 1700 and 2000 and the flowers fall until 1000 the next morning (Itino *et al.* 1991). The female flowers have a life span of up to three days. Flowers are available throughout the year in both species.

8.9.2 Strelitziaceae

Strelitzia

Strelitzia is a genus of five species from southern Africa. Some species are popular ornamental plants ('bird-of-paradise') because of their large, spectacular flowers adapted to bird-pollination (Scott-Elliot 1890, Rowan 1974, S. & P. Frost 1981). In *S. reginae* each inflorescence is produced on a long, stout, upright axis (Fig. 8.57.1). It carries a rigid, bluish, boat-shaped bract, which is horizontally oriented. Usually four flowers in monochasial arrangement are included in this bract. The flowers open sequentially. The first flower rises out of the bract, then the second, until all four are exposed. Each flower is open for about a week in *S. reginae* (three days in *S. nicolai*; S. & P. Frost 1981). Since a plant produces several inflorescences in sequence, the flowering period may be quite extended (*S. nicolai* flowers throughout the year; S. & P. Frost 1981). The flowers are almost scentless. The floral architecture resembles that of a papilionate flower in several respects (Fig. 8.57.2). The three sepals are orange; the two upper ones are directed upwards and form a flag (standard), the lower one is at first still included in the bract. The three petals are blue; the two lower ones form a keel and enclose the five stamens and the style, the upper one is much shorter and forms a protective roof over the entrance towards the nectar space (Fig. 8.59.1–2). The anthers are 5 cm long (the longest I have seen in angiosperms!). Kronestedt & Walles (1986) mention that there are no visible remnants of a sixth stamen. However, there is a small appendage on the inner side of the upper petal, which may be a rudiment of the missing stamen. The stigma is exposed at the tip of the keel; it is up to 3.5 cm long, three-lobed, papillate and very sticky.

In *S. nicolai*, studied by S. & P. Frost (1981), sunbirds are pollinators. They sit on the keel, on the stigma, on the flag or on the subtending bract and may take nectar from different positions. They pollinate the flowers with their feet; the same was found for *S. reginae* by Rowan (1974). In addition, bat visits are mentioned for *S. nicolai* (Faegri & van der Pijl 1979).

The two keel petals are tightly postgenitally united by their lower margins, except at the very base (as in Fabaceae). They form a median groove, which is protected by two tightly overlapping seams. The outer flanks of

Figure 8.57. *Strelitzia reginae*. 1. Inflorescence with three flowers raised out of the spathe, the flower on the right side in anthesis, the other two past anthesis (× 0.7). 2. A single flower, showing the three more or less equal sepals and the three very unequal petals; the two lateral petals form a keel enclosing the pollination organs, while the short median petal covers the access to the nectar (× 1).

the keel are directed obliquely upwards. If they are pressed downwards, the groove opens. When relaxed, the groove closes again. Thus, the keel apparatus is firm but elastic. The anthers of each side lie upon one another in the closed keel. When it opens they come into a more flattened position and are well exposed. Nectar is produced in large quantities from highly convoluted septal nectaries with a large surface area. In *S. nicolai* nectar production is highest in a second-day flower with 118 μl h^{-1} of a thin nectar containing glucose and fructose (S. & P. Frost 1981).

Aspects of the particular functional anatomy of the flowers of *S. reginae* have been studied by Wagner (1894), Johnson (1977), and Kronestedt & Walles (1986). The following description also contains original observations (Fig. 8.59.1–2). The firmness and elasticity of the keel apparatus is due to three thick longitudinal ribs of fibrous tissue (representing massive vascular bundle sheaths) in each of the two petals and thin, membranous, flexible parts between the ribs. The thin, membranous parts between the ribs consist almost solely of epidermis, with large intercellular spaces in between. However, the cells of the epidermis are thick-walled and collenchyma-like. The thin parts are also somewhat reinforced by small vascular bundles that obliquely connect the longitudinal ribs. The seams of the keel, which cover the pollination organs, are also thin (3–5-layered) but collenchymatic and papillate; near the keel they contain several sclerenchymatic strands. The wings of the keel are without sclerenchyma but have a thick-walled, papillate epidermis as well. The upper, short petal also has sclerified bundle sheaths, which are, however, much smaller than in the keel petals.

The mature anthers have a sclerified connective with endothecium-like cells. The thecal walls are two-layered and both layers, endothecium and epidermis, have lignified cell walls. However, in a large part of the region where the two pollen sacs meet they are not lignified. Consequently, each pollen sac has its own dehiscence line, which develops at the border between the lignified and the unlignified part (Fig. 8.59.4). The clumps of pollen grains are held together by viscin threads, which are derived from the unlignified epidermis cells between the two pollen sacs of a theca (Kronestedt & Bystedt 1981).

The thin style is made rigid and wire-like by extremely thick-walled sclerification of its entire periphery. Sclerification is weaker (with thinner cell walls) in the stigma and in the basal stylar parts. The ovary is 3-locular. The large, anatropous ovules are in two longitudinal rows along the placentae (Fig. 2.24). The micropyle is near the placental surface and is connected with it by a secretion. The three septa contain elaborated septal nectaries. Nectar is released by three canals into the nectar chamber in the floral centre.

Ravenala

Among the Zingiberales the flowers of the monotypic genus *Ravenala* are outstanding in having six fertile stamens. As in *Strelitzia* the two lower petals are united, forming a sheath for stamens and style. In bud the six stamens are under tension in this sheath. When the flower emerges between the rigid edges of the inflorescence bract, a touch sets the two united petals free and the pollination organs spring out of their sheath. They then remain in this exposed position (Scott-Elliot 1890).

Thus, *Ravenala* shows a further parallel to the Fabaceae in having flag-flowers with an explosive mechanism, as opposed to the valvular mechanism in *Strelitzia*.

The principal pollinators of *Ravenala madagascariensis* are lemurs (Kress *et al.* 1992), although it may also be pollinated by sunbirds, as observed by Scott-Elliot (1890). Kress *et al.* (1992) suggest that pollination by non-flying mammals is ancestral in the Strelitziaceae, while sunbird-pollination in *Strelitzia* and phyllostomid bat-pollination in *Phenakospermum* is more derived.

8.9.3 Heliconiaceae

Heliconia

Heliconia is a pantropical genus of often large banana-like herbs with perhaps 250 species. Many species are used as ornamentals for their conspicuous and elegant inflorescences (Berry & Kress 1991). The inflorescences consist of a number of large, boat-shaped bracts, each containing a cincinnus (monochasium) of several or numerous (up to 50) flowers that flower sequentially over a longer period (Kress 1984b) (Fig. 8.58). An inflorescence may flower for 40–200 days (Dobkin 1984). Each flower is open for only a single day (Stiles 1979). While the anthetic flowers are exposed, the floral buds are enclosed by the bract. The inflorescences are erect or pendent. The flowers have three sepals and three petals, which are postgenitally fused, except for the median sepal and the median margins of the two lateral petals. These free parts open the way to the nectar, which is released in the floral centre from a septal nectary. The five united organs enclose the pollination organs. The distal, arching parts of the two free flanks of the lateral petals are hard and contain sclerified tissue (Fig. 8.59.3). Of the six stamens, only five are fertile; the sixth is a staminode and may function in some species as a guide leading the pollinator's tongue to the nectar (Kress 1984a). The anthers have massive endothecium-like tissue on the connective side of the thecae but endothecium is lacking in a broad zone beside the median furrow of each theca, and the style is heavily sclerified, all much as in *Strelitzia* (personal observation on *H. bihai*) (Fig. 8.59.5). Each of the three ovary locules contains a single ovule in median position. In most species the flowers are presented upside-down with regard to their position in early development. This is achieved either by the pendent inflorescence or, in erect inflorescences, by a curvature of the individual flowers. Thus, in contrast to *Strelitzia*, the flowers are lip-flowers with the pollination organs in the upper lip. In some species with more hidden flowers and erect inflorescences, the flowers are not resupinate (Andersson 1985).

Species with erect inflorescences may contain a liquid in the inflorescence bracts in which the lower floral parts are immersed. The liquid is actively secreted and serves as protection of the flowers against herbivores (Wootton & Sun 1990).

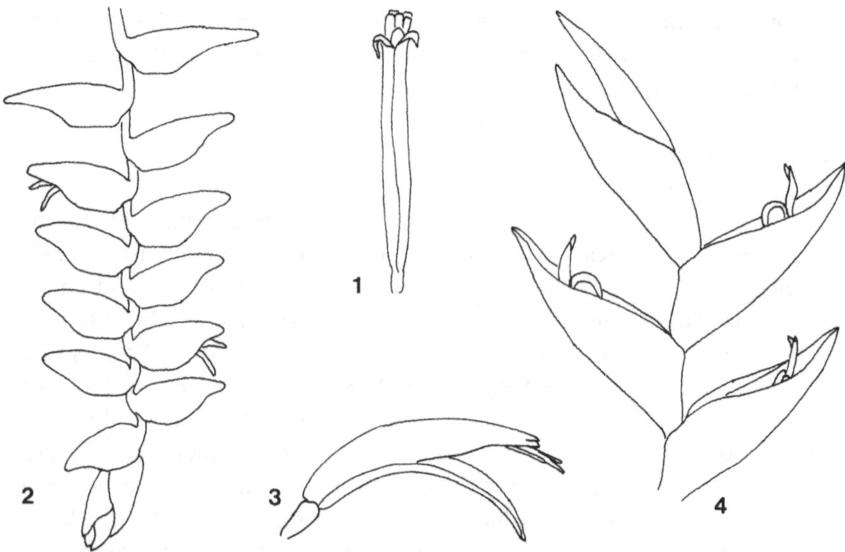

Figure 8.58. Musaceae and Heliconiaceae. 1. *Musa textilis*; male flower with five protruding stamens, from adaxial side (× 0.7). 2. *Heliconia rostrata*; pendent inflorescence with large bracts, two flowers visible, each with upper and lower lip (× *ca.* 0.3). 3. *Heliconia collinsiana*; flower from pendent inflorescence, with stamens protruding out of 'upper' lip (× *ca.* 1). 4. *Heliconia bihai*; terminal part of upright inflorescence with flowers resupinate (recurved) (× 0.4).

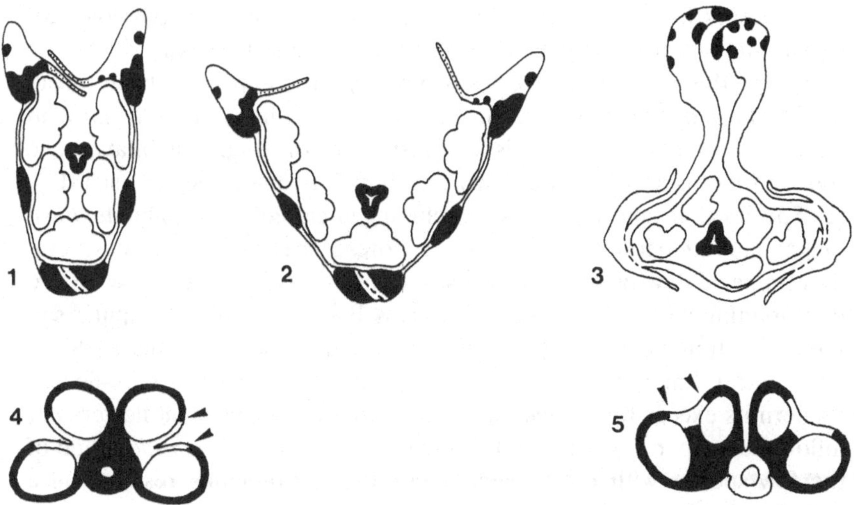

Figure 8.59. Strelitziaceae and Heliconiaceae. 1–3. Transverse sections of pollination organs and of those perianth parts that enclose them. Black, sclerified tissues; stippling, collenchyma; interrupted lines, postgenitally fused parts. 1–2. *Strelitzia reginae*; keel formed by two postgenitally united petals; in anthers sclerified tissue not indicated (× 6). 1. In closed condition. 2. As opened by the weight of a visiting bird. 3. *Heliconia bihai*; 'upper' lip of flower, formed by postgenital fusion of the two lateral sepals and the three petals, drawn in unresupinate position; sclerified tissue in anthers not indicated (× 12). 4–5. Transverse section of anthers. Black, sclerified tissue; arrows mark the two stomia of each theca. 4. *Strelitzia reginae* (× 22). 5. *Heliconia bihai* (× 45).

In the Neotropics, where the bulk of the species occur, most species have red and yellow inflorescences (most commonly red bracts and yellow flowers) and long, tubular, diurnal odourless flowers that produce copious nectar; the brilliant blue fruits add to the colour contrasts. Hummingbirds are the primary and mostly the exclusive pollinators (Kress 1985). Nectar is produced mainly very early in the morning (to escape nectar thieves) (Stiles 1979). In contrast, in the Palaeotropical Pacific region most species have green bracts and greenish flowers that open for one night; the fruits are red or orange; some species, especially those with pendent inflorescences, are pollinated by bats (Kress 1985, 1990a). Adaptation to bat-pollination is probably secondary (Andersson 1989).

Most *Heliconia* species in Central America are self-compatible and autogamy occurs in some species. In self-incompatible species, physiological regulation acts at the stigmatic surface and in the style (Kress 1983).

Among the Neotropical species two biological groups have been recognized. (1) Plants visited by territorial, non-hermit hummingbirds grow in large, often monoclonal stands at forest edges or in open habitats, have short perianths and produce many flowers with copious nectar each day. (2) Species that tend to form small clumps in closed forest habitats, have longer curved perianths, and have inflorescences that produce several flowers per day with small amounts of dilute nectar, are generally visited by traplining hermit hummingbirds (Kress 1985). The role of trapliners in the pollination biology of *Heliconia* was first discussed by Linhart (1973). All Central American species with pendent inflorescences are pollinated by hermits except for *H. mariae*, which produces many small flowers per inflorescence per day with a relatively short perianth; *H. mariae* is one of the few species with red flowers (Stiles 1975). Traplining resulting in a high level of outcrossing may compensate for the low level of self-incompatibility within the genus. However, hummingbird flower mites may effect self-pollination (Dobkin 1984). They inhabit *Heliconia* inflorescences, where they feed on nectar, and are transported in the nasal cavities of hermit hummingbirds. Within an inflorescence they move within flowers and from flower to flower and disperse pollen.

8.9.4 Zingiberaceae

The Zingiberaceae are a prominent family of more than a thousand species of mainly tropical herbs. Besides their use as spices, many species with showy inflorescences are grown as ornamentals. The inflorescences are usually thyrses, sometimes with large coloured bracts. The floral groundplan is pronouncedly monosymmetric: there is only a single fertile stamen, while the other stamens are replaced by petal-like organs (Fig. 8.60). Some genera have asymmetric flowers (e.g., *Hedychium*, *Kaempferia*). Another peculiarity is the frequent presence of two stamens in the second flowers of a partial inflorescence in *Alpinia* species (Nelson 1954, Kunze 1985).

The three sepals are congenitally united and form a tube. The three petals are congenitally fused with the androecium. The flowers are usually differentiated as complicated lip-flowers. The lip is formed by the two united lateral inner staminodes (the median inner staminode is lost). The stamen filament is firm and the anther large with a broad connective. The connective contains three or more vascular bundles. There are several layers of endothecium-like tissue on the connective side of each theca (Fig. 8.61.4). The endothecium may also be two or more layers thick. In addition, the thecal epidermis may have lignified cell walls. In contrast to the massive stamen, the hollow style is slender and weak. It is supported by the anther, which embraces it from above, so that the stigma is exposed exactly above the anther (Fig. 8.62.1). This complex formed by anther and stigma is usually in the median plane but it may also be directed to the side (as in *Hedychium* species). The anther has a conspicuous crest in some genera but, in contrast to Costaceae, it is not broadened into petaloid flanks. In some genera the anther has firm basal appendages that are located at the floral entrance. A pollinator pushing into the flower moves them backwards, and the hinged anther together with the clasped stigma comes down towards the back of the visitor (*Curcuma*, *Camptandra*, *Roscoea*) (Troll 1929, 1951, Holttum 1950). In *Globba* species, firm appendages occur on the side of the anther (see below). In *Hedychium* each theca is flanked by a band of secretory hairs. They secrete a glue, which covers the pollen as the thecae open and serves to attach the pollen mass to pollinating lepidopter wings (Vogel 1984).

Stamen and lip are congenitally fused and form a narrower or broader tube or cup at the base. The free part of the lip may be relatively large and expanded (e.g. *Alpinia*, *Globba*) or smaller and involved in the tube formation (e.g. *Amomum*). In *Hornstedtia*, in addition, the petals are involved in the tube formation; the tube may be very long, e.g. 6–7 cm in *H. incana* (Smith 1985). The ovary is trilocular with axile placentae or unilocular with parietal placentae. There are no septal nectaries. At the base of the tube two lateral conical nectaries are situated, which may be seen as evolutionarily derived from septal nectaries, as demonstrated by Costaceae (see section 8.9.5) (Fig. 8.61.3). *Costus* has septal nectaries that flare out at the upper end into two short fingerlike processes (Rao 1963) (Fig. 8.61.2). Apparently, in Zingiberaceae these processes have been further elaborated with a concomitant loss of the septal cavities. The presence of unilocular ovaries with parietal placentae in some Zingiberaceae is also associated with the loss of septal nectaries.

As in other families of the Zingiberales, flowers last less than a day. In some species of *Zingiber* they open in the afternoon and last only a few hours (Holttum 1950). Self-incompatibility is reported for *Hedychium* (Holttum 1950). In *Alpinia hookeriana* the flowers are protandrous (Porsch 1924b). Reports on pollinators are scant for the family. *Hedychium coronarium* with white and heavily scented flowers (and the lip turned upwards) is pollinated by hawkmoths, the red-flowered *H. coccineum* by butterflies (Knuth *et al.* 1904). The flowers of *Alpinia zerumbet* are pollinated by large bees (euglossines, *Centris*, *Bombus*) (F. Müller 1888); those of *A. malaccensis* and *A. hookeriana* by *Xylocopa* (Porsch 1924b, van der

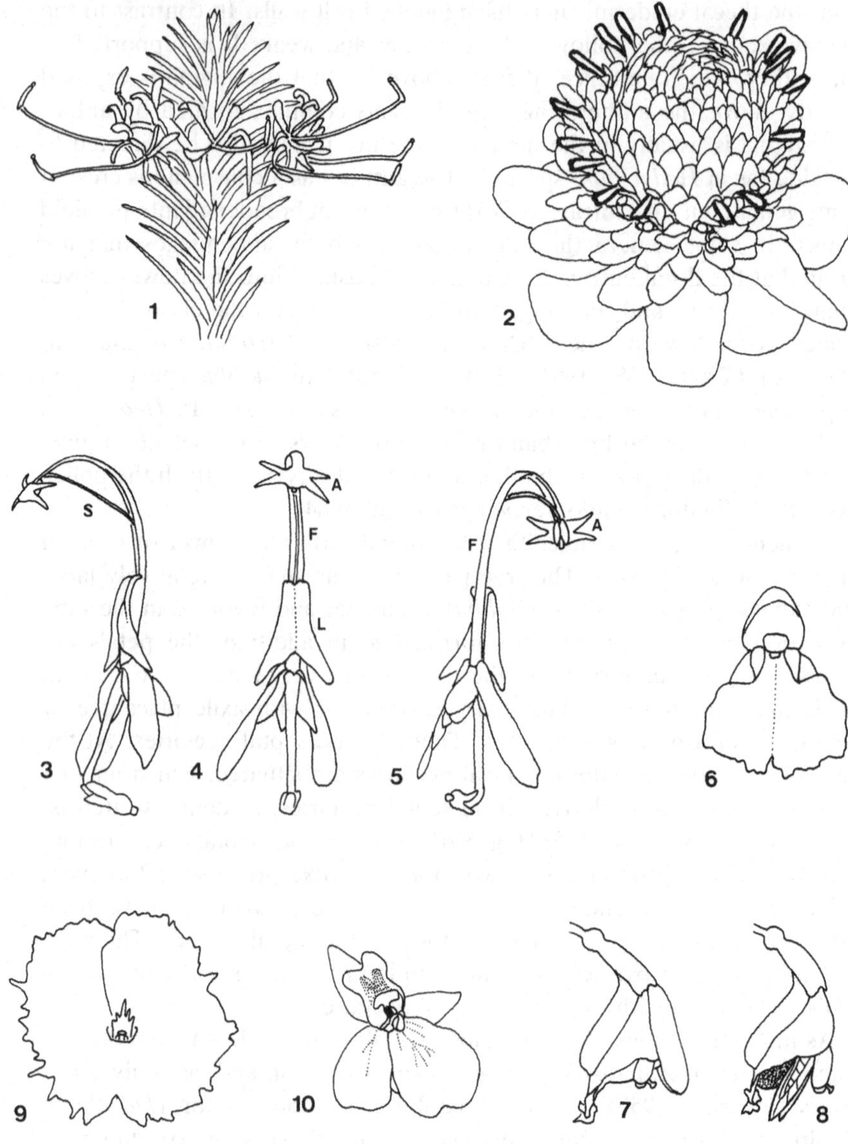

Figure 8.60. Zingiberaceae (1–8), Costaceae (9), and Marantaceae (10). 1. *Hedychium thyrsiforme*; inflorescence with probably hawkmoth-pollinated flowers (× *ca.* 0.4). 2. *Etlingera elatior*; compact inflorescence with bird-pollinated flowers; lower lips of flowers drawn with thick contours (× *ca.* 0.3). 3–5. *Globba winitii*; flower. A, anther clasping the upper part of the style; F, stamen filament with ventral

Pijl 1954) (Fig. 8.60.8). Tubular flowers of *Renealmia* species are pollinated by hummingbirds (Maas 1977), those of *Hornstedtia* by honeyeaters (Ippolito & Armstrong 1990).

In some groups the bracts of the partial inflorescences contain a liquid, which protects flower buds and young fruits, a parallel to *Heliconia*. In some genera the inflorescences are buried underground (e.g. *Achasma*, *Elettariopsis*, *Elettaria*, *Siphonochilus*) (Holttum 1950, Gordon-Gray *et al.* 1989). In extreme cases only the expanded floral parts are above ground, while the floral tube and the inferior ovary are buried (Holttum 1950).

Globba (Globbeae)

Globba is an Eastern Asian and Malesian genus of almost a hundred species. It is outstanding by reason of its delicate flowers with long, arching filaments and lateral anther appendages (Smith 1988) (Fig. 8.60.3–5, 8.62.1). The inflorescences are lax, upright or pendent thyrses and the flower buds are not included by bracts. The inflorescence bracts are green, orange, pink or red. They often contrast in colour with the flowers, which are white, yellow, orange or purple. Several flowers may be open in an inflorescence at any given time. The flowers have a very narrow tube of a few centimetres length (Fig. 8.61.3). The inner surface of the tube is provided with hairs, which act as a capillary system for the nectar (Müller 1931). The ovary is unilocular and has three parietal placentae. As usual in Zingiberaceae, the style is clasped by the anther and rests in the ventral furrow of the filament in bud. At anthesis, however, the style is shorter than the arching filament and it is now stretched like the string of a bow. The stigma, immediately above the anther, is concave and fringed by stiff hairs. The anther is hinged by a relatively thin zone, which is collenchymatic (Müller 1931). This hinge may act as a shock absorber against pushes by pollinators. Since the style is held by the anther, it is slightly compressed by such a movement; secretion comes out of the stigmatic chamber and may then be sucked in again together with deposited pollen

Fig. 8.60 (*cont.*)
furrow enclosing the style; L, lower lip; S, style (× 1.5). 3. Lateral view. 4. Frontal view. 5. From the back. 6. *Alpinia* cf. *coerulea*; flower in frontal view (× 2). 7–8. *Alpinia hookeriana*; protandrous flower in male phase (modified after Porsch 1924) (× 0.5). 7. Flower in lateral view. 8. Same flower with visiting *Xylocopa* bee, dorsal petal removed. 9. *Costus igneus*; flower in frontal view, lower lip surrounding the pollination organs (× 0.5). 10. *Maranta leuconeura*; asymmetric flower in frontal view; black, style (× 1.5).

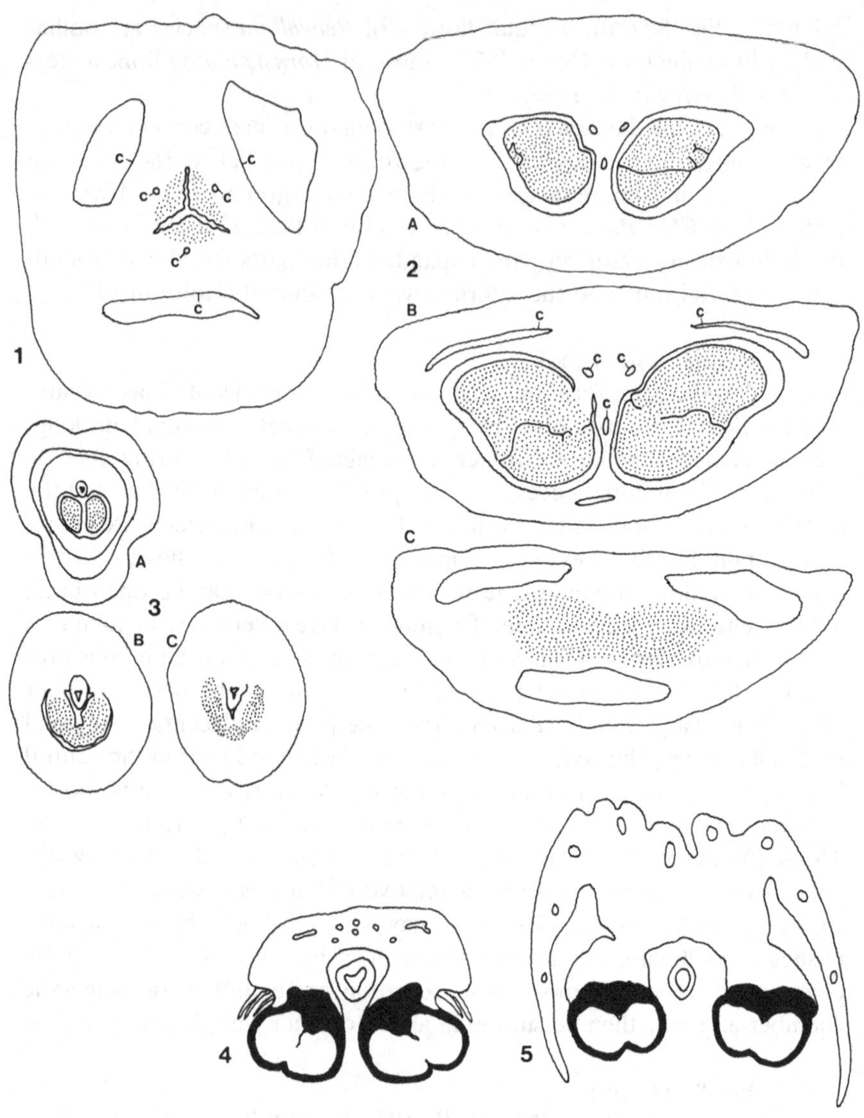

Figure 8.61. Heliconiaceae (1), Costaceae (2,5) and Zingiberaceae (3,4) (× 16). 1–3. Forms of nectaries as seen in transverse sections. Stippling, nectariferous regions; c, inner surfaces of carpels. 1. *Heliconia bihai*; septal nectary. 2. *Costus igneus*; three successive levels. A. Free protruding part of the two nectaries in the two septal cavities. B. Attachment level of nectaries in septal cavities. C. Below the septal cavities. 3. *Globba winitii*; three successive levels. A. Free protruding part of the two nectaries in the floral tube

(Müller 1931). The lateral anther appendages may act as a lever to put the anther in a favourable position, even if the flower is approached laterally by a pollinator. Leinfellner (1956) noted the late formation of the appendages. They arise by extreme transverse cell elongation (personal observations).

Surprisingly, there seem to be no reports on pollination biology. From the floral architecture one could suppose that butterflies were pollinators (Müller 1931).

Etlingera (Alpinieae)

This Malesian genus contains a number of species with underground inflorescences (Smith 1986). However, the best known species is *Etlingera* (*Nicolaia*) *elatior*, in which the inflorescences are well raised above ground. They are conspicuous because of their brilliant red, densely arranged bracts, the lower ones radiating, so that a sort of a pseudanthium is formed (Fig. 8.60.2). The flowers are also red but the lips have a contrasting yellow margin. The lips are held upright forming a cup, which is filled with copious aqueous nectar. Several flowers in a ring are open at the same time. An inflorescence flowers for about three weeks. Since there are always several inflorescences of different ages on an individual, nectar may be continuously available.

Classen (1987) observed Nectarinidae as pollinators in the Botanical Garden, Singapore; Knuth (in Knuth *et al.* 1904) in the Botanical Garden, Bogor (Java). They perch in the centre of the stout inflorescence and push their bill downward into the open flowers. The underside of the head may take up pollen and transfer it to stigmas. The relatively wide and long (up to 5 cm) floral tube is provided with a hair ring at the base of the lip, which may prevent the entrance of nectar thieves. Butterflies were also observed as pollinators (Classen 1987). Vogel (1981a) mentions the function of stigmatic secretion as glue for the pollen transport.

8.9.5 Costaceae

The Costaceae is a family of pantropical herbs with about 150 species. The flowers, in contrast to those of Zingiberaceae, have all five

Fig. 8.61 (*cont.*)
formed by congenitally fused petals and androecial elements. B. Attachment level of nectaries at base of floral tube. C. Below attachment. 4–5. Transverse sections of anthers clasping the hollow style; black, sclerenchymatic tissue; vascular bundles indicated. 4. *Hedychium coronarium*. 5. *Costus igneus*.

staminodia united and transformed into a broad labellum (Costerus 1916, Troll 1928a) (Fig. 8.60.9). This is confirmed by the development (Kirchoff 1988a,b). The stamen is broad and petaloid. The petaloid parts arise early in ontogeny (in contrast to the appendages of the zingiberaceous *Globba*) (Leinfellner 1956). Further, there are septal nectaries. However, they consist of two secretory protrusions in the septal cavities and may well be seen as an intermediate stage between normal septal nectaries and the peculiar conical nectaries in Zingiberaceae (see section 8.9.4) (Fig. 8.61.2) (also Newman & Kirchoff 1992). In some *Costus* species the sepals have pointed subapical spurs with presumably protective function (Weber 1980).

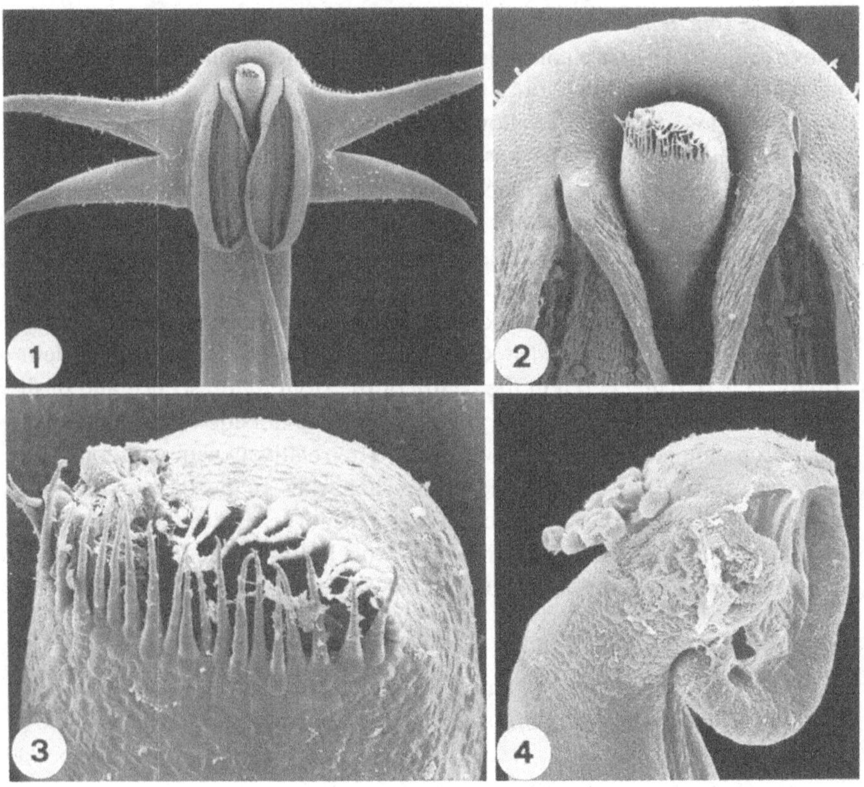

Figure 8.62. 1–3. *Globba winitii* (Zingiberaceae); winged anther clasping the slender style. 1. Overview (× 13). 2. Upper part with stigma (× 50). 3. Hollow stigma with non-receptive bristles (× 200). 4. *Ctenanthe* sp. (Marantaceae); upper part of style with stigma (to the right) and pollen grains attached at dorsal side (× 40).

The stigma is concave as in Zingiberaceae. It is either bilamellate or cup-shaped (Maas 1972). The ovary is trilocular or bilocular (*Dimerocostus*).

Floral proportions are diverse. Some groups have narrow, thick-walled floral tubes, the lip of which is yellow, orange or red; they are pollinated by hummingbirds (*Costus* sect. *Ornithophilus*) (Stiles 1978a). In other species the floral tube is short and wide, the lip is more flat and white to yellow, often striped with red, and bee-pollination occurs (*Costus* sect. *Costus*) (Maas 1972, 1977). *Costus mallortieanus* is intermediate in its characters and is pollinated by euglossine bees and hummingbirds. *Dimerocostus* is pollinated by euglossine bees (Vogel 1966a). In the Old World tropics *Costus* is predominantly pollinated by carpenter bees (Cammerloher 1931, van der Pijl 1954). The flowers of *C. speciosus* open early in the morning and wilt in the afternoon (van der Pijl 1954).

In *Costus woodsonii* (and other species) floral buds and young fruits are protected by thick bracts. The bracts produce nectar and the inflorescence is protected by ants against predators (Schemske 1980). Self-compatibility was found in three *Costus* species from Panama, although combined with inbreeding depression (Schemske 1983a). However, the production of a single flower per day and pollination by euglossine bees suggest a high probability of outbreeding. Schemske (1981) also found two species (*C. allenii* and *C. laevis*) with almost the same flower characteristics in the same region flowering synchronously and pollinated by the same *Euglossa* species but with strong internal hybridization barriers. This was interpreted as a convergence due to selection for pollinator sharing.

8.9.6 Marantaceae

The family is predominantly Neotropical and contains about 550 species. The flowers are asymmetric (Fig. 8.60.10). The single (upper median) stamen is fertile on only one side; the other side is petaloid. Four other androecial members (two outer and two inner) are sterile and petaloid, the sixth (lower median) is lacking. In *Calathea* there is only one outer staminode. The petals are congenitally united. Stamen and staminodes are fused with the corolla tube; the style is also fused with the tube on one side. The ovary is trilocular or unilocular with a single ovule per locule. The unilocular condition has evolved independently several times (Andersson 1981). Septal nectaries are present. The concave stigma is wet and non-papillate (Fig. 2.28.4). The main pollinators of Marantaceae in the Neotropics are euglossine bees (Vogel 1966a, Kennedy 1978).

The Marantaceae have an unique floral mechanism. The style is thick and is under tension at the beginning of anthesis. When triggered, it moves explosively, which is important for pollination. Pollen is presented secondarily on the style (Fig. 8.62.4). The mechanism has long been known. It was described in detail for *Calathea* by Kennedy (1978) and revised for other genera by Kunze (1984) and Classen-Bockhoff (1991).

In *Calathea* in the newly opened flower the style is surrounded by a cucullate staminode. Already in bud pollen is deposited in a shallow depression on the convex side of the style, since the anther is also in the cucullate staminode and pressed toward this stylar depression. But the empty anther moves away from the cucullate staminode before the flower opens. Triggering occurs when a pollinator presses its head to the floral centre to obtain nectar and touches a slightly curved, finger-like projection of the cucullate staminode. The cucullate staminode moves backwards and the style bends explosively forward in an arc, coming to rest against the callous staminode. Thus pollen on the pollinator's body is taken up by the stigmatic cavity; at the same time, the pollen presented on the style is attached to the pollinator together with a sticky secretion produced at the edge of the stylar depression (Kennedy 1978, Vogel 1984). Kunze (1984) and Classen-Bockhoff (1991) showed that in *Maranta*, *Calathea*, and *Thalia* species the style is sensitive in a small region below the finger-like projection of the staminode. If this region is touched, the explosion occurs. Thus the style is not held back by the cucullate staminode but vice versa.

In *Calathea insignis*, pollen is deposited on the style between 1700 and 2100 in the bud, and the next morning the newly opened flowers are generally pollinated between 0600 and 1000, while the unpollinated flowers drop off between 1030 and 1200 (Kennedy 1983). An inflorescence contains about 20–25 bracts. Each bract contains eight to ten pairs of flowers. Anthesis starts in the lowermost bract with the first flower pair. The next day the first flower pair of the second bract is in anthesis. After six to nine days the second flower pair of the lowermost bract comes to flower. Thus there are waves of flowering. When the third pair of flowers in the basal bracts opens, the bracts halfway up the inflorescence are in their second flowering, while the uppermost bracts show the first pair in anthesis (Kennedy 1983). With this pattern an inflorescence may flower for a few months. Among the flowers of an inflorescence nectar production is quite unequal; some have almost no nectar. Within a population two flower colour morphs may be found, one yellow, the other pinkish purple (Kennedy 1983).

In contrast, *C. gymnocarpa* exhibits mass flowering. Flowering of an individual lasts only about a week. This is unusual in Marantaceae (and Zingiberales in general) (Kennedy 1977).

In some species of *Calathea* the flowers remain closed and are forced open by their pollinators (Kennedy 1978). Thus they are not cleistogamous. Nectar is produced as in the open-flowered species. However, the (single) outer staminode is reduced compared with those of the open-flowered species, where it is larger and attractive. This closed-flower syndrome was found in 15 species of five species groups of the series *Scapifoliae*. It has probably arisen more than once (Kennedy 1978). The plants are pollinated by a single species or a few closely related species of euglossines (*Eulaema* and *Euglossa*). In contrast, the bees are polytropic. The evolution of this closed-flower behaviour may be understood by observations on *Ischnosiphon*, where the flowers normally open but euglossines frequently forcibly open still closed flowers and pollinate them (Kennedy 1978).

The explosive mechanism has been thoroughly analysed by slow-motion pictures in *Thalia geniculata* by Classen-Bockhoff (1991). The revolving movement of the style occurs in three phases in approximately 0.03 seconds. The flowers of cultivated specimens in the Botanical Garden, Bogor, were pollinated by *Xylocopa caerulea*. *Thalia geniculata* observed in Costa Rica had carpenter bees, hummingbirds and butterflies as floral visitors (Davis 1987). Carpenter bees triggered almost all flowers visited, hummingbirds about half, and butterflies none. The species was also found to produce seeds autogamously.

In *Calathea ovandensis*, floral visitors were also Hymenoptera and Lepidoptera but only bees tripped the flowers (Schemske & Horvitz 1984).

The flowers of *Maranta* subg. *Maranta* are usually nocturnal and partly autogamous. Pollinators have not been observed (Andersson 1986).

8.10 Orchidales (Liliidae)

8.10.1 Orchidaceae

The orchids, with probably more than 20 000 species, are unsurpassed by any other group in their floral elaboration and diversity. For a long time they were subdivided into the three subfamilies: Apostasioideae, Cypripedioideae and Orchidoideae. Their floral groundplan in a conventional sense (with number and position of the structural elements) is very stable. The flowers are monosymmetric; they have three sepals, two petals, a lip and a gynoecium with a tricarpellate, inferior ovary; only stamen number varies between three and one. In flowers of Orchidoideae there is a single median stamen of the outer of originally two whorls ('monandrous orchids'), in Cypripedioideae two lateral stamens of the inner whorl ('diandrous orchids') and in Apostasioideae two to three stamens ('triandrous orchids') (Fig. 8.63). However, more recently the monandrous orchids have been subdivided into three or four subfamilies

(Neottioideae, Orchidoideae incl. Spiranthoideae, Epidendroideae incl. Vandoideae) (Dressler 1981, 1986a); similar classifications with deviations in detail are used by Rasmussen (1985a) and Burns-Balogh & Funk (1986). The flowers are generally resupinate: the lip is in an adaxial position but during development the flower is twisted through 180°. Since the majority of tropical orchids are epiphytes and many of them have pendent inflorescences the flowers attain the lip-downward position without twisting.

Although there are excellent comprehensive works on the structure and biology of orchid flowers (van der Pijl & Dodson 1966, Dressler 1981) a short account in the framework of this book seems appropriate. The treatment by Dressler (1993) appeared after completion of the manuscript.

Floral groundplan

The lip is often much more complicated than the two lateral petals, and it has been debated whether it represents the median petal or an organ complex (see below). Often the lip is differentiated into a basal narrow portion ('hypochile'), a middle portion often with two lateral lobes ('mesochile') and a distal portion, which may also be subdivided ('epichile') (Fig. 8.64.5). Various stout ridges ('callus') may appear on the lip (Fig. 8.65).

The stamen in monandrous orchids is congenitally fused with the gynoecium to form a 'gynostemium' (Fig. 8.66). Partial fusion of other floral organs is also common. They may originate as common primordia (Fig. 8.67). Two lateral outgrowths are often present on the gynostemium, which are interpreted as two lateral staminodes that are also fused with

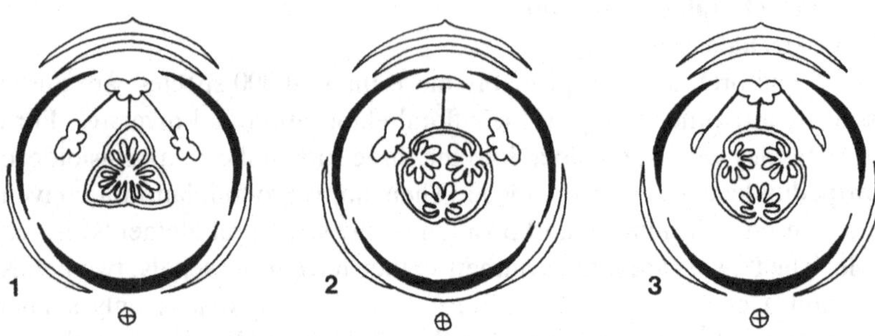

Figure 8.63. Floral diagrams of main groups of Orchidaceae (modified after Rasmussen 1985a). Black, petals; crossed circle, inflorescence axis; keeled leaf, subtending bract; stamens and staminodes congenitally fused with gynoecium and with each other. 1. Apostasioideae. 2. Cypripedioideae. 3. Monandrous orchids.

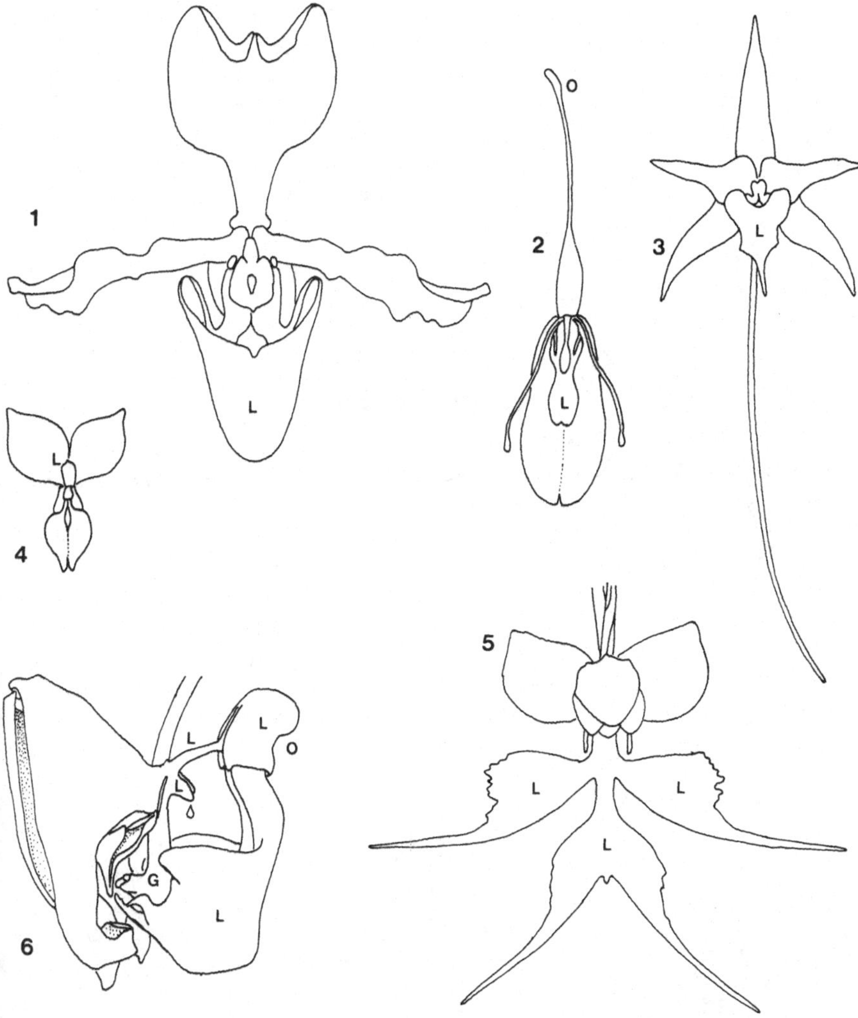

Figure 8.64. Flower forms of Orchidaceae. L, lip (if a lip is differentiated into several parts, each part is indicated with 'L'); O, osmophore. 1. *Paphiopedilum insigne* (× 1). 2. *Restrepia elegans;* antenna-like median sepal differentiated as osmophore, lateral sepals united (× 1.5). 3. *Angraecum sesquipedale*; lip with long spur (× 0.25). 4. *Ponthieva maculata*; flower unresupinate, lateral petals united to form a functional lip, while original lip is minute and not functional (× 1). 5. *Habenaria crinifera*; lip flat (× 1.5). 6. *Coryanthes speciosa*; lip 3-dimensional and extremely complicated (× 0.5) (modified after Dodson 1965).

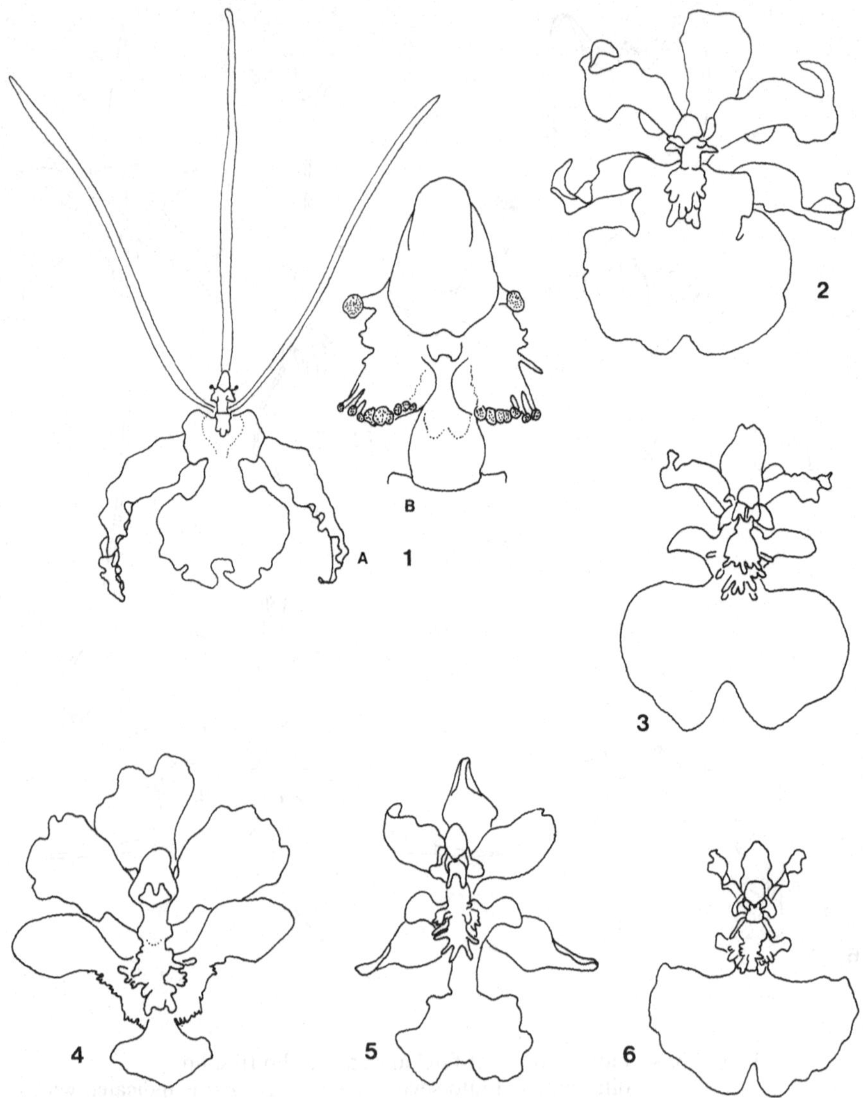

Figure 8.65. Diversity of floral forms in *Oncidium* and relatives (Orchidaceae). 1. *Psychopsis papilio*. A. Flower (× 0.5). B. Gynostemium with dark brown knobs at several protrusions (× 3). 2. *Oncidium sphacelatum* (× 1.5). 3. *O. varicosum* (× 1.5). 4. *O. ciliatum* (× 1.5). 5. *O. baueri* (× 1.5). 6. *O. flexulosum* (× 1.5).

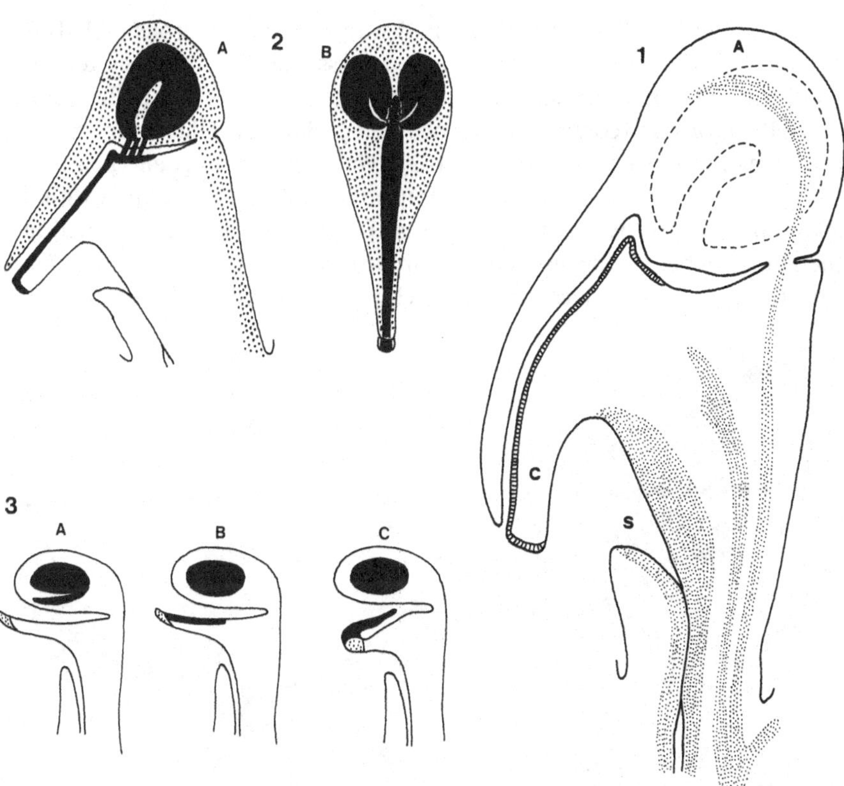

Figure 8.66. Pollinarium development in Orchidaceae. 1–2. *Oncidium ornithorrhynchum*; pollinarium with stipes. 1. Median longitudinal section of gynostemium of old floral bud. A, anther; C, distal part of median carpel; S, stigma; stippling, vascular bundle of stamen and of median carpel, and secretory tissue of stigma and stylar canal; hatching, part of the median carpel that will give rise to the stipe and viscidium of the pollinarium; interrupted lines, pollinium in paramedian position (× 45). 2 A. Schematic figure of the same. Black, components of the pollinarium; the later connection between stipes and pollinium is marked with three thick lines; stippling, stamen; white, gynoecium. B. The same as seen from above to show the entire pollinarium below the hood-like periphery of the anther. 3. Different forms of pollinaria in Orchidaceae as seen in schematic median longitudinal sections (modified after Rasmussen 1985a). Components of the pollinarium black, except for viscidium, which is stippled. A. Stalk formed in the anther. B–C. Stalk formed by the median carpel (stipe). B. Stipe in the form of a tegula. C. Stipe in the form of a hamulus.

the gynostemium (Fig. 8.68, 8.69). Pollen in most monandrous orchids is coherent in pollinia (or aggregations of massulae, i.e. little packets of coherent pollen that are differentiated in larger numbers in a pollen sac). At pollination the (commonly two or four) pollinia of a flower are transported as a pollinarium (Fig. 8.70.1–2). The pollinarium is an apparatus with a viscid basal part for attachment to the pollinator and often with a stalk upon which the pollinia are sitting. Pollinia and stalk are formed in the anther, whereas the viscid basal part is formed by the adjacent median (mostly non-receptive) carpel apex. Thus the pollinarium is formed by

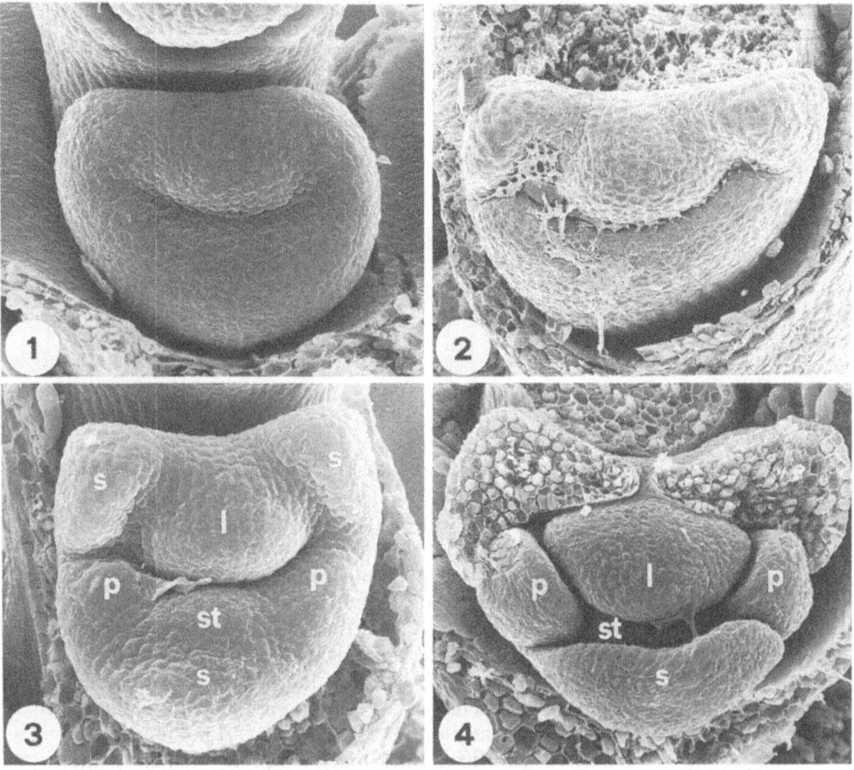

Figure 8.67. *Oncidium ornithorrhynchum* (Orchidaceae); early floral development; flowers in non-resupinate position. l, Lip; p, petals; s, sepals; st, stamen. 1. Young flower with ring primordium of the floral organs (× 200). 2. Adaxial organ primordia apparent on the ring primordium (× 200). 3. All organ primordia present except for carpels (× 200). 4. The expanded lip covering the stamen primordium, lateral sepals removed (× 150).

intimate synorganization of androecium and gynoecium. This is a noteworthy convergence with the Asclepiadaceae, where the pollinaria are also formed by both androecium and gynoecium. In contrast to Asclepiadaceae, androecium and gynoecium are congenitally fused in orchids (gynostemium) and not postgenitally (gynostegium). Further, the translator in orchids is not a secretion as in Asclepiadaceae, but derived from tissues. Another convergence is that after removal from the flower the stalk of the pollinarium dries out and bends in such a way that the pollinia will be exactly inserted into the sticky stigmatic region of another flower. The structure of the gynostemium is diverse within the orchids, and will be discussed later.

Whereas the gynostemium is an unequivocal organ complex, the nature of the lip is more controversial. Although the lip seems to represent the median petal at first sight (Vermeulen 1966), it has been interpreted to be an organ complex by some authors. According to them it is (1) a congenital fusion product of the median petal and three adjacent staminodia (Brown 1833, Darwin 1862), or (2) a congenital fusion product of three lower staminodia without participation of the median petal, the median petal being lost (Nelson 1967). These interpretations were mainly based on the study of teratological flowers. Extensive comparative developmental studies (Kurzweil 1987) tend toward the interpretation of the lip as a simple petal, since it is initiated as a single primordium (Fig. 8.67.3–4). This is also suggested by some genera with a lip not different in structure from the lateral petals (e.g. *Thelymitra*). However, there is no ready answer. One has to take into account that complex primordia occur in orchid flowers, e.g. the median sepal and the two lateral petals in *Bletia* originate from a common crescent-shaped primordium (Kurzweil 1987) (see Fig. 8.67.1–2 for *Oncidium*). It is also possible that the bauplan of the lip is not as uniform within the family as it has always been supposed. A parallel situation may be present in the gynostemium whose lateral appendages do not seem to correspond to staminodes in every case. In some African Orchidinae the staminodial primordia 'disappear' in ontogeny and the lateral appendages develop much later (Kurzweil & Weber 1991). Also in the gingers the lip is not uniform. It corresponds to two united staminodes in Zingiberaceae but to five united staminodes in Costaceae (see sections 8.9.4 and 8.9.5).

The ovary is trilocular or more often unilocular with three parietal placentae. The placentae are protruding-diffuse and more or less branched. They produce an enormous number of miniature tenuinucellate, unitegmic ovules (Fig. 8.70.5). For *Cycnoches chlorochilon* 3.7 million seeds have been recorded from a single fruit (Walter 1983). Especially in epiphytic

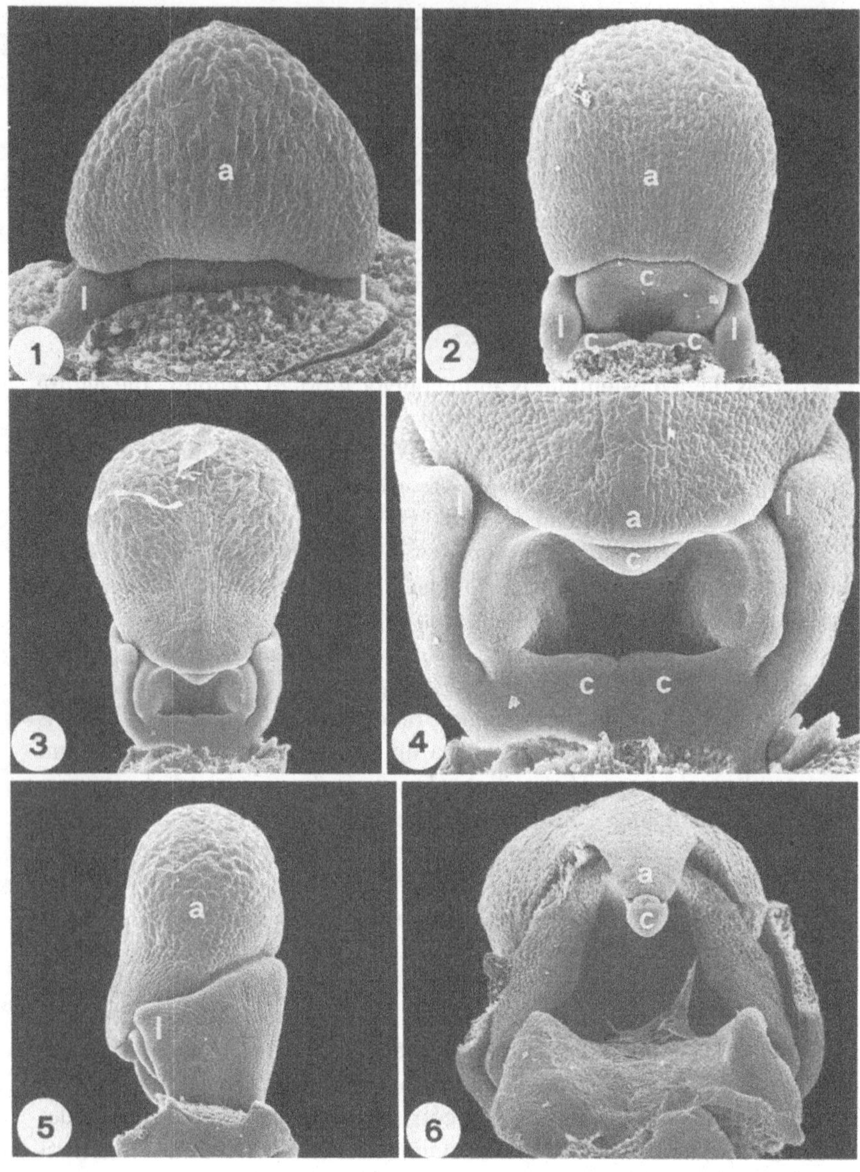

Figure 8.68. *Oncidium ornithorrhynchum* (Orchidaceae); later floral development; flowers in resupinate position, all outer floral organs removed to show androecium and gynoecium. a, Anther; c, carpels; l, lateral staminodes. 1. Young androecium with anther and lateral staminodes (× 150). 2. Gynoecium with the three carpels visible (× 80).

orchids the ovules are only formed after pollination has taken place (Dressler 1981). Thus ovary differentiation is retarded at anthesis. Therefore, pollen tubes may take from several days to several months to reach the ovary (Walter 1983).

Diversity of the pollination apparatus

The most intriguing part of the flower is certainly the pollination apparatus, the gynostemium, whose general structure has been described above. It exhibits fascinating diversity in detail, which will be discussed now. The plethora of terms used for its different parts is confusing and will not be discussed here (see Dressler 1981, 1986b, 1989a; Rasmussen 1982, 1985a, 1986a). I use the terms in the sense of Dressler (1986b, 1989a).

The anther is erect in the less specialized flowers. It bends forward ('incumbent') in many Epidendroideae, extremely so ('hyperincumbent') in Vanilleae and Vandeae. In contrast, in some Orchidoideae the anther is bent backwards (Burns-Balogh & Bernhardt 1985).

Pollen of a pollen sac is usually coherent in a pollinium. The number of pollinia of a pollinarium is originally four. In many groups two pollinia are formed by sterilization of two pollen sacs. In two clades (within the Orchidoideae and Epidendroideae) eight pollinia are formed by secondary subdivision. In some Epidendroideae there is secondary reduction in number to six, four and two. In several groups of the Orchidoideae and lower Epidendroideae with relatively soft pollinia they are each subdivided into small portions ('massulae'). This condition has probably evolved independently several times (Dressler 1986b). In primitive orchids there are no pollinia. In Apostasioideae and Cypripedioideae pollen is in single grains, forming sticky masses in Cypripedioideae. In some monandrous orchids mature pollen occurs as tetrads (Dressler 1986b).

The stigma is three-lobed and convex in its least specialized condition according to the apices of the three carpels; it is papillate or non-papillate (e.g. Apostasioideae or *Corymborkis*). In more specialized stigmas the median lobe is receptive only at the base, which, however, is enlarged, or it may even be unreceptive (Dressler 1981), while the apical part is

Fig. 8.68 (*cont.*)

3–5. Median carpel beginning to elongate distally to form the rostellum. 3. Frontal view (× 35). 4. Same, central part (× 90). 5. Lateral view (× 35). 6. Frontal view of a later stage to show distal part of median carpel (rostellum) covered by the extended anther, and stigmatic cavity; lateral staminodes removed (× 30).

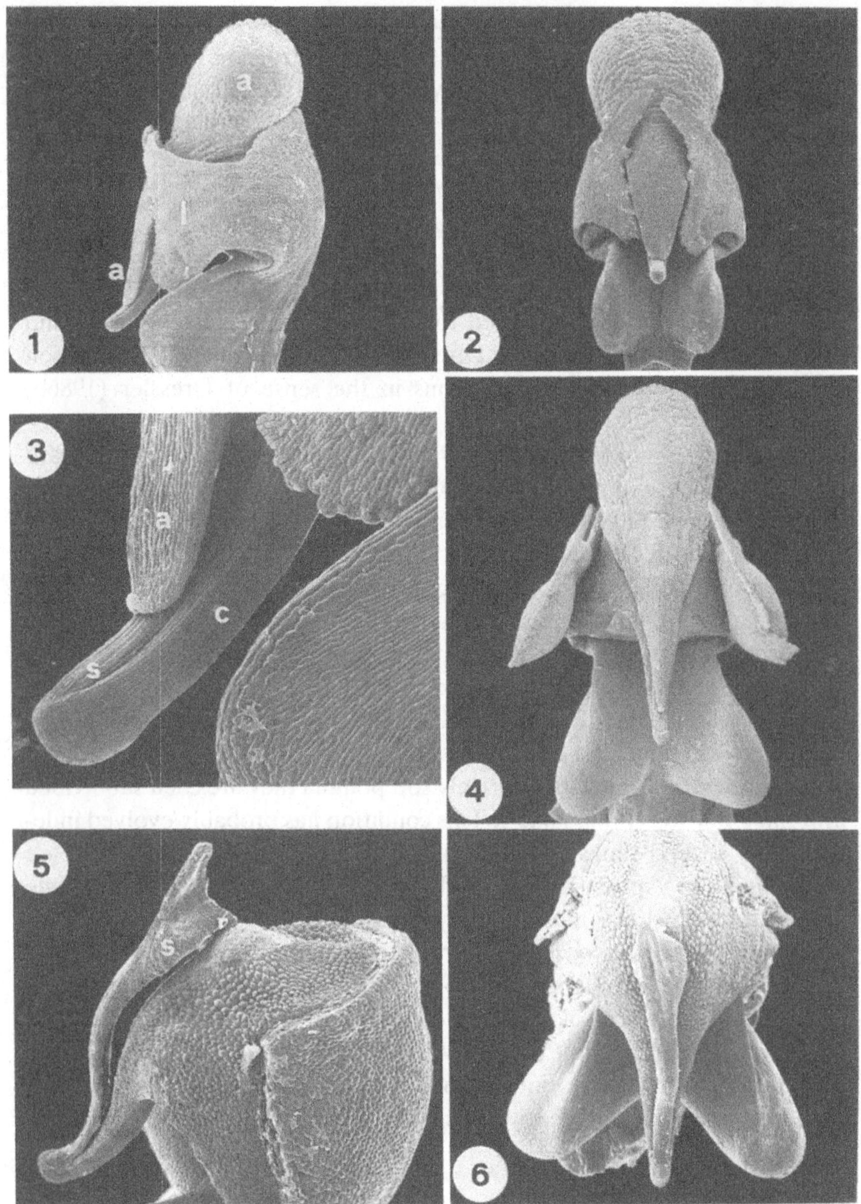

Figure 8.69. *Oncidium ornithorrhynchum* (Orchidaceae); gynostemia shortly before and at anthesis. a, Anther; c, carpel; l, lateral staminode; s, stipe. 1–3. Shortly before anthesis. 1. Lateral view, lateral staminodes partly covering the protracted part of the anther (× 13). 2. Same in frontal

differentiated as a rostellum (Figs 8.68, 8.69). The stigmatic surface is large, concave (or convex in a cavity), smooth and covered by a sticky secretion to take up the pollinia (Rasmussen 1982). The secretion may contain detached cells in some groups (see section 2.4.4) (Calder & Slater 1985). The secretion may be covered by a detached cuticle (Dannenbaum *et al.* 1989).

The rostellum may be short or quite long (e.g. *Oncidium*, Fig. 8.66). An extreme form occurs in the oil-flowers of *Zygostates pustulata*, where it is asymmetrically twisted (Vogel 1974).

The part of the pollinarium that serves for its attachment to the pollinator, the viscidium, is produced at the end of the rostellum, which is the modified apical part of the median carpel. This part is not receptive but involved with pollen transport. The viscidium is a localized body of glue developed by holocrine secretion (secretion associated with cell disintegration) at the rostellum surface. It may or may not involve a firm platelet part above the glue (Schick 1988, 1989). In more basal groups (e.g. *Vanilla*, a primitive genus of the Epidendroideae) there is no special viscidium. The median stigmatic lobe is still receptive. A floral visitor receives some stigmatic secretion on its back, and this serves to attach the pollen masses to the insect as it pushes under the anther (Dressler 1986b). It is therefore hypothesized that the site of viscidium formation is an evolutionarily modified part of the stigmatic surface. Dressler (1986b) estimates that the viscidium has evolved at least 23 times in the Orchidaceae.

Between the viscidium and the pollinia there is usually a kind of stalk of various origins (Rasmussen 1986a) (Fig. 8.66.3). In the lower groups the stalk is differentiated from anther tissue and called a caudicle; it may be hard or it may consist of mealy parts (histogenetically derived from sterile sporogeneous tissue) held together by elastoviscin, a very elastic material (probably derived from the tapetum) (Schill & Wolter 1986, Yeung 1987). Also, the pollinia or massulae of a pollinarium are held together by elastoviscin. At maturity the anther wall opens in a way that the caudicle comes into contact with the viscidium either before or at a visit by a pollinator. The caudicle may be 16 mm long in *Cynorkis uniflora*

Fig. 8.69 (*cont.*)
view (× 13). 3. Same, showing distal part of median carpel (rostellum) with the preformed stipe covered by the protracted part of the anther (× 70). 4–6. At anthesis. 4. Frontal view, lateral staminodes expanded (× 13). 5–6. Anther with pollinia removed to show upper part of median carpel with the partly detached stipe. 5. Lateral view (× 25). 6. from above (× 18).

(Nilsson *et al.* 1992). In advanced groups (mainly Epidendroideae) the stalk of the pollinarium is more complicated. It is derived, in addition, from tissue of the rostellum. This part of the stalk derived from the rostellum is called a stipe (stipes). The stipe is formed either by the surface layers of the upper rostellum part ('tegula') or by the entire recurved apex of the rostellum ('hamulus') (Rasmussen 1982, 1985a). A hamulus may be

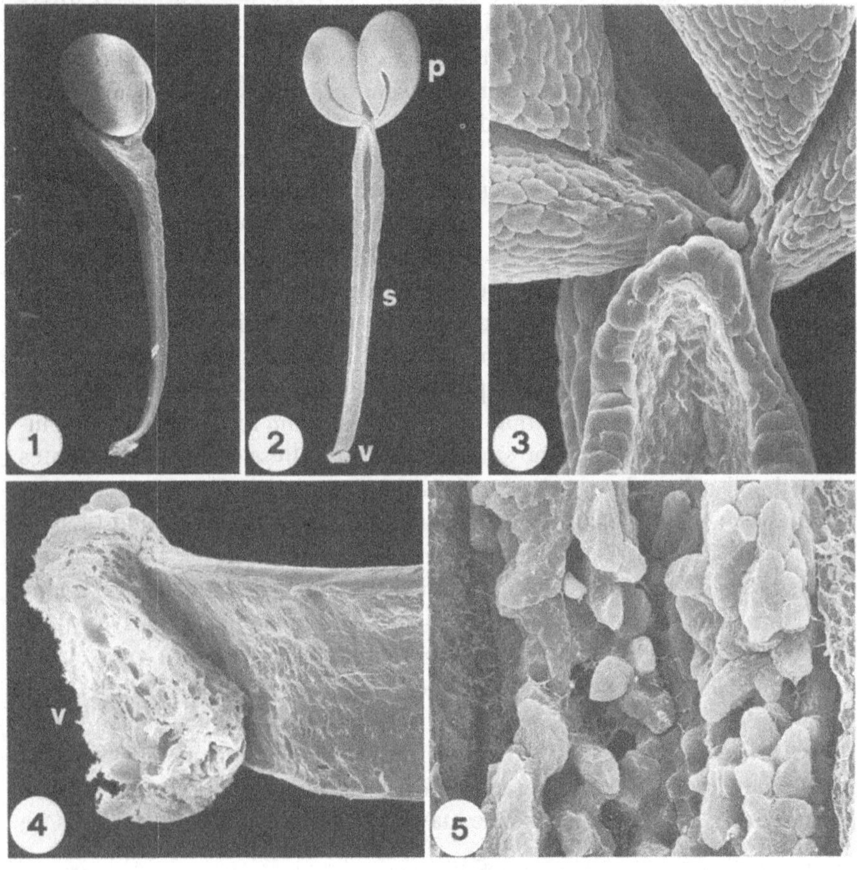

Figure 8.70. *Oncidium ornithorrhynchum* (Orchidaceae); floral parts at anthesis. p, Pollinia; s, stipe; v, viscidium. 1–4. Pollinarium. 1. Lateral view, dorsal side at left (× 18). 2. Ventral view (× 18). 3. Attachment of pollinia to stipe, ventral view (magnified part of 2) (× 200). 4. Distal part of stipe with viscidium, ventral view (magnified part of 2) (× 250). 5. Ovary in longitudinal section, showing placentae with numerous young ovules (× 150).

present or absent among closely related species of *Bulbophyllum* (Rasmussen 1985b). Hamuli have arisen independently several times (Rasmussen 1985b). Between the stipe and the pollinia there is still a short caudicle (Schick 1988). In *Leochilus* the stipe part adjoining the viscidium is also viscous (Chase 1986a). Dressler (1983) estimates that stipes have evolved at least five times within the Epidendroideae.

Depending on the degree of anther curvature and the length of the rostellum the connection between stalk (or viscidium) and pollinia is more at the upper end of the anther ('acrotonic') or at the lower end ('basitonic') (Dressler 1986b, 1990a).

Thus it is impressive how originally different parts have become topographically adjacent and by synorganization are involved in the formation of a new organ, the pollinarium (Fig. 8.66). This synorganization is at different levels: morphological (stamen and carpel) and histological (sporogenous tissue, tapetum, stigmatic surface, and dorsal part of carpel apex). It is also important to see the variations in detail of this theme. The figures in Schill & Pfeiffer (1977) and Dressler (1981) give an idea of the diversity of pollinaria within the family.

It is becoming more and more apparent that these trends of specialization occur with variations on a broad front in the monandrous orchids. Certain levels of organization of the pollinarium are not characteristic of clades but rather of grades (Dressler 1983, 1986a,b, 1990b). For instance, the Epidendroideae, the most diverse subfamily (containing about 80% of the orchid species) shows the entire spectrum from simple pollinaria to the most advanced ones. The former Vandoideae with elaborate stipes have been included in the Epidendroideae (Dressler 1983). They probably represent three clades independently derived from epidendroid ancestors (Dressler 1989b).

Synorganizations and other elaborations of perianth parts

Perianth parts including the lip of orchids are evolutionarily plastic in the highest degree and it is almost unbelievable what bizarre forms may originate.

The sepals may be united in various ways. In Cypripedioideae (Fig. 8.64.1) and in various monandrous orchids (e.g. *Restrepia*, Fig. 8.64.2) the two lateral sepals are congenitally united into a flat organ behind the lip. In some monandrous orchids (e.g. *Masdevallia*, *Monosepalum*) all three sepals are congenitally united and may form a floral tube, which contains the small petals and the pollination apparatus. In *Cirrhopetalum* (*Bulbophyllum*) (Epidendroideae) the two lateral sepals are postgenitally united on their adaxial (!) margins (Fig. 8.71.1). In *Cryptophoranthus*

(*Pleurothallis*) the sepals are united at the tip forming lateral entrances. In *Ponthieva* (Spiranthoideae), with unresupinate flowers, the two lateral petals are postgenitally united and form a landing platform, while the lip is reduced (and turned upwards) (Fig. 8.64.4).

In many groups the lip has a basal spur, which may or may not secrete nectar. In the extreme case of *Angraecum sesquipedale* the spur may reach a length of 30 cm (see also section 4.1.4) (Fig. 8.64.3). A spur may also be formed by synorganization of several floral parts, e.g. by one-sided congenital union of the lateral sepals and their folding over the nectary at the base of the lip (e.g. *Rodriguezia batemanii*). A similar spur is more elaborated in *Comparettia* and *Neokoehleria*. In *Cryptocentrum* there is even a double structure: a lip spur inside a spur by the lateral sepals (Vogel 1959, 1969d). Nectar may also be produced on the exposed surface of the lip (e.g. *Bulbophyllum*).

The lip may also be congenitally fused with the gynostemium (e.g. *Epidendrum*).

Elaiophores were found on the lip of *Zygostates pustulata* (trichomes), *Oncidium ornithorrhynchum* (Fig. 5.7), *Oncidium* sect. *Cyrtochilum*, and *Sigmatostalix bicallosa* (epithelium) (Vogel 1974). Osmophores are sometimes very conspicuous, especially if sepals and petals have the shape of antennae, e.g. *Restrepia antennifera* (median sepal), *Pleurothallis palpigera* and *Dendrobium minax* (lateral petals), *Pleurothallis fuegii* (all three sepals) (Vogel 1963a, Stern *et al.* 1986). *Restrepia* osmophores have epidermis cells with mushroom-shaped papillae (Vogel 1963a, Pridgeon & Stern 1983). In the perfume flowers of Catasetinae and Stanhopeinae scent production is excessive and the osmophores are histologically still more elaborated. The secretory tissue is massive and contains a large intercellular space system (Vogel 1963a).

Bizarre emergences, almost like Christmas tree decorations, occur on the sepals and petals of fly-pollinated *Bulbophyllum* spp. They include vibratile hairs, vibratile platelets, mobile club-shaped parts or parts that look like miniature armchairs dangling on mobile strings (Seidenfaden 1979, Dressler 1981) (Fig. 8.71.3). Furthermore, *Bulbophyllum* has hinged lips that work like seesaws and so move the visitor to the pollination organs. As in other cases of movable floral parts an extensive collenchyma is differentiated in the hinge (Pohl 1935).

Floral modes and pollination

As in Asclepiadaceae, which are in many ways the dicotyledonous counterparts of the orchids, the peak of pollination biological adaptations is with Hymenoptera and Diptera. However, within these limits the range of these adaptations is fantastic. In the primitive *Apostasia* with powdery pollen the flowers are probably buzz-pollinated by bees (Vogel 1981a). Some sphingid-pollinated groups, including the famous *Angraecum sesquipedale*, are mainly known from Madagascar (e.g. Nilsson *et al.* 1987), somewhat less from other parts of the tropics. Birds play an insubstantial role (van der Pijl & Dodson 1966). *Sacoila lanceolata* var. *lanceolata* with reddish to orange tubular flowers is an example of a hummingbird-pollinated orchid (Catling 1987). Mammals are not known to act as pollinators: large, imprecise animals do not fit the elaborations of orchid flowers. A list of orchid species with their observed pollinators is given by van der Pijl & Dodson (1966).

Perhaps the most spectacular and relatively best studied orchid flowers in the tropics are those that are pollinated by euglossine bees for their perfume, especially Stanhopeinae and Catasetinae (see below and section 4.1.1). Darwin (1862) made accurate observations on the structure and function of the flowers of several representatives, although the biological relationships between these flowers and their bees were not known at the time. It has been argued that in addition to the structure of the (often large) flowers the specific combinations of fragrance compounds are also important isolation factors (Dodson *et al.* 1969, Williams & Dodson 1972). Pollinations have been observed in many species and fragrance analyses have been carried out for many examples. Pollinator sharing because of the same fragrance but different floral shape may maintain isolation by attaching pollinia at different parts of the pollinator's body (Dressler 1968). Nevertheless, the degree of this specifity is still uncertain, since the data for orchid and bee species are still too scant for generalizations, and further since the bees also collect scent substances from sources other than flowers (Dressler 1968, Williams & Whitten 1983, Ackerman 1983, Roubik & Ackerman 1987, Young 1991; see also section 4.1.1). Results on some interspecific variation of fragrance production were presented for two species of *Cycnoches* (Gregg 1983). Oil-flowers of *Oncidium* are pollinated by *Centris*, those of *Sigmatostalix* and *Zygostates* perhaps by smaller bees (Vogel 1974).

Many genera seem to be pollinated predominantly by flies, although pollination has been studied in only a few species (e.g. the very large genera *Bulbophyllum*, *Pleurothallis*, and *Stelis*). There are also fly-pollinated species in many genera that are predominantly bee-pollinated (van der Pijl & Dodson 1966).

Floral mimicry and pollination by deceit abound in orchids (Vogel 1975, Dafni 1984, Ackerman 1986a, Nilsson 1992). Many flowers have no reward for pollinators. They are visited because they mimic flowers of other

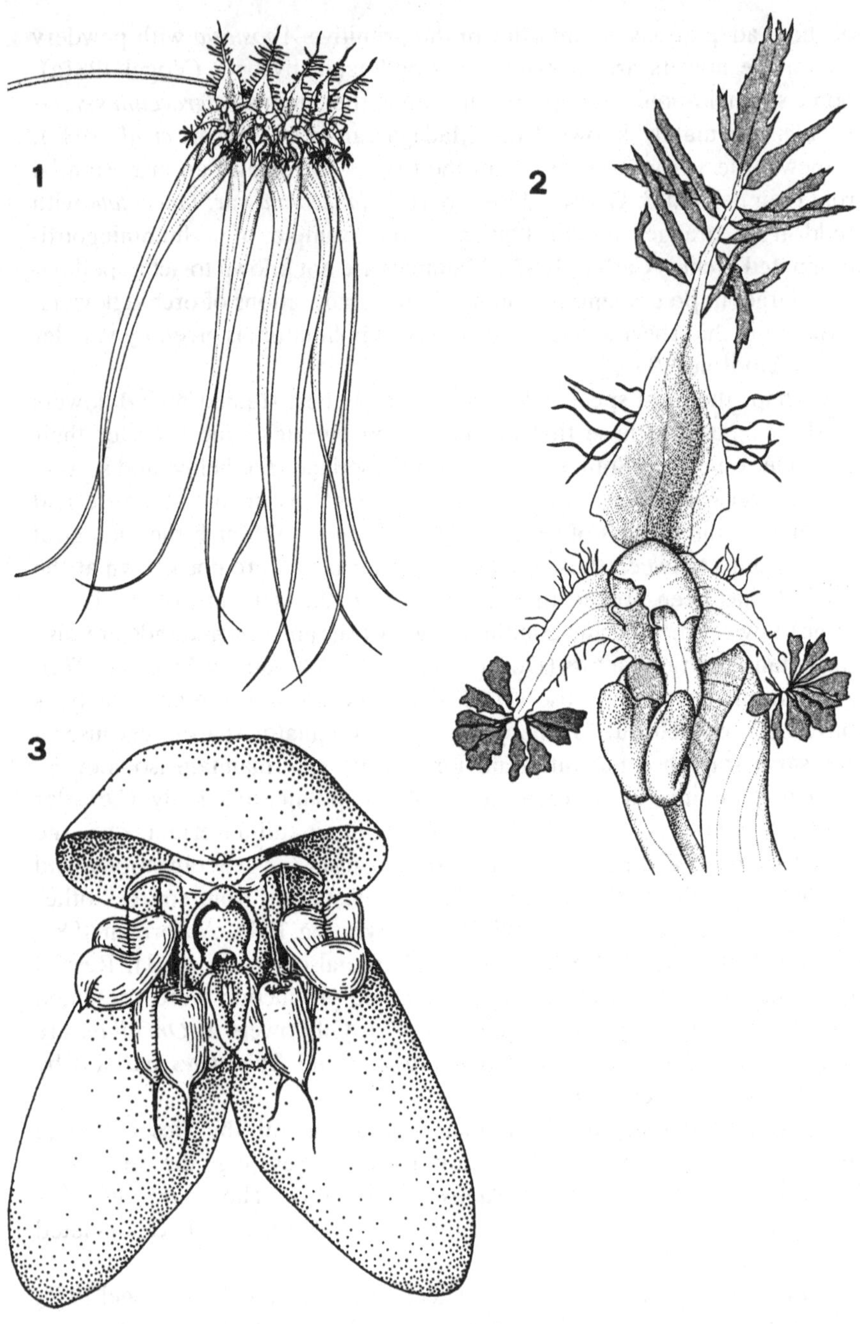
1
2
3

groups. An example are the two deceiving orchids *Cymbidium insigne* and *Dendrobium infundibulum*, both similar in floral habit to *Rhododendron lyi*, all growing together in northern Thailand. All three species are pollinated by *Bombus eximius* (Kjellsson *et al.* 1985, du Puy & Cribb 1988). Another trio involving an orchid with nectarless flowers together with two species of other families with nectariferous flowers is *Epidendrum radicans* (*E. ibaguense*), *Asclepias curassavica* and *Lantana camara*. They are all pollinated by the same butterflies. The significance of this trio as a guild has been amply disputed (see section 7.6). The nectarless *Cymbidiella flabellata* in Madagascar is pollinated by sphecid wasps that otherwise forage for nectar in similar flowers (Nilsson *et al.* 1986). In the blue bee-pollinated flowers of some species of the Australasian genus *Thelymitra* the gynostemium has two tufts of bright yellow moniliform hairs; these flowers seem to mimic (buzz-pollinated) pollen-flowers (Bernhardt & Burns-Balogh 1986, Dafni & Calder 1987). Similar yellow dummies occur on the gynostemium or on the lip of many other orchids (Vogel 1978a).

In other orchid groups, flowers intrude into the sexual cycle of pollinators in mimicking mates and are pollinated by pseudocopulation by males. This is known especially from extratropical groups, where some Hymenoptera, bees and wasps, are involved (survey in Dafni & Bernhardt 1990), or ants in the unique case of *Leporella fimbriata* (Peakall *et al.* 1991). Neotropical orchids pollinated by pseudocopulation include *Trichoceros antennifera*, pollinated by male tachinid flies (Dodson 1962a), and *Trigonidium obtusum*, pollinated by male *Trigona* bees (Kerr & Lopez 1963, Vogel 1963b). It may be expected that pseudocopulation also plays a role in many unstudied tropical groups, if the peculiar flowers of such genera as, for example, *Haraella*, *Luisia*, or *Lepanthes* are considered.

Many orchids that are pollinated by flies also work in some way by deceit. The common occurrence of dark brown spots, often with hairs, may be seen as fly mimicry. In species of *Lepanthes* and *Pleurothallis* a single flower is open at a time and is held immediately above an expanded

Figure 8.71. Excessively adorned flowers in *Bulbophyllum* (Orchidaceae); petals with various appendages. 1–2. *B. wendlandianum* (from Seidenfaden 1979, with permission). 1. Inflorescence; lateral sepals very elongated and postgenitally united by their adaxial (distant!) margins (× 0.6). 2. Flower (only bases of lateral sepals drawn) (× 3.4). 3. *B.* cf. *macrorhopalon*; each petal with three mobile appendages (× 5) (from Dressler 1981, with permission).

foliage leaf. It looks like a fly sitting on a leaf. However, their pollination biology is unknown. Extracellular refractile crystals (raphides) on the flowers of nectarless *Stelis* species are supposed to act as pseudonectaries (Chase & Peacor 1987, Christensen 1992) (see section 5.7.2). Sapromyophilous species seem to be numerous. A urinous odour occurs in *Paphiopedilum* and *Phragmipedium* species (see below). The flowers of *Masdevallia fractiflexa* have a foul odour (Dodson 1962a). In other species of *Masdevallia* that have now been separated as a genus *Dracula* the lip mimics a mushroom with its lamellate surface that is turned towards the substrate and emits a mushroom-like odour (e.g. *D. chestertoni*). These flowers are pollinated by Diptera (fungus gnats or drosophilids) that normally lay their eggs in mushrooms (Vogel 1978b).

In spite of the extreme specializations of orchid flowers in relation to their pollinators, autogamy and cleistogamy are not absent among tropical orchids. On the contrary, the proximity of pollinia and stigma on the gynostemium is a good precondition for autogamy or cleistogamy. The occurrence of autogamy has been reported for 350 species of 130 genera out of most of the tribes of all subfamilies (except Apostasioideae). Commonly it occurs together with chasmogamy in a genus. As an example, in *Nervilia* a wide spectrum of chasmogamy (including pollination by deceit), autogamy and cleistogamy occurs (Pettersson 1989). Thus autogamy has evolved independently many times within the orchids (Catling 1990). Geographically, autogamy is more concentrated in higher latitudes and in the tropics at higher altitudes.

Life span of flowers

Orchids are well known for having especially long-lived flowers. This is true for many groups. It may be seen as a compensation for the rarity of pollination events, especially in species with deceptive flowers. However, it should not be surprising for such a large and diverse family that extremely short-lived flowers also occur. The following enumeration gives an idea of this spectrum: *Paphiopedilum*, one to three months (Cribb 1987); *Oncidium ornithorrhynchum*, about two weeks (Vogel 1974); *Dendrobium crumenatum*, nine days (Smith 1925); *Bulbophyllum ecornutum*, about a week (Rasmussen 1985b); *Catasetum maculatum*, female flowers, six days (Janzen 1981a); *Stanhopea*, two to five days (Williams & Whitten 1983); *Fregea amabilis* and many *Sobralia* species, one day (Walter 1983); species of *Bulbophyllum* and *Stelis*, less than one day (van der Pijl & Dodson 1966); *Dendrobium appendiculatum*, five minutes (van der Pijl & Dodson 1966). Considering flowers that are only open for a short time, perhaps the transition to cleistogamy is continuous. Gregarious flowering occurs in some of the species with short-lived flowers (van der Pijl & Dodson 1966). In some genera the flowers disintegrate by autolysis after anthesis: e.g. *Sobralia*, *Stanhopea*, *Gongora*, *Coryanthes* (van der Pjil & Dodson 1966).

Paphiopedilum (Cypripedioideae)

Paphiopedilum with about 60 species occurs in tropical Eastern Asia and Malesia (Cribb 1987) (Fig. 8.64.1). The flowers are large and in many species only one is produced in an inflorescence. In *P. sanderianum* the petals are up to 90 cm long. In *P. rothschildianum* with spreading petals the flower diameter is up to 30 cm. As in all Cypripedioideae the lip has the shape of a slipper. It is three-lobed with the mid-lobe saccate and the side lobes involute. There are two lateral stamens. The anthers release sticky or granular pollen masses. A large median staminode covers the lip base like a shield. The lip is a trap for pollinators. They slide into the trap on the slippery walls and can only escape along a ladder of hairs through the base of the lip through a gap on either side of the gynostemium, where they pass the stigma and one of the two anthers. The two exit gaps are formed by the involute lateral lobes of the lip and the shield-like staminode (Cribb 1987). Pollination has only been studied in *P. rothschildianum* (Atwood 1985). A syrphid fly was found to pollinate the flowers. The flowers are considered to mimic brood sites and attract female flies to the staminode where they lay their eggs. Atwood found 76 eggs on one staminode. He suggests that glandular hairs on the staminode mimic aphids.

That pollination by flies also plays an important role in some other Cypripedioideae was suggested by Vogel (1963a) for Neotropical *Phragmipedium*. In *Ph. grande*, glandular hairs on the long petals emit a urinous scent. Delpino (1873) mentioned this scent for some *Paphiopedilum* species. Dodson (1966) observed syrphid flies and bees as pollinators of *Phragmipedium longifolium*. In addition, the predominantly greenish and brown floral colours with dark stripes and dark hairy spots and the often narrow, tail-like petals suggest the importance of fly-pollination in the group.

Anthesis of *Paphiopedilum* flowers is exceedingly long. It is between one and three months (Cribb 1987).

Oncidium (Epidendroideae–Maxillarieae–Oncidiinae)

The Neotropical genus *Oncidium* contains some 400 species (Figs 5.7, 8.65–8.70). They are epiphytes and vary between miniature plants with a single flower per inflorescence (e.g. *O. glossomystax*) and plants with giant twining inflorescences of up to more than 3 m length and with hundreds of flowers (e.g. *O. volvox*). The flowers are commonly yellow and often have brown spots. The usually flat lip is often distinctly differentiated into meso- and epichile and has a prominent callus at the base. In *O. heteranthum* the inflorescence contains many small sterile, though yellow, flowers in addition to the larger bisexual ones.

In the large-flowered *O.* (*Psychopsis*) *papilio* the two lateral appendages of the gynostemium have a dark brown knob at the end and look like the feelers of an insect. In *O. baccatum* the apex of the lip is seamed by hard translucent globular blisters. The function of these embellishments has not been studied.

Oncidium flowers are commonly pollinated by bees of different groups (Dressler 1981). In some species of *Oncidium* the flowers are mimics of the oil-flowers of some Malpighiaceae; they deceive female bees of the genus *Centris* (Vogel 1974). The story becomes more complicated in that the flowers of a few *Oncidium* species, such as *O. ornithorrhynchum* (Fig. 5.7) and *O.* sect. *Cyrtochilum*, are real oil-flowers, a rare feature among orchids (see above). In *O. ornithorrhynchum* the flowers are pink and the flanks of the mesochile containing the elaiophores are reflexed. This allows the oil-harvesting bees (also *Centris*) to take up a position on the flower similar to that in Malpighiaceae, where the elaiophores are behind the corolla on the calyx (Vogel 1974) (see section 5.3). The flowers of other species seem to mimic bees, since they are attacked by male *Centris* bees, perhaps for territorial defence (Dodson 1962a, Nierenberg 1972, Vogel 1974).

It has been argued that *Oncidium* is not a monophyletic group. Convergence due to similar pollination biology has masked cladistic relationships (Dodson 1962a, Dressler 1981). Chase (1986b) and Chase & Palmer (1992) made an attempt to disentangle the group by a combination of morphological and molecular biological methods. This is a fascinating problem to follow up.

Stanhopea (Epidendroideae–Maxillarieae–Stanhopeinae)

Stanhopea, with about 45 species in the Neotropics, has heavily (often vanilla-like) scented, sometimes large flowers of up to 20 cm diameter. The inflorescences bear only few, pendent flowers. Primitive species of the genus *Stanhopea* (e.g. *S. tricornis*, *S. ecornuta*, *S. pulla*) have smaller, relatively closed flowers in a more or less horizontal position. Male euglossine bees enter the flower, collect scent, then back out, touching the apex of the gynostemium (Dodson & Frymire 1961, van der Pijl & Dodson 1966). In more advanced species (e.g. *S. oculata*, *S. tigrina*) the flowers are larger and more open. Lip and gynostemium are long and face downward. The mesochile has two arms that are directed in such a way as to embrace the upper part of the gynostemium. The hypochile has a pouch, which contains the osmophore. The osmophore surface is convoluted or contains large multicellular projections (Stern *et al.* 1987, Curry *et al.* 1991). Male euglossine bees approach the flowers from the side. They collect scent from the pouch. Then they slide down, since the surface of the lip is slippery with an oily secretion (Pohl 1927). They fall through the two arms of the mesochile, and touch the apex of the gynostemium on the way. Thereby they either remove the pollinarium or insert pollinia into the stigmatic cleft. In most species all flowers of an inflorescence open at the same time and persist for two to five days. Fragrance production is strongest between 0800 and 1300 hours (Williams & Whitten 1983). This is also the main activity period of fragrance-collecting bees. Compounds of the perfume have been analysed by Williams & Whitten (1983).

Dodson (1962b) suggested that, during evolution within the genus *Stanhopea*, changes in pollination from small to large euglossines or vice versa occurred ('leap-frog speciation'), and some lines within the genus may have shifted two or three times.

Coryanthes (Epidendroideae–Maxillarieae–Stanhopeinae)

Coryanthes, with around 30 species in the Neotropics, has flowers of such a bizarre form and biology that Darwin cited observations by H. Crüger on its pollination biology in his '*Origin of species*' (1859) (Fig. 8.64.6). As in *Stanhopea* they are perfume-flowers pollinated by male euglossine bees. The last comparative account of the floral biology is by Dodson (1965), who studied six species and found quite specific pollination by different *Euglossa* and *Eulaema* species. Fragrance diversity was studied by Gerlach & Schill (1989). The few-flowered inflorescence is pendent and the flowers face downward. The complicated lip gives the large flower an extremely three-dimensional appearance. The flowers are trap-flowers somewhat like those of Cypripedioideae but they work by different means. The sepals and petals are more or less reflexed. The lip is most complicated. The hypochile is hood-shaped and contains the osmophore; the mesochile is involute-tubular; the epichile is large and bucket-shaped. The long gynostemium is bent outwards at the end; it has two basal lateral appendages, which secrete large amounts of water during anthesis. Lip and gynostemium are highly synorganized and form a complicated apparatus. Water secreted from the gynostemium drips constantly into the bucket formed by the lip. It gives the impression of a bath tub with two water taps. Male euglossine bees are attracted by the perfume at the hood of the hypochile. They swarm around the hypochile and try to get perfume below the hood. As soon as their wings touch a water drop hanging at the water gland the bees fall into the water-filled bucket, because the water has a low surface-tension. The only exit is at the tip of the lip, where a bee automatically comes into contact with the viscidium of the pollinarium or with the stigma. However, the bee usually has to struggle for 15–30 minutes until it reaches the exit out of the bath. After removal of the pollinarium it becomes easier for subsequent bees to pass through. The bee visits observed by Dodson (1965) were between 0700 and 1000 hours.

It would certainly be most interesting to study the phylogenetic relationships of *Coryanthes* with the neighbouring genera *Stanhopea* and *Gongora*, whose pollination mechanism is so different in detail.

Catasetum (Epidendroideae–Cymbidieae–Catasetinae)

Catasetum has *ca.* 70 Neotropical species. Surprisingly, it has unisexual flowers, commonly on different inflorescences, which may be quite heteromorphic. Bisexual flowers have repeatedly been observed but they are not functional (Romero 1992). Unique among orchids is, furthermore, the shooting mechanism of pollinaria application to the pollinator's back. Catapulting of the pollinarium is triggered if one of two long lateral appendages of the gynostemium is touched (Dodson 1962b, Ebel 1974).

Not only the floral structure but also the floral behaviour may be quite different in male and female flowers as impressively shown by Janzen (1981a,b) for *C. maculatum* in Costa Rica. In June–July 1979 two five-flowered female inflorescences attracted at least 407 male *Eulaema polychroma* euglossine bees. The bees arrived between 0600 and 0930 hours. Only two bees had pollinaria and pollinated three flowers. The unpollinated flowers lasted for six days. The pollinated flowers ceased odour production shortly after pollination, and they were no longer visited by bees (Janzen 1981a). In June 1980 another individual produced a five-flowered male inflorescence about 150 m apart from the female. The flowers began producing scent at about 0700 hours. By 45 minutes later the pollinaria of all five flowers had been removed by *Eulaema polychroma*. The flowers ceased odour production as soon as the pollinaria were removed and were no longer visited by the bees (Janzen 1981b). Possibly each of the five pollinaria resulted in the pollination of a female flower at another location. This would be an extremely high pollen-donation success for an outcrossing plant. According to Vogel (1963b) scent emission may start only on the second or third day of anthesis in some species, and anthesis may last for several weeks.

Romero & Nelson (1986) showed for *C. ochraceum* that after pollinarium attachment *Euglossa* bees avoided male flowers (because of the weight of the big pollinarium) but not female ones. Possibly this behaviour promoted the selection of pronouncedly dimorphic male and female flowers within the genus. Floral fragrance specificity in many species of the genus seems to be the most direct factor of congruence of male and female flowers as well as an isolating factor between species (Hills *et al.* 1972). Compounds of floral fragrance were also analysed by Williams & Whitten (1983) and Gerlach & Schill (1991).

In the neighbouring genus *Cycnoches* the same character combination of dimorphic male and female flowers and catapulting of the pollinarium occurs. Unisexual flowers also occur in some *Mormodes* species. Evidence from various sources shows that evolution was most probably from bisexual flowers in Cyrtopodiinae to unisexual flowers in Catasetinae (*Catasetum*, *Cycnoches*) and again to bisexual ones in other Catasetinae (*Clowesia*, *Dressleria*) but with protandry, in that a flower can only be pollinated if the pollinarium has been removed (Chase & Hills 1992; see also Romero 1990 for a different viewpoint). *Cyrtopodium*, probably the sister group of the Catasetinae, does not have perfume-flowers. However, they attract both male and female euglossines (and other bees) by deceit. This may be a good evolutionary starting point for the scenario proposed by Chase & Hills (1992).

Conclusion: Peculiarities of orchid flower evolution

Orchid flowers are extremely diverse, although the floral bauplan in a conventional sense is extremely stable. One is tempted to apply different criteria to define the floral bauplan in orchids compared with other angiosperms: androecium and gynoecium are united forming a single complex organ, the gynostemium. Perhaps also the lip is a complex organ formed by a petal and parts of the androecium. These complex organs are exceedingly plastic in many respects. Other organ combinations may also form synorganized complexes (Vogel 1959).

A prominent theme in orchid flowers is deceit and mimicry. This strategy is so successful because a majority of orchid species are rare in scattered populations and because floral features are so evolutionarily plastic (Nilsson 1992). Pollinator specificity is pronounced in many orchids, and there are trends towards increase of specificity within the family (Tremblay 1992). Different features may work towards pollinator specificity and relatively easy advent of isolation in populations and subsequent speciation: (1) the presence of pollinaria and their precise attachment at insects' bodies in conjunction with plasticity of floral shapes (Dressler 1968, 1981, Benzing 1987); and (2) the presence of specific fragrances, besides shapes and colours, that may act on the instincts of specific pollinators. These features favour prepollination isolation mechanisms, whereas postpollination isolation mechanisms are weakly developed in orchids. Pollination by pollinia is efficient because of the large number of pollen grains in a compact mass, which allows synchronous pollen tube emergence and so enhances pollen tube selection (Benzing 1987). This efficiency may also compensate for the generally rare floral visits owing to the scattered distribution of the flowers. Rare though the visits may be, the probability that a pollinarium attached to a pollinator reaches the right species is relatively high (Wagenitz 1981). Rare visits may also be responsible for the long anthesis of many species.

The most extreme forms in terms of floral construction are perhaps the large-flowered Stanhopeinae and Catasetinae. The most extreme forms in terms of floral adornment diversity are species of *Bulbophyllum*. The most extreme group in terms of species diversity is the Pleurothallidinae (Dressler 1990b).

The occurrence of parallel evolution strikes every worker in a large plant group. Perhaps the phenomenon is at a peak in the orchids with their plastic flowers (see Dressler 1981, 1986a, Rasmussen 1986a, Benzing 1987). The synorganized 'new' organs are notable; in particular, the gynostemium with the pollinarium is especially impressive in this respect. It acts as a unit, and the pollinarium stalk is formed according to topography irrespective of morphology: it is differentiated in various ways from former gynoecium or androecium parts, which can still be recognized. Certain combinations have arisen many times in parallel (see above).

9

Salient aspects of flower evolution

9.1 What is special in tropical flowers?

The question 'What is special in tropical flowers?' is asked here, because the study of flowers began in temperate regions and some general concepts on flowers have been developed based on flowers of temperate plants. A general answer may be: the diversity of tropical flowers encompasses almost the entire range of flower diversity on earth, while the diversity of flowers in other parts of the world is much more restricted. Owing to the great diversity of tropical flowers, most of the phenomena of flowers in general occur in the tropics. Specialities of tropical flowers are more in the emphasis of particular phenomena than in the absence of certain traits. Tropical flowers, therefore, provide a basis for the study of flower evolution in general.

The greater diversity of angiosperms in the tropics, as compared with other regions, is directly related to the greater diversity of flower forms and to the greater diversity of pollinators available. Many angiosperm families or genera that are exclusively tropical may also have survived to the present for other reasons but, in turn, they form the raw material for further diversification, and, indirectly, for increased floral diversity.

Although the majority of tropical plants have small flowers, there are more large and deep flowers in the tropics than in other regions. This is largely due to the fact that the largest animal pollinators are all exclusively or predominantly tropical, such as bats, birds, primates, large moths and butterflies. The largest flowers on earth (*Rafflesia* and *Aristolochia*) are tropical, although, in this case, not because of large pollinators, but probably due to the climate, which allows excessive growth. In contrast, there are growth limitations in temperate climates.

Although bees are the most important pollinator group, since the entire range of pollinator groups is higher, the relative diversity of flower-pollinating bees in the tropics is lower than might be expected (Heithaus 1979a,b). However, the diversity and abundance of wasps that visit flowers is higher in the seasonal tropical communities than in the temperate communities that have been studied (Heithaus 1979a,b). Apart from some extreme situations, there is less wind-pollination in the tropics, probably because a rich fauna is available for pollination throughout the year, and not limited by cold periods. There are many one-day flowers in tropical lowlands. This is due to the availability of a diversity of traplining animals, which are largely absent in temperate regions. These pollinators are active year-round and not hindered by longer adverse weather conditions.

A number of tropical flowers open rapidly (explosively) because the sepals (or tepals) are postgenitally or congenitally united in bud. Before the flower opens they come under tension. The flowers may open within seconds or a few minutes. In some groups an explosion is even effected by the pollinators, which actively open them (some Fabaceae, Loranthaceae, Proteaceae, Lamiaceae and Marantaceae) (see chapter 8). Two aspects may be important here: (1) efficient bud protection, and (2), at the community level, timing in floral opening because of strong competition among foraging animals.

Although the organization of inflorescences is not covered here, the occurrence of an especially broad range in the architecture of floral display in tropical plants should be mentioned. Flowers on trunks (cauliflory), on larger branches (ramiflory), on long, pendulous peduncles (flagelliflory) or the protrusion of flowering branches out of the crowns are characteristic of some tropical plants (especially, but not exclusively, some bat-pollinated and bat-dispersed trees); however, these are not really common. The presence of flowers on foliage leaves or flowers that appear from underground shoots, separate from the vegetative parts, is more rare but does hardly occur in extratropical plants. All this diversity is an expression of the greater potential of architectural forms in the tropics (see Hallé *et al.* 1978).

9.2 Recent advances in the study of flower evolution

9.2.1 Scanning electron microscopy and flower development

The SEM has become an important tool for the comparative investigation of floral development, since its first application in this field on the carpels of Laurales (Endress 1972a). After a relatively slow initial phase, its use for the study of flowers has explosively expanded in the last decade. It allows a more precise analysis and much better visualization of forms and developmental events than the light microscope. But it also reveals the great quality of many earlier studies, beginning with the classical work by Payer (1857).

The SEM is also an ideal tool for the study of differentiation and function of floral surfaces at the histological level (e.g. stigmas and other secretory surfaces, complicated surface patterns at the cellular level) (Y. Heslop-Harrison 1981, Barthlott 1990).

A new dimension that has been opened by the application of the SEM is the clearer recognition of the amount of variation of developmental patterns. On the other hand, it only encompasses the developmental changes at the surface of the floral apex. The final number and position of the organs may be the same, irrespective of whether the organs in each whorl originate sequentially or simultaneously or even if the origin is unidirectional (e.g. Fabales, Tucker 1992c). This suggests that the position of organs is determined at an earlier time than they originate at the surface. Thus, the notion of organ initiation needs critical revision.

This also sheds new light on floral symmetry. What is the difference between polysymmetric and monosymmetric flowers? Monosymmetric flowers may arise from polysymmetric early stages by differential growth of floral sectors. This change may be late in development but it may also be protracted far back toward organ initiation (Mair 1977) (see section 3.5). As a rule of thumb, late monosymmetry is a feature of groups where floral monosymmetry occurs more exceptionally among otherwise polysymmetric taxa, while early monosymmetry occurs in groups where monosymmetry is the predominant or sole condition. These two conditions also coincide more or less with positional and constitutional monosymmetry, as discussed in chapter 3.

The situation is, however, more complicated in detail. The degree of monosymmetry in a flower may change in different directions during development. Floral symmetry should also be seen from a more holistic point of view, since it may be influenced by the symmetry of the entire inflorescence. In racemose inflorescences the upper (adaxial) half of each flower is often retarded compared with the lower (adaxial) half in young stages. The young stages may be pronouncedly monosymmetric irrespective of the final symmetry. Monosymmetry may, therefore, merely be an episodic feature and it may be absent at maturity (e.g. in the flowers of *Arabidopsis* and other Brassicaceae). It has been argued that the differentiation gradient of the inflorescence axis as well as that of the floral bract may be directly expressed in the young flower and thus cause transient early monosymmetry (Hagemann 1963). The adaxial half of the flower primordium is on the less differentiated side of the inflorescence axis.

There may even be more than one symmetry change in the development of a flower. In *Couroupita* (Lecythidaceae) early floral development up to completion of petal formation is pronouncedly monosymmetric (Fig.

8.12). It is followed by a short polysymmetric phase up to early androecium and gynoecium development (Fig. 8.13). The late flower development is again pronouncedly monosymmetric by differentiation of the peculiar one-sided androecium (Fig. 8.14).

Although our knowledge about the diversity of patterns of floral development has greatly increased, we are still far from knowing the entire range within the angiosperms and far from understanding all the evolutionary relationships of all these patterns. Certain evolutionary changes may be formally described by the terms homoeosis (replacement of an organ by another kind of organ) and heterochrony (alteration in developmental timing) (Lord & Hill 1987, Guerrant 1988, Sattler 1988, Kirchoff 1991, Lord 1991), although the application of these terms is limited to flowers with a constant number and position of floral organs, and the genetic background is unknown in almost all cases.

A feature that has not been comparatively studied is the occurrence of common primordia between the organs of subsequent (adjacent) whorls. It appears that this is more prominent in the monocots than in the dicots. Examples include tepals and stamens in Liliaceae (Sattler 1973), petals and stamens in Zingiberaceae (Weber 1980), and sepals, petals and stamens in Orchidaceae (Kurzweil 1987) (Fig. 8.67), or, in dicots, petals and stamens in *Kandelia* (Rhizophoraceae) (Juncosa & Tomlinson 1987).

There is an impressive amount of regulatory capacity in floral development. Flowers that are similar at maturity may show considerable variations in some developmental phases. They are then capable of reaching the same final form, which no longer contains any clues to differences in the former development. This may also be a source of misinterpretation of differences in youngest stages. Young stages should, therefore, be evaluated cautiously and always with comparisons of both adult stages and closely related groups.

9.2.2 Early fossil angiosperm flowers

The past decade has brought unprecedented finds of well-preserved early fossil angiosperm flowers, especially since the spectacular discovery of charcoalified, three-dimensionally preserved Cretaceous flowers by Friis & Skarby (1981). These finds and the subsequent expansion of fossil flower studies have considerably changed our views on the primitive angiosperm flowers. They include a large number of small to very small flowers; large flowers are not present among the earliest floral fossils of the Lower Cretaceous. Although their absence may be an artifact due to differential preservation of different-sized flowers, the present state of knowledge shows that small flowers at least co-occurred with larger ones, if they were not the only first flowers (see also Dilcher 1979, Krassilov 1991). However, the new fossils clearly show the primitiveness of the Magnoliidae (and Hamamelididae) (Friis *et al.* 1991), which has long been suggested from the characteristics of extant relatives and of fossil pollen (Doyle 1969). One of the best represented groups of early fossil angiosperms, documented by reproductive material, pollen and flowers, is the Chloranthaceae (Crane *et al.* 1989).

The critical evaluation of the origin of flowers showed that flowers are preangiospermous. An anthophyte clade including Bennettitales, Pentoxylales, Gnetales and angiosperms, all containing 'flowers', has been recognized by the comparative work of palaeobotanists (Crane 1985, Doyle & Donoghue 1986).

Most of the extant angiosperms that are closest to the early fossils are tropical groups. This emphasizes the role of the tropics as conservation regions for archaic flowering plants.

9.2.3 Molecular genetics of flower development

The advent of new molecular biological techniques made possible the study of specific gene activities during floral development. The field has considerably developed in the past five years (Meyerowitz *et al.* 1989, 1991, Coen 1991, Coen & Meyerowitz 1991). These studies require the use of particular model organisms and are, at present, not applicable for broader comparative purposes. However, the two prominent model species, *Arabidopsis thaliana* (Brassicaceae) and *Antirrhinum majus* (Scrophulariaceae), allow at least limited comparisons. Although they are only very distantly related at the level of the angiosperms, both have in common the presence of four kinds of floral organs (sepals, petals, stamens, carpels) in constant numbers and positions.

Analysis concentrates on the events in the young floral meristem at organ initiation. A cascade of regulatory events has been recognized with the interaction of three kinds of genes: (1) meristem identity (heterochronic) genes, (2) cadastral genes, and (3) organ identity (homeotic) genes. A model of organ identity determination involves the interaction of three genes. If one of them is lacking, a particular simplification of the floral structure occurs. If all are lacking, uniform leaf-like organs are produced by a floral meristem. The changes are the same in both *Arabidopsis* and *Antirrhinum* (Coen & Meyerowitz 1991).

Since a majority of angiosperms have flowers with sepals, petals, stamens, and carpels in constant numbers and arrangement, this model may turn out to have a wide application. However, there are many tropical plant groups with either more simple or more complicated flowers. In the primitive Magnoliidae only stamens and carpels, or only sepals ('tepals'), stamens and carpels occur; in several subclasses flowers with many stamens that develop centrifugally occur; in some families unisexual flowers without any rudiments of the other sex occur (Fig. 5.5.1). These groups may help to expand the model into an evolutionary context in the future.

In addition to the earliest stages of flower development, the differentiation processes within organs have also been studied in model organisms, particularly in *Nicotiana*, where the sequential activity of genes has been demonstrated in stamen development (Koltunow *et al.* 1990, Gasser 1991). However, it is too early for comparisons among different groups.

9.3 Major events in the evolution of angiosperm flowers

A basic event in the origin of angiosperm flowers was the formation of closed carpels. It is still unclear what structures preceded the carpels. A characteristic attribute of closed carpels are anatropous ovules (vs. orthotropous ovules in gymnosperms). It was believed that ovule curvature was a consequence of carpel closure. However, the finds of Tomlinson (1991, 1992) and Tomlinson *et al.* (1991) in podocarps (with 'anatropous' ovules) of 'pollen scavenging' on the bracts where the ovules are attached show functional significance of anatropy without the ovules being enclosed in a carpel. Thus the advent of anatropous ovules before that of closed carpels in angiosperm ancestors cannot be excluded. Lloyd & Wells (1992) suggest the evolutionary transference of the landing position of pollen from a pollination drop on ovules to a wet stigma on the outside of a closed carpel. An intermediate stage would have been the reception of pollen on the adaxial surface of open carpels with anatropous ovules (Lloyd & Wells 1992). This, however, does not explain the origin of carpels. A number of hypotheses on the origin of angiosperm carpels from different potential ancestral groups have been advanced, none of which, however, may be distinctly favoured over the other ones (Ophioglossaceae, Kato 1990, 1991; pteridosperms, Long 1984; glossopterids, Retallack & Dilcher 1981; Caytoniales, Stebbins 1974a, 1976, Doyle 1978; Bennettitales, Meyen 1988, Crane 1986b; Gnetales, Meeuse 1990; see also Friis & Endress 1990, Brückner 1991).

In contrast to carpels, angiosperm stamens are less different from the male organs of related gymnosperms. The most distinctive feature of angiosperm stamens is the occurrence of a pair of thecae, each consisting of two pollen sacs that have a common dehiscence region. Two evolutionary pathways for the origin of thecae may be hypothesized: (1) differentiation of a single pollen sac into a theca by longitudinal septation, or (2) collateral association and synorganization of two pollen sacs into a theca (Fig. 2.5). Models of precursor stages are to be found in Gnetales: for (1) as in *Gnetum*, for (2) as in *Ephedra*.

The role of neoteny (paedomorphosis) (Takhtajan 1976) and more rapid individual development (Stebbins 1974a, Doyle & Hickey 1976, Haig & Westoby 1991b), as discussed for the origin of angiosperms, may also have been crucial at the flower level.

It seems that there were not many major, irreversible steps during flower evolution (Friis & Endress 1990). This may not be surprising, since it is also a general feature for other organs and organisms (Stebbins 1974b).

From the present knowledge it appears that in the first angiosperm flowers stamens and carpels were the only major components. A perianth was absent or protective bracts were loosely integrated in the flower, forming a simple perianth (J. & A. Walker 1985, Endress 1987c). Flowers were small or of moderate size and had a low or moderate number of organs. The position of organs in spirals or whorls was not fixed. Bisexual flowers were protogynous (Endress 1990b). All attractive devices were displayed by stamens and carpels. Stamens were scented and optically attractive. Secretions of the wet stigma formed major rewards for pollinators (Lloyd & Wells 1992), and pollen may have played a role as an additional reward. Scent may also have played a major role (Pellmyr & Thien 1986, Pellmyr *et al.*, 1991). The first pollinators were probably flies, perhaps also beetles, micropterigid moths (Thien *et al.* 1985) and wasps (Crepet & Friis 1987, Willemstein 1987). Wind-pollination is almost absent in extant Magnoliidae and may have been so in the first angiosperms (Endress 1990b). Efficient animal-pollination resulting in stronger selection of pollen tubes may have been important for the initial radiation of the angiosperms (Regal 1977, Mulcahy 1979), in concert with the advent of self-incompatibility (Whitehouse 1950, 1960, Olmstead 1989, Lloyd & Wells 1992). Midgley & Bond (1991) emphasize the competitive significance of rapid growth in angiosperms. However, this may be a consequence of enhanced selection capacity in angiosperms. Animal-pollination also opened the way to highly discriminating pollination modes (Crepet 1984), which, in extreme cases, led to 'hyperdiversity' (Ehrlich & Wilson 1991).

An important step in early angiosperm flower evolution was the stronger integration and functional differentiation of the perianth into protective sepals and attractive petals. Petals may have evolved from the outer part of the androecium or, under the influence of the androecium, from the inner tepals (Hiepko 1965, Ehrendorfer 1977, 1989, Weber 1980, J. & A. Walker 1985, Crepet *et al.* 1992). The emergence of petals released the stamens from the burden of being the main attractive floral parts. In the monocots and at the middle evolutionary levels of the dicots the stamens differentiated into anthers and filaments that are well articulated from each other. The frequent occurrence of dehiscence by valves in the primitive, relatively massive anthers was lost; simple longitudinal dehiscence prevails throughout, apart from Magnoliidae and Lower Hamamelididae. A further step is congenital union of the petals (and sepals) by basal intercalary meristems, and often, in addition, fusion of stamens with the petals; this is prominent in the Asteridae. This opened the way for more elaborate floral architectures, especially tubular flowers, forms of monosymmetric flowers, and large flowers, as well as more floral modes, especially bird-, lepidopter- and bee-pollinated flowers.

One of the most general evolutionary advancements is the congenital union of carpels into syncarpous gynoecia (Endress 1982), which is a major feature of all angiosperms except for Magnoliidae and most Alismatidae. This allows the formation of a compitum, a common pollen tube transmitting tract for all carpels. Another step in the gynoecium, albeit less general, is the change from a superior to an inferior ovary by an intercalary meristem in the floral base and, therefore, increased protection of the developing fruit and seeds. Progression from crassinucellar to tenuinucellar ovules opened the way for an increased seed production per fruit, which was explored by several parasitic and mycotrophic groups and other specialists (Endress 1990b). Likewise, the progression from anatropous to campylotropous ovules allowed the production of larger embryos (Takaso & Bouman 1984, Bouman & Boesewinkel 1991).

The originally flexible floral phyllotaxis was restricted to a whorled arrangement (Endress 1987a, 1990a, Kubitzki 1987) and the number of floral organs was restricted to a fixed pattern. This greatly favoured increased synorganization of parts (e.g. union; see also Robinson 1985), because whorls and constant positional relations are a precondition for precise interactions of different parts. Conversely, it opened the way for flexibility at other levels. Families such as Orchidaceae and Asclepiadaceae, with extreme synorganizations, also have extremely stable numerical patterns in flowers but incredibly plastic organ forms. However, the potential to increase organ number was not completely lost at the middle and higher evolutionary level of the dicots. This is best seen in the androecium, where a fixed and low number of primordia could give rise to a large number of stamens by secondary subdivision (e.g. Stebbins 1967).

The main evolutionary events discussed above are traits of elaboration. However, diversification and increase in flexibility are also general themes of evolution in flowers. Another principle is the decoupling of conflicting double functions of the same organ by the advent of new organs (e.g.

advent of petals and of separate nectaries in addition to stamens and carpels). However, secondary pollen flowers with loss of nectaries also evolved and show that the principle is not universal.

All this results in increased 'evolutionary competence' (Stebbins 1988). The advent of great flexibility in acquiring diverse floral modes and breeding system traits at low systematic levels, such as, for example, in Gesneriaceae and Orchidaceae, may also be seen as important advancements. Features related to breeding systems are rarely involved as stable parts of major evolutionary trends, perhaps with the exception of heterostyly, which occurs only at the middle and higher evolutionary level of the dicots and monocots. This is due to its dependence on a floral organization that is not too open but shows a fixed and low number of floral organs (Ganders 1980).

Genetic repertoires can be used in concerted action, for example with changing floral colours (Paige & Whitham 1985) or increased nectar secretion (Koptur 1989).

9.4 Principles and traits of flower evolution and their role in phylogenetic reconstructions

9.4.1 Evolutionary tinkering, constraints and compromises

Evolution works with the raw material at hand, which Jacob (1977) called 'evolutionary tinkering'. Similar ends are reached from different starting points, and the result is never perfect but is burdened by all sorts of earlier events (Riedl 1977). Diversification proceeds in the directions of the least resistance.

As an example, in floral organs of Scrophulariales there is a trend to form efficient secretory hairs of a particular kind: they have a unicellular stalk and a multicellular head. They are plastic in shape: the unicellular stalk may also become broad and multicellular and the head may develop into a highly specialized epithelium. These hairs are versatile in function and can be used for all kinds of purposes. (1) They are the glands that secrete the fluid for the water calyces (e.g. Bignoniaceae, Verbenaceae). (2) They are, in the large form just described, the extranuptial nectaries on the outer surface of the calyx that serve as ant guards (e.g. Bignoniaceae, Verbenaceae). (3) They are the elaiophores on the corollas and stamens of Scrophulariaceae. (4) They may also occur inside the ovary and become involved in fruit differentiation as juice-producing devices (e.g. in *Clerodendrum minahassae*, Verbenaceae, Bremekamp 1914). In other angiosperm groups nectaries, elaiophores or juice-producing devices in fruits have other structural origins.

Flowers of Scrophulariales have a constant number of floral organs with very limited possibility of change, because they are constrained by their bauplan with its synorganization between corolla, androecium and gynoecium. Increase in the amount of pollen production is not possible by stamen number increase per flower but only by an increase in flower number or increase in anther size.

Different levels of pollinator specificity may be seen as examples of evolutionary compromises, such as extreme specialization of a plant group to one pollinator as well as generalistic behaviour with many pollinators and pollinator sharing with other sympatric species. Extreme specialization depends completely on a reliable pollinator. Conversely, in certain plant communities selection in favour of similar flower characteristics seems to be strong (e.g. Schemske 1981), although this may be more widespread in temperate regions (Brown & Kodric-Brown 1979, Koptur 1983). The negative consequences of pollinator sharing, including stigma clogging and pollen load dilution on pollinators, may be outweighed by the benefits of greater pollinator attraction and more interspecific transfer of pollen (Koptur 1983).

9.4.2 The role of the different evolutionary levels

As shown in section 9.2, there are not many major evolutionary directions in flowers that would hold for the entire angiosperms. In the concept of different aspects of flowers, as used in this book: (1) organization, (2) construction, (3) mode, (4) breeding systems, the number of major directions decreases in the sequence from (1) to (4). The greatest amount of stability is in the organization, the least amount in the breeding systems and in some superficial traits of form that oscillate within populations. Thus, the direction from (1) to (4) decreases in importance at the macroevolutionary level, and at the same time increases at the microevolutionary level. It has to be pointed out again that these levels are not independent, closed systems in the biology of the plants. They are evolutionarily interlinked and they do not show clear borderlines against each other. For instance, past adaptations expressed in the floral mode may have been evolutionarily stabilized and become part of the organization, such as sympetaly in the Asteridae. This may also be subsumed under 'evolutionary tinkering'. The action of selection and the expression of fitness is most directly seen at the superficial levels of (3) and (4) but is ultimately not independent of the organization. How should the organization be defined? Is it the floral groundplan (as expressed in a floral diagram) plus (stable) synorganizations between the parts, or is it the sum of relatively stable features in a group? If the first definition is applied, it would be difficult to define the floral organization of some Magnoliidae.

From the point of view of macroevolution, features of breeding systems and floral mode largely appear as 'evolutionary noise', as they are highly labile, and attempts at character polarization and the use of parsimony criteria would hardly make sense from this angle (e.g. Armbruster 1992). It would be like trying to trace the origin of a molecule in Brownian motion. Conversely, from the point of view of microevolution, breeding systems and floral modes are the central elements, while the bauplan is too static to be of relevance. Nevertheless, the spectrum of reproductive strategies in larger groups is not completely random but shaped by bauplan constraints to some extent (Webb 1984).

In many instances floral features are crucial for speciation in that they effectively cause reproductive isolation. Slight variations in superficial, flexible characters, such as floral tube depth, colour or scent, or in breeding systems, may provide effective isolation within populations (Kay 1984; see also Baker 1959, 1963, Dronamraju 1960, Cruden 1976b, Ornduff 1978, Raven 1980b). Striking floral variations of potential evolutionary significance may even occur within an individual, depending on age or environmental conditions, and thus be epigenetic rather than genetic (Porsch 1908/09). An example of variation in the floral mode at low levels is *Drymonia* (Gesneriaceae) (Wiehler 1983) (see section 8.8.2) with oscillation between hummingbird-, euglossine bee-, moth-, and bat-pollination. An example of variation at the breeding system level is *Rhododendron* sect. *Vireya*. In this group changes in style length lead to failure of pollen tubes to reach the ovary, thus effecting isolation between different morphs (Williams & Rouse 1988).

9.4.3 Homology and convergence: the evolutionary interpretation of peculiar floral features

This mainly concerns the bauplan level. Floral parts are not always easy to homologize. The nature of special organs bearing nectaries (staminodes or organs sui generis?) or of a simple perianth (calyx or corolla?) is often disputed and not always easy to resolve. The use of 'homology criteria', such as (1) position of an organ in the entire system, or (2) form and differentiation (quality) of an organ, or (3) the occurrence of transitions between organs, is helpful (Eckardt 1964, Kaplan 1984, Tomlinson *et al.* 1984). But the most important approach is an analysis of the flowers in the entire systematic context. Thus the central question that should be asked is: How is the organ in question related to the organs of the plant's closest relatives? Three examples may illustrate this.

(1) The male structure of *Hedyosmum* (Chloranthaceae) (see section 8.2.1) appears as an elongate axis bearing numerous stamens. From a formalistic morphological point of view it has been interpreted as a flower (Leroy 1983). However, if it is viewed in a systematic and evolutionary context, it clearly appears to be an inflorescence consisting of extremely reduced (unistaminate, bractless) flowers (Endress 1987b). This example leads to the old question of how common camouflaged inflorescences, generally taken as flowers, may be in the angiosperms.

(2) The female structure of *Balanophora* (Balanophoraceae) (see section 2.4.6) appears as a minute, compact organ bearing an embryo sac in the centre. From a formalistic morphological point of view it has been interpreted as a reduced ovule (Shivamurthy & Arekal 1982). However, if viewed in a systematic context, it clearly appears to represent an entire reduced gynoecium (e.g. Fagerlind 1948).

(3) The outermost perianth organ of Loranthaceae (see section 2.2.5) has been interpreted as a new formation termed 'calyculus' (Venkata Rao 1963). However, if viewed in a broader systematic context it appears that this organ represents rather a calyx (Baillon 1862, Stauffer 1961).

These results may, in turn, be used in phylogenetic reconstructions. The danger of circular reasoning is overcome by repeated testing of different possibilities.

9.4.4 Parallel evolution and the evolution of unique traits in flowers

Full phylogenetic analysis of plant groups involves a concerted search for the shape of grades and of clades. This is generally a long-term procedure, in which a continuous mutual refinement and correction of earlier hypotheses is envisaged to avoid circular reasoning. Each group has its taxonomic and biological foundations, which are the basis of this refinement. The recognition of biological characteristics and trends (anagenesis) and the recognition of clades (cladogenesis) go hand in hand and depend on each other. Together, they will eventually lead to a better understanding of phylogeny.

Such concerted studies inevitably lead to the recognition of a large amount of parallel evolution (homoplasy), based on similar genetic (biological) constitution and, therefore, similar possibilities of change of a group under consideration. This had been recognized by Darwin (1868), was elaborated by Vavilov (1922), and has recently attracted much attention once again (Sanderson & Donoghue 1989, Donoghue & Sanderson 1992, Kubitzki *et al.* 1991, Szalay & Bock 1991, Wake 1991, Endress 1992b). The potential to form the same sets of alternative features may directly be present in the genome of related groups, but may not be apparent under normal circumstances (Kubitzki *et al.* 1991). The genome of an individual plant may have a repertoire of different possibilities of character state expression, only a small part of which is actually expressed during the lifetime of the individual. What appears then as a parallelism or a reversal may have already been present as a 'silent' potential in the original genome. Detailed examples of parallel evolutionary trends at different taxonomic levels are the studies by Armbruster (1992), McDade (1992), Stein (1992), Chase & Hills (1992) and Chase & Palmer (1992).

Character evaluation for phylogenetic reconstruction only makes sense if a character complex is understood in its biology as fully as possible (Stebbins 1970). For flowers this means the entire range from the bauplan down to the mode and the oscillations at population level. Different characters are often linked either extrinsically by ecological demands or intrinsically by bauplan constraints. Such a complex of linked characters should therefore not be atomized into single characters that have equal weight to other, more independent characters (Szalay & Bock 1991). Even without a fully 'perfect' taxonomic classification at hand it can be recognized that not all characters are equally prone to showing parallel evolutionary trends in closely related taxa. If it turns out that a character or character complex shows multiple parallel evolutionary trends in a group, it should have less weight for the judgement of clades within the group (Endress 1992b). Special weight has to be given to unique characters and characters that are relatively constant at the taxonomic level under study. An example of an unique character is anthers with bilocular thecae and each theca opening with a single ventrally attached valve. This anther type occurs exclusively in all hamamelidaceous genera of the Southern Hemisphere. It is not known from any other angiosperms. This character alone, in concert with its exclusive geographic distribution, is a strong argument that all these genera form a monophyletic group (Endress 1989). Further support comes from other, less unique characters. Weighting of conflicting taxonomic properties must not be done *a posteriori* (Szalay & Bock 1991). It should be done both at the outset and repeatedly during the process of study.

Therefore, parallel evolution should not be regarded as 'noise'. It is an essential component of evolutionary plasticity in each group. At present we can recognize the pervasive occurrence of parallel evolution. Numerous examples have been discussed in chapter 8, especially for large groups, such as Fabales, Scrophulariales and Orchidales, at all levels from species up to groups of families. Only future broad and detailed comparative biological studies can lead to a sensible quantification of parallel evolution for particular traits in particular groups. Such studies require a great effort. This is a challenge that makes the work more difficult, but all the more fascinating.

10

Prospects

10.1 The need for synthesis of research at all levels

To receive a balanced view of the evolution of organisms or organs a focus on all levels of the systematic hierarchy is necessary, from the entire group down to populations and individuals. Therefore, a broad spectrum of approaches has to be evaluated. More specifically, there is a need for a clarification of the relationships between the evolutionary implications of the morphological tradition and the population and molecular genetic approach including the fields in between (Riedl 1977, Wagner 1986). Such an approach is still in its infancy in botany, and, as zoological attempts have shown, leads to intricate difficulties (Wagner 1986).

This difficulty is an intrinsic trait of diversity but it is reinforced by the trend in contemporaneous biology to specialize into ever more narrow fields. A result of this splitting is increasingly deeper insight in narrow compartments but an increasing lack of coherence in the insight into nature as a whole. In the evolutionary biology of flowers, examples of such important specializations are the study of the energetics of pollination (Heinrich & Raven 1972) and the study of gender allocation (e.g. Lloyd 1980). The increasing specialization also reveals that nature is more complicated and more flexible than one imagined a few decades ago and that a complete understanding is impossible. The recognition of this failure of a more comprehensive understanding is especially acute at present in the light of the biodiversity crisis. An example of the impossibility of perceiving the entire web of tropical flower biology is given by the attempt to provide a synthesis on Fabales by Schrire (1989). He demonstrates the lacunae that are bigger than the impressive amount of knowledge at hand (see also J. Heslop-Harrison 1976a).

We need more data; even more, we need more attempts at syntheses. We should try to combine different approaches into a new whole (see also Wanntorp *et al.* 1990, Lawton 1991). Attempts at syntheses may at least direct the attention of researchers more closely to those areas most badly in need of study.

10.2 The diversity of tropical flowers: what should we know?

A better understanding of the biology and diversity of tropical organisms is urgently required (Raven 1980a, Soulé 1990, Ehrlich & Wilson 1991, Wilson 1992). The design of conservation actions and setting priorities largely depends on broad knowledge of the organisms.

The following are specific questions to be asked. (1) Are there keystone taxa, taxa with flowers and fruits providing food for pollinators and dispersers that play key roles in communities, and which taxa are they? (e.g. Gilbert 1980, Wheelwright 1983, Howe 1984, Prance 1985, Lovejoy *et al.* 1986, Cox *et al.* 1991, Fujita & Tuttle 1991, Terborgh 1992, Woinarski *et al.* 1992). The extinction of which taxa will cause 'domino effects' by breakdown of biological interactions? (Janzen 1979, 1986a,b, 1987, Dressler 1982, Terborgh 1986, Myers 1987, Prance 1990, 1991). An example are Malpighiaceae with oil-flowers, which are a key resource for *Centris* bees. Conversely, *Centris* bees are the major pollinators of a number of orchids, which attract them by other means (Frankie *et al.* 1989) (see section 8.10.1). (2) Which taxa are phylogenetically 'old' and, therefore, deserve special attention? (Vane-Wright *et al.* 1991).

Large-scale comparative research has its innate difficulties. If a larger plant group is studied, living material has to be studied at, and assembled from, different places, often different continents, at different times and under different circumstances. This is more difficult for field studies than for bauplan studies, both of which are important components of profound comparative biological investigations that can give answers to the questions posed above.

If aspects of biology are studied in the field, one has to take into account that there are wide-ranging ecological phenomena, long-term ecological phenomena, and rare events, which may be pivotal in the biology and evolution of organisms. They may be missed, and the results may be misleading, if the study time is short, as is often the case in tropical field studies. Such studies, as accurate as they may be in their narrow scope, are then of only limited value.

Examples of wide-ranging ecological relationships include those involving migratory species. Perry (1990) points out that the flowers of *Norantea sessilis* (Marcgraviaceae) attract many species of North American migratory birds, such as warblers and orioles, in addition to tropical species. This shows that flowers of the tropical rain forests may also play a role in the life cycle of temperate migratory birds, and that destruction of tropical forests may so produce short-circuits in the global balance of life. Wide-ranging ecological relationships on a much smaller and more local scale are those of traplining pollinators, large insects, birds or bats, that may fly routes of more than 20 km every day (Janzen 1974b, Start & Marshall 1976). Since a large number of animal species belongs to this biological class, its consideration in devising nature reserves is important (Janzen 1986b, 1987, Bawa & Ashton 1991). Flower-pollinating monkeys may also need very large territories (Terborgh & Stern 1987). Australian honeyeaters have complex and unpredictable seasonal patterns of movement (Woinarski *et al.* 1992).

Examples of long-term ecological phenomena or rare events may be the occurrence of excessive drought or cyclones in some years, which may affect flowering and pollinators in a way that cannot be recognized in the more frequent 'normal' years or which may cause long-term fluctuations that are not apparent in short-term studies (e.g. Roubik & Ackerman 1987, Magnuson 1990, Feinsinger & Tiebout 1991, Feinsinger *et al.* 1991, Bellingham *et al.* 1992).

One of the most valuable and important things to do is to observe plants and their interactions in the tropics with an open mind as to the potential meaning of the observations and the role the organisms play. The observation of the flowering behaviour of a plant during a short period, combined with the study of floral structure, may give a preliminary idea about its ecological relationships and help in designing more profound studies, which, in turn, will reveal chapters of its evolutionary biology.

REFERENCES

Ackerman, J.D. 1983. Specificity and mutual dependency of the orchid-euglossine bee interaction. *Biol. J. Linn. Soc.* **20**: 301–14.

Ackerman, J.D. 1985. Euglossine bees and their nectar hosts. In D'Arcy, W.G. & Correy, M.D.A. (eds), *The Botany and Natural History of Panama*, 225–33. St. Louis: Missouri Botanical Garden.

Ackerman, J.D. 1986a. Mechanisms and evolution of food-deceptive pollination systems in orchids. *Lindleyana* **1**: 108–13.

Ackerman, J.D. 1986b. Coping with the epiphytic existence: Pollination strategies. *Selbyana* **9**: 52–60.

Ackerman, J.D. 1989. Geographic and seasonal variation in fragrance choice and preferences of male euglossine bees. *Biotropica* **21**: 340–47.

Ackerman, J.D. & Montalvo, A.M. 1985. Longevity of euglossine bees. *Biotropica* **17**: 79–81.

Agren, J. & Schemske, D.W. 1991. Pollination by deceit in a neotropical monoecious herb, *Begonia involucrata. Biotropica* **23**: 235–41.

Ali, T. & Ali, S.I. 1989. Pollination biology of *Calotropis procera* subsp. *hamiltonii* (Asclepiadaceae). *Phyton (Horn)* **29**: 175–88.

Alon, G., Dulberger, R., Heyn, C.C. & Werker, E. 1987. Pollination mechanism, breeding system, and stigma structure and development in six species of Papilionoideae. *Israel J. Bot.* **36**: 199–213.

Altenburger, R. & Matile, P. 1988. Circadian rhythmicity of fragrance emission in flowers of *Hoya carnosa* R. Br. *Planta* **174**: 248–52.

Aluri, R.J.S. 1990. Studies on pollination ecology in India: A review. *Proc. Indian Natl Sci. Acad. B* **56**: 375–88.

Ampornpan, L.-A. & Armstrong, J.E. 1991. In quest of the oblique ovary in Solanaceae, an adventure in floral development. *Am. J. Bot.* **78** (6), Suppl., 164.

Anderson, G.J. & Symon, D.E. 1989. Functional dioecy and andromonoecy in *Solanum. Evolution* **43**: 204–19.

Anderson, W.R. 1980. Cryptic self-fertilization in the Malpighiaceae. *Science* **207**: 892–3.

Anderson, W.R. 1990. The origin of the Malpighiaceae – The evidence from morphology. *Mem. New York Bot. Gard.* **64**: 210–24.

Andersson, L. 1981. The neotropical genera of Marantaceae. Circumscription and relationships. *Nord. J. Bot.* **1**: 218–45.

Andersson, L. 1985. Revision of *Heliconia* subgen. *Stenochlamys* (Musaceae-Heliconioideae). *Opera Bot.* **82**: 1–123.

Andersson, L. 1986. Revision of *Maranta* subgen. *Maranta* (Marantaceae). *Nord. J. Bot.* **6**: 729–56.

Andersson, L. 1989. An evolutionary scenario for the genus *Heliconia*. In Holm-Nielsen, L.B., Nielsen, I.C. & Balslev, H. (eds), *Tropical Forests. Botanical Dynamics, Speciation and Diversity*, 173–84. London: Academic Press.

Appanah, S. 1981. Pollination in Malaysian primary forests. *Malays. Forest.* **44**: 37–42.

Appanah, S. 1985. General flowering in the climax rain forests of South-east Asia. *J. Trop. Ecol.* **1**: 225–40.

Appanah, S. 1990. Plant-pollinator interactions in Malaysian rain forests. In Bawa, K.S. & Hadley, M. (eds), *Reproductive Ecology of Tropical Forest Plants*, 85–101. Paris: Unesco.

Appanah, S. & Chan, H.T. 1981. Thrips: The pollinators of some dipterocarps. *Malay. Forest.* **44**: 234–52.

Appanah, S., Willemstein, S.C. & Marshall, A.G. 1986. Pollen foraging by two *Trigona* colonies in a Malaysian rain forest. *Malay. Nat. J.* **39**: 177–91.

Arber, A. 1937. The interpretation of the flower: A study of some aspects of morphological interest. *Biol. Rev.* **12**: 157–84.

Armbruster, W.S. 1984. The role of resin in angiosperm pollination: Ecological and chemical considerations. *Am. J. Bot.* **71**: 1149–60.

Armbruster, W.S. 1992. Phylogeny and the evolution of plant-animal interactions. *BioSci.* **42**: 12–20.

Armbruster, W.S. & Edwards, M.E. 1992. Pollination ecology and patterns of floral variation in Australian trigger plants (*Stylidium*, Stylidiaceae). *Am. J. Bot.* **79** (6), Suppl., 3.

Armbruster, W.S. & McCormick, K.D. 1990. Diel foraging patterns of male euglossine bees: Ecological causes and evolutionary response by plants. *Biotropica* **22**: 160–71.

Armbruster, W.S. & Webster, G.L. 1979. Pollination of two species of *Dalechampia* (Euphorbiaceae) in Mexico by euglossine bees. *Biotropica* **11**: 278–83.

Armstrong, J.E. 1985. The delimitation of Bignoniaceae and Scrophulariaceae based on floral anatomy, and the placement of problem genera. *Am. J. Bot* **72**: 755–66.

Armstrong, J.E. 1992 Lever action anthers and the forcible shedding of pollen in *Torenia*. *Am. J. Bot.* **79**: 34–40.

Armstrong, J.E. & Douglas, A.W. 1989. The ontogenetic basis for corolla aestivation in Scrophulariaceae. *Bull. Torrey Bot. Club* **116**: 378–89.

Armstrong, J.E. & Irvine, H.K. 1990. Functions of staminodia in the beetle-pollinated flowers of *Eupomatia laurina*. *Biotropica* **22**: 429–31.

Arroyo, M.T.K. 1976. Geitonogamy in animal pollinated tropical angiosperms: A stimulus for the evolution of self-incompatibility. *Taxon* **25**: 543–48.

Arroyo, M.T.K. 1979. Comments on breeding systems in neotropical forests. In Larsen, K. & Holm-Nielsen, L.B. (eds), *Tropical Botany*, 371–80. London: Academic Press.

Arroyo, M.T.K. 1981. Breeding systems and pollination biology in Leguminosae. In Polhill, R.M. & Raven, P.H. (eds), *Advances in Legume Systematics*. Part 2, 723–69. Kew: Royal Botanic Gardens.

Arroyo, M.T.K., Rozzi, R., Squeo, F. & Belmonte, E. 1990. Pollination in tropical and temperate high elevation ecosystems: Hypotheses and the Asteraceae as a test case. In Winiger, M., Wiesmann, U. & Rheker, J. (eds), *Mount Kenya Area: Differentiation and Dynamics of a Tropical Mountain Ecosystem*, 21-31. *Geographica Bernensia*, African Studies, Vol. 8A. Berne: Institute of Geography, University of Berne.

Ashton, P.S. 1969. Speciation among tropical forest trees: some deductions in the light of recent evidence. *Biol. J. Linn. Soc.* **1**: 155–96.

Ashton, P.S. 1976. An approach to the study of breeding systems, population structure and taxonomy of tropical trees. In Burley, J. & Styles, B.T. (eds), *Tropical Trees. Variation, Breeding and Conservation*, 35–42. London: Academic Press.

Ashton, P.S. 1977. A contribution of rain forest ecology to evolutionary theory. *Ann. Missouri Bot. Gard.* **64**: 694–705.

Ashton, P.S. 1988. Dipterocarp biology as a window to the understanding of tropical forest structure. *Ann. Rev. Ecol. Syst.* **19**: 347–70.

Ashton, P.S., Givnish, T.J. & Appanah, S. 1988. Staggered flowering in the Dipterocarpaceae: New insights into floral induction and the evolution of mast fruiting in the aseasonal tropics. *Amer. Nat.* **132**: 44–66.

Askenasy, E. 1879. Ueber das Aufblühen der Gräser. *Verh. Naturh. Medic. Ver. Heidelberg*, n.s., **2**: 261–73.

Asker, S.E. & Jerling, L. 1992. *Apomixis in Plants*. Boca Raton, Florida: CRC Press.

Atwood, J.T. 1985. Pollination of *Paphiopedilum rothschildianum*: Brood-site deception. *Natl Geogr. Res.* Spring 1985: 247–54.

von Aufsess, A. 1960. Geruchliche Nahorientierung der Biene bei entomophilen und ornithophilen Blüten. *Z. Vergl. Physiol.* **43**: 469–98.

Augspurger, C.K. 1982. A cue for synchronous flowering. In Leigh, E.G., Jr., Rand, A.S. & Windsor, D.M. (eds), *The Ecology of a Tropical Forest. Seasonal Rhythms and Long-term Changes*, 133–50. Washington, DC: Smithsonian Press.

Augspurger, C.K. 1985. Flowering synchrony of Neotropical plants. In D'Arcy, W.G. & Correa, M.D.A. (eds), *The Botany and Natural History of Panama*, 235–43. St. Louis: Missouri Botanical Garden.

Autrum, H. & von Zwehl, V. 1964. Die spektrale Empfindlichkeit einzelner Sehzellen des Bienenauges. *Z. Vergl. Physiol.* **48**: 357–84.

Backer, C.A. 1916. *Thunbergia grandiflora*. *Trop. Natuur* **5**: 113–21.

Bahadur, B., Reddy, N.P., Rao, M.M. &

Farooqui, S.M. 1984. Corolla handedness in Oxalidaceae, Linaceae and Plumbaginaceae. *J. Indian Bot. Soc.* **63**: 408–11.

Bailey, I.W. & Smith, A.C. 1942. Degeneriaceae, a new family of flowering plants from Fiji. *J. Arn. Arb.* **23**: 356–65.

Bailey, I.W. & Swamy, B.G.L. 1951. The conduplicate carpel of dicotyledons and its initial trends of specialization. *Am. J. Bot.* **38**: 373–9.

Baillon, H. 1862. Mémoire sur les Loranthacées. *Adansonia* **2**: 330–81.

Baker, H.G. 1948. Corolla-size in gynodioecious and gynomonoecious species of flowering plants. *Proc. Leeds Philos. Lit. Soc.*, Sci. Sect., **5, II**: 136–9.

Baker, H.G. 1959. Reproductive methods as factors in speciation in flowering plants. *Cold Spring Harbor Symp. Quant. Biol.* **24**: 177–90.

Baker, H.G. 1961. The adaptation of flowering plants to nocturnal and crepuscular pollinators. *Quart. Rev. Biol.* **36**: 64–73.

Baker, H.G. 1963. Evolutionary mechanisms in pollination biology. *Science* **139**: 877–83.

Baker, H.G. 1964. Opportunities for evolutionary studies in the tropics. *Taxon* **13**: 121–4.

Baker, H.G. 1970. Evolution in the tropics. *Biotropica* **2**: 101-11.

Baker, H.G. 1973. Evolutionary relationships between flowering plants and animals in American and African tropical forests. In Meggers, B.J., Ayensu, E.S. & Duckworth, W.D. (eds), *Tropical Forest Ecosystems of Africa and South America: A Comparative Review*, 145–59. Washington: Smithsonian Institution Press.

Baker, H.G. 1976. Mistake pollination as a reproductive system, with special reference to the Caricaceae. In Burley, J. & Stiles, B.T. (eds), *Tropical Trees: Variation, Breeding and Conservation*, 161–70. London: Academic Press.

Baker, H.G. 1978. Chemical aspects of the pollination biology of woody plants in the tropics. In Tomlinson, P.B. & Zimmermann, M.H. (eds), *Tropical Trees as Living Systems*, 57–82. Cambridge University Press.

Baker, H.G. 1980. Anthecology: Old Testament, New Testament, Apocrypha. *New Zeal. J. Bot.* **17**: 431–40.

Baker, H.G. 1983. An outline of the history of anthecology, or pollination biology. In Real, L. (ed.), *Pollination Biology*, 7–28. Orlando, Florida: Academic Press.

Baker, H.G. 1984. Some functions of dioecy in seed plants. *Am. Nat.* **124**: 149–58.

Baker, H.G. 1986a. Trends in pollination biology. *Aliso* **11**: 213–29.

Baker, H.G. 1986b. Yuccas and yucca moths – A historical commentary. *Ann. Missouri Bot. Gard.* **73**: 556–64.

Baker, H.G. & Baker, I. 1973. Amino acids in nectar and their evolutionary significance. *Nature* **241**: 543–5.

Baker, H.G. & Baker, I. 1975. Nectar constitution and pollinator-plant coevolution. In Gilbert, L.E. & Raven, P.H. (eds), *Animal and Plant Coevolution*, 100–40. Austin: University of Texas Press.

Baker, H.G. & Baker, I. 1979. Starch in angiosperm pollen grains and its evolutionary significance. *Am. J. Bot.* **66**: 591–600.

Baker, H.G. & Baker, I. 1983. A brief historical review of the chemistry of floral nectar. In Bentley, B. & Elias, T. (eds), *The Biology of Nectaries*, 126–52. New York: Columbia University Press.

Baker, H.G. & Baker, I. 1990. The predictive value of nectar chemistry to the recognition of pollinator types. *Israel J. Bot.* **39**: 157–66.

Baker, H.G. & Harris, B.J. 1957. The pollination of *Parkia* by bats and its attendant evolutionary problems. *Evolution* **11**: 449–60.

Baker, H.G. & Hurd, P.D. 1968. Intrafloral ecology. *Ann. Rev. Entomol.* **13**: 385–414.

Baker, H.G., Baker, I. & Opler, P.A. 1973. Stigmatic exudates and pollination. In Brantjes, N.B.M. (ed.), *Pollination and Dispersal*, 47–60. Nijmegen: Department of Botany, University of Nijmegen.

Baker, H.G., Bawa, K.S., Frankie, G.W. & Opler, P.A. 1982. Reproductive biology of plants in tropical forests. In Golley, F.B. (ed.), *Tropical Rain Forest Ecosystems: Structure and Function*, 183–215. Ecosystems of the World, Vol. 14 A. Amsterdam: Elsevier.

Baker, I. & Baker, H.G. 1982. Some chemical constituents of floral nectars of *Erythrina* in relation to pollinators and systematics. *Allertonia* **3**: 25–37.

Bamert, U. 1990. Aspekte der Struktur, Funktion und Diversität der Blüten bei Polygalaceae. Ph.D. Dissertation, University of Zurich. Zurich: ADAG.

Bänziger, H. 1991. Stench and fragrance: Unique pollination lure of Thailand's largest flower, *Rafflesia kerrii* Meijer. *Nat. Hist. Bull. Siam Soc.* **39**: 19–52.

Barabé, D., Chrétien, L. & Forget, S. 1987. On the pseudomonomerous gynoecia of the Araceae. *Phytomorphology* **37**: 139–43.

Barneby, R.C. 1991. Sensitivae censitae. A description of the genus *Mimosa* Linnaeus (Mimosaceae) in the New World. *Mem. New York Bot. Gard.* **65**: 1–835.

Barneby, R.C. & Grimes, J.W. 1990. *Orphanodendron*, a new caesalpinioid Leguminosae from northwestern Colombia. *Brittonia* **42**: 249–53.

Barrett, S.C.H. 1988. The evolution, maintenance, and loss of self-incompatibility systems. In Lovett Doust, J. & Lovett Doust, L. (eds), *Plant Reproductive Ecology. Patterns and Strategies*, 98–124. Oxford University Press.

Barrett, S.C.H. 1989. The evolutionary breakdown of heterostyly. In Bock, J.H. & Linhart, Y.B. (eds), *The Evolutionary Ecology of Plants*, 151–69. Boulder, Colorado: Westview Press.

Barrett, S.C.H. 1992. Heterostylous genetic polymorphisms: Model systems for evolutionary analysis. In Barrett, S.C.H. (ed.), *Evolution and Function of Heterostyly*, 1–29. Berlin: Springer.

Barrett, S.C.H. & Eckert, C.G. 1990a. Variation and evolution of mating systems in seed plants. In Kawano, S. (ed.), *Biological Approaches and Evolutionary Trends in Plants*, 229–54. London: Academic Press.

Barrett, S.C.H. & Eckert, C.G. 1990b. Current issues in plant reproductive ecology. *Israel J. Bot.* **39**: 5–12.

Barrett, S.C.H. & Richards, J.H. 1990. Heterostyly in tropical plants. *Mem. New York Bot. Gard.* **55**: 35–61.

Barrows, E.M. 1976. Nectar robbing and pollination of *Lantana camara* (Verbenaceae). *Biotropica* **8**: 132–35.

Barth, F.G. 1991. *Insects and Flowers. The Biology of a Partnership.* Princeton University Press.

Barthlott, W. 1990. Scanning electron microscopy of the epidermal surface in plants. In Claugher, D. (ed.), *Scanning Electron Microscopy in Taxonomy and Functional Morphology*, 69–94. Oxford: Clarendon Press.

Barthlott, W. & Ehler, N. 1977. Raster-Elektronenmikroskopie der Epidermis-Oberflächen von Spermatophyten. *Trop. Subtrop. Pflanzenwelt* **19**: 1–105. Wiesbaden: Steiner.

Bateman, A.J. 1952. Self-incompatibility systems in angiosperms. *Heredity* **6**: 285–310.

Bateson, W. & Bateson, A. 1891. On variations in the floral symmetry of certain plants having irregular corollas. *J. Linn. Soc., Bot.*, **28**: 386–424.

Batra, S.W.T. 1967. Crop pollination and the flower relationships of the wild bees of Ludhiana, India (Hymenoptera: Apoidea). *J. Kansas Entomol. Soc.* **40**: 164–77.

Baum, H. 1950. Unifaziale und subunifaziale Strukturen im Bereich der Blütenhülle und ihre Verwendbarkeit für die Homologisierung der Kelch- und Kronblätter. *Oesterr. Bot. Z.* **97**: 1–43.

Baum, H. 1951. Die Frucht von *Ochna multiflora*, ein Fall von ökologischer Apokarpie. *Oesterr. Bot. Z.* **98**: 383–94.

Baum, H. 1952. Ueber die 'primitivste' Karpellform. *Oesterr. Bot. Z.* **99**: 632–4.

Baum, H. 1953. Die Peltationsnomenklatur der Karpelle. *Oesterr. Bot. Z.* **100**: 424–6.

Baumberger, R. 1987. Floral Structure, Coloration, and Evolution of Bird-pollinated Plants; Correlation with Functional Traits in Nectarivorous Birds. Ph.D. Dissertation, University of Zurich. Zurich: ADAG.

Baumgartner, P. 1913. Untersuchungen an Bananenblütenständen. *Beitr. Biol. Centralbl.*, I, **30**: 1–132.

Bawa, K.S. 1974. Breeding systems of tree species of a lowland tropical community. *Evolution* **28**: 85–92.

Bawa, K.S. 1977. The reproductive biology of *Cupania guatemalensis* Radlk. (Sapindaceae). *Evolution* **31**: 52–63.

Bawa, K.S. 1980a. Breeding systems of trees in a tropical wet forest. *New Zeal. J. Bot.* **17**: 521–4.

Bawa, K.S. 1980b. Evolution of dioecy in flowering plants. *Ann. Rev. Ecol. Syst.* **11**: 15–39.

Bawa, K.S. 1980c. Mimicry of male by female flowers and intrasexual competition for pollinators in *Jacaratia dolichaula* (D. Smith) Woodson (Cariaceae). *Evolution* **34**: 467–74.

Bawa, K.S. 1983. Patterns of flowering in tropical plants. In Jones, C.E. & Little, R.J. (eds), *Handbook of Experimental Pollination Biology*, 394–410. New York: Van Nostrand Reinhold.

Bawa, K.S. 1990. Plant-pollinator interactions in tropical rainforests. *Ann. Rev. Ecol. Syst.* **21**: 399–422.

Bawa, K.S. & Ashton, P.S. 1991. Conservation of rare trees in tropical rain forests: A genetic perspective. In Falk, D.A. & Holsinger, K.E. (eds), *Genetics and Conservation of Rare Plants*, 62–71. New York: Oxford University Press.

Bawa, K.S. & Beach, J.H. 1981. Evolution of sexual systems in flowering plants. *Ann. Missouri Bot. Gard.* **68**: 254–74.

Bawa, K.S. & Buckley, D.P. 1989. Seed:ovule ratios, selective seed abortion, and mating systems in Leguminosae. *Monogr. Syst. Bot. Missouri Bot. Gard.* **29**: 243–62.

Bawa, K.S. & Crisp, J.E. 1980. Wind-pollination in the understory of a rain forest in Costa Rica. J. Ecol. **68**: 871–6.

Bawa, K.S. & Opler, P.A. 1975. Dioecism in tropical forest trees. *Evolution* **29**: 167–79.

Bawa, K.S. & Webb, C.J. 1984. Flower, fruit and seed abortion in tropical trees: Implications for the evolution of paternal and maternal reproduction patterns. *Am. J. Bot.* **71**: 736–51.

Bawa, K.S., Ashton, P.S. & Nor, S.M. 1990. Reproductive ecology of tropical forest plants: Management issues. In Bawa, K.S. & Hadley, M. (eds), *Reproductive Ecology of Tropical Forest Plants*, 3–13. Paris: UNESCO.

Bawa, K.S., Perry, D. & Beach, J.H. 1985a. Reproductive biology of tropical lowland rain forest trees. I. Sexual systems and incompatibility mechanisms. *Am. J. Bot.* **72**: 331–45.

Bawa, K.S., Perry, D.R., Bullock, S.H., Covile, R.E. & Grayum, M.H. 1985b. Reproductive biology of tropical lowland rain forest trees. II. Pollination mechanisms. *Am. J. Bot.* **72**: 346–56.

Bayer, C. & Hoppe, J.R. 1990. Die Blütenentwicklung von *Theobroma cacao* L. (Sterculiaceae). *Beitr. Biol. Pfl.* **65**: 301–12.

Beach, J.H. 1982. Beetle pollination of *Cyclanthus bipartitus* (Cyclanthaceae). *Am. J. Bot.* **69**: 1074–81.

Beach, J.H. 1984. The reproductive biology of the peach or 'pejibaye' palm (*Bactris gasipaes*) and a wild congener (*Bactris porschiana*) in the Atlantic lowlands of Costa Rica. *Principes* **28**: 107–19.

Beaman, R.S., Decker, P.J. & Beaman, J.H. 1988. Pollination of *Rafflesia* (Rafflesiaceae). *Am. J. Bot.* **75**: 1148–62.

Beccari, O. 1904. *Wanderings in the Great Forests of Borneo*. London: Archibald Constable.

Behrens, W.J. 1875. Untersuchungen über den anatomischen Bau des Griffels und der Narbe einiger Pflanzenarten. Ph.D. Dissertation, University of Göttingen. Göttingen: Dieterich.

Behrens, W.J. 1879. Die Nectarien der Blüten. *Flora* **62**: 2–11, 17–27, 49–54, 81–90, 113–23, 145–53, 233–40, 241–7, 305–14, 369–75, 433–57.

Bell, G. 1986. The evolution of empty flowers. *J. Theor. Biol.* **118**: 253–8.

Bellingham, P.J., Kapos, V., Varty, N., Healey, J.R., Tanner, E.V.J., Kelly, D.L., Dalling, J.W., Burns, L.S., Lee, D. & Sidrak, G. 1992. Hurricanes need not cause high mortality: The effects of hurricane Gilbert on forests in Jamaica. *J. Trop. Ecol.* **8**: 217–23.

Belt, T. 1874. *The Naturalist in Nicaragua*. London: Murray.

Bené, F. 1941. Experiments on the color preference of black-chinned hummingbirds. *Condor* **43**: 237–42.

Benson, W.W., Brown, K.S., Jr. & Gilbert, L.E. 1976. Coevolution of plants and herbivores: Passion flower butterflies. *Evolution* **29**: 659–80.

Bentley, B.L. 1977. Extrafloral nectaries and protection by pugnacious bodyguards. *Ann. Rev. Ecol. Syst.* **8**: 407–27.

Benzing, D.H. 1987. Major patterns and processes in orchid evolution: A critical synthesis. In Arditti, J. (ed.), *Orchid Biology. Reviews and Perspectives* IV: 33–77. Ithaca: Comstock.

Benzing, D.H. 1990. *Vascular Epiphytes. General Biology and Related Biota*. Cambridge University Press.

Berg, C.C. 1990a. Differentiation of flowers and inflorescences of Urticales in relation to their protection against breeding insects and to pollination. *Sommerfeltia* **11**: 13–34.

Berg, C.C. 1990b. Reproduction and evolution in *Ficus* (Moraceae): Traits connected with the adequate rearing of pollinators. *Mem. New York Bot. Gard.* **55**: 169–85.

Bergström, G., Groth, I., Pellmyr, O., Endress, P.K., Thien, L.B., Hübener, A. & Francke, W. 1991. Chemical basis of a highly specific mutualism: Chiral esters attract pollinating beetles in Eupomatiaceae. *Phytochemistry* **30**: 3221–5.

Bernhardt, P. 1989. The floral ecology of Australian *Acacia*. *Monogr. Syst. Bot. Missouri Bot. Gard.* **29**: 263–81.

Bernhardt, P. & Burns-Balogh, P. 1986.

Floral mimesis in *Thelymitra nuda* (Orchidaceae). *Pl. Syst. Evol.* **151**: 187–202.

Bernhardt, P. & Thien, L.B. 1987. Self-isolation and insect pollination in the primitive angiosperms: New evaluations of older hypotheses. *Pl. Syst. Evol.* **156**: 159–76.

Bernhardt, P. & Walker, K. 1984. Bee foraging on three sympatric species of Australian *Acacia*. *Int. J. Entomol.* **26**: 322–30.

Bernier, G., Kinet, J.-M. & Sachs, R.M. 1981/1985. *The Physiology of Flowering*, I-III. Boca Raton, Florida: CRC Press.

Berry, F. & Kress, W.J. 1991. *Heliconia. An Identification Guide*. Washington: Smithsonian Institution Press.

Bhatnagar, S.P. & Uma, M.C. 1969. The structure of style and stigma in some Tubiflorae. *Phytomorphology* **19**: 99–109.

Bierzychudek, P. 1981. *Asclepias*, *Lantana*, and *Epidendrum*: A floral mimicry complex? *Biotropica* **13** (2), Suppl., 54–8.

Björnstad, I.N. 1970. Comparative embryology of Asparagoideae-Polygonateae, Liliaceae. *Nytt Mag. Bot.* **17**: 169–207.

Blatter, E. & Millard, W.S. 1977. *Some Beautiful Indian Trees* (Ed. 2). Bombay: Natural History Society.

Bobisud, L.B. & Neuhaus, R.J. 1975. Pollinator constancy and survival of rare species. *Oecologia* **21**: 263–72.

Bocquet, G. 1960. The campylotropous ovule. *Phytomorphology* **9**: 222–7.

Bocquillon, H. 1863. Revue du groupe des Verbénacées. Paris: Germer Baillière.

Boesewinkel, F.D. & Bouman, F. 1984. The seed: Structure. In Johri, B.M. (ed.), *Embryology of Angiosperms*, 567–610. Berlin: Springer.

Boesewinkel, F.D. & Bouman, F. 1991. The development of bi- and unitegmic ovules and seeds in *Impatiens* (Balsaminaceae). *Bot. Jahrb. Syst.* **113**: 87–104.

Böhme, S. 1988. Bromelienstudien III. Vergleichende Untersuchungen zu Bau, Lage und systematischer Verwertbarkeit der Septalnektarien von Bromeliaceen. *Trop. Subtrop. Pflanzenwelt* **62**: 1–154. Stuttgart: Steiner.

Bokhari, M.H. & Hedge, I.C. 1971. Observations on the tribe Meriandreae of the Labiatae. *Notes Roy. Bot. Gard. Edinburgh* **31**: 53–67.

Bonner, L.J. & Dickinson, H.G. 1989. Anther dehiscence in *Lycopersicon esculentum* Mill. I. Structural aspects. *New Phytol.* **113**: 97–115.

Bor, N.L. & Raizada, M.B. 1990. *Some Beautiful Indian Climbers and Shrubs* (rev. ed. 2). Bombay: Oxford University Press.

Borchert, R. 1980. Phenology and ecophysiology of tropical trees: *Erythrina poeppigiana* O.F. Cook. *Ecology* **61**: 1065–74.

Borchert, R. 1983. Phenology and control of flowering in tropical trees. *Biotropica* **15**: 81–9.

Bouman, F. 1984. The ovule. In Johri, B.M. (ed.), *Embryology of Angiosperms*, 123–57. Berlin: Springer.

Bouman, F. & Boesewinkel, F.D. 1991. The campylotropous ovules and seeds, their structure and functions. *Bot. Jahrb. Syst.* **113**: 255–70.

Bouman, F. & Calis, J.I.M. 1977. Integumentary shifting - a third way to unitegmy. *Ber. Deutsch. Bot. Ges.* **90**: 15–28.

Bouman, F. & Devente, N. 1986. Seed micromorphology. In Maas, P.J.M. & Ruyters, P.: *Voyria* and *Voyriella* (saprophytic Gentianaceae), 9–25. *Flora Neotropica Monogr.* **41**: 1–93.

Bouman, F. & Meijer, W. 1986. Comparative seed morphology in Rafflesiaceae. *Acta Bot. Neerl.* **35**: 521–8.

Boyden, T.C. 1980. Floral mimicry by *Epidendrum ibaguense* (Orchidaceae) in Panama. *Evolution* **34**: 135–6.

Brantjes, N.B.M. 1973. Sphingophilous flowers, function of their scent. In Brantjes, N.B.M. (ed.), *Pollination and Dispersal*, 27–46. Nijmegen: Department of Botany, University of Nijmegen.

Brantjes, N.B.M. 1978. Sensory responses to flowers in night-flying moths. In Richards, A.J. (ed.), *The Pollination of Flowers by Insects*, 13–19. London: Academic Press.

Brantjes, N.B.M. 1980. Flower morphology of *Aristolochia* species and the consequences for pollination. *Acta Bot. Neerl.* **29**: 212–13.

Brantjes, N.B.M. 1981a. Mechanical aspects of the corolla, a quantitative and functional approach. *Acta Bot. Neerl.* **30**: 317–18.

Brantjes, N.B.M. 1981b. Floral mechanics in *Phlomis* (Lamiaceae). *Ann. Bot.* **47**: 279–82.

Brantjes, N.B.M. 1981c. Nectar and pollination of bread fruit, *Artocarpus altilis* (Moraceae). *Acta Bot. Neerl.* **30**: 345–52.

Brantjes, N.B.M. 1982. Pollen placement and reproductive isolation between Brazilian *Polygala* species (Polygalaceae). *Pl. Syst. Evol.* **141**: 45–52.

Brantjes, N.B.M. 1983a. Regulated pollen issue in *Isotoma*, Campanulaceae, and evolution of secondary pollen presentation. *Acta Bot. Neerl.* **32**: 213–22.

Brantjes, N.B.M. 1983b. Mechanical tuning of flowers. *Acta Bot. Neerl.* **32**: 343.

Brantjes, N.B.M. & de Vos, O.C. 1981. The explosive release of pollen in flowers of *Hyptis* (Lamiaceae). *New Phytol.* **87**: 425–30.

Braun, A. 1831. Vergleichende Untersuchung über die Ordnung der Schuppen an den Tannenzapfen als Einleitung zur Untersuchung der Blattstellung überhaupt. *Nova Acta Acad. Leop.-Carol.* **15I**: 195–402.

Breindl, M. 1934. Zur Kenntnis der Baumechanik des Blütenkelches der Dikotylen. *Bot. Arch.* **36**: 191–268.

Bremekamp, C.E.B. 1914. Eine besondere Funktion der Drüsenschuppen im Fruchtknoten von *Clerodendron minahassae* Miq. *Ann. Jard. Bot. Buitenzorg*, Sér. 2, **13**: 93–7.

Bremekamp, C.E.B. 1953. The delimitation of the Acanthaceae. *Proc. Kon. Nederl. Akad. Wet. C* **56**: 533–46.

Bremekamp, C.E.B. 1955. The *Thunbergia* species of the Malesian area. *Verh. Kon. Nederl. Akad. Wet.*, Afd. Nat., R. 2, **50** (4): 3–90.

Bremekamp, C.E.B. 1965. Delimitation and subdivision of the Acanthaceae. *Bull. Bot. Surv. India* **7**: 21–30.

Bremer, B. 1987. The sister group of the paleotropical tribe Argostemmateae: A redefined neotropical tribe Hamelieae (Rubiaceae, Rubioideae). *Cladistics* **3**: 35–51.

Brewbaker, J.L. & Majumder, S.K. 1961. Cultural studies of the pollen population effect and the self-incompatibility inhibition. *Am. J. Bot.* **48**: 457–64.

Brongniart, A. 1854. Mémoire sur les glandes nectarifères de l'ovaire dans diverses familles de plantes monocotylédones. *Ann. Sci. Nat., Bot.*, Sér. 4, **2**: 5–23.

Bronstein, J.L. 1989. A mutualism at the edge of its range. *Experientia* **45**: 622–37.

Bronstein, J.L. & McKey, D. 1989. The fig/pollinator mutualism: A model system for comparative biology. *Experientia* **45**: 601–4.

Brown, J.H. & Kodric-Brown, A. 1979. Convergence, competition, and mimicry in a temperate community of hummingbird-pollinated flowers. *Ecology* **60**: 1022–35.

Brown, R. 1833. On the organs and mode of fecundation in Orchideae and Asclepiadeae. *Trans. Linn. Soc.* **16**: 685–745.

Brown, W.H. 1938. The bearing of nectaries on the phylogeny of flowering plants. *Proc. Amer. Phil. Soc.* **71**: 549–95.

Brückner, C. 1991. Zur Interpretation des Karpells – eine Uebersicht. *Gleditschia* **19**: 3–14.

Brues, C.T. 1928. Some Cuban Phoridae which visit the flowers of *Aristolochia elegans*. *Psyche* **35**: 160–1.

Bruyns, P.V. &. Forster, P.I. 1991. Recircumscription of the Stapelieae (Asclepiadaceae). *Taxon* **40**: 381–91.

Buchmann, S.L. 1980. Preliminary anthecological observations on *Xiphidium caeruleum* Aubl. (Monocotyledoneae: Haemodoraceae) in Panama. *J. Kansas Entomol. Soc.* **53**: 685–99.

Buchmann, S.L. 1983. Buzz pollination in angiosperms. In Jones, C.E. & Little, R.J. (eds), *Handbook of Experimental Pollination Biology*, 73–113. New York: Scientific and Academic Editions.

Buchmann, S.L. 1985. Bees use vibration to aid pollen collection from non-poricidal anthers. *J. Kansas Entomol. Soc.* **58**: 517–25.

Buchmann, S.L. 1986. Vibratile pollination in *Solanum* and *Lycopersicon*: A look at pollen chemistry. In D'Arcy, W.G. (ed.), *Solanaceae. Biology and Systematics*, 237–52. New York: Columbia University Press.

Buchmann, S.L. 1987. The ecology of oil flowers and their bees. *Ann. Rev. Ecol. Syst.* **18**: 343–69.

Buchmann, S.L. & Cane, J.H. 1989. Bees assess pollen returns while sonicating *Solanum* flowers. *Oecologia* **81**: 289–94.

Buchmann, S.L. & Hurley, J.P. 1978. A biophysical model for buzz pollination in angiosperms. *J. Theor. Biol.* **72**: 639–57.

Bullock, S.H. 1985. Breeding systems in the flora of a tropical deciduous forest in Mexico. *Biotropica* **17**: 287–301.

Bünning, E. 1959. Die seismonastischen Reaktionen. In Ruhland, W. (ed.), *Handbuch der Pflanzenphysiologie* **XVII(1)**: 184–238. Berlin: Springer.

Burck, W. 1890. Ueber Kleistogamie im weiteren Sinne und das Knight-Darwin'sche Gesetz. *Ann. Jard. Bot. Buitenzorg* **8**: 122–63.

Burck, W. 1891. Beiträge zur Kenntniss der myrmecophilen Pflanzen. *Ann. Jard. Bot. Buitenzorg* **10**: 75–144.

Burckhardt, D. 1977. On the vision of insects. *J. Comp. Physiol.* **120**: 33–50.

Burkill, I.H. 1906. The pollination of *Thunber-*

gia grandiflora. *J. Proc. Asiat. Soc. Bengal* **2**: 511–14.

Burns-Balogh, P. & Bernhardt, P. 1985. Evolutionary trends in the androecium of the Orchidaceae. *Pl. Syst. Evol.* **149**: 119–34.

Burns-Balogh, P. & Funk, V.A. 1986. A phylogenetic analysis of the Orchidaceae. *Smithsonian Contr. Bot.* **61**: 1–79.

Burquez, A. & Corbet, S.A. 1991. Do flowers reabsorb nectar? *Funct. Ecol.* **5**: 369–79.

Burtt, B.L. 1958. *Opithandra*, a genus with sterile anticous stamens. *Notes Roy. Bot. Gard. Edinburgh* **22**: 301–3.

Burtt, B.L. 1961. Interpretive morphology. *Notes Roy. Bot. Gard. Edinburgh* **23**: 569–72.

Burtt, B.L. 1963. Studies on the Gesneriaceae of the Old World XXIV: Tentative keys to the tribes and genera. *Notes Roy. Bot. Gard. Edinburgh* **24**: 205–20.

Burtt, B.L. 1964. Angiosperm taxonomy in practice. In Heywood, V.H. & McNeill, J. (eds), *Phenetic and Phylogenetic Classification*, 5–16. London: The Systematics Association.

Burtt, B.L. 1965. The transfer of *Cyrtandromoea* from Gesneriaceae to Scrophulariaceae, with notes on the classification of that family. *Bull. Bot. Surv. India* **7**: 73–88.

Burtt, B.L. 1970. Studies in the Gesneriaceae of the Old World XXXI: Some aspects of functional evolution. *Notes Roy. Bot. Gard. Edinburgh* **30**: 1–10.

Burtt, B.L. 1977. Classification above the genus, as exemplified by Gesneriaceae, with parallels from other groups. *Pl. Syst. Evol., Suppl.* **1**, 97–109.

Burtt, B.L. 1978. A preliminary revision of *Monophyllaea*. *Notes Roy. Bot. Gard. Edinburgh* **37**: 1–59.

Burtt, B.L. & Smith, R.M. 1964. A new genus of Acanthaceae from Sarawak. *Notes Roy. Bot. Gard. Edinburgh* **26**: 325–9.

Buttrose, M.S., Grant, W.J.R. & Lott, J.N.A. 1977. Reversible curvature of style branches of *Hibiscus trionum* L., a pollination mechanism. *Austral. J. Bot.* **25**: 567–70.

Buzato, S. & Franco, A.L.M. 1992. *Tetrastylis ovalis*: A second case of bat-pollinated passionflower (Passifloraceae). *Pl. Syst. Evol.* **181**: 261–7.

Bystedt, P.-A. 1990. The transmitting tract in *Trimezia fosteriana* (Iridaceae). I. Ultrastructure in the stigma, style and ovary. *Nord. J. Bot.* **9**: 507–18.

Bystedt, P.-A. & Vennigerholz, F. 1991a/b. The transmitting tract in *Trimezia fosteriana* (Iridaceae). II. Development of secretory cells in the stigma, style and ovary./III. Pollen tube growth in the stigma, style and ovary. *Nord. J. Bot.* **11**: 345–57/459–64.

Calder, D.M. & Slater, A.T. 1985. The stigma of *Dendrobium speciosum* Sm. (Orchidaceae): A new stigma type comprising detached cells within a mucilaginous matrix. *Ann. Bot.* **55**: 297–307.

Cammerloher, H. 1923. Zur Biologie der Blüte von *Aristolochia grandiflora* Swartz. *Oesterr. Bot. Z.* **72**: 180–98.

Cammerloher, H. 1929a. Zur Kenntnis von Bau und Funktion extrafloraler Nektarien. *Biol. Gen.* **5**: 281–302.

Cammerloher, H. 1929b. Blütenökologische Beobachtungen an den Blüten einer *Bauhinia*. *Bul. Facult. Stiinte diu Cernauti* **3**: 171–4.

Cammerloher, H. 1931. *Blütenbiologie* I. Berlin: Borntraeger.

Cammerloher, H. 1933. Die Bestäubungseinrichtungen der Blüten von *Aristolochia lindneri* Berger. *Planta* **19**: 351–65.

Campbell, D.H. 1930. The relationships of *Paulownia*. Bull. *Torrey Bot. Club* **57**: 47–50.

Cantino, P.D. 1992. Evidence for a polyphyletic origin of the Labiatae. *Ann. Missouri Bot.Gard.* **79**: 361–379.

Capus, G. 1879. Anatomie du tissu conducteur. *Ann. Sci. Nat., Bot.*, Sér. 6, **7**: 209–91.

Carlquist, S. 1970. Toward acceptable evolutionary interpretations of floral anatomy. *Phytomorphology* **19**: 332–62.

Carlquist, S. 1981. Wood anatomy of *Zygogynum* (Winteraceae); field observations. *Bull. Mus. Natl Hist. Nat., Paris*, Sér. 4, Sect. B, *Adansonia*, **3**: 281–92.

Carlquist, S. 1982. *Exospermum stipitatum* (Winteraceae): Observations on wood, leaves, flowers, pollen, and fruit. *Aliso* **10**: 277–89.

Carlquist, S. 1983. Wood anatomy of *Belliolum* (Winteraceae) and a note on flowering. *J. Arn. Arb.* **64**: 161–9.

Carr, S.G.M. & Carr, D.J. 1961. The functional significance of syncarpy. *Phytomorphology* **11**: 249–56.

de Carvalho, C.T. 1960. Das visitas de morcegos as flôres (Mammalia, Chiroptera). *An. Acad. Bras. Cienc.* **32**: 359–77.

Caspary, J.X.R. 1848. *De nectariis*. Elberfeld: Schellhoff.

Catling, P.M. 1987. Notes on the breeding systems of *Sacoila lanceolata* (Aublet) Garay (Orchidaceae). *Ann. Missouri Bot. Gard.* **74**: 58–68.

Catling, P.M. 1990. Auto-pollination in the Orchidaceae. In Arditti, J. (ed.), *Orchid Biology. Reviews and Perspectives* V: 121–58. Portland, Oregon: Timber Press.

Čelakovský, L.J. 1876. Vergleichende Darstellung der Placenten in den Fruchtknoten der Phanerogamen. *Abh. Böhm. Ges. Wiss.*, Math.-Naturwiss. Cl, 6. Folge, 8 (2): 1–74.

Čelakovský, L.J. 1895. Das Reductionsgesetz der Blüthen, das Dédoublement und die Obdiplostemonie. *Sitzber. Königl. Böhm. Ges. Wiss.*, Math.-Naturwiss. Cl., 1894 (3): 1–142.

Čelakovský, L.J. 1896/1900. Ueber den phylogenetischen Entwicklungsgang der Blüthe und über den Ursprung der Blumenkrone 1/2. *Sitzber. Königl. Böhm. Ges. Wiss.*, Math.-Naturwiss. Cl., 1896 (40): 1–91/1900 (3): 1–221.

Chan, H.T. & Appanah, S. 1980. Reproductive biology of some Malaysian dipterocarps. I. Flowering biology. *Malay. Forester* **43**: 132–43.

Charlesworth, D. & Charlesworth, B. 1979. A model for the evolution of distyly. *Am. Nat.* **114**: 467–98.

Chase, M.W. 1986a. A monograph of *Leochilus*. *Syst. Bot. Monogr.* **14**: 1–97.

Chase, M.W. 1986b. A reappraisal of the oncidioid orchids. *Syst. Bot.* **11**: 477–91.

Chase, M.W. & Hills, H.G. 1992. Orchid phylogeny, flower sexuality, and fragrance seeking. *BioSci.* **42**: 43–9.

Chase, M.W. & Palmer, J.D. 1992. Floral morphology and chromosome number in subtribe Oncidiinae (Orchidaceae): Evolutionary insights from a phylogenetic analysis of chloroplast DNA restriction site variation. In Soltis, P.S., Soltis, D.E. & Doyle, J.J. (eds), *Molecular Systematics of Plants*, 324–39. New York: Chapman and Hall.

Chase, M.W. & Peacor, D.R. 1987. Crystals of calcium oxalate hydrate on the perianth of *Stelis* Sw. *Lindleyana* **2**: 91–4.

Chaw, S.-M. 1992. Pollination, breeding syndromes, and systematics of *Trochodendron aralioides* Sieb. & Zucc. (Trochodendraceae), a relictual species in Eastern Asia. In Peng, C.-I. (ed.), *Phytogeography and Botanical Inventory of Taiwan*, 63–77. Institute of Botany, Academia Sinica Monograph Series 12. Taipei: ROC.

Chen, D.-M., Collins, J.S. & Goldsmith, T.H. 1984. The ultraviolet receptor of bird retinas. *Science* **225**: 337–40.

Cheplik, G.P. 1987. The ecology of amphicarpic plants. *Trends Ecol. Evol.* **2**: 97–101.

Chittka, L. & Menzel, R. 1992. The evolutionary adaptation of flower colours and the insect pollinators' colour vision. *J. Comp. Physiol.* A **171**: 171–81.

Christ, P. & Schnepf, E. 1985. The nectaries of *Cynanchum vincetoxicum* (Asclepiadaceae). *Israel J. Bot.* **34**: 79–90.

Christ, P. & Schnepf, E. 1988. Zur Struktur und Funktion von Asclepiadaceen-Nektarien. *Beitr. Biol. Pfl.* **63**: 55–79.

Christensen, D.E. 1992. Notes on the reproductive biology of *Stelis argentata* Lindl. (Orchidaceae: Pleurothallidinae) in Eastern Ecuador. *Lindleyana* **7**: 28–33.

Church, A.H. 1908. *Types of Floral Mechanism*. Oxford: Clarendon Press.

Clarke, A.E., Anderson, M.A., Bernatsky, R., Cornish, E.C. & Mau, S.-L. 1989. Molecular aspects of self-incompatibility. In Goldberg, R. (ed.), *The Molecular Basis of Plant Development*, 87–98. New York: Alan R. Liss.

Classen, R. 1986. Organisation und Funktion der blumenbildenden Hochblatthüllen bei Symphoremoideae (Verbenac.). *Beitr. Biol. Pfl.* **60**: 383–402.

Classen, R. 1987. Morphological adaptations for bird pollination in *Nicolaia elatior* (Jack) Horan (Zingiberaceae). *Gard. Bull. Singapore* **40**: 37–43.

Classen-Bockhoff, R. 1990. Pattern analysis in pseudanthia. *Pl. Syst. Evol.* **171**: 57–88.

Classen-Bockhoff, R. 1991. Untersuchungen zur Konstruktion des Bestäubungsapparates von *Thalia geniculata* (Marantaceen). *Bot. Acta* **104**: 183–93.

Clifford, H.T. 1987. Spikelet and floral morphology. In Soderstrom, T.R., Hilu, K.W., Campbell, C.S. & Barkworth, M.E. (eds), *Grass Systematics and Evolution*, 21–30. Washington, D.C.: Smithsonian Institution Press.

Cocucci, A.A. 1989. El mecanismo floral de *Schizanthus* (Solanaceae). *Kurtziana* **20**: 113–32.

Cocucci, A.A. 1991. Pollination biology of *Nierembergia* (Solanaceae). *Pl. Syst. Evol.* **174**: 17–35.

Cocucci, A.A., Galetto, L. & Sersic, A. 1992. El sindrome floral de *Caesalpinia gilliesii* (Fabaceae-Caesalpinioideae). *Darwiniana* **31**: 111–35.

Coen, E.S. 1991. The role of homeotic genes in flower development and evolution. *Ann. Rev. Pl. Physiol. Pl. Mol. Biol.* **42**: 241–79.

Coen, E.S. & Meyerowitz, E.M. 1991. The war of the whorls: Genetic interactions controlling flower development. *Nature* **353**: 31–7.

Collins, B.G. & Rebelo, T. 1987. Pollination biology of the Proteaceae in Australia and southern Africa. *Austral. J. Ecol.* **12**: 387–421.

Colwell, R.K. 1983. *Rhinoseius colwelli* (acaro floral del colibri, totolate floral de colibri, hummingbird flower mite). In Janzen, D.H. (ed.), *Costa Rican Natural History*, 767–8. Chicago: University of Chicago Press.

Connor, H.E. 1980. Breeding systems in the grasses: A survey. *New Zeal. J. Bot.* **17**: 547–74.

Cook, C.D.K. 1982. Pollination mechanisms in the Hydrocharitaceae. In Symoens, J.J., Hooper, S.S. & Compère, P. (eds), *Studies on Aquatic Vascular Plants*, 1–15. Brussels: Royal Botanical Society of Belgium.

Cook, C.D.K. 1988. Wind pollination in aquatic angiosperms. *Ann. Missouri Bot. Gard.* **75**: 768–77.

Cook, C.D.K. & Guo, Y.-H. 1990. A contribution to the natural history of *Althenia filiformis* Petit (Zannichelliaceae). *Aquat. Bot.* **38**: 261–81.

Corbet, S.A. 1990. Pollination and the weather. *Israel J. Bot.* **39**: 13–30.

Corbet, S.A. & Willmer, P.G. 1980. Pollination of the yellow passionfruit: Nectar, pollen and carpenter bees. *J. Agric. Sci.* **95**: 655–66.

Corbet, S.A., Beament, J.W.L. & Eisikowitch, D. 1982. Are electrostatic forces involved in pollen transfer? *Plant Cell Environm.* **5**: 125–9.

Corbet, S.A., Chapman, H. & Saville, N. 1988. Vibratory pollen collection and flower form: Bumble-bees on *Actinidia*, *Symphytum*, *Borago* and *Polygonatum*. *Funct. Ecol.* **2**: 147–55.

Corner, E.J.H. 1940. *Wayside Trees of Malaya* (Ed. 1). Singapore: Government Printers.

Corner, E.J.H. 1946. Centrifugal stamens. *J. Arn. Arb.* **27**: 423–37.

Corner, E.J.H. 1949. The durian theory or the origin of the modern tree. *Ann. Bot.* **52**: 367–414.

Corner, E.J.H. 1958. Transference of function. *J. Linn. Soc., Bot.*, **56**: 33–40.

Corner, E.J.H. 1964. *The Life of Plants*. London: Weidenfeld & Nicolson.

Corner, E.J.H. 1976. *The Seeds of Dicotyledons*, Vols 1/2. Cambridge University Press.

Corner, E.J.H. 1988. *Wayside Trees of Malaya*, Vols 1/2 (Ed. 3). Kuala Lumpur: Malayan Nature Society.

del Coro-Arizmendi, M. & Ornelas, J.F. 1990. Hummingbirds and their floral resources in a tropical dry forest in Mexico. *Biotropica* **22**: 172–80.

Costa, E. de L. & Hime, N. da C. 1982. Biologia floral de *Aristolochia gigantea* Mart. et Zucc. (Aristolochiaceae) I. *Rodriguesia* **56**: 23–44.

Costa, E. de L. & Hime, N. da C. 1983. Observações sobre a biologia floral de *Aristolochia macroura* Gomez (Aristolochiaceae). *Atas Soc. Bot. Bras. RJ* **1** (11): 63–6.

Coster, C. 1926. Periodische Blüteerscheinungen in den Tropen. *Ann. Jard. Bot. Buitenzorg* **35**: 125–62.

Costerus, J.C. 1916. Das Labellum und das Diagramm der Zingiberaceen. *Ann. Jard. Bot. Buitenzorg*, Sér. II, **14**: 95–108.

Cowan, R.S. & Polhill, R.M. 1981. Tribe 4. Detarieae DC. (1825). In Polhill, R.M. & Raven, P.H. (eds), *Advances in Legume Systematics* Part 1: 117–34. Kew: Royal Botanic Gardens.

Cox, P.A. 1983. Search theory, random motion, and the convergent evolution of pollen and spore morphologies in aquatic plants. *Amer. Nat.* **121**: 9–31.

Cox, P.A. 1988. Hydrophilous pollination. *Ann. Rev. Ecol. Syst.* **19**: 261–80.

Cox, P.A. 1990. Pollination and the evolution of breeding systems in Pandanaceae. *Ann. Missouri Bot. Gard.* **77**: 816–40.

Cox, P.A. 1991a. Abiotic pollination: An evolutionary escape for animal-pollinated angiosperms. *Phil. Trans. Roy. Soc. Lond. B* **333**: 217–24.

Cox, P.A. 1991b. Hydrophilous pollination of a dioecious seagrass, *Thalassodendron ciliatum* (Cymodoceaceae) in Kenya. *Biotropica* **23**: 159–65.

Cox, P.A. & Knox, R.B. 1988. Pollination postulates and two-dimensional pollination in hydrophilous monocotyledons. *Ann. Missouri Bot. Gard.* **75**: 811–18.

Cox, P.A. & Knox, R.B. 1989. Two-dimensional pollination in hydrophilous plants: Convergent evolution in the genera *Halodule* (Cymodoceaceae), *Halophila*

(Hydrocharitaceae), *Ruppia* (Ruppiaceae), and *Lepilaena* (Zannichelliaceae). *Am. J. Bot.* **76**: 164–75.

Cox, P.A. & Tomlinson, P.B. 1988. Pollination ecology of a seagrass, *Thalassia testudinum* (Hydrocharitaceae), in St. Croix. *Am. J. Bot.* **75**: 958–65.

Cox, P.A., Elmqvist, T. & Tomlinson, P.B. 1990. Submarine pollination and reproductive morphology in *Syringodium filiforme* (Cymodoceaceae). *Biotropica* **22**: 259–65.

Cox, P.A., Elmqvist, T., Pierson, E.D. & Rainey, W.E. 1991. Flying foxes as strong interactors in South Pacific island ecosystems: A conservation hypothesis. *Conserv. Biol.* **5**: 448–54.

Crane, E. 1991. *Apis* species of tropical Asia as pollinators, and some rearing methods for them. *Acta Horticult.* **288**: 29–48.

Crane, P.R. 1985. Phylogenetic analysis of seed plants and the origin of angiosperms. *Ann. Missouri Bot. Gard.* **72**: 716–93.

Crane, P.R. 1986a. Form and function in wind dispersed pollen. In Blackmore, S. & Ferguson, I.K. (eds.), *Pollen and Spores. Form and Function*, 179–202. London: Academic Press.

Crane, P.R. 1986b. The morphology and relationships of the Bennettitales. In Spicer, R.A. & Thomas, B.A. (eds), *Systematic and Taxonomic Approaches in Palaeobotany*, 163- 75. Oxford: Clarendon Press.

Crane, P.R., Friis, E.M. & Pedersen, K.R. 1989. Reproductive structure and function in Cretaceous Chloranthaceae. *Pl. Syst. Evol.* **165**: 211–26.

Crepet, W.L. 1984. Advanced (constant) insect pollination mechanisms: Pattern of evolution and implications vis-à-vis angiosperm diversity. *Ann. Missouri Bot. Gard.* **71**: 607–30.

Crepet, W.L. & Friis, E.M. 1987. The evolution of insect pollination in angiosperms. In Friis, E.M., Chaloner, W.G. & Crane, P.R. (eds), *The Origin of Angiosperms and Their Biological Consequences*, 181–201. Cambridge University Press.

Crepet, W.L. & Taylor, D.W. 1985. The diversification of the Leguminosae: First fossil evidence of the Mimosoideae and Papilionoideae. *Science* **228**: 1087–9.

Crepet, W.L., Friis, E.M. & Nixon, K.C. 1991. Fossil evidence for the evolution of biotic pollination. *Phil. Trans. Roy. Soc. Lond. B* **333**: 187–95.

Crepet, W.L., Nixon, K.C., Friis, E.M. & Freudenstein, J.V. 1992. Oldest fossil flowers of hamamelidaceous affinity, from the late Cretaceous of New Jersey. *Proc. Natl Acad. Sci. USA* **89**: 8986–9.

Cresens, E.M. & Smets, E.F. 1992. On the character 'carpel-form'. Trends in the development of the Magnoliatae pistil. *Candollea* **47**: 373–90.

Cresti, M., van Went, J.L., Pacini, E. & Willemse, M.T.M. 1976. Ultrastructure of transmitting tissue of *Lycopersicon peruvianum* style: Development and histochemistry. *Planta* **132**: 305–12.

Cribb, P. 1987. *The Genus Paphiopedilum*. Kew: Royal Botanic Gardens.

Croat, T.B. 1978. *Flora of Barro Colorado Island*. Stanford: Stanford University Press.

Croat, T.B. 1979. The sexuality of the Barro Colorado Island flora. *Phytologia* **42**: 319–48.

Croat, T.B. 1980. Flower behavior of the Neotropical genus *Anthurium* (Araceae). *Am. J. Bot.* **67**: 888–904.

Crome, F.H.J. & Irvine, A.K. 1986. 'Two bob each way': The pollination and breeding system of the Australian rain forest tree *Syzygium cormiflorum* (Myrtaceae). *Biotropica* **18**: 115–25.

Crone, W. & Lord, E.M. 1991. A kinematic analysis of gynoecial growth in *Lilium longiflorum*: Surface growth patterns in all floral organs are triphasic. *Develop. Biol.* **143**: 408–17.

Cronquist, A. 1957. Outline of a new system of families and orders of dicotyledons. *Bull. Jard. Bot. Natl Belg.* **27**: 13–40.

Cronquist, A. 1968. *The Evolution and Classification of Flowering Plants* (Ed. 1). New York: New York Botanical Garden.

Cronquist, A. 1981. *An Integrated System of Classification of Flowering Plants*. New York: Columbia University Press.

Cronquist, A. 1988. *The Evolution and Classification of Flowering Plants* (Ed. 2). Bronx: New York Botanical Garden.

Cruden, R.W. 1972. Pollinators in high-elevation ecosystems: Relative effectiveness of birds and bees. *Science* **176**: 1439–40.

Cruden, R.W. 1976a. Fecundity as a function of nectar production and pollen-ovule ratios. In Burley, J. & Styles, B.T. (eds), *Tropical Trees. Variation, Breeding and Conservation*, 171–8. New York: Academic Press.

Cruden, R.W. 1976b. Intraspecific variation in pollen-ovule ratios and nectar secretion. Preliminary evidence of ecotypic adaptation. *Ann. Missouri Bot. Gard.* **63**: 277–89.

Cruden, R.W. 1977. Pollen-ovule ratios: a conservative indicator of breeding systems in flowering plants. *Evolution* **31**: 32–46.

Cruden, R.W. 1988. Temporal dioecism: systematic breadth, associated traits, and temporal patterns. *Bot. Gaz. (Crawfordsville)* **149**: 1–15.

Cruden, R.W. & Hermann-Parker, S.M. 1977. Temporal dioecism: an alternative to dioecism? *Evolution* **31**: 863–6.

Cruden, R.W. & Hermann-Parker, S.M. 1979. Butterfly pollination of *Caesalpinia pulcherrima*, with observations on a psychophilous syndrome. *J. Ecol.* **67**: 155–68.

Cruden, R.W. & Jensen, K.G. 1979. Viscin threads, pollination efficiency and low pollen-ovule ratios. *Am. J. Bot.* **66**: 875–9.

Cruden, R.W. & Lyon, D.L. 1985. Correlations among stigma depth, style length, and pollen grain size: Do they reflect function or phylogeny? *Bot. Gaz. (Crawfordsville)* **146**: 143–9.

Cruden, R.W. & Lyon, D.L. 1989. Facultative xenogamy: Examination of a mixed mating system. In Bock, J.H. & Linhart, Y.B. (eds), *The Evolutionary Ecology of Plants*, 173–207. Boulder, Colorado: Westview Press.

Cruden, R.W. & Miller-Ward, S. 1981. Pollen-ovule ratio, pollen size, and the ratio of stigmatic area to the pollen-bearing area of the pollinator: an hypothesis. *Evolution* **35**: 964–74.

Cruden, R.W. & Toledo, V.M. 1977. Oriole pollination of *Erythrina breviflora* (Leguminosae): Evidence for a polytypic view of ornithophily. *Pl. Syst. Evol.* **126**: 393–403.

Cruden, R.W., Hermann, S.M. & Peterson, S. 1983. Patterns of nectar production and plant-pollinator coevolution. In Bentley, B. & Elias, T. (eds), *The Biology of Nectaries*, 80–125. New York: Columbia University Press.

Cruden, R.W., Kinsman, S., Stockhouse II, R.E. & Linhart, Y.B. 1976. Pollination, fecundity, and the distribution of moth-flowered plants. *Biotropica* **8**: 204–10.

Cruzan, M.B. 1986. Pollen tube distribution in *Nicotiana glauca*: evidence for density dependent growth. *Am. J. Bot.* **73**: 902–7.

Curry, K.J., McDowell, L.M., Judd, W.S. & Stern, W.L. 1991. Osmophores, floral features, and systematics of *Stanhopea* (Orchidaceae). *Am. J. Bot.* **78**: 610–23.

Cusick, F. 1966. On phylogenetic and ontogenetic fusions. In Cutter, E.G. (ed.), *Trends in Plant Morphogenesis*, 170–83. London: Longmans Green.

Dafni, A. 1984. Mimicry and deception in pollination. *Ann. Rev. Ecol. Syst.* **15**: 259–78.

Dafni, A. 1986. Floral mimicry – mutualism and unidirectional exploitation of insects by plants. In Juniper, B. & Southwood, R. (eds), *Insects and the Plant Surface*, 81–90. London: Arnold.

Dafni, A. 1992. *Pollination Ecology*. Oxford University Press.

Dafni, A. & Bernhardt, P. 1990. Pollination of terrestrial orchids of southern Australia and the mediterranean region. Systematic, ecological, and evolutionary implications. *Evol. Biol.* **24**: 193–252.

Dafni, A. & Calder, D.M. 1987. Pollination by deceit and floral mimesis in *Thelymitra antennifera* (Orchidaceae). *Pl. Syst. Evol.* **158**: 11–22.

Dafni, A., Bernhardt, P., Shmida, A., Ivri, Y., Greenbaum, S., O'Toole, C. & Losito, L. 1990. Red bowl-shaped flowers: convergence for beetle pollination in the mediterranean region. *Israel J. Bot.* **39**: 81–92.

Dahlgren, R. & Rao, V.S. 1971. The genus *Oftia* and its systematic position. *Bot. Not.* **124**: 451–72.

Dahlgren, R. & Thorne,R.F. 1984. The order Myrtales: Circumscription, variation and relationships. *Ann. Missouri Bot. Gard.* **71**: 633–99.

Dahlgren, R., Clifford, H.T. & Yeo, P. 1985. *The Families of the Monocotyledons*. Berlin: Springer.

Dalmer, M. 1880. Ueber die Leitung der Pollenschläuche bei den Angiospermen. *Jenaische Z. Naturwiss.*, n.F., **7**: 530–66.

Daniel, T.F. 1983. *Carlowrightia* (Acanthaceae). *Flora Neotropica Monogr.* **34**: 1–116.

Daniel, T.F. 1986. Systematics of *Tetramerium* (Acanthaceae). *Syst. Bot. Monogr.* **12**: 1–134.

Daniel, T.F. 1990. Systematics of *Henrya* (Acanthaceae). *Contr. Univ. Mich. Herb.* **17**: 99–131.

Dannenbaum, C. & Schill, R. 1991. Die Entwicklung der Pollentetraden und Pollinien bei den Asclepiadaceae. *Bibl. Bot.* **141**: 1–138.

Dannenbaum, C., Wolter, M. & Schill, R. 1989. Stigma morphology of the orchids. *Bot. Jahrb. Syst.* **110**: 441–60.

D'Arcy, W.G., D'Arcy, N.S. & Keating, R.C. 1990. Scented anthers in the Solanaceae. *Rhodora* **92**: 50–3.

Darwin, C. 1858. On the agency of bees in the fertilisation of the papilionaceous flowers, and on the crossing of kidney beans. *Ann. Mag. Nat. Hist.*, Ser. 3, **2**: 459–65.

Darwin, C. 1859. *On the Origin of Species by Means of Natural Selection.* London: Murray.

Darwin, C. 1862. *On the Various Contrivances by which British and Foreign Orchids Are Fertilised by Insects.* London: Murray.

Darwin, C. 1868. *The Variation of Animals and Plants under Domestication.* London: Murray.

Darwin, C. 1877. *The Different Forms of Flowers on Plants of the Same Species.* London: Murray.

Daumann, E. 1930. Das Blütennektarium von *Nepenthes. Beitr. Bot. Centralbl.*, Abt. I, **47**: 1–14.

Daumann, E. 1931. Nektarabscheidung in der Blütenregion einiger Araceen. *Planta* **12**: 38–52.

Daumann, E. 1959. Zur Kenntnis der Blütennektarien von *Aristolochia. Preslia* **31**: 359–72.

Daumann, E. 1970. Das Blütennektarium der Monocotyledonen unter besonderer Berücksichtigung seiner systematischen und phylogenetischen Bedeutung. *Fedde Rep.* **80**: 463–590.

Daumer, K. 1958. Blumenfarben wie sie die Bienen sehen. *Z. Vergl. Physiol.* **41**: 49–110.

Davis, A.R. & Gunning, B.E.S. 1992. The modified stomata of the floral nectary of *Vicia faba* L. 1. Development, anatomy and ultrastructure. *Protoplasma* **166**: 134–52.

Davis, M.A. 1987. The role of flower visitors in the explosive pollination of *Thalia geniculata* (Marantaceae), a Costa Rican marsh plant. *Bull. Torrey Bot. Club* **114**: 134–8.

Dayanandan, S., Attygalla, D.N.C., Abeygunasekera, A.W.W.L., Gunatilleke, I.A.U.N. & Gunatilleke, C.V.S. 1990. Phenology and floral morphology in relation to pollination of some Sri Lankan dipterocarps. In Bawa, K.S. & Hadley, M. (eds), *Reproductive Ecology of Tropical Forest Plants*, 103–33. Paris: UNESCO.

Delpino, F. 1867. Sugli apparecchi della fecondazione nelle piante antocarpee. Firenze: Cellini.

Delpino, F. 1868/1869a/1869b/1870/1873/1874. Ulteriori osservazioni e considerazioni sulla dicogamia nel regno vegetale I,1/ I,2/ I,3/ II,1/ II,2/ II,3. *Atti Soc. Ital. Sci. Nat. Milano* **11**: 265–332/ **12**: 21–141/ **12**: 179–233/ **13**: 167–205/ **16**: 151–319/ **17**: 266–407.

Delpino, F. 1890. Significazione biologica dei nettarostegii florali. *Malpighia* **4**: 21–23.

Demeter, K. 1922. Vergleichende Asclepiadeenstudien. *Flora* **15**: 130–76.

Deroin, T. 1985. Contribution à la morphologie comparée du gynécée des Monodoroidées (Annonacées). *Bull. Mus. Natl Hist. Nat.*, Sér. 4, B, *Adansonia*, **7**: 167–76.

Deroin, T. 1988a. Aspects anatomiques et biologiques de la fleur des Annonacées. Ph.D. Thesis, Université de Paris-Sud, Centre d'Orsay.

Deroin, T. 1988b. Biologie florale d'une Annonacée introduite en Côte d'Ivoire: *Cananga odorata* (Lam.) Hook.f. & Thoms. *Bull. Mus. Natl Hist. Nat., Paris*, Sér. 4, Sect. B, *Adansonia*, **10**: 377–93.

Deroin, T. 1989. Evolution des modalités de la pollinisation au cours du développement des axes aériens chez une Annonacée savanicole soumise aux feux annuels: *Annona senegalensis. C.R. Acad. Sci. Paris* **308**, Sér. III, 307–11.

Deroin, T. 1991a. La vascularisation florale des Magnoliales: première approche expérimentale de son rôle au cours de la pollinisation. *C.R. Acad. Sci. Paris* **312**, Sér. III, 355–60.

Deroin, T. 1991b. La répartition des modèles de plateaux stigmatiques et l'évolution des Annonacées. *C.R. Acad. Sci. Paris* **312**, Sér. III, 561–6.

Deroin, T. & Le Thomas, A. 1989. Sur la systématique et les potentialités évolutives des Annonacées: Cas d'*Ambavia gerrardii* (Baill.) Le Thomas, espèce endémique de Madagascar. *C.R. Acad. Sci. Paris* **309**, Sér. III, 647–52.

Derstine, K.S. & Tucker, S.C. 1991. Organ initiation and development of inflorescences and flowers of *Acacia baileyana*. *Am. J. Bot.* **78**: 816–32.

DeVries, P.J. 1979. Pollen-feeding rain forest *Parides* and *Battus* butterflies in Costa Rica. *Biotropica* **11**: 237–8.

DeVries, P.J. 1983a/1983b. *Papilio cresphontes* (lechera, papilio grande, giant swallowtail)/ *Phoebis philea* (sulfurea quemada, organe-barred sulfur). In Janzen, D.H. (ed.), *Costa Rican Natural History*, 751–2/ 755–6. Chicago: University of Chicago Press.

DeVries, P.J. 1987. *The Butterflies of Costa Rica and Their Natural History.* Princeton: Princeton University Press.

Dickison, W.C. 1981. The evolutionary relationships of the Leguminosae. In Polhill, R.M. & Raven, P.H. (eds), *Advances in Legume Systematics* Part 1: 35–54. Kew: Royal Botanic Gardens.

Diels, L. 1916. Käferblumen bei den Ranales und ihre Bedeutung für die Phylogenie der Angiospermen. *Ber. Deutsch. Bot. Ges.* **34**: 758–74.

Diggle, P.K. 1991a. Labile sex expression in andromonoecious *Solanum hirtum*: Floral development and sex determination. *Am. J. Bot.* **78**: 377–93.

Diggle, P.K. 1991b. Labile sex expression in the andromonoecious *Solanum hirtum*: Pattern of variation in floral structure. *Can. J. Bot.* **69**: 2033–43.

Dilcher, D.L. 1979. Early angiosperm reproduction: An introductory report. *Rev. Palaeobot. Palynol.* **27**: 291–328.

Dobat, K. & Peikert-Holle, T. 1985. *Blüten und Fledermäuse. Bestäubung durch Fledermäuse und Flughunde (Chiropterophilie).* Frankfurt: Kramer.

Dobkin, D.S. 1984. Flowering patterns of long-lived *Heliconia* inflorescences – Implications for visiting and resident nectarivores. *Oecologia* **64**: 245–54.

Dobkin, D.S. 1987. Synchronous flower abscission in plants pollinated by hermit hummingbirds and the evolution of one-day flowers. *Biotropica* **19**: 90–3.

Dobson, H.E.M. 1988. Survey of pollen and pollenkitt lipids – chemical cues to flower visitors? *Am. J. Bot.* **75**: 170–82.

Dobson, H.E.M. 1989. Pollenkitt in plant reproduction. In Bock, J.H. & Linhart, Y.B. (eds), *The Evolutionary Ecology of Plants*, 227–246. Boulder, Colorado: Westview Press.

Docters van Leeuwen, W.M. 1938. Observations about the biology of tropical flowers. *Ann. Jard. Bot. Buitenzorg* **48**: 27–68.

Dodson, C.H. 1962a. The importance of pollination in the evolution of the orchids of tropical America. *Am. Orch. Soc. Bull.* **31**: 525–34, 641–9, 731–5.

Dodson, C.H. 1962b. Pollination and variation in the subtribe Catasetinae (Orchidaceae). *Ann. Missouri Bot. Gard.* **49**: 35–56.

Dodson, C.H. 1965. Studies in orchid pollination: The genus *Coryanthes*. *Am. Orch. Soc. Bull.* **34**: 680–7.

Dodson, C.H. 1966. Studies in orchid pollination – *Cypripedium*, *Phragmipedium* and allied genera. *Am. Orch. Soc. Bull.* **35**: 125–8.

Dodson, C.H. & Frymire, G.P. 1961. Preliminary studies in the genus *Stanhopea* (Orchidaceae). *Ann. Missouri Bot. Gard.* **48**: 137–72.

Dodson, C.H., Dressler, R.L., Hills, H.G., Adams, R.M. & Williams, N.H. 1969. Biologically active compounds in orchid fragrances. *Science* **164**: 1243–9.

Donoghue, M.J. & Doyle, J.A. 1989. Phylogenetic analysis of angiosperms and the relationships of Hamamelidae. In Crane, P.R. & Blackmore, S. (eds), *Evolution, Systematics, and Fossil History of the Hamamelidae*, Vol. 1, 17–45. Oxford: Clarendon Press.

Donoghue, M.J. & Sanderson, M.J. 1992. The suitability of molecular and morphological evidence in resconstructing plant phylogeny. In Soltis, P.S., Soltis, D.E. & Doyle, J.J. (eds), *Molecular Systematics of Plants*, 340–68. New York: Chapman & Hall.

Dowding, P. 1987. Wind pollination mechanisms and aerobiology. *Int. Rev. Cytol.* **107**: 421–38.

Downie, S.R. & Palmer, J.D. 1992. Restriction site mapping of the chloroplast DNA inverted repeat: A molecular phylogeny of the Asteridae. *Ann. Missouri Bot. Gard.* **79**: 266–83.

Doyle, J.A. 1969. Cretaceous angiosperm pollen of the Atlantic coastal plain and its evolutionary significance. *J. Arn. Arb.* **50**: 1–35.

Doyle, J.A. 1978. Origin of angiosperms. *Ann. Rev. Ecol. Syst.* **9**: 365–92.

Doyle, J.A. & Donoghue, M.J. 1986. Seed plant phylogeny and the origin of angiosperms: An experimental cladistic approach. *Bot. Rev.* **52**: 321–431.

Doyle, J.A. & Hickey, L.J. 1976. Pollen and leaves from the mid-Cretaceous Potomac Group and their bearing on early angiosperm evolution. In Beck, C.B. (ed.), *Origin and Early Evolution of Angiosperms*, 139–206. New York: Columbia University Press.

Doyle, J.A. & Hotton, C.L. 1991. Diversification of early angiosperm pollen in a cladistic context. In Blackmore, S. & Barnes, S.H. (eds), *Pollen and Spores*, 169–95. Oxford: Clarendon Press.

Dressler, R.L. 1968. Pollination by euglossine bees. *Evolution* **22**: 202–10.

Dressler, R.L. 1981. *The Orchids. Natural History and Classification*. Cambridge, Massachusetts: Harvard University Press.

Dressler, R. 1982. Biology of the orchid bees (Euglossini). *Ann. Rev. Ecol. Syst.* **13**: 373–94.

Dressler, R. 1983. Classification of the Orchidaceae and their probable origin. *Telopea* **2**: 413–24.

Dressler, R.L. 1986a. Recent advances in orchid phylogeny. *Lindleyana* **1**: 5–20.

Dressler, R.L. 1986b. Features of pollinaria and orchid classification. *Lindleyana* **1**: 125–30.

Dressler, R.L. 1989a. Rostellum and viscidium: divergent definitions. *Lindleyana* **4**: 48–9.

Dressler, R. 1989b. The vandoid orchids: a polyphyletic grade? *Lindleyana* **4**: 89–93.

Dressler, R. 1990a. The Spiranthoideae: Grade or subfamily? *Lindleyana* **5**: 110–16.

Dressler, R. 1990b. The major clades of the Orchidaceae-Epidendroideae. *Lindleyana* **5**: 117–25.

Dressler, R.L. 1993. *Phylogeny and Classification of the Orchid Family*. Cambridge University Press.

Drinnan, A.N. & Ladiges, P.Y. 1989. Operculum development in the Eudesmieae B eucalypts and *Eucalyptus caesia* (Myrtaceae). *Pl. Syst. Evol.* **165**: 227–37.

Dronamraju, K.R. 1960. Selective visits of butterflies to flowers: a possible factor in sympatric speciation. *Nature* **186**: 178.

Ducker, S.C. & Knox, R.B. 1976. Submarine pollination in seagrasses. *Nature* **263**: 705–6.

Dukas, R. 1987. Foraging behavior of three bee species in a natural mimicry system: Female flowers which mimic male flowers in *Ecballium elaterium* (Cucurbitaceae). *Oecologia* **74**: 256–63.

Dulberger, R. 1981. The floral biology of *Cassia didymobotrya* and *C. auriculata* (Caesalpiniaceae). *Am. J. Bot.* **68**: 1350–60.

Dulberger, R. 1992. Floral polymorphisms and their functional significance in the heterostylous syndrome. In Barrett, S.C.H. (ed.), *Evolution and Function of Heterostyly*, 41–84. Berlin: Springer.

Dumas, C. & Russell, S.D. 1992. Plant reproductive biology: Trends. *Int. Rev. Cytol.* **140**: 565–92.

Dumas, C., Knox, R.B. & Gaude, T. 1984. Pollen-pistil recognition: New concepts from electron microscopy and cytochemistry. *Int. Rev. Cytol.*, Suppl., **90**: 239–72.

Dumas, C., Rougher, M., Zandonella, P., Ciampolini, F., Cresti, M. & Pacini, E. 1978. The secretory stigma in *Lycopersicum peruvianum* Mill.: Ontogenesis and glandular activity. *Protoplasma* **96**: 173–87.

Dumas, C., Bowman, R.B., Gaude, T., Guilly, C.M., Heizmann, P., Roeckel, P. & Rougier, M. 1988. Stigma and stigmatic secretion reexamined. *Phyton* **28**: 193–200.

Durkee, L.T. 1982. The floral and extra-floral nectaries of *Passiflora*. II. The extrafloral nectary. *Am. J. Bot.* **69**: 1420–8.

Durkee, L.T., Gaal, D.J. & Reisner, W.H. 1981. The floral and extra-floral nectaries of *Passiflora*. I. The floral nectary. *Am. J. Bot.* **68**: 453–62.

Dyer, R.A. 1983. *Ceropegia, Brachystelma and Riocreuxia in Southern Africa*. Rotterdam: Balkema.

Eames, A.J. 1961. *Morphology of the Angiosperms*. New York: McGraw-Hill.

East, E.M. 1940. The distribution of self-sterility in the flowering plants. *Proc. Amer. Philos. Soc.* **82**: 449–518.

Ebel, F. 1974. Beobachtungen über das Bewegungsverhalten des Pollinariums von *Catasetum fimbriatum* Lindl. während Abschuss, Flug und Landung. *Flora* **163**: 342–56.

Eckardt, T. 1937. Untersuchungen über Morphologie, Entwicklungsgeschichte und systematische Bedeutung des pseudomonomeren Gynoeceums. *Nova Acta Leopold.*, n.F., **5**: 1–112.

Eckardt, T. 1964. Das Homologieproblem und Fälle strittiger Homologien. *Phytomorphology* **14**: 79–92.

Eckert, G. 1966. Entwicklungsgeschichtliche und blütenanatomische Untersuchungen zum Problem der Obdiplostemonie. *Bot. Jahrb. Syst.* **85**: 523–604.

Edwards, J. & Jordan, J.R. 1992. Reversible anther opening in *Lilium philadelphicum* (Liliaceae): A possible means of enhancing male fitness. *Am. J. Bot.* **79**: 144–8.

Ehler, N. 1975. Beitrag zur Kenntnis der Mikromorphologie der Coroll-Epidermen von Stapelieen und ihre taxonomische Verwendbarkeit. *Trop. Subtrop. Pflanzenwelt* **14**: 83–139. Wiesbaden: Steiner.

Ehler, N. 1976. Struktur und Funktion der Oberflächen von Orchideenblüten. In Senghas, K. (ed.), *Tagungsbericht der 8. Welt-Orchideen-Konferenz, Palmengarten Frankfurt, 1975*: 456–62. Frankfurt/M.: Deutsche Orchideen-Gesellschaft.

Ehrendorfer, F. 1977. New ideas about the early differentiation of angiosperms. *Pl. Syst. Evol., Suppl.* **1**: 227–34.

Ehrendorfer, F. 1989. The phylogenetic position of the Hamamelidae. In Crane, P.R. & Blackmore, S. (eds), *Evolution, Systematics, and Fossil History of the Hamamelidae*, Vol. 1, 1–7. Oxford: Clarendon Press.

Ehrlich, P.R. & Raven, P.H. 1964. Butterflies and plants: A study in coevolution. *Evolution* **18**: 586–608.

Ehrlich, P.R. & Wilson, E.O. 1991. Biodiversity studies: Science and policy. *Science* **253**: 758–62.

Eichler, A.W. 1875/1878. *Blüthendiagramme* I/II. Leipzig: Engelmann.

Eisikowitch, D. 1986. Morpho-ecological aspects on the pollination of *Calotropis procera* (Asclepiadaceae) in Israel. *Pl. Syst. Evol.* **152**: 185–94.

Eisikowitch, D. & Rotem, R. 1987. Flower orientation and color change in *Quisqualis indica* and their possible role in pollinator partitioning. *Bot. Gaz. (Crawfordsville)* **148**: 175–9.

Eisner, T., Eisner, M. & Aneshansley, D. 1973. Ultraviolet patterns on rear of flowers: Basis of disparity of buds and blossoms. *Proc. Natl Acad. Sci. USA* **70**: 1002–4.

Elias, T.S. 1981. Tribe 1. Parkieae (Wight & Arn.) Benth. (1842). In Polhill, R.M. & Raven, P.H. (eds), *Advances in Legume Systematics*, Part 1, 153. Kew: Royal Botanic Gardens.

Elias, T.S. 1983. Extrafloral nectaries. Their structure and distribution. In Bentley, B. & Elias, T. (eds), *The Biology of Nectaries*, 174–203. New York: Columbia University Press.

Elisens, W.J. 1985. Monograph of the Maurandyinae (Scrophulariaceae-Antirrhineae). *Syst. Bot. Monogr.* **5**: 1–97.

Endress, P.K. 1971. Bau der weiblichen Blüten von *Hedyosmum mexicanum* Cordemoy. *Bot. Jahrb. Syst.* **91**: 37–60.

Endress, P.K. 1972a. Zur vergleichenden Entwicklungsmorphologie, Embryologie und Systematik bei Laurales. *Bot. Jahrb. Syst.* **92**: 331–428.

Endress, P.K. 1972b. Aspekte der Karpellontogenese. *Verh. Schweiz. Naturf. Ges.* **152**: 126–30.

Endress, P.K. 1975. Nachbarliche Formbeziehungen mit Hüllfunktion im Infloreszenz- und Blütenbereich. *Bot. Jahrb. Syst.* **96**: 1–44.

Endress, P.K. 1977a. Blütenmorphologie – Rückblick und aktuelle Probleme. *Ber. Deutsch. Bot. Ges.* **90**: 1–13.

Endress, P.K. 1977b. Ueber Blütenbau und Verwandtschaft der Eupomatiaceae und Himantandraceae. *Ber. Deutsch. Bot. Ges.* **90**: 83–103.

Endress, P.K. 1977c. Evolutionary trends in the Hamamelidales-Fagales-group. *Pl. Syst. Evol.*, Suppl. 1, 321–47.

Endress, P.K. 1978. Blütenontogenese, Blütenabgrenzung und systematische Stellung der perianthlosen Hamamelidoideae. *Bot. Jahrb. Syst.* **100**: 249–317.

Endress, P.K. 1980a. Ontogeny, function and evolution of extreme floral construction in Monimiaceae. *Pl. Syst. Evol.* **134**: 79–120.

Endress, P.K. 1980b. Floral structure and relationships of *Hortonia* (Monimiaceae). *Pl. Syst. Evol.* **133**: 199–221.

Endress, P.K. 1980c. The reproductive structures and systematic position of the Austrobaileyaceae. *Bot. Jahrb. Syst.* **101**: 393–433.

Endress, P.K. 1982. Syncarpy and alternative modes of escaping disadvantages of apocarpy in primitive angiosperms. *Taxon* **31**: 48–52.

Endress, P.K. 1983. The early floral development of *Austrobaileya*. *Bot. Jahrb. Syst.* **103**: 481–97.

Endress, P.K. 1984a. The flowering process in the Eupomatiaceae (Magnoliales). *Bot. Jahrb. Syst.* **104**: 297–319.

Endress, P.K. 1984b. The role of inner staminodes in the floral display of some relic Magnoliales. *Pl. Syst. Evol.* **146**: 269–82.

Endress, P.K. 1985. Stamenabszission und Pollenpräsentation bei Annonaceae. *Flora* **176**: 95–8.

Endress, P.K. 1986a. Reproductive structures and phylogenetic significance of extant primitive angiosperms. *Pl. Syst. Evol.* **152**: 1–28.

Endress, P.K. 1986b. Floral structure, systematics and phylogeny in Trochodendrales. *Ann. Missouri Bot. Gard.* **73**: 297–324.

Endress, P.K. 1987a. Floral phyllotaxis and floral evolution. *Bot. Jahrb. Syst.* **108**: 417–38.

Endress, P.K. 1987b. The Chloranthaceae: reproductive structures and phylogenetic position. *Bot. Jahrb. Syst.* **109**: 153–226.

Endress, P.K. 1987c. The early evolution of the angiosperm flower. *Trends Ecol. Evol.* **2**: 300–4.

Endress, P.K. 1989. Phylogenetic relationships in the Hamamelidoideae. In Crane,P.R. & Blackmore, S. (eds), *Evolution, Systematics, and Fossil History of the Hamamelidae, 1: Introduction and 'Lower' Hamamelidae*, 227–48. Oxford: Clarendon Press.

Endress, P.K. 1990a. Patterns of floral construction in ontogeny and phylogeny. *Biol. J. Linn. Soc.* **39**: 153–75.

Endress, P.K. 1990b. Evolution of reproductive structures and functions in primitive angiosperms (Magnoliidae). *Mem. New York Bot. Gard.* **55**: 5–34.

Endress, P.K. 1992a. Protogynous flowers in Monimiaceae. *Pl. Syst. Evol.* **181**: 227–32.

Endress, P.K. 1992b. Evolution and floral diversity: The phylogenetic surroundings of *Arabidopsis* and *Antirrhinum*. *Int. J. Pl. Sci.* **153**: s106–22.

Endress, P.K. & Hufford, L.D. 1989. The diversity of stamen structures and dehiscence patterns among Magnoliidae. *Bot. J. Linn. Soc.* **100**: 45–85.

Endress, P.K. & Lorence, D.H. 1983. Diversity and evolutionary trends in the floral structure of *Tambourissa* (Monimiaceae). *Pl. Syst. Evol.* **143**: 53–81.

Endress, P.K. & Sampson, F.B. 1983. Floral structure and relationships of the Trimeniaceae (Laurales). *J. Arn. Arb.* **64**: 447–73.

Endress, P.K. & Stumpf, S. 1990. Non-tetrasporangiate stamens in the angiosperms: structure, systematic distribution and evolutionary aspects. *Bot. Jahrb. Syst.* **112**: 193–240.

Endress, P.K. & Stumpf, S. 1991. The diversity of stamen structures in 'Lower' Rosidae (Rosales, Fabales, Proteales, Sapindales). *Bot. J. Linn. Soc.* **107**: 217–93.

Endress, P.K., Jenny, M. & Fallen, M.E. 1983. Convergent elaboration of apocarpous gynoecia in higher advanced dicotyledons (Sapindales, Malvales, Gentianales). *Nord. J. Bot.* **3**: 293–300.

Engler, A. 1931. Reihe Geraniales. In Engler, A. & Prantl, K. (eds), *Die natürlichen Pflanzenfamilien*, Ed. 2, 19a: 4–10. Leipzig: Engelmann.

Erbar, C. 1983. Zum Karpellbau einiger Magnoliiden. *Bot. Jahrb. Syst.* **104**: 3–31.

Erbar, C. 1986. Untersuchungen zur Entwicklung der spiraligen Blüte von *Stewartia pseudocamellia* (Theaceae). *Bot. Jahrb. Syst.* **106**: 391–407.

Erbar, C. 1991. Sympetaly – A systematic character? *Bot. Jahrb. Syst.* **112**: 417–51.

Erbar, C. 1992. Floral development of two species of *Stylidium* (Stylidiaceae) and some remarks on the systematic position of the family Stylidiaceae. *Can. J. Bot.* **70**: 258–71.

Erbar, C. & Leins, P. 1981. Zur Spirale in Magnolien-Blüten. *Beitr. Biol. Pfl.* **56**: 225–41.

Erbar, C. & Leins, P. 1983. Zur Sequenz von Blütenorganen bei einigen Magnoliiden. *Bot. Jahrb. Syst.* **103**: 433–49.

Erickson, E.H. 1975. Surface electric potentials on worker honeybees leaving and entering the hive. *J. Apicult. Res.* **14**: 141–7.

Erickson, E.H. & Buchmann, S.L. 1983. Electrostatics and pollination. In Jones, C.E. & Little, R.J. (eds), *Handbook of Experimental Pollination Biology*, 173–84. New York: Scientific and Academic Editions.

Ernst, A. & Schmid, E. 1913. Ueber Blüte und Frucht von *Rafflesia*. *Ann. Jard. Bot. Buitenzorg*, Sér. 2, **12**: 1–58.

Ernst-Schwarzenbach, M. 1944. Zur Blütenbiologie einiger Hydrocharitaceen. *Ber. Schweiz. Bot. Ges.* **54**: 33–69.

Esch, H. 1967. Evolution of bee language. *Sci. Am.* **216 (4)**: 96–105.

Eyde, R.H. & Morgan, J.T. 1973. Floral structure and evolution in Lopezieae (Onagraceae). *Am. J. Bot.* **60**: 771–87.

Eyde, R.H. & Tseng, C.C. 1969. Flower of *Tetraplasandra gymnocarpa*. Hypogyny with epigynous ancestry. *Science* **166**: 506–8.

Exner, F. & Exner, S. 1910. Die physikalischen Grundlagen der Blütenfärbungen. *Sitzber. Akad. Wiss. Wien*, Math.- Naturwiss. Kl., Abt. I, **119**: 191–245.

Faden, R.B. 1992. Floral attraction and floral hairs in the Commelinaceae. *Ann. Missouri Bot. Gard.* **79**: 46–52.

Faegri, K. 1986. The solanoid flower. *Trans. Bot. Soc. Edinburgh*, 150th Anniv. Suppl., 51–9.

Faegri, K. & van der Pijl, L. 1979. *The Principles of Pollination Ecology* (Ed. 3). Oxford: Pergamon.

Fagerlind, F. 1948. Beiträge zur Kenntnis der Gynäceummorphologie und Phylogenie der Santalales-Familien. *Svensk Bot. Tidskr.* **42**: 195–229.

Fahn, A. 1949. Studies in the ecology of nectar secretion. *Palest. J. Bot., Ser.* **4**, 207–24.

Fahn, A. 1952. On the structure of floral nectaries. *Bot. Gaz.* **113**: 464–70.

Fahn, A. 1954. The topography of the nectary in the flower and its phylogenetical trend. *Phytomorphology* **3**: 424–6.

Fahn, A. 1979. Ultrastructure of nectaries in relation to nectar secretion. *Am. J. Bot.* **66**: 977–85.

Fahn, A. 1988. Secretory tissues in vascular plants. *New Phytol.* **108**: 229–57.

Fahn, A. 1990. *Plant Anatomy* (Ed. 4). Oxford: Pergamon Press.

Fahn, A. & Benouaiche, P. 1979. Ultrastructure, development and secretion in the nectary of banana flowers. *Ann. Bot.* **44**: 85–93.

Fallen, M.E. 1985. The gynoecial development and systematic position of *Allamanda* (Apocynaceae). *Am. J. Bot.* **72**: 575–9.

Fallen, M.E. 1986. Floral structure in the Apocynaceae: morphological, functional, and evolutionary aspects. *Bot. Jahrb. Syst.* **106**: 245–86.

Feehan, J. 1985. Explosive flower opening in ornithophily: A study of pollination mechanisms in some central African Loranthaceae. *Bot. J. Linn. Soc.* **90**: 129–44.

Feil, J.P. 1992. Reproductive ecology of dioecious *Siparuna* (Monimiaceae) in Ecuador – A case of gall midge pollination. *Bot. J. Linn. Soc.* **110**: 171–203.

Feinsinger, P. 1976. Organization of a tropical guild of nectarivorous birds. *Ecol. Monogr.* **46**: 257–91.

Feinsinger, P. 1983a. Coevolution and pollination. In Futuyma, D.J. & Slatkin, M. (eds), *Coevolution*, 282–310. Sunderland, Massachusetts: Sinauer.

Feinsinger, P. 1983b. Variable nectar secretion in a *Heliconia* species pollinated by hermit hummingbirds. *Biotropica* **15**: 48–52.

Feinsinger, P. & Colwell, R.K. 1978. Community organization among Neotropical nectar-feeding birds. *Am. Zool.* **18**: 779–95.

Feinsinger, P. & Tiebout III, H.M. 1991. Competition among plants sharing hummingbird pollinators: Laboratory experiments on a mechanism. *Ecology* **72**: 1946–52.

Feinsinger, P., Tiebout III, H.M. & Young, B.E. 1991. Do tropical bird-pollinated plants exhibit density-dependent interaction? Field experiments. *Ecology* **72**: 1953–63.

Ferguson, I.K. & Tucker, S.C. (eds) 1994. *Advances in Legume Systematics 6. Structural Botany*. Kew: Royal Botanic Gardens.

Findlay, N. 1982. Secretion of nectar. In Loewus, F.A. & Tanner, W. (eds), *Plant Carbohydrates I. Intercellular Carbohydrates*, 677–83. Berlin: Springer.

Fischer, E. 1989. Contributions to the flora of central Africa II. *Crepidorhopalon*, a new genus within the relationship of *Craterostigma*, *Torenia* and *Lindernia* (Scrophulariaceae) with two new or noteworthy species from Central and South Central Africa (Zaire, Zambia). *Fedde Rep.* **100**: 439–50.

Ford, H.A., Paton, D.C. & Forde, N. 1980. Birds as pollinators of Australian plants. *New Zeal. J. Bot.* **17**: 509–19.

Forster, P.I. 1992. Pollination of *Hoya australis* (Asclepiadaceae) by *Ocybadistes walkeri* Sothis (Lepidoptera: Hesperiidae). *Austral. Entomol. Mag.* **19**: 39–43.

Frankel, R. & Galun, E. 1977. *Pollination Mechanisms, Reproduction and Plant Breeding*. Berlin: Springer.

Frankie, G.W. 1975. Tropical forest phenology and pollinator plant coevolution. In Gilbert, L.E. & Raven, P.H. (eds), *Coevolution of Animals and Plants*, 192–209. Austin: University of Texas Press.

Frankie, G.W. 1976. Pollination of widely dispersed trees by animals in Central America, with an emphasis on bee pollination systems. In Burley, J. & Styles, B.T. (eds), *Tropical Trees: Variation, Breeding and Conservation*, 151–9. New York: Academic Press.

Frankie, G.W. & Vinson, S.B. 1977. Scent marking of passion flowers in Texas by females of *Xylocopa viginica texana* (Hymenoptera: Anthophoridae). *J. Kansas Entomol. Soc.* **50**: 613–25.

Frankie, G.W., Baker, H.G. & Opler, P.A. 1974. Tropical plant phenology: Applications for studies in community ecology. In Lieth, H. (ed.), *Phenology and Seasonality Modelling*, 287–96. New York: Springer.

Frankie, G.W., Vinson, S.B. & Williams, H. 1989. Ecological and evolutionary sorting of 12 sympatric species of *Centris* bees in Costa Rican dry forest. In Bock, J.H. & Linhart, Y.B. (eds), *The Evolutionary Ecology of Plants*, 535–49. Boulder, Colorado: Westview Press.

Frankie, G.W., Haber, W.A., Opler, P.A. & Bawa, K.S. 1983. Characteristics and organization of the large bee pollination system in the Costa Rican dry forest. In Jones, C.E. & Little, R.J. (eds), *Handbook of Experimental Pollination Biology*, 411–47. New York: Scientific and Academic Editions.

Frankie, G.W., Vinson, S.B., Newstrom, L.E., Barthell, J.F., Haber, W.A. & Frankie, J.K. 1990. Plant phenology, pollination ecology, pollinator behaviour and conservation of pollinators in Neotropical dry forest. In Bawa, K.S. & Hadley, M. (eds), *Repro-*

ductive Ecology of Tropical Forest Plants, 37–47. Paris: UNESCO.

Free, J.B. 1970. *Insect Pollination of Crops*. London: Academic Press.

Frei, E. 1955. Die Innervierung der floralen Nektarien dikotyler Pflanzenfamilien. *Ber. Schweiz. Bot. Ges.* **65**: 60–114.

French, J.C. 1985. Patterns of endothecial wall thickenings in Araceae: subfamilies Pothoideae and Monsteroideae. *Am. J. Bot.* **72**: 472–86.

Frey-Wyssling, A. 1933. Ueber die physiologische Bedeutung der extrafloralen Nektarien von *Hevea brasiliensis* Müll. *Ber. Schweiz. Bot. Ges.* **42**: 109–22.

Friedman, W.E. 1990a. Double fertilization in *Ephedra*, a nonflowering seed plant: Its bearing on the origin of angiosperms. *Science* **247**: 951–4.

Friedman, W.E. 1990b. Sexual reproduction in *Ephedra nevadensis* (Ephedraceae): further evidence of double fertilization in a nonflowering seed plant. *Am. J. Bot.* **77**: 1582–98.

Friedman, W.E. 1992. Evidence of a pre-angiosperm origin of endosperm: Implications for the evolution of flowering plants. *Science* **225**: 336–9.

Friis, E.M. & Endress, P.K. 1990. Origin and evolution of angiosperm flowers. *Adv. Bot. Res.* **17**: 99–162.

Friis, E.M. & Skarby, A. 1981. Structurally preserved angiosperm flowers from the Upper Cretaceous of southern Sweden. *Nature* **291**: 485–6.

Friis, E.M., Crane, P.R. & Pedersen, K.R. 1991. Stamen diversity and in situ pollen of Cretaceous angiosperms. In Blackmore, S. & Barnes, S.H. (eds), *Pollen and Spores. Patterns of Diversification*, 197–224. Oxford: Clarendon Press.

von Frisch, K. 1919. Ueber den Geruchssinn der Bienen und seine blütenbiologische Bedeutung. *Zool. Jahrb. (Physiol.)* **37**: 1–238.

Froebe, H.A., Magin, N., Jöhlinger, H. & Netz, M. 1983. A re-evaluation of the inflorescence of *Dalechampia spathulata* (Scheidw.) Baillon (Euphorbiaceae). *Bot. Jahrb. Syst.* **104**: 249–60.

Frost, S.K. & Frost, P.G.H. 1981. Sunbird pollination of *Strelitzia nicolai*. *Oecologia* **49**: 379–84.

Fujita, M.S. & Tuttle, M.D. 1991. Flying foxes (Chiroptera: Pteropodidae): Threatened animals of key ecological and economic importance. *Conserv. Biol.* **5**: 455–63.

Galil, J. 1977. Fig biology. *Endeavour*, n.S., **1**: 52–6.

Galil, J. & Zeroni, M. 1965. Nectar system of *Asclepias curassavica*. *Bot. Gaz.* **126**: 144–8.

Galil, J. & Zeroni, M. 1969. On the organization of the pollinium in *Asclepias curassavica*. *Bot. Gaz.* **130**: 1–4.

Galle, P. 1977. Untersuchungen zur Blütenentwicklung der Polygonaceen. *Bot. Jahrb. Syst.* **98**: 449–89.

Gamerro, J.C. 1986. Dimorphismo y viabilidad del polen en *Tripogandra diuretica* (Commelinaceae). *Darwiniana* **28**: 143–52.

Ganders, F.R. 1980. The biology of heterostyly. *New Zeal. J. Bot.* **17**: 607–35.

Garber, P.A. 1988. Foraging decisions during nectar feeding by tamarin monkeys (*Saguinus mystax* and *Saguinus fuscicollis*, Callitrichidae, Primates) in Amazonian Peru. *Biotropica* **20**: 100–6.

Garcia, M.U. 1975. Floral biology of *Amherstia*. *Pterocarpus* **1**: 26–35.

Gasser, C.S. 1991. Molecular studies on the differentiation of floral organs. *Ann. Rev. Pl. Physiol. Pl. Mol. Biol.* **42**: 621–49.

Geitler, L. 1934. Blütenfärbung durch einen Membranfarbstoff bei *Leonotis*. *Oesterr. Bot. Z.* **83**: 284–7.

Gemmeke, V. 1982. Entwicklungsgeschichtliche Untersuchungen an Mimosaceen-Blüten. *Bot. Jahrb. Syst.* **103**: 185–210.

Gentry, A.H. 1974a. Coevolutionary patterns in Central American Bignoniaceae. *Ann. Missouri Bot. Gard.* **61**: 728–59.

Gentry, A.H. 1974b. Flowering phenology and diversity in tropical Bignoniaceae. *Biotropica* **6**: 64–8.

Gentry, A.H. 1976. Relationships of the Madagascar Bignoniaceae: A striking case of convergent evolution. *Pl. Syst. Evol.* **126**: 255–66.

Gentry, A.H. 1978. Anti-pollinators for mass-flowering plants? *Biotropica* **10**: 68–9.

Gentry, A.H. 1982. Patterns of neotropical plant species diversity. *Evol. Biol.* **15**: 1–84.

Gentry, A.H. 1988. Distribution and evolution of the Madgascar Bignoniaceae. *Monogr. Syst. Bot. Missouri Bot. Gard.* **25**: 175–85.

Gentry, A.H. 1990a. Evolutionary patterns in neotropical Bignoniaceae. *Mem. New York Bot. Gard.* **55**: 118–29.

Gentry, A.H. 1990b. *Sphingiphila* (Bignoniaceae), a new genus from the Paraguayan Chaco. *Syst. Bot.* **15**: 277–9.

Gerenday, A. & French, J.C. 1988. Endothecial thickenings in anthers of porate monocotyledons. *Am. J. Bot.* **75**: 22–5.

Gerlach, G. & Schill, R. 1989. Fragrance analyses, an aid to taxonomic relationships of the genus *Coryanthes* (Orchidaceae). *Pl. Syst. Evol.* **168**: 159–65.

Gerlach, G. & Schill, R. 1991. Composition of orchid scents attracting euglossine bees. *Bot. Acta* **104**: 379–91.

Ghouse, A.K.M. & Hashmi, S. 1981. A note on the shoot growth activity of *Delonix regia*. *Indian J. Bot.* **4**: 99–101.

Gibbs, P.E. 1988. Self-incompatibility mechanisms in flowering plants: some complications and clarifications. *Lagascalia* **15** (Extra): 17–28.

Gifford, E.M., Jr. 1963. Developmental studies of vegetative and floral meristems. *Brookhaven Symp. Biol.* **16**: 126–37.

Gilbert, F.S., Haines, N. & Dickson, K. 1991. Empty flowers. *Funct. Ecol.* **5**: 29–39.

Gilbert, L.E. 1972. Pollen feeding and reproductive biology of *Heliconius* butterflies. *Proc. Natl Acad. Sci. USA* **69**: 1403–7.

Gilbert, L.E. 1980. Food web organization and the conservation of Neotropical diversity. In Soulé, M.E. & Wilcox, B.A. (eds), *Conservation Biology. An Evolutionary-Ecological Perspective*, 11–33. Sunderland, Massachusetts: Sinauer.

Gilbert, L.E. & Raven, P.H. (eds), 1975. *Coevolution of Animals and Plants*. Austin: University of Texas Press.

Gill, F.B. & Conway, C.A. 1979. Floral biology of *Leonotis nepetifolia* (L.) R.Br. (Labiatae). *Proc. Acad. Nat. Hist. Philadelphia* **131**: 244–56.

Gill, F.B., Mack, A.L. & Ray, R.T. 1982. Competition between hermit hummingbirds Phaethorninae and insects for nectar in a Costa Rican rain forest. *Ibis* **124**: 44–9.

Giurfa, M. & Nunez, J.A. 1992. Honeybees mark with scent and reject recently visited flowers. *Oecologia* **89**: 113–7.

Godley, E.J. & Smith, D.H. 1981. Breeding systems in New Zealand plants. 5. *Pseudowintera colorata* (Winteraceae). *New Zeal. J. Bot.* **19**: 151–6.

Goebel, K. 1933. *Organographie der Pflanzen 3* (Ed. 3). Jena: Fischer.

von Goethe, J.W. 1790. *Versuch die Metamorphose der Pflanzen zu erklären*. Gotha: Ettinger.

Goldingay, R.L., Carthew, S.M. & Whelan, R.J. 1991. The importance of non-flying mammals in pollination. *Oikos* **61**: 79–87.

Goldsmith, G.W. & Hafenrichter, A.L. 1932. *Anthokinetics: The Physiology and Ecology of Floral Movements*. Washington, D.C.: Carnegie Institution.

Gopal, K.G. & Puri, V. 1962. Morphology of the flower of some Gentianaceae with special reference to placentation. *Bot. Gaz.* **124**: 42–57.

Gordon-Gray, K.D., Cunningham,A.B. & Nichols, G.R. 1989. *Siphonochilus aethiopicus* (Zingiberaceae): observations on floral and reproductive biology. *South Afr. J. Bot.* **55**: 281–7.

Gori, D.F. 1983. Post-pollination phenomena and adaptive floral changes. In Jones, C.E. & Little, L.J. (eds), *Handbook of Experimental Pollination Biology*, 31–49. New York: Scientific and Academic Editions.

Goss, G.J. 1977. The reproductive biology of the epiphytic orchids of Florida. **6**: *Polystachya flavescens* (Lindley) J.J. Smith. *Am. Orch. Soc. Bull.* **46**: 990–4.

Gottsberger, G. 1970. Beiträge zur Biologie von Annonaceen-Blüten. *Oesterr. Bot. Z.* **118**: 237–79.

Gottsberger, G. 1971. Colour change of petals in *Malvaviscus arboreus* flowers. *Acta Bot. Neerl.* **20**: 381–8.

Gottsberger, G. 1974. The structure and function of the primitive angiosperm flower – a discussion. *Acta Bot. Neerl.* **23**: 461–71.

Gottsberger, G. 1977. Some aspects of beetle pollination in the evolution of flowering plants. *Pl. Syst. Evol.*, Suppl. **1**, 211–26.

Gottsberger, G. 1986. Some pollination strategies in neotropical savannas and forests. *Pl. Syst. Evol.* **152**: 29–45.

Gottsberger, G. 1988. The reproductive biology of primitive angiosperms. *Taxon* **37**: 630–43.

Gottsberger, G. 1989a. Beetle pollination and flowering rhythm of *Annona* spp. (Annonaceae) in Brazil. *Pl. Syst. Evol.* **167**: 165–87.

Gottsberger, G. 1989b. Comments on flower evolution and beetle pollination in the genera *Annona* and *Rollinia* (Annonaceae). *Pl. Syst. Evol.* **167**: 189–94.

Gottsberger, G. 1990. Flowers and beetles in the South American tropics. *Bot. Acta* **103**: 360–5.

Gottsberger, G. 1991. Pollination of some species of the Carludovicoideae, and remarks on the origin and evolution of the Cyclanthaceae. *Bot. Jahrb. Syst.* **113**: 221–35.

Gottsberger, G. & Amaral, A. 1984. Pollination strategies in Brazilian *Philodendron* species. *Ber. Deutsch. Bot. Ges.* **97**: 391–410.
Gottsberger, G. & Gottlieb, O.R. 1981. Blue flower pigmentation and evolutionary advancement. *Biochem. Syst. Ecol.* **9**: 13–8.
Gottsberger, G. & Silberbauer-Gottsberger, I. 1988. Evolution of flower structures and pollination in Neotropical Cassiinae (Caesalpiniaceae) species. *Phyton (Horn)* **28**: 293–320.
Gottsberger, G. & Silberbauer-Gottsberger, I. 1991. Olfactory and visual attraction of *Erioscelis emarginata* (Cyclocephalini, Dynastinae) to the inflorescences of *Philodendron selloum* (Araceae). *Biotropica* **23**: 23–8.
Gottsberger, G., Camargo, J.M.F. & Silberbauer-Gottsberger, I. 1988. A bee-pollinated tropical community: The beach dune vegetation of Ilha de São Luis, Maranhão, Brazil. *Bot. Jahrb. Syst.* **109**: 469–500.
Gottsberger, G., Silberbauer-Gottsberger, I. & Ehrendorfer, F. 1980. Reproductive biology in the primitive relic angiosperm *Drimys brasiliensis* (Winteraceae). *Pl. Syst. Evol.* **135**: 11–39.
Gould, E. 1978. Foraging behavior of Malaysian nectar-feeding bats. *Biotropica* **10**: 184–93.
Gould, K.S. & Lord, E.M. 1988. Growth of anthers in *Lilium longiflorum*. A kinematic analysis. *Planta* **173**: 161–71.
Gould, K.S. & Lord, E.M. 1989. A kinematic analysis of tepal growth in *Lilium longiflorum*. *Planta* **177**: 66–73.
Gould, S.J. & Lewontin, R.C. 1979. The spandrels of San Marco and the Panglossian paradigm: A critique of the adaptationist programme. *Proc. Roy. Soc. Lond. B* **205**: 581–98.
Gracie, C. 1991. Observation of dual function of nectaries in *Ruellia radicans* (Nees) Lindau (Acanthaceae). *Bull. Torrey Bot. Club* **118**: 188–190.
Graham, A., Barker, G. & Freitas da Silva, M. 1980. Unique pollen types in the Caesalpinioideae (Leguminosae). *Grana* **19**: 79–84.
Des Granges, J.L. 1979. Organization of a tropical nectar feeding bird guild in a variable environment. *Living Bird* **17**: 199–236.
Grant, K.A. 1966. A hypothesis concerning the prevalence of red coloration in California hummingbird flowers. *Am. Nat.* **100**: 85–98.
Grant, K. & Grant, V. 1968. *Hummingbirds and Their Flowers*. New York: Columbia University Press.
Grant, V. 1949. Pollination systems as isolating mechanisms in the angiosperms. *Evolution* **3**: 82–97.
Grant, V. 1950a. The protection of the ovules in flowering plants. *Evolution* **4**: 179–201.
Grant, V. 1950b. The flower constancy of bees. *Bot. Rev.* **16**: 379–98.
Grant, V. & Grant, K.A. 1965. *Flower Pollination in the Phlox Family*. New York: Columbia University Press.
Grant, V. & Grant, K.A. 1983a. Behavior of hawkmoths on flowers of *Datura meteloides*. *Bot. Gaz. (Crawfordsville)* **144**: 280–4.
Grant, V. & Grant, K.A. 1983b. Hawkmoth pollination of *Mirabilis longiflora* (Nyctaginaceae). *Proc. Natl Acad. Sci. USA, Biol. Sci.*, **80**: 1298–9.
Greathead, D.J. 1983. The multi-million dollar weevil that pollinates oil-palms. *Antenna* **7 (3)**: 105–7.
Gregg, K.B. 1983. Variation in floral fragrances and morphology: Incipient speciation in *Cycnoches*? *Bot. Gaz.* **144**: 566–76.
Grey-Wilson, C. 1980. *Impatiens of Africa. Morphology, Pollination and Pollinators, Ecology, Phytogeography, Hybridisation, Keys and Systematic Treatment of All African Species*. Rotterdam: Balkema.
Gribel, R. 1988. Visits of *Caluromys lanatus* (Didelphidae) to flowers of *Pseudobombax tomentosum* (Bombacaceae): A probable case of pollination by marsupials in central Brazil. *Biotropica* **20**: 344–7.
Grünmeier, R. 1990. Pollination by bats and non-flying mammals of the African tree *Parkia bicolor* (Mimosaceae). *Mem. New York Bot. Gard.* **55**: 83–104.
Guédès, M. 1966. Stamen, carpel and ovule. The teratological approach to their interpretation. *Adv. Front. Pl. Sci.* **14**: 43–108.
Guédès, M. 1968. La feuille végétative et le périanthe de quelques Aristoloches. *Flora B* **158**: 167–79.
Guédès, M. 1979. *Morphology of Seed-Plants*. Cramer: Vaduz.
Guédès, M. & Le Thomas, A. 1981. Le gynécée syncarpe de *Monodora* (Annonacées-Monodoroidées). *C.R. Acad. Sci. Paris* **292**: 1025–8.
Guéguen, F. 1901. *Anatomie comparée du tissu conducteur du style et du stigmate des phanérogames. I. Monocotylédones, Apétales et Gamopétales*. Paris: Mersch.
Guerrant, E.O., Jr. 1988. Heterochrony in plants. The intersection of evolution, eco-

logy and ontogeny. In McKinney, M.L. (ed.), *Heterochrony in Evolution. A Multidisciplinary Approach*, 111–33. New York: Plenum Press.

Guerrant, E.O., Jr. 1989. Early maturity, small flowers and autogamy: a developmental connection? In Bock, J.H. & Linhart, Y.B. (eds), *The Evolutionary Ecology of Plants*, 61–84. Boulder, Colorado: Westview Press.

Guinet, P. 1981. Mimosoideae: The characters of their pollen grains. In Polhill, R.M. & Raven, P.H. (eds), *Advances in Legume Systematics*, Part 2, 835–57. Kew: Royal Botanic Gardens.

Guo, Y.-H. & Cook, C.D.K. 1990. The floral biology of *Groenlandia densa* (L.) Fourreau (Potamogetonaceae). *Aquat. Bot.* **38**: 283–8.

von Guttenberg, H. 1959. Die physiologische Anatomie seismonastisch reaktionsfähiger Organe. In Ruhland, W. (ed.), *Handbuch der Pflanzenphysiologie* XVII (1): 168–83. Berlin: Springer.

Haber, W.A. & Frankie, G.W. 1989. A tropical hawkmoth community: Costa Rican dry forest Sphingidae. *Biotropica* **21**: 155–72.

Haberlandt, G. 1893. Eine botanische Tropenreise. Leipzig: Engelmann.

Hagemann, W. 1963. Die morphologische Sprossdifferenzierung und die Anordnung des Leitgewebes. *Ber. Deutsch. Bot. Ges.* **76**: (113)-(20).

Hagemann, W. 1970. Studien zur Entwicklungsgeschichte der Angiospermenblätter. *Bot. Jahrb. Syst.* **90**: 297–413.

Hagemann, W. 1984. Morphological aspects of leaf development in ferns and angiosperms. In White, R.A. & Dickison, W.C. (eds), *Contemporary Problems in Plant Anatomy*, 301–50. Orlando, Florida: Academic Press.

Haig, D. & Westoby, M. 1991a. Genomic imprinting in endosperm: Its effect on seed development in crosses between species, and between different ploidies of the same species, and its implications for the evolution of apomixis. *Phil. Trans. R. Soc. Lond. B* **333**: 1–13.

Haig, D. & Westoby, M. 1991b. Seed size, pollination costs and angiosperm success. *Evol. Ecol.* **5**: 231–47.

Halevy, A.H. 1986. Pollination induced corolla senescence. *Acta Horticult.* **181**: 25–32.

Hallé, F. 1961. Contribution à l'étude biologique et taxonomique des Mussaendeae (Rubiaceae) d'Afrique tropicale. *Adansonia*, Sér. 2, **1**: 266–98.

Hallé, F. 1967. Etude biologique et morphologique de la tribu des Gardéniées (Rubiacées). *Mém. ORSTOM* **22**: 1–146.

Hallé, F., Oldeman, R.A.A. & Tomlinson, P.B. 1978. *Tropical Trees and Forests. An Architectural Analysis.* Berlin: Springer.

Hallier, H. 1897. Ueber *Leea amabilis* und ihre Wasserkelche. *Ann. Jard. Bot. Buitenzorg* **14**: 241–7.

Hamann, U. 1966. Embryologische, morphologisch-anatomische und systematische Untersuchungen an Philydraceen. *Willdenowia, Beih.* **4**: 1–178.

Hamilton, A.G. 1898. On the fertilisation of *Eupomatia laurina* R.Br. *Proc. Linn. Soc. New South Wales* **22**: 48–55.

Hamrick, J.L. & Loveless, M.D. 1989. The genetic structure of tropical tree populations: Associations with reproductive biology. In Bock, J.H. & Linhart, Y.B. (eds), *The Evolutionary Ecology of Plants*, 129–46. Boulder, Colorado: Westview Press.

Hamrick, J.L. & Murawski, D.A. 1990. The breeding structure of tropical tree populations. *Pl. Spec. Biol.* **5**: 157–65.

Hanf, M. 1936. Vergleichende und entwicklungsgeschichtliche Untersuchungen über Morphologie und Anatomie der Griffel und Griffeläste. *Beih. Bot. Centralbl. A* **54**: 99–141.

Hansen, B. & Engell, K. 1978. Inflorescences in Balanophoroideae, Lophophytoideae and Scybalioideae (Balanophoraceae). *Bot. Tidsskr.* **72**: 177–88.

Harborne, J.B. (ed.) 1993. The Flavonoids: *Advances in Research since 1986*. London: Chapman & Hall.

Haring, V., Gray, J.E., McClure, B.A., Anderson, M.A. & Clarke, A.E. 1990. Self-incompatibility: A self-recognition system in plants. *Science* **250**: 937–41.

Harley, R.M. 1971. An explosive pollination mechanism in *Eriope crassipes*, a Brazilian labiate. *Biol. J. Linn. Soc.* **3**: 159–64.

Harley, R.M. & Reynolds, T. (eds) 1992. *Advances in Labiate Science*. Kew: Royal Botanic Gardens.

Harms, H. 1935. Rafflesiaceae. In Engler, A. & Prantl, K. (eds), *Die natürlichen Pflanzenfamilien* 16 b, 243–81 (Ed. 2). Leipzig: Engelmann.

Harris, J.A. 1905. The dehiscence of anthers by apical pores. *Ann. Rep. Missouri Bot. Gard.* **16**: 167–257.

Hart, I.W. 1990. *Plant Tropisms and Other Growth Movements*. London: Unwin Hyman.

Hartl, D. 1955. Das Vorkommen rhinanthoider Knospendeckung bei *Lindenbergia* Lehm., einer Gattung der Scrophulariaceae-Antirrhinoideae. *Oesterr. Bot. Z.* **102**: 80–3.

Hartl, D. 1956a. Morphologische Studien am Pistill der Scrophulariaceen. *Oesterr. Bot. Z.* **103**: 185–242.

Hartl, D. 1956b. Die Beziehungen zwischen den Plazenten der Lentibulariaceen und Scrophulariaceen nebst einem Exkurs über die Spezialisationsrichtungen der Plazentation. *Beitr. Biol. Pfl.* **32**: 471–90.

Hartl, D. 1957. Die Pseudosympetalie von *Correa speciosa* (Rutaceae) und *Oxalis tubiflora* (Oxalidaceae). *Akad. Wiss. Lit. Mainz*, Abh. Math.-Naturwiss. Kl., 1957 (2): 53–63.

Hartl, D. 1962. Die morphologische Natur und die Verbreitung des Apikalseptums. *Beitr. Biol. Pfl.* **37**: 241–330.

Hartl, D. 1963. Das Placentoid der Pollensäcke, ein Merkmal der Tubifloren. *Ber. Deutsch. Bot. Ges.* **76**: (70)-(2).

Hartl, D. 1965–1974. Scrophulariaceae. In Hegi, G. (ed.), *Illustrierte Flora von Mitteleuropa* (Ed. 2) VI (1), 1–631. München: Hanser.

Hartl, D. & Severin, I. 1981. Verwachsungen im Umfeld des Griffels bei *Allium*, *Cyanastrum* und *Heliconia* und den Monokotylen allgemein. *Beitr. Biol. Pfl.* **55**: 235–60.

Hedberg, O. 1955. A taxonomic revision of the genus *Sibthorpia* L. *Bot. Not.* **108**: 161–83.

Hedge, I.C. 1972. The pollination mechanism of *Aeollanthus njassae*. *Notes Roy. Bot. Gard. Edinburgh* **32**: 45–8.

Hedrén, M. 1989. *Justicia* sect. *Harnieria* (Acanthaceae) in tropical Africa. *Symb. Bot. Upsal.* **29** (1): 1–141.

Hedström, I. & Thulin, M. 1986. Pollination by a hugging mechanism in *Vigna vexillata* (Leguminosae-Papilionoideae). *Pl. Syst. Evol.* **154**: 275–83.

van Heel, W.A. 1966. Morphology of the androecium in Malvales. *Blumea* **13**: 177–394.

van Heel, W.A. 1981. A S.E.M.-investigation on the development of free carpels. *Blumea* **27**: 499–552.

van Heel, W.A. 1983. The ascidiform early development of free carpels, a S.E.M. investigation. *Blumea* **28**: 231–70.

van Heel, W.A. 1984. Variation in the development of ascidiform carpels, an S.E.M. investigation. *Blumea* **29**: 443–52.

van Heel, W.A. 1988. On the development of some gynoecia with septal nectaries. *Blumea* **33**: 477–504.

Heinrich, B. & Raven R.H. 1972. Energetics and pollination ecology. *Science* **176**: 597–602.

Heithaus, E.R. 1979a. Community structure of neotropical flower visiting bees and wasps: Diversity and phenology. *Ecology* **60**: 190–202.

Heithaus, E.R. 1979b. Flower visitation records and resource overlap of bees and wasps in northwest Costa Rica. *Brenesia* **16**: 9–52.

Heithaus, E.R. 1982. Coevolution between bats and plants. In Kunz, T.H. (ed.), *Ecology of Bats,* 327–67. New York: Plenum Press.

Heithaus, E.R., Opler, P.A. & Baker, H.G. 1974. Bat activity and pollination of *Bauhinia pauletia*: Plant-pollinator coevolution. *Ecology* **55**: 412–19.

Hemsley, A.J. & Ferguson, I.K. 1985. Pollen morphology of the genus *Erythrina* (Leguminosae: Papilionoideae) in relation to floral structure and pollinators. *Ann. Missouri Bot. Gard.* **72**: 570–90.

Henderson, A. 1986. A review of pollination studies in the Palmae. *Bot. Rev.* **52**: 221–59.

Hepher, A. & Boulter, M.E. 1987. Pollen tube growth and fertilization efficiency in *Salpiglossis sinuata*: Implications for the involvement of chemotropic factors. *Ann. Bot.* **60**: 595–601.

Herendeen, P.S. & Dilcher, D.L. (eds), 1992. *Advances in Legume Systematics*. Part 4. *The Fossil Record*. Kew: Royal Botanic Gardens.

Herrera, J. 1991. The reproductive biology of a riparian mediterranean shrub, *Nerium oleander* L. (Apocynaceae). *Bot. J. Linn. Soc.* **106**: 147–72.

Hertz, M. 1935. Zur Physiologie des Formen- und Bewegungssehens. II. Auflösungsvermögen des Bienenauges und optomotorische Reaktion. *Z. Vergl. Physiol.* **21**: 579–603.

Heslop-Harrison, J. 1952. A reconsideration of plant teratology. *Phyton (Horn)* **4**: 19–34.

Heslop-Harrison, J. 1972. Sexuality of angiosperms. In Steward, F.C. (ed.), *Plant Physiology, a Treatise* 6 C: 133–289. New York: Academic Press.

Heslop-Harrison, J. 1975. Incompatibility and the pollen stigma interaction. *Ann. Rev. Pl. Physiol.* **26**: 403–25.

Heslop-Harrison, J. 1976a. Reproductive physiology. In Simmons, J.B. *et al.* (eds),

Conservation of Threatened Plants, 199–205. NATO Conference, Ser. 1, Ecology, Vol. 1. New York: Plenum Press.

Heslop-Harrison, J. 1976b. A new look at pollination. *Rep. E. Malling Res. Stn* 1975: 141–57.

Heslop-Harrison, J. 1978. Genetics and physiology of angiosperm incompatibility systems. *Proc. Roy. Soc. Lond.* B **202**: 73–92.

Heslop-Harrison, J. 1980. Pollen-stigma interactions in grasses: A brief review. *New Zeal. J. Bot.* **17**: 537–46.

Heslop-Harrison, J. 1983. Self-incompatibility: Phenomenology and physiology. *Proc. Roy. Soc. Lond. B,* **218**: 371–395.

Heslop-Harrison, J. & Heslop-Harrison, Y. 1982a. The specialized cuticles of the receptive surfaces of angiosperm stigmas. In Cutter, K.L., Alvin, K.L. & Price, C.E. (eds), *The Plant Cuticle*, 99–120. London: Academic Press.

Heslop-Harrison, J. & Heslop-Harrison, Y. 1982b. Pollen-stigma interaction in the Leguminosae: The secretory system of the style in *Trifolium pratense* L. *Ann. Bot.* **50**: 635–45.

Heslop-Harrison, J. & Heslop-Harrison, Y. 1982c. Pollen-stigma interaction in the Leguminosae: Constituents of the stylar fluid and stigma secretion of *Trifolium pratense* L. *Ann. Bot.* **49**: 729–35.

Heslop-Harrison, J. & Heslop-Harrison, Y. 1985. Surfaces and secretions in the pollen-stigma interaction: A brief review. *J. Cell Sci.*, Suppl. 2, 287–300.

Heslop-Harrison, J. & Heslop-Harrison, Y. 1986. Pollen tube chemotropism: fact or delusion? In Cresti, M. & Dallai, R. (eds), *Biology of Reproduction and Cell Motility in Plants and Animals*, 169–74. Siena: University of Siena.

Heslop-Harrison, J., Heslop-Harrison, Y. & Reger, B.J. 1987. Anther-filament extension in *Lilium*: Potassium ion movement and some anatomical features. *Ann. Bot.* **59**: 505–15.

Heslop-Harrison, Y. 1981. Stigma characteristics and angiosperm taxonomy. *Nord. J. Bot.* **1**: 401–20.

Heslop-Harrison, Y. & Shivanna, K.R. 1977. The receptive surface of the angiosperm stigma. *Ann. Bot.* **41**: 1233–58.

Heslop-Harrison, Y., Heslop-Harrison, J. & Reger, B.J. 1985. The pollen-stigma interaction in the grasses. 7. Pollen tube guidance and the regulation of tube number in *Zea mays* L. *Acta Bot. Neerl.* **34**: 193–211.

Hess, D. 1990. *Die Blüte. Eine Einführung in Struktur und Funktion, Oekologie und Evolution der Blüten.* (Ed. 2). Stuttgart: Ulmer.

Hesse, M. 1981. Pollenkitt and viscin threads: Their role in cementing pollen grains. *Grana* **20**: 145–52.

Hesse, M. 1984a. An exine architecture model for viscin threads. *Grana* **23**: 69–75.

Hesse, M. 1984b. Form and function of *Delonix* pollen surface. *Mikroskopie* **41**: 70–2.

Hesse, M. 1986. Nature, form and function of pollen-connecting threads in angiosperms. In Blackmore, S., Ferguson, I.K. (eds), *Pollen and Spores. Form and Function*, 109–18. London: Academic Press.

van Heusden, E.C.H. 1992. Flowers of Annonaceae: Morphology, classification and evolution. *Blumea*, Suppl. 7, 1–218.

Heyneman, A.J., Colwell, R.K., Naeem, S., Dobkin, D.S. & Hallet, B. 1991. Host plant discrimination: Experiments with hummingbird flower mites. In Price, P.W., Lewinsohn, T.M., Fernandes, G.W. & Benson, W.W. (eds), *Plant-Animal Interactions. Evolutionary Ecology in Tropical and Temperate Regions*, 455–85. New York: Wiley.

Hiepko, P. 1965. Vergleichend-morphologische und entwicklungsgeschichtliche Untersuchungen über das Perianth bei den Polycarpicae. *Bot. Jahrb. Syst.* **84**: 359–508.

Hiepko, P. 1966. Das Blütendiagramm von *Drimys winteri* J.R. et G. Forst. (Winteraceae). *Willdenowia* **4**: 221–6.

Hills, H.G., Williams, N.H. & Dodson, C.H. 1972. Floral fragrances and isolating mechanisms in the genus *Catasetum* (Orchidaceae). *Biotropica* **4**: 61–76.

Hilty, S.L. 1980. Flowering and fruiting periodicity in a premontane forest in Pacific Columbia. *Biotropica* **12**: 292–306.

Hirmer, M. 1918. Beiträge zur Morphologie der polyandrischen Blüten. *Flora* **110**: 140–92.

Hirmer, M. 1931. Zur Kenntnis der Schraubenstellungen im Pflanzenreich. *Planta* **14**: 132–206.

Hirouchi, T. & Suda, S. 1975. Thigmotropism in the growth of pollen tubes of *Lilium longiflorum*. *Pl. Cell Physiol.* **16**: 377–81.

Hofmann, U. & Specht, A.K. 1986. Der morphologische Charakter der Asclepiadaceencorona. *Beitr. Biol. Pfl.* **61**: 79–85.

Hofmeister, W. 1868. Allgemeine Morphologie der Gewächse. In Hofmeister, W. (ed.),

Handbuch der Physiologischen Botanik, 405–664. Leipzig: Engelmann.

Hoggart, R.M. & Clarke, A.E. 1984. Arabinogalactans are common components of angiosperm styles. *Phytochemistry* **23**: 1571–3.

Hokche, O. & Ramirez, N. 1990. Pollination ecology of seven species of *Bauhinia* L. (Leguminosae: Caesalpinioideae). *Ann. Missouri Bot. Gard.* **77**: 559–72.

Holm, E. 1988. *On Pollination and Pollinators in Western Australia*. Gedved, Denmark: Holm.

Holm-Nielsen, L.B., Jörgensen, P.M. & Lawesson, J.E. 1988. Passifloraceae. In Harling, G. & Andersson, L. (eds), *Flora of Ecuador* **31**: 1–130. Arlöv: Berlings.

Holttum, R.E. 1940. On periodic leaf-change and flowering of trees at Singapore. II. *Gard. Bull. Singapore* **11**: 119–75.

Holttum, R.E. 1950. The Zingiberaceae of the Malay Peninsula. *Gard. Bull. Singapore* **13**: 1–250.

Holttum, R.E. 1954. *Plant Life in Malaya*. London: Longman.

Hopkins, H.C. 1983. The taxonomy, reproductive biology and economic potential of *Parkia* (Leguminosae: Mimosoideae) in Africa and Madagascar. *Bot. J. Linn. Soc.* **87**: 135–67.

Hopkins, H.C. 1984. Floral biology and pollination ecology of the neotropical species of *Parkia*. *J. Ecol.* **72**: 1–23.

Hopkins, H.C. 1992. The radiation of *Mucuna* in New Guinea and the role of birds, bats and possums as floral visitors. In Schrire, B. (ed.), *International Legume Conference, Poster Sessions*, Abstr. 14. Kew: Royal Botanic Gardens.

Horner, H.T., Jr. & Wagner, B.L. 1980. The association of druse crystals with the developing stomium of *Capsicum annuum* (Solanaceae) anthers. *Am. J. Bot.* **67**: 1347–60.

Hotchkiss, A.T. 1959. Pollen and pollination in the Eupomatiaceae. *Proc. Linn. Soc. New South Wales* **83**: 86–91.

Howard, R.A. 1948. The morphology and systematics of the West Indian Magnoliaceae. *Bull. Torrey Bot. Club* **75**: 335–57.

Howe, H.F. 1984. Implications of seed dispersal by animals for tropical reserve management. *Biol. Conserv.* **30**: 261–81.

Huber, E. 1953. Beitrag zur anatomischen Untersuchung der Antheren von *Saintpaulia*. Sitzber. *Oesterr. Akad. Wiss.*, Math.-Naturwiss. Kl., Abt. 1, **162**: 227–34.

Huber, H. 1957. Revision der Gattung *Ceropegia*. *Mem. Soc. Brot.* **12**: 1–203.

Huber, H. 1985. Samenmerkmale und Gliederung der Aristolochiaceen. *Bot. Jahrb. Syst.* **107**: 277–320.

Huber, K.A. 1980. Morphologische und entwicklungsgeschichtliche Untersuchungen an Blüten und Blütenständen von Solanaceen und von *Nolana paradoxa* Lindl. (Nolanaceae). *Dissertationes Botanicae* **55**: 1–252. Vaduz: Cramer.

Hudson, P. & Sugden, A.M. 1982. Floral biology of *Brownea rosa-de-monte*. In Richards, P.W. (ed.), *Symposium – The Tropical Rain Forest*, Leeds, Poster Abstracts. Department of Pure and Applied Zoology, University of Leeds.

Hufford, L.D. 1988. Roles of early ontogenetic modifications in the evolution of floral form of *Eucnide* (Loasaceae). *Bot. Jahrb. Syst.* **109**: 289–333.

Hufford, L.D. 1990. Androecial development and the problem of monophyly of Loasaceae. *Can. J. Bot.* **68**: 402–19.

Hufford, L.D. 1992. Floral structure of *Besseya* and *Synthyris* (Scrophulariaceae). *Int. J. Pl. Sci.* **153**: 217–29.

Hufford, L.D. & Endress, P.K. 1989. The diversity of anther structures and dehiscence patterns among Hamamelididae. *Bot. J. Linn. Soc.* **99**: 301–46.

Hurd, P.D. 1978. *An Annotated Catalog of the Carpenter Bees (Genus Xylocopa Latreille) of the Western Hemisphere (Hymenoptera: Anthophoridae)*. Washington, D.C.: Smithsonian Institution Press.

Hyland, B.P.M. 1989. Revision of Lauraceae in Australia (excluding *Cassytha*). *Austral. Syst. Bot.* **2**: 135–367.

Iltis, H.H. 1959. Studies in the Capparidaceae. VI. *Cleome* sect. *Physostemon*: Taxonomy, geography and evolution. *Brittonia* **11**: 123–62.

Inouye, D.W. 1980. The terminology of floral larceny. *Ecology* **61**: 1251–3.

Ippolito, A. & Armstrong, J.E. 1990. Flowering and pollination of *Hornstedtia scottiana* (Zingiberaceae). *Am. J. Bot.* **77** (6), Suppl., 181.

Irvine, A.K. & Armstrong, J.E. 1988. Beetle pollination in Australian tropical rainforests. In Kitching, R.L. (ed.), *Ecology of Australia's Wet Tropics. Proc. Ecol. Soc. Australia* **15**: 107–14.

Irvine, A.K. & Armstrong, J.E. 1990. Beetle pollination in tropical forests of Australia. In Bawa, K.S. & Hadley, M. (eds), *Reproductive Ecology of Tropical Forest Plants*, 135–49. Paris: Unesco.

Irwin, H.S. & Barneby, R.C. 1981. Tribe 2. Cassieae Bronn (1822). In Polhill, R.M. & Raven P.H. (eds), *Advances in Legume Systematics*, Part 1: 97–106. Kew: Royal Botanic Gardens.

Irwin, H.S. & Barneby, R.C. 1982. The American Cassiinae. A synoptical revision of Leguminosae tribe Cassieae subtribe Cassiinae in the New World. *Mem. New York Bot. Gard.* **35**: 1–918.

Itino, T., Kato, M. & Hotta, M. 1991. Pollination ecology of the two wild bananas, *Musa acuminata* subsp. *halabanensis* and *M. salaccensis*: Chiropterophily and ornithophily. *Biotropica* **23**: 151–8.

Iwanami, Y. 1953. Physiological researches of pollen. 5. On the conductive tissue and the growth of the pollen tube in the style. *Bot. Mag. (Tokyo)* **66**: 189–96.

Jacob, F. 1977. Evolution and tinkering. *Science* **196**: 1161–6.

Jaeger, P. 1954. Les aspects actuels du problème de la cheiroptérogamie. *Bull. Inst. Franç. Afr. Noire A* **16**: 796–821.

Jaeger, P. 1959. *La vie étrange des fleurs*. Paris: Horizons de France.

Jain, S.K. 1976. The evolution of inbreeding in plants. *Ann. Rev. Ecol. Syst.* **7**: 469–95.

Janson, C.A., Terborgh, J. & Emmons, L.H. 1981. Non-flying mammals as pollinating agents in the Amazonian forest. *Biotropica* **13** (2), Suppl, 1–6.

Janzen, D.H. 1967a. Interaction of the bull-horn acacia (*Acacia cornigera*) with an ant inhabitant (*Pseudomyrmex ferruginea*) in eastern Mexico. *Kans. Univ. Bull.* **47**: 315–58.

Janzen, D.H. 1967b. Synchronization of sexual reproduction of trees within the dry season in Central America. *Evolution* **21**: 620–37.

Janzen, D.H. 1968. Reproductive behavior in the Passifloraceae and some of its pollinators in Central America. *Behavior* **32**: 33–48.

Janzen, D.H. 1971. Euglossine bees as long distance pollinators of tropical plants. *Science* **171**: 203–5.

Janzen, D.H. 1974a. Tropical blackwater rivers, animals, and mast fruiting by the Dipterocarpaceae. *Biotropica* **6**: 69–103.

Janzen, D.H. 1974b. The deflowering of Central America. *Nat. Hist.* **83** (4): 48–53.

Janzen, D.H. 1975. *Ecology of Plants in the Tropics*. London: Arnold.

Janzen, D.H. 1976. Why bamboos wait so long to flower. *Ann. Rev. Ecol. Syst.* **7**: 347–91.

Janzen, D.H. 1977. Promising directions of study in tropical animal-plant interactions. *Ann. Missouri Bot. Gard.* **64**: 706–45.

Janzen, D.H. 1979. How to be a fig. *Ann. Rev. Ecol. Syst.* **10**: 13–51.

Janzen, D.H. 1980. When is it coevolution? *Evolution* **34**: 611-2.

Janzen, D.H. 1981a. Bee arrival at two Costa Rican female *Catasetum* orchid inflorescences, and a hypothesis on euglossine population structure. *Oikos* **36**: 177–83.

Janzen, D.H. 1981b. Differential visitation of *Catasetum* orchid male and female flowers. *Biotropica* **13** (2), Suppl., 77.

Janzen, D.H. (ed.) 1983a. *Costa Rican Natural History*. Chicago: University of Chicago Press.

Janzen, D.H. 1983b. Insects. Introduction. In Janzen, D.H. (ed.), *Costa Rican Natural History*, 619–45. Chicago: University of Chicago Press.

Janzen, D.H. 1983c. *Sapranthus palanga* (palanco, guineo, platano, turru). In Janzen, D.H. (ed.), *Costa Rican Natural History*, 320–2. Chicago: Chicago University Press.

Janzen, D.H. 1986a. The future of tropical ecology. *Ann. Rev. Ecol. Syst.* **17**: 305–24.

Janzen, D.H. 1986b. The eternal external threat. In Soulé, M.E. (ed.), *Conservation Biology. The Science of Scarcity and Diversity*, 286–303. Sunderland, Massachusetts: Sinauer.

Janzen, D.H. 1987. Insect diversity of a Costa Rican dry forest: Why keep it, and how? *Biol. J. Linn. Soc.* **30**: 343–56.

Janzen, D.H. 1989. Natural history of a wind-pollinated Central American dry forest legume tree (*Ateleia herbert-smithii* Pittier). *Monogr. Syst. Bot. Missouri Bot. Gard.* **29**: 293–376.

Jenny, M. 1985. Struktur, Funktion und systematische Bedeutung des Gynoeciums bei Sterculiaceen. Ph.D. Dissertation, University of Zurich. Zurich: ADAG.

Jenny, M. 1988. Different gynoecium types in Sterculiaceae: Ontogeny and functional aspects. In Leins, P., Tucker, S.C. & Endress, P.K. (eds), *Aspects of Floral Development*, 225–36. Berlin: Cramer.

Jensen, W.A. 1969. Cotton embryogenesis: Pollen tube development in the nucellus. *Can. J. Bot.* **47**: 383–5.

Jensen, W.A. & Fisher, D.B. 1968. Cotton embryogenesis: The entrance and discharge of the pollen tube in the embryo sac. *Planta* **78**: 158–83.
Jérémie, J. 1989. Autogamie dans le genre *Utricularia* L. (Lentibulariaceae). *Bull. Mus. Natl Hist. Nat., Paris*, Sér. 4, B, *Adansonia*, **11**: 17–28.
Joel, D.M. & Eisenstein, D. 1980. A bridge between the ovule and ovary wall in *Mangifera indica* L. (Anacardiaceae). *Acta Bot. Neerl.* **29**: 203–6.
Johnsgard, P.A. 1983. *The Hummingbirds of North America*. Washington, D.C.: Smithsonian Institution.
Johnson, C.T. 1977. 'N anatomiese ondersoek van die blom van *Strelitzia reginae* Banks. *J. South Afr. Bot.* **43**: 81–91.
Johnson, D.M. 1989. Revision of *Disepalum* (Annonaceae). *Brittonia* **41**: 356–78.
Johnson, S.D. 1992. Buzz pollination of *Orpheum frutescens*. *Veld and Flora* **78** (2): 36–7.
Johri, B.M. 1963. Embryology and taxonomy. In Maheshwari, P. (ed.), *Recent Advances in the Embryology of Angiosperms*, 395–444. Delhi: International Society of Plant Morphologists.
Johri, B.M. & Bhatnagar, S.P. 1961. Embryology and taxonomy of the Santalales – I. *Proc. Natl Inst. Sci. India* **26** B, Suppl., 199–220.
Jones, C.E. & Buchmann, S.L. 1974. Ultraviolet floral patterns as functional orientation cues in hymenopterous pollination systems. *Anim. Behav.* **22**: 481–5.
Joshi, A.C. 1932. Dedoublement of stamens in *Achyranthes aspera*, Linn. *J. Indian Bot. Soc.* **11**: 335–9.
Jost, L. 1907. Ueber die Selbststerilität einiger Blüten. *Bot. Zeitg., Abt. 2*, **65**: 77–117.
Juncosa, A.M. 1988. Floral development and character evolution in Rhizophoraceae. In Leins, P., Tucker, S.C. & Endress, P.K. (eds), *Aspects of Floral Development*, 83–101. Berlin: Cramer.
Juncosa, A.M. & Tomlinson, P.B. 1987. Floral development in mangrove Rhizophoraceae. *Am. J. Bot.* **74**: 1263–79.
Juncosa, A.M. & Tomlinson, P.B. 1989. Systematic comparison and some biological characteristics of Rhizophoraceae and Anisophylleaceae. *Ann. Missouri Bot. Gard.* **75**: 1296–318.
Juncosa, A.M. & Webster, B.D. 1989. Pollination in *Lupinus nanus* subsp. *latifolius* (Leguminosae). *Am. J. Bot.* **76**: 59–66.
Junell, S. 1934. Zur Gynäzeummorphologie und Systematik der Verbenaceen und Labiaten nebst Bemerkungen über ihre Samenentwicklung. *Symbolae Bot. Upsal.* **1**: 1–219.
Junell, S. 1961. Ovarian morphology and taxonomical position of Selagineae. *Svensk Bot. Tidskr.* **55**: 168–92.
Jung, J. 1956. Sind Narbe und Griffel Eintrittspforten für Pilzinfektionen? *Phytopathol. Z.* **27**: 405–26.
Kampny, C.M. & Canne-Hilliker, J.M. 1988. Aspects of floral development in Scrophulariaceae. Striking early differences in three tribes. In Leins, P., Tucker, S.C. & Endress, P.K. (eds), *Aspects of Floral Development*, 147–57. Berlin: Cramer.
Kania, W. 1973. Entwicklungsgeschichtliche Untersuchungen an Rosaceenblüten. *Bot. Jahrb. Syst.* **93**: 175–246.
Kaplan, D.R. 1967. Floral morphology, organogensis and interpretation of the inferior ovary in *Downingia bacigalupii*. *Am. J. Bot.* **54**: 1274–90.
Kaplan, D.R. 1968. Histogenesis of the androecium and gynoecium in *Downingia bacigalupii*. *Am. J. Bot.* **55**: 933–50.
Kaplan, D.R. 1971. On the value of comparative development in phylogenetic studies – a rejoinder. *Phytomorphology* **21**: 134–40.
Kaplan, D.R. 1984. The concept of homology and its central role in the elucidation of plant systematic relationships. In Duncan, T. & Stuessy, T.F. (eds), *Cladistics: Perspectives on the Reconstruction of Evolutionary History*, 51–70. New York: Columbia University Press.
Kaplan, D.R. & Hagemann, W. 1991. The relationships of cell and organism in vascular plants. *BioSci.* **41**: 693–703.
Karrer, A. 1991. Blütenentwicklung und systematische Stellung der Papaveraceae und Capparaceae. Ph.D. Dissertation, University of Zurich. Zurich: ADAG.
Kato, M. 1990. Ophioglossaceae: A hypothetical archetype for the angiosperm carpel. *Bot. J. Linn. Soc.* **102**: 303–11.
Kato, M. 1991. Further comments on an ophioglossoid archetype for the angiosperm carpel: Ovular paedomorphosis. *Taxon* **40**: 189–94.
Kaur, A., Ha, C.O., Jong, K., Sands, V.E., Cain, H.T., Soepadmo, E. & Ashton, P.S. 1978. Apomixis may be widespread among

trees of the climax rain forest. *Nature* **271**: 440–2.

Kay, Q.O.N. 1984. Variation, polymorphism and gene-flow within species. In Heywood, V.H. & Moore, D.M. (eds), *Current Concepts in Plant Taxonomy*, 181–99. London: Academic Press.

Kay, Q.O.N. 1987. Ultraviolet patterning and ultraviolet-absorbing pigments in flowers of the Leguminosae. In Stirton, C.H. (ed.), *Advances in Legume Systematics*, Part 3, 317–53. Kew: Royal Botanic Gardens.

Kay, Q.O.N. 1988. More than the eye can see: the unexpected complexity of petal structure. *Plants today* **1**: 109–14.

Kay, Q.O.N., Daoud, H.S. & Stirton, C.H. 1981. Pigment distribution, light reflection and cell structure in petals. *Bot. J. Linn. Soc.* **83**: 57–84.

Keeler, K.H. 1981. Function of *Mentzelia nuda* (Loasaceae) postfloral nectaries in seed defense. *Am. J. Bot.* **68**: 295–9.

Keijzer, C.J. 1987a. The processes of anther dehiscence and pollen dispersal. I. The opening mechanism of longitudinally dehiscing anthers. *New Phytol.* **105**: 487–98.

Keijzer, C.J. 1987b. The processes of anther dehiscence and pollen dispersal: 2. The formation and the transfer mechanism of pollenkitt, cell-wall development of the loculus tissues and a function of orbicules in pollen dispersal. *New Phytol.* **105**: 499–507.

Kelbessa, E. 1990. *Justicia* sect. *Ansellia* (Acanthaceae). *Symb. Bot. Upsal.* **29** (2): 1–96.

Keller, S. & Armbruster, S. 1989. Pollination of *Hyptis capitata* by eumenid wasps in Panama. *Biotropica* **21**: 190–2.

Kennedy, H. 1977. An unusual flowering strategy and new species in *Calathea*. *Bot. Not.* **130**: 333–9.

Kennedy, H. 1978. Systematics and pollination of the 'closed-flowered' species of *Calathea* (Marantaceae). *Univ. Calif. Publ. Bot.* **71**: 1–90.

Kennedy, H. 1983. *Calathea insignis* (hoya negra, hoja de sal, bijagua, rattlesnake plant). In Janzen, D.H. (ed.), *Costa Rican Natural History*, 204–7. Chicago: University of Chicago Press.

Kenrick, J. & Knox, R.B. 1982. Function of the polyad in reproduction of *Acacia*. *Ann. Bot.* **50**: 721–7.

Kenrick, J. & Knox, R.B. 1989. Pollen-pistil interactions in Leguminosae (Mimosoideae). *Monogr. Syst. Bot. Missouri Bot. Gard.* **29**: 127–56.

Kenrick, J., Kaul, V. & Knox, R.B. 1985. Self-incompatibility in the nitrogen-fixing tree legume, *Acacia retinodes*, pre- or post-zygotic mechanism? In Willemse, M.T.M. & van Went, J.L. (eds), *Sexual Reproduction in Seed Plants, Ferns and Mosses,* 111. Wageningen: Pudoc.

Kephart, S.R. 1983. The partitioning of pollinators among three species of *Asclepias*. *Ecology* **64**: 120–33.

Kerner, A. 1905. *Pflanzenleben 2* (Ed. 2). Leipzig: Bibliographisches Institut.

Kerr, W.E. & Lopez, C.R. 1963. Biologia de reprodução de *Trigona* (*Plebeja*) *Droryana* F. Smith. *Rev. Brasil. Biol.* **22**: 335–41.

Kers, L.E. 1969. Studies in *Cleome* III. Morphology and distribution of some African species. *Bot. Not.* **122**: 549–88.

Kessler, P.J.A. 1988. Revision der Gattung *Orophea* Blume (Annonaceae). *Blumea* **33**: 1–80.

Kevan, P.G. 1978. Floral coloration, its colorimetric analysis and significance in anthecology. In Richards, A.J. (ed.), *The Pollination of Flowers by Insects*, 51–78. London: Academic Press.

Kevan, P.G. & Baker, H.G. 1983. Insects as flower visitors and pollinators. *Ann. Rev. Ent.* **28**: 407–53.

Kevan, P.G. & Lane, M.A. 1985. Flower petal microtexture is a tactile cue for bees. *Proc. Natl Acad. Sci. USA* **82**: 4750–2.

Kirchoff, B.K. 1988a. Inflorescence and flower development in *Costus scaber* (Costaceae). *Can. J. Bot.* **66**: 339–45.

Kirchoff, B.K. 1988b. Floral ontogeny and evolution in the ginger group of the Zingiberales. In Leins, P., Tucker, S.C. & Endress, P.K. (eds), *Aspects of Floral Development*, 45–56. Berlin: Cramer.

Kirchoff, B.K. 1991. Homeosis in the flowers of the Zingiberales. *Am. J. Bot.* **78**: 833–7.

Kirk, W.D.J. 1984. Pollen-feeding in thrips (Insecta: Thysanoptera). *J. Zool.* **204**: 107–17.

Kirk, W.D.J. 1985. Effect of some floral scents on host finding by thrips (Insecta: Thysanoptera). *J. Chem. Ecol.* **11**: 35–43.

Kjellsson, G., Rasmussen, F.N. & Dupuy, D. 1985. Pollination of *Dendrobium infundibulum*, *Cymbidium insigne* (Orchidaceae) and *Rhododendron lyi* (Ericaceae) by *Bombus eximius* (Apidae) in Thailand: A possible case of floral mimicry. *J. Trop. Ecol.* **1**: 289–302.

Klekowski, E.J., Jr., 1988. *Mutation, Develop-*

mental Selection and Plant Evolution. New York: Columbia University Press.

Knoll, F. 1921. Insekten und Blumen. *Abh. Zool.-Bot. Ges. Wien* **12**: 17–119.

Knoll, F. 1922. Fettes Oel auf den Blütenepidermen der Cypripedilinae. *Oesterr. Bot. Z.* **71**: 120–9.

Knoll, F. 1956. *Die Biologie der Blüte*. Berlin: Springer.

Knox, R.B. 1984. Pollen-pistil interactions. In Linskens, H.F. & Heslop-Harrison, J. (eds), *Encyclopedia of Plant Physiology*, n.s., **17**: 508–608. Berlin: Springer.

Knox, R.B. & Kenrick, J. 1983. Polyad function in relation to the breeding system of *Acacia*. In Mulcahy, D.L. & Ottaviano, E. (eds), *Pollen: Biology and Implications for Plant Breeding*, 411–7. New York: Elsevier.

Knox, R.B. & Singh, M.B. 1987. New perspectives in pollen biology and fertilization. *Ann. Bot.* **60**, Suppl. 4, 15–37.

Knox, R.B. & Singh, M.B. 1990. Reproduction and recognition phenomena in the Poaceae. In Chapman, G.P. (ed.), *Reproductive Versatility in the Grasses*, 220–39. Cambridge University Press.

Knox, R.B., Williams, E.G. & Dumas, C. 1986. Pollen, pistil and reproductive function in crop plants. *Pl. Breed. Rev.* **4**: 9–79.

Knuth, P. 1898/1899. *Handbuch der Blütenbiologie* I, II,1 / II,2. Leipzig: Engelmann.

Knuth, P., Appel, O. & Loew, E. 1904/1905. *Handbuch der Blütenbiologie* III,1 / III,2. Leipzig: Engelmann.

Koepcke, H.-W. 1974. *Die Lebensformen II*. Krefeld: Goecke & Evers.

Kölreuter, J.G. 1761. *Vorläufige Nachricht von einigen das Geschlecht der Pflanzen betreffenden Versuchen und Beobachtungen*. Leipzig: Gleditsch.

Koltunow, A.M., Truettner, J., Cox, K.H., Wallroth, M. & Goldberg, R.B. 1990. Different temporal and spatial gene expression patterns occur during anther development. *Pl. Cell* **2**: 1201–24.

Koorders, S.H. 1897. *Ueber die Blüthenknospen-Hydathoden einiger tropischen Pflanzen*. Leiden: Brill.

Koptur, S. 1983. Flowering phenology and floral biology of *Inga* (Fabaceae: Mimosoideae). *Syst. Bot.* **8**: 354–68.

Koptur, S. 1984. Outcrossing and pollinator limitation of fruit set: Breeding systems of Neotropical *Inga* trees (Fabaceae: Mimosoideae). *Evolution* **38**: 1130–43.

Koptur, S. 1989. Is extrafloral nectar production an inducible defense? In Bock, J.H. & Linhart, Y.B. (eds), *The Evolutionary Ecology of Plants*, 323–39. Boulder, Colorado: Westview Press.

Krassilov, V.A. 1991. Origin of angiosperms: New and old problems. *Trends Ecol. Evol.* **6**: 215–20.

Kraus, G. 1895. Wasserhaltige Kelche bei *Parmentiera cerifera* Seem. *Flora* **81**: 435–7.

Krebs, A.S. 1990. Eine Ameisenpflanze im Regenwald Malaysias: *Thunbergia grandiflora* und ihre Symbiosepartner. *Palmengarten* **90** (2): 126–30.

Kress, W.J. 1983. Self-incompatibility in central American *Heliconia*. *Evolution* **37**: 735–44.

Kress, W.J. 1984a. Systematics of Central American *Heliconia* (Heliconiaceae) with pendent inflorescences. *J. Arn. Arb.* **65**: 429–532.

Kress, W.J. 1984b. Pollination and reproductive biology of *Heliconia*. In D'Arcy, W.G. & Correa A., M.D. (eds), *The Botany and Natural History of Panama. Monogr. Syst. Bot. Missouri Bot. Gard.* **10**: 267–71.

Kress, W.J. 1985. Bat pollination of an Old World *Heliconia*. *Biotropica* **17**: 302–8.

Kress, W.J. 1990a. The taxonomy of Old World *Heliconia* (Heliconiaceae). *Allertonia* **6** (1): 1–58.

Kress, W.J. 1990b. The phylogeny and classification of the Zingiberales. *Ann. Missouri Bot. Gard.* **77**: 698–721.

Kress, W.J., Schatz, G.E. & Andrianifihanana, M. 1992. Lemur pollination of *Ravenala madagascariensis* and the evolution of floral biology in the bird-of-paradise family (Strelitziaceae). *Am. J. Bot.* **79** (6), Suppl., 184.

Kroh, M. & Helsper, J.P.F.G. 1974. Transmitting tissue and pollen tube growth. In Linskens, H.F. (ed.), *Fertilization in Higher Plants*, 167–75. Amsterdam: North-Holland.

Kronestedt, E. & Bystedt, P.-A. 1981. Thread-like formations in the anthers of *Strelitzia reginae*. *Nord. J. Bot.* **1**: 523–9.

Kronestedt, E. & Walles, B. 1986. Anatomy of the *Strelitzia reginae* flower (Strelitziaceae). *Nord. J. Bot.* **6**: 307–20.

Kubitzki, K. 1978. *Caraipa* and *Mahurea* (Bonnetiaceae). *Mem. New York Bot. Gard.* **29**: 82–138.

Kubitzki, K. 1987. Origin and significance of trimerous flowers. *Taxon* **36**: 21–8.

Kubitzki, K. & Amaral, M.C.E. 1991. Transference of function in the pollination system of the Ochnaceae. *Pl. Syst. Evol.* **177**: 77–80.

Kubitzki, K. & Kurz, H. 1984. Synchronized dichogamy and dioecy in neotropical Lauraceae. *Pl. Syst. Evol.* **147**: 253–66.

Kubitzki, K., von Sengbusch, P. & Poppendiek, H.-H. 1991. Parallelism, its evolutionary origin and systematic significance. *Aliso* **13**: 191–206.

Kugler, H. 1947. Hummeln und die UV-Reflexion an Kronblättern. *Naturwissenschaften* **34**: 315–6.

Kugler, H. 1955. Zum Problem der Dipterenblumen. *Oesterr. Bot. Z.* **102**: 529–41.

Kugler, H. 1956. Ueber die optische Wirkung von Fliegenblumen auf Fliegen. *Ber. Deutsch. Bot. Ges.* **69**: 387–98.

Kugler, H. 1963. UV-Musterungen auf Blüten und ihr Zustandekommen. *Planta* **59**: 296–329.

Kugler, H. 1966. UV-Male auf Blüten. *Ber. Deutsch. Bot. Ges.* **79**: 57–70.

Kugler, H. 1970. *Einführung in die Blütenökologie* (Ed. 2). Stuttgart: Fischer.

Kugler, H. 1971. Zur Bestäubung grossblumiger *Datura*-Arten. *Flora* **160**: 511–7.

Kugler, H. 1980. Zur Bestäubung von *Lantana camara. Flora* **169**: 524–9.

Kühn, G. 1928. Beiträge zur Kenntnis der intraseminalen Leitbündel bei den Angiospermen. *Bot. Jahrb. Syst.* **61**: 325–79.

Kujit, J. 1969. *The Biology of Parasitic Flowering Plants.* Berkeley: University of California Press.

Kunze, H. 1982a. Morphogenese und Synorganisation des Bestäubungsapparates einiger Asclepiadaceen. *Beitr. Biol. Pfl.* **56**: 133–70.

Kunze, H. 1982b. Aspekte der Blütengestalt. II. Innenraumbildung. III. Die Bildung von Apparaten. *Elemente Naturwiss.* **37**: 19–30, 31–41.

Kunze, H. 1984. Vergleichend-morphologische Studien an Cannaceen- und Marantaceenblüten. *Flora* **175**: 301–18.

Kunze, H. 1985. Die Infloreszenzen der Marantaceen und ihr Zusammenhang mit dem Typus der Zingiberales-Synfloreszenz. *Beitr. Biol. Pfl.* **60**: 93–140.

Kunze, H. 1990. Morphology and evolution of the corona in Asclepiadaceae and related families. *Trop. Subtrop. Pflanzenwelt* **76**: 1–51. Stuttgart: Steiner.

Kunze, H. 1991. Structure and function in asclepiad pollination. *Pl. Syst. Evol.* **176**: 227–53.

Kurzweil, H. 1987. Developmental studies in orchid flowers I: Epidendroid and vandoid species. *Nord. J. Bot.* **7**: 427–42.

Kurzweil, H. & Weber, A. 1991. Floral morphology of southern African Orchideae. I. Orchidinae. *Nord. J. Bot.* **11**: 155–78.

Lack, A.J. & Kevan, P.G. 1984. On the reproductive biology of a canopy tree, *Syzygium syzygioides* (Myrtaceae), in a rain forest in Sulawesi, Indonesia. *Biotropica* **16**: 31–6.

LaFrankie, J.V., Jr. & Chan, H.T. 1991. Confirmation of sequential flowering in *Shorea. Biotropica* **23**: 200–3.

de Lagerheim, G. 1891. Zur Biologie der *Iochroma macrocalyx* Benth. *Ber. Deutsch. Bot. Ges.* **9**: 348–51.

Lamond, M. & Vieth, J. 1972. L'androcée synanthéré du *Rechsteinera cardinalis* (Gesnériacées). Une contribution au problème des fusions. *Can. J. Bot.* **50**: 1633–7.

Lamont, B. 1985. The significance of flower colour change in eight co-occurring shrub species. *Bot. J. Linn. Soc.* **90**: 145–55.

Lamoureux, C.H. 1975. Phenology and floral biology of *Monodora myristica* (Annonaceae) in Bogor, Indonesia. *Ann. Bogor.* **6**: 1–25.

Lasseigne, A. 1979. Studies in *Cassia* (Leguminosae-Caesalpinioideae) III. Anther morphology. *Iselya* **1**: 141–60.

Lavin, M. & Delgado S., A. 1990. Pollen brush of Papilionoideae (Leguminosae): Morphological variation and systematic utility. *Am. J. Bot.* **77**: 1294–312.

Lawton, J. 1991. Warbling in different ways. *Oikos* **60**: 273–4.

Leach, L.C. 1978. A contribution towards a new classification of Stapelieae (Asclepiadaceae) with a preliminary review of *Orbea* Haw. and descriptions of three new genera. *Excelsa* **1**: 1–75.

Léandri, J. 1933. Sur la station d'origine de *Poinciana regia* Boj. *Bull. Mus. Natl Hist. Nat.*, Paris, II, **5**: 413–14.

Léandri, J. 1936. Le milieu et la végétation de la Réserve Naturelle d'Antsingy (Madagascar). *Bull. Mus. Natl Hist. Nat.*, Paris, II, **8**: 557–72.

Ledin, R.B. & Menninger, E.A. 1956. *Bauhinia*, the so-called orchid trees. *Natl Hortic-ult. Mag.* **35**: 183–200.

Lee, D.W. 1991. Ultrastructural basis and function of iridescent blue colour of fruits in *Elaeocarpus. Nature* **349**: 260–1.

Lee, T.D. 1988. Patterns of fruit and seed production. In Lovett Doust, J. & Lovett

Doust, L. (eds), *Reproductive Ecology of Plants*, 179–263. Oxford University Press.

Lehmann, E. 1919. Die Pentasepalie in der Gattung *Veronica* und die Vererbungsweise der pentasepalen Zwischenrassen. *Ber. Deutsch. Bot. Ges.* **36**: (28)-(46).

Lehmann, N.L. & Sattler, R. 1992. Irregular floral development in *Calla palustris* (Araceae) and the concept of homeosis. *Am. J. Bot.* **79**: 1145–57.

Leinfellner, W. 1950. Der Bauplan des synkarpen Gynoeceums. *Oesterr. Bot. Z.* **97**: 403–36.

Leinfellner, W. 1951. Die U-förmige Plazenta als der Plazentationstypus der Angiospermen. *Oesterr. Bot. Z.* **98**: 338–58.

Leinfellner, W. 1956. Die blattartig flachen Staubblätter und ihre gestaltlichen Beziehungen zum Bautypus des Angiospermen-Staubblattes. *Oesterr. Bot. Z.* **103**: 247–90.

Leinfellner, W. 1964. Ueber die falsche Sympetalie bei *Lonchostoma* und anderen Gattungen der Bruniaceen. *Oesterr. Bot. Z.* **111**: 345–53.

Leinfellner, W. 1969a. Ueber die Karpelle verschiedener Magnoliales. VIII. Ueberblick über alle Familien der Ordnung. *Oesterr. Bot. Z.* **117**: 107–27.

Leinfellner, W. 1969b. Zur Kenntnis der Karpelle der Leguminosen. I. Papilionaceae. *Oesterr. Bot. Z.* **117**: 332- 47.

Leinfellner, W. 1970. Zur Kenntnis der Karpelle der Leguminosen. 2. Caesalpiniaceae und Mimosaceae. *Oesterr. Bot. Z.* **118**: 108–20.

Leinfellner, W. 1973a. Das Gynözeum der Bignoniaceen. II. Die U-förmige Plazenta von *Schlegelia* (Crescentieae). *Oesterr. Bot. Z.* **121**: 13–22.

Leinfellner, W. 1973b. Das Gynözeum der Bignoniaceen. III. Crescentieae (*Amphitecna*, *Colea*, *Rhodocolea*, *Ophiocolea*, *Phyllarthron*, *Phylloctenium*, *Parmentiera*, *Enallagma* und *Crescentia*). *Oesterr. Bot. Z.* **122**: 59–73.

Leins, P. 1964. Entwicklungsgeschichtliche Studien an Ericales-Blüten. *Bot. Jahrb. Syst.* **83**: 57–88.

Leins, P. 1972a. Das Karpell im ober- und unterständigen Gynoeceum. *Ber. Deutsch. Bot. Ges.* **85**: 291–4.

Leins, P. 1972b. Das zentrifugale Androeceum von *Couroupita guianensis* (Lecythidaceae). *Beitr. Biol. Pfl.* **48**: 313–9.

Leins, P. & Erbar, C. 1980. Die Entwicklung der Blüten von *Monodora crispata* (Annonaceae). *Beitr. Biol. Pfl.* **55**: 11–22.

Leins, P. & Erbar, C. 1982. Das monokarpellate Gynoeceum von *Monodora crispata* (Annonaceae). *Beitr. Biol. Pfl.* **57**: 1–13.

Leins, P. & Erbar, C. 1985. Ein Beitrag zur Blütenentwicklung der Aristolochiaceen, einer Vermittlergruppe zu den Monokotylen. *Bot. Jahrb. Syst.* **107**: 343–68.

Leins, P. & Erbar, C. 1990. On the mechanisms of secondary pollen presentation in the Campanulales-Asterales-complex. *Bot. Acta* **103**: 87–92.

Leins, P. & Erbar, C. 1991. Fascicled androecia in Dilleniidae and some remarks on the *Garcinia* androecium. *Bot. Acta* **104**: 336–44.

Leins, P. & Metzenauer, G. 1979. Entwicklungsgeschichtliche Untersuchungen an *Capparis*-Blüten. *Bot. Jahrb. Syst.* **100**: 542–54.

Leins, P. & Schwitalla, S. 1986. Studien an Cactaceen-Blüten I. Einige Bemerkungen zur Blütenentwicklung von *Pereskia*. *Beitr. Biol. Pfl.* **60**: 313–23.

Leins, P. & Stadler, P. 1973. Entwicklungsgeschichtliche Untersuchungen am Androeceum der Alismatales. *Oesterr. Bot. Z.* **121**: 51–63.

Leins, P., Erbar, C. & van Heel, W.A. 1988. Note on the floral development of *Thottea* (Aristolochiaceae). *Blumea* **33**: 357–70.

Leppik, E.E. 1953. The ability of insects to distinguish number. *Am. Nat.* **87**: 229–36.

Leppik, E.E. 1966. Floral evolution and pollination in the Leguminosae. *Ann. Bot. Fenn.* **3**: 299–308.

Leppik, E.E. 1972. Origin and evolution of bilateral symmetry in flowers. *Evol. Biol.* **5**: 49–85.

Leppik, E.E. 1977. *Floral Evolution in Relation to Pollination Ecology*. New Delhi: Today & Tomorrow's Printers and Publishers.

Leroy, J.-F. 1977. A compound ovary with open carpels in Winteraceae (Magnoliales). Evolutionary implications. *Science* **196**: 977–8.

Leroy, J.-F. 1983. The origin of angiosperms: An unrecognized ancestral dicotyledon, *Hedyosmum* (Chloranthales), with a strobiloid flower is living today. *Taxon* **32**: 169–75.

Les, D.H. 1988. Breeding systems, population structure, and evolution in hydrophilous angiosperms. *Ann. Missouri Bot. Gard.* **75**: 819–35.

Lewis, D. 1966. The genetic integration of breeding systems. In Hawkes, J.G. (ed.), *Reproductive Biology and Taxonomy of Vascular Plants*, 20–5. Oxford: Pergamon Press.

Lewis, D. 1979. *Sexual Incompatibility in Plants*. London: Arnold.

Lewis, D. 1980. Genetic versatility of incompatibility in plants. *New Zeal. J. Bot.* **17**: 637–44.

Lewis, D. 1982. Incompatibility, stamen movement and pollen economy in a heterostylous tropical forest tree, *Cratoxylum formosum* (Guttiferae). *Proc. Roy. Soc. Lond. B* **214**: 273–283.

Lewis, D. & Jones, D.A. 1992. The genetics of heterostyly. In Barrett, S.C.H. (ed.), *Evolution and Function of Heterostyly*, 129–50. Berlin: Springer.

Lex, T. 1954. Duftmale an Blüten. *Z. Vergl. Physiol.* **36**: 212-34.

Liede, S. & Whitehead, V. 1991. Studies in the pollination biology of *Sarcostemma viminale* R.Br. sensu lato. *South Afr. J. Bot.* **57**: 115–22.

Lindman, C.A.M. 1902. Die Blüteneinrichtungen einiger südamerikanischer Pflanzen. I. Leguminosae. *Bihang till Svenska Vet.-Akad. Handl.* 27 III, **14**: 1–63.

Lindman, C.A.M. 1906. Zur Kenntnis der Corona einiger Passifloren. In Sernander, R., Svedelius, N. & Norén, C.O. (eds), *Botaniska Studier Tillägnade F.R. Kjellman*, 55–79. Uppsala: Almqvist & Wiksell.

Lindner, E. 1928. *Aristolochia lindneri* Berger und ihre Bestäubung durch Fliegen. *Biol. Zentralbl.* **48**: 93–101.

Linhart, Y.B. 1973. Ecological and behavioral determinants of pollen dispersal in hummingbird-pollinated *Heliconia*. *Am. Nat.* **107**: 511–23.

Link, D.A. 1992. The floral nectaries of the Geraniales and their systematic implications. VI. Ixonanthaceae Exell & Mendonça. *Bot. Jahrb. Syst.* **114**: 81–90.

Linnaeus, C. 1751. *Philosophia Botanica*. Stockholm: Kiesewetter.

Little, R.J. 1983. A review of floral food deception mimicries with comments on floral mutualism. In Jones, C.E. & Little, R.J. (eds), *Handbook of Experimental Pollination Biology*, 294–309. New York: Scientific and Academic Editions.

Lloyd, D.G. 1979. Some reproductive factors affecting the selection of self-fertilization in plants. *Am. Nat.* **113**: 67–97.

Lloyd, D.G. 1980. Parental strategies of angiosperms. *New Zeal. J. Bot.* **17**: 595–606.

Lloyd, D.G. 1982. Selection of combined versus separate sexes in seed plants. *Am. Nat.* **120**: 571–85.

Lloyd, D.G. & Bawa, K.S. 1984. Modification of the gender of seed plants in varying conditions. *Evol. Biol.* **17**: 255–338.

Lloyd, D.G. & Schoen, D.J. 1992. Self- and cross-fertilization in plants. I. Functional dimensions. *Int. J. Pl. Sci.* **153**: 358–69.

Lloyd, D.G. & Webb, C.J. 1986. The avoidance of interference between the presentation of pollen and stigmas in angiosperms. I. Dichogamy. *New Zeal. J. Bot.* **24**: 135–62.

Lloyd, D.G. & Webb, C.J. 1992. The evolution of heterostyly. In Barrett, S.C.H. (ed.), *Evolution and Function of Heterostyly*, 151–78. Berlin: Springer.

Lloyd, D.G. & Wells, M.S. 1992. Reproductive biology of a primitive angiosperm, *Pseudowintera colorata* (Winteraceae), and the evolution of pollination systems in the Anthophyta. *Pl. Syst. Evol.* **181**: 77–95.

Loconte, H. & Stevenson, D.W. 1991. Cladistics of the Magnoliidae. *Cladistics* **7**: 267–96.

Long, A.G. 1984. The cupule-carpel theory. A defence. *Trans. Bot. Soc. Edinburgh* **44**: 281–5.

Longman, K.A. & Jenik, J. 1987. *Tropical Forest and Its Environment* (Ed. 2). New York: Longman.

Lorch, J. 1978. The discovery of nectar and nectaries and its relation to views on flowers and insects. *Isis* **69**: 514–33.

Lord, E.M. 1981. Cleistogamy: a tool for the study of floral morphogenesis, function and evolution. *Bot. Rev.* **47**: 421–49.

Lord, E.M. 1991. The concepts of heterochrony and homeosis in the study of floral morphogenesis. *Flowering Newsl.* **11**: 4–13.

Lord, E.M. & Hill, J.P. 1987. Evidence for heterochrony in the evolution of plant form. In Raff, R.A. & Raff, E.C. (eds), *Development as an Evolutionary Process*, 47–70. New York: Liss.

Lorence, D.H. 1985. A monograph of the Monimiaceae (Laurales) in the Malagasy region (Southwest Indian Ocean). *Ann. Missouri Bot. Gard.* **72**: 1–165.

Lovejoy, T.E., Bierregaard, R.O., Jr., Rylands, A.B., Malcolm, J.R., Quintela, C.E., Harper, L.H., Brown, K.S., Jr., Powell, A.H., Powell, G.V.N., Schubart, H.O.R. & Hays, M.B. 1986. Edge and other effects of isolation on Amazon forest fragments. In Soulé, M.E. (ed.), *Conserva-*

tion Biology, 257–85. Sunderland, Massachusetts: Sinauer.

Lovett Doust, J. & Lovett Doust, L. 1988. Sociobiology of plants: An emerging synthesis. In Lovett Doust, J. & Lovett Doust, L. (eds), *Plant Reproductive Ecology. Patterns and Strategies*, 5–29. New York: Oxford University Press.

Lu, J., Mayer, A. & Pickersgill, B. 1990. Stigma morphology and pollination in *Arachis* L. (Leguminosae). *Ann. Bot.* **66**: 73–82.

Ludwig, F. 1897. Ueber das Leben und die botanische Thätigkeit Dr. Fritz Müller's. *Bot. Centralbl.* **18** (II): 291–302, 347–63, 401–8.

Lumer, C. 1980. Rodent pollination of *Blakea* (Melastomataceae) in a Costa Rican cloud forest. *Brittonia* **32**: 512–7.

Lumer, C. & Schoer, R.D. 1986. Pollination of *Blakea austin-smithii* and *B. penduliflora* (Melastomataceae) by small rodents in Costa Rica. *Biotropica* **18**: 363–4.

Lunau, K. 1990. Colour saturation triggers innate reactions to flower signals: Flower dummy experiments with bumblebees. *J. Comp. Physiol. A* **166**: 827–34.

Lunau, K. 1991. Innate flower recognition in bumblebees (*Bombus terrestris*, *B. lucorum*; Apidae): Optical signals from stamens as landing reaction releasers. *Ethology* **88**: 203–14.

Lundblad, H. 1922. *Ueber die baumechanischen Vorgänge bei der Entstehung von Anomomerie bei homochlamydeischen Blüten sowie damit zusammenhängende Fragen*. Lund: Lindstedt.

Luza, J.G. & Polito, V.S. 1991. Porogamy and chalazogamy in walnut (*Juglans regia* L.) *Bot. Gaz. (Crawfordsville)* **152**: 100–6.

Lyndon, R.F. 1979. Rates of growth and primordial initiation during flower development in *Silene* at different temperatures. *Ann. Bot.* **43**: 539–51.

Maas, P.J.M. 1972. Costoideae (Zingiberaceae). In Stafleu, F. (ed.), *Flora Neotropica Monograph* **8**: 1–139. New York: Hafner.

Maas, P.J.M. 1977. *Renealmia* (Zingiberaceae-Zingiberoideae), Costoideae (Additions) (Zingiberaceae). In Rogerson, C.T. (Ed.), *Flora Neotropica Monograph* **18**: 1–218. Bronx: New York Botanical Garden.

Mabberley, D.J. 1987. *The Plant-Book*. Cambridge University Press.

Mabberley, D.J. 1992. *Tropical Rain Forest Ecology* (Ed. 2). Glasgow: Blackie.

Macior, L.W. 1965. Insect adaptation and behavior in *Asclepias* pollination. *Bull. Torrey Bot. Club* **92**: 114–26.

Magin, N., Classen, R. & Gack, C. 1989. The morphology of false anthers in *Craterostigma plantagineum* and *Torenia polygonoides* (Scrophulariaceae). *Can. J. Bot.* **67**: 1931–7.

Magnuson, J.J. 1990. Long-term ecological research and the invisible present. Uncovering the processes hidden because they occur slowly or because effects lay years behind causes. *BioSci.* **40**: 495–501.

Maheshwari, P. 1931. Contributions to the morphology of *Albizzia lebbek*. *J. Indian Bot. Soc.* **10**: 241–64.

Mair, O. 1977. Zur Entwicklungsgeschichte monosymmetrischer Dicotylen-Blüten. *Dissertationes Botanicae* **38**: 1–90. Vaduz: Cramer.

Manning, A. 1956. The effect of honey guides. *Behaviour* **9**: 114–39.

Manning, A. 1957. Some evolutionary aspects of the flower constancy of bees. *Proc. Roy. Phys. Soc. Edinburgh* **25**: 67–71.

Markgraf, F. 1936. Blütenbau und Verwandtschaft bei den einfachsten Helobiae. *Ber. Deutsch. Bot. Ges.* **54**: 191–229.

Marshall, A.G. 1983. Bats, flowers and fruit: evolutionary relationships in the Old World. *Biol. J. Linn. Soc.* **20**: 115–35.

Marshall, D.L. & Folsom, M.W. 1991. Mate choice in plants: An anatomical to population perspective. *Ann. Rev. Ecol. Syst.* **22**: 37–63.

Martinez del Rio, C. & Burquez, A. 1986. Nectar production and temperature dependent pollination in *Mirabilis jalapa* L. *Biotropica* **18**: 28–31.

Marubashi, W. & Nakajima, T. 1981. Pollen tube behavior in the ovary of *Nicotiana tabacum* L. *Jap. J. Breed.* **31**: 133–40.

Massart, J. 1895. *Un botaniste en Malaisie*. Gand: Annoot-Braeckman.

Masters, M.T. 1871. Contribution to the natural history of the Passifloraceae. *Trans. Linn. Soc. Lond., Bot.*, **27**: 593–645.

Masters, M.T. 1873. On the development of the androecium in *Cochliostema* Lem. *J. Linn. Soc. Lond., Bot.*, **13**: 204–9.

Mathä A. 1936. Der Seidenglanz der Kakteenblüten. *Oesterr. Bot. Z.* **85**: 81–115.

Mathur, G. & Mohan Ram, H.Y. 1986. Floral biology and pollination of *Lantana camara*. *Phytomorphology* **36**: 79- 100.

Matile, P. 1978. Entwicklung einer Blüte. *Neujahrsbl. Naturforsch. Ges. Zürich* **180**: 1–40.

Matile, P. & Altenburger, R. 1988. Rhythms of fragrance emission in flowers. *Planta* **174**: 242–7.

Matthews, J.R. & Knox, E.M. 1926. The comparative morphology of the stamen in the Ericaceae. *Trans. Bot. Soc. Edinburgh* **29**: 243–81.

Matthews, J.R. & Maclachlan, C.M. 1929. The structure of certain poricidal anthers. *Trans. Proc. Bot. Soc. Edinburgh* **30**: 104–22.

Mattsson, D. 1982. The morphogenesis of dimorphic pollen and anthers in *Tripogandra amplexicaulis* – Light microscopy and growth analysis. *Opera Bot.* **66**: 1–46.

Mau, S.-L., Williams, E.G., Andersson, M.A., Cornish, E.C., Grego, B., Simpson, R.J., Kheyr-Pour, A. & Clarke, A.E. 1986. Style proteins of a wild tomato (*Lycopersicon peruvianum*) associated with expression of self-incompatibility. *Planta* **169**: 184–91.

Mayak, S. & Halevy, A.H. 1980. Flower senescence. In Thimann, K.V. (ed.), *Senescence in Plants*, 131–56. Boca Raton, Florida: CRC Press.

Mayer, S.S. & Charlesworth, D. 1991. Cryptic dioecy in flowering plants. *Trends Ecol. Evol.* **6**: 320–5.

McCann, C. 1943. 'Light-windows' in certain flowers (Asclepiadaceae and Araceae). *J. Bombay Nat. Hist. Soc.* **44**: 182–4.

McConchie, C.A. 1982. The diversity of hydrophilous pollination in monocotyledons. In Williams, E.G., Knox, R.B., Gilbert, J.H., Bernhardt, P. (eds), *Pollination '82*, 148–66. Melbourne: University of Melbourne Press.

McConchie, C.A. 1983. Floral development of *Maidenia rubra* Rendle (Hydrocharitaceae). *Austral. J. Bot.* **31**: 585–603.

McConchie, C.A. & Knox, R.B. 1989. Pollen-stigma interaction in the seagrass *Posidonia australis*. *Ann. Bot.* **63**: 235–48.

McConchie, C.A., Ducker, S.C. & Knox, R.B. 1982. Biology of Australian seagrasses: Floral development and morphology in *Amphibolis* (Cymodoceaceae). *Austral. J. Bot.* **30**: 251–64.

McConchie, C.A., Hough, T., Singh, M.B. & Knox, R.B. 1986. Pollen presentation on petal combs in the geoflorous heath *Acrotriche serrulata* (Epacridaceae). *Ann. Bot.* **57**: 155–64.

McDade, L.A. 1984. Systematics and reproductive biology of the Central American species of the *Aphelandra pulcherrima* complex (Acanthaceae). *Ann. Missouri Bot. Gard.* **71**: 104–65.

McDade, L.A. 1992. Pollinator relationships, biogeography, and phylogenetics. *BioSci.* **42**: 21–6.

McDade, L.A. & Kinsman, S. 1980. The impact of floral parasitism in two Neotropical hummingbird-pollinated plant species. *Evolution* **34**: 944–58.

Meeuse, A.D.J. 1990. *Flowers and Fossils*. Delft: Eburon.

Meeuse, B.J.D. 1961. *The Story of Pollination*. New York: Ronald Press.

Meeuse, B.J.D. & Morris, S. 1984. *The Sex Life of Flowers*. London: Faber and Faber.

Meeuse, B.J.D. & Raskin, 1988. Sexual reproduction in the arum lily family, with emphasis on thermogenicity. *Sex Pl. Reprod.* **1**: 3–15.

Meeuse, B.J.D., Schneider, E.L., Hess, C.M., Kirkwood, K. & Patt, J.M. 1984. Activation and possible role of the 'food-bodies' of *Sauromatum* (Araceae). *Acta Bot. Neerl.* **33**: 483–96.

Meijer, W. 1984. New species of *Rafflesia* (Rafflesiaceae). *Blumea* **30**: 209–15.

Meijer, W. 1985. Saving the world's largest flower. *Natl Geogr.* **168** (1): 136–40.

Meinhardt, H. 1982. *Models of Biological Pattern Formation*. London: Academic Press.

Menzel, R. & Backhaus, W. 1991. Colour vision in insects. In Gouras, P. (ed.), *The Perception of Colour*, 262–93. Houndsmills: Macmillan.

Meyen, S.V. 1988. Origin of the angiosperm gynoecium by gamoheterotopy. *Bot. J. Linn. Soc.* **97**: 171–8.

Meyerowitz, E.M., Smyth, D.R. & Bowman, J.L. 1989. Abnormal flowers and pattern formation in floral development. *Development* **106**: 209–17.

Meyerowitz, E.M., Bowman, J.L., Brockman, L.L., Drews, G.N., Jack, T., Sieburth, L.E. & Weigel, D. 1991. A genetic and molecular model for flower development in *Arabidopsis thaliana*. *Development*, Suppl. 1, 157–67.

Michener, C.D. 1962. An interesting method of pollen collecting by bees from flowers with tubular anthers. *Rev. Biol. Trop.* **10**: 167–75.

Midgley, J.J. & Bond, W.J. 1991. How important is biotic pollination and dispersal to the success of the angiosperms? *Phil. Trans. Roy. Soc. Lond. B* **333**: 187–95.

Miller, J.M. 1988. A new species of *Degeneria* (Degeneriaceae) from the Fiji Archipelago. *J. Arn. Arb.* **69**: 275–80.

Miller, J.M. 1989. The archaic flowering plant family Degeneriaceae: Its bearing on an old enigma. *Natl Geogr. Res.* **5**: 218–31.

Mitchison, G.J. 1977. Phyllotaxis and the Fibonacci series. Science **196**: 270–5.

Mohan Ram, H.Y. & Rao, I.V.R. 1984. Physiology of flower bud growth and opening. *Proc. Indian Acad. Sci., Pl. Sci.*, **93**: 253–74.

Molau, U. 1988. Scrophulariaceae – Part I. Calceolarieae. *Flora Neotropica Monogr.* **47**: 1–326.

Möller, A. (ed.) 1915, 1920, 1921. *Fritz Müller. Werke, Briefe und Leben*, 3 Vols. Jena: Fischer.

Moncur, M.W. 1988. *Floral Development of Tropical and Subtropical Fruit and Nut Species. An Atlas of Scanning Electron Micrographs.* Canberra: CSIRO.

Morawetz, W. 1982. Morphologisch-ökologische Differenzierung, Biologie, Systematik und Evolution der neotropischen Gattung *Jacaranda* (Bignoniaceae). *Denkschr. Oesterr. Akad. Wiss.*, Math.-Naturwiss. Kl., **123**: 1–184.

Morawetz, W. 1988. Karyosystematics and evolution of Australian Annonaceae as compared with Eupomatiaceae, Himantandraceae, and Austrobaileyaceae. *Pl. Syst. Evol.* **159**: 49–79.

Morawetz, W. & Waha, M. 1991. Zur Entstehung und Funktion pollenverbindender Fäden bei *Porcelia* (Annonaceae). *Beitr. Biol. Pfl.* **66**: 145–54.

Mori, S.A. & Boeke, J.D. 1987. Pollination. In Mori, S.A. (ed.), *The Lecythidaceae of a Lowland Neotropical Forest: La Fumée Mountain, French Guiana. Mem. New York Bot. Gard.* **44**: 137–55.

Mori, S.A. & Kallunki, J. 1976. Phenology and floral biology of *Gustavia superba* (Lecythidaceae) in central Panama. *Biotropica* **8**: 184–92.

Mori, S.A. & Prance, G.T. 1990. Taxonomy, ecology and economic botany of the Brazil nut (*Bertholletia excelsa* Humb. & Bonpl.: Lecythidaceae). *Adv. Econ. Bot.* **8**: 130–50.

Mori, S.A., Orchard, J.E. & Prance, G.T. 1980. Intrafloral pollen differentiation in the New World Lecythidaceae, subfamily Lecythidoideae. *Science* **209**: 400–3.

Mori, S.A., Prance, G.T. & Bolten, A.B. 1978. Additional notes on the floral biology of neotropical Lecythidaceae. *Brittonia* **30**: 113–30.

Morse, D.H. 1985. Milkweeds and their visitors. *Sci. Am.* **53** (1): 90–6C, 104.

Morse, D.H. & Fritz, R.S. 1985. Variation in the pollinaria, anthers, and alar fissures of common milkweed (*Asclepias syriaca* L.). *Am. J. Bot.* **72**: 1032–8.

Moseley, H.N. 1879. *Notes by a Naturalist on the 'Challenger'.* London: Macmillan.

Mulcahy, D.L. 1979. The rise of the angiosperms: a genecological factor. *Science* **206**: 20–3.

Mulcahy, D.L. & Mulcahy, G.B. 1983. Pollen selection: An overview. In Mulcahy, D.L. & Ottaviano, E. (eds), *Pollen: Biology and Implications for Plant Breeding*, 15–7. New York: Elsevier.

Mulcahy, D.L. & Mulcahy, G.B. 1987. The effects of pollen competition. *Am. Sci.* **75**: 44–50.

Mulcahy, G.B. & Mulcahy, D.L. 1982. The two phases of growth of *Petunia hybrida* (Hort. Vilm.-Andz.) pollen tubes through compatible styles. *J. Palynol.* **18**: 61–4.

Mulcahy, G.B. & Mulcahy, D.L. 1986. More evidence on the preponderant influence of the pistil on pollen tube growth. In Cresti, M. & Dallai, R. (eds), *Biology of Reproduction and Cell Motility in Plants and Animals*, 139–44. Siena: University of Siena Press.

Mulcahy, G.B. & Mulcahy, D.L. 1987. Induced pollen tube directionality. *Am. J. Bot.* **74**: 1458–9.

Mulcahy, G.B. & Mulcahy, D.L. 1988. Induced polarity as an index of pollination-triggered stylar activation. In Cresti, M., Gori, P. & Pacini, E. (eds), *Sexual Reproduction in Higher Plants*, 327–32. Berlin: Springer.

Müller, F. 1877. Flowers and insects. *Nature* **17**: 78–9.

Müller, F. 1888. Zweimännige Zingiberaceenblumen. *Ber. Deutsch. Bot. Ges.* **6**: 95–100.

Müller, F. 1896. Einige Bemerkungen über Bromeliaceen. *Flora* **82**: 314–28.

Müller, H. 1873. *Die Befruchtung der Blumen durch Insekten und die gegenseitigen Anpassungen beider.* Leipzig: Engelmann.

Müller, H. 1876. On the relation between flowers and insects. *Nature* **15**: 178–80.

Müller, H. 1879. Die Wechselbeziehungen zwischen den Blumen und den ihre Kreuzung vermittelnden Insekten. In Schenk, A. (ed.), *Handbuch der Botanik I*: 1–112. Breslau: Trewent.

Müller, H. 1883. Arbeitstheilung bei Staubgefässen von Pollenblumen. *Kosmos* **7**: 241–59.

Müller, L. 1926. Zur biologischen Anatomie der Blüte von *Ceropegia woodii* Schlechter. *Biol. Gen.* **2**: 799–814.

Müller, L. 1929. Anatomisch-biomechanische Studien an maskierten Scrophulariaceenblüten. *Oesterr. Bot. Z.* **78**: 193–214.

Müller, L. 1931. Ueber den Bau und die Mechanik der Blüte von *Globba atrosanguinea*. *Oesterr. Bot. Z.* **80**: 149–61.

Müller-Doblies, D. 1970. Ueber die Verwandtschaft von *Typha* und *Sparganium* im Infloreszenz- und Blütenbau. *Bot. Jahrb. Syst.* **89**: 451–562.

Murbeck, S. 1914. Ueber die Baumechanik bei Aenderungen im Zahlenverhältnis der Blüte. *Lunds Univ. Aarsskr., n.F.*, **26** (3): 1–35.

Murray, N.A. & Johnson, D.M. 1987. Synchronous dichogamy in a Mexican anonillo *Rollinia jimenezii* var. *nelsonii* (Annonaceae). *Contr. Univ. Mich. Herb.* **16**: 173–8.

Musselman, L.J. & Visser, J.H. 1989. Taxonomy and natural history of *Hydnora* (Hydnoraceae). *Aliso* **12**: 317–26.

Myers, N. 1987. The extinction spasm impending: Synergisms at work. *Conserv. Biol.* **1**: 14–21.

Nair, N.C. & Kahate, S. 1961. Floral morphology and embryology of *Parkinsonia aculeata* L. *Phyton (Buenos Aires)* **17**: 77–90.

Nakamura, N., Fukushima, A., Iwayama, H. & Suzuki, H. 1991. Electrotropism of pollen tubes of *Camellia* and other plants. *Sex Pl. Reprod.* **4**: 138–43.

Neill, D.A. 1987. Trapliners in the trees: Hummingbird pollination of *Erythrina* sect. *Erythrina* (Leguminosae: Papilionoideae). *Ann. Missouri Bot. Gard.* **74**: 27–41.

Nelson, B.W., Absy, M.L., Barbosa, E.M. & Prance, G.T. 1985. Observations on flower visitors to *Bertholletia excelsa* H.B.K. and *Couratari tenuicarpa* A.C. Sm. (Lecythidaceae). *Acta Amaz., Supl.*, **15**: 225–34.

Nelson, E. 1954. *Gesetzmässigkeiten der Gestaltwandlung im Blütenbereich, ihre Bedeutung für das Problem der Evolution*. Nelson: Chernex-Montreux.

Nelson, E. 1967. Das Orchideenlabellum ein Homologon des einfachen medianen Petalums oder ein zusammengesetztes Organ? *Bot. Jahrb. Syst.* **87**: 22–35.

de Nettancourt, D. 1977. *Incompatibility in Angiosperms*. Berlin: Springer.

Newcombe, F.C. 1922/1924. Significance of the behavior of sensitive stigmas. I./ II. *Am. J. Bot.* **9**: 99–120/ **11**: 85–93.

Newman, S.W.H. & Kirchoff, B.K. 1992. Ovary structure in the Costaceae (Zingiberales). *Int. J. Pl. Sci.* **153**: 471–87.

Newstrom, L.E., Frankie, G.W. & Baker, H.G. 1991. Survey of long-term flowering patterns in lowland tropical rain forest trees at La Selva, Costa Rica. In Edelin, C. (ed.), *L'arbre. Biologie et developpement. Nat. Monspel.*, hors série, 345–66.

Nicholls, M.S. & Cook, C.D.K. 1986. The function of pollen tetrads in *Typha*. *Veröff. Geobot. Inst. ETH, Stiftung Rübel*, **87**: 112–19.

Nielsen, I. 1981. Tribe 5. Ingeae Benth. (1865). In Polhill, R.M. & Raven, P.H. (eds), *Advances in Legume Systematics*, Part 1, 173–90. Kew: Royal Botanic Gardens.

Nierenberg, L. 1972. The mechanism for the maintenance of species integrity in sympatrically occurring equitant oncidiums in the Caribbean. *Amer. Orch. Soc. Bull.* **41**: 873–82.

Nieuwenhuis – von Uexküll-Güldenbandt, M. 1907. Extraflorale Zuckerausscheidungen und Ameisenschutz. *Ann. Jard. Bot. Buitenzorg, Sér. 2*, **6**: 195–328.

Niklas, K.J. 1985. The aerodynamics of wind pollination. *Bot. Rev.* **51**: 328–86.

Niklas, K.J. 1987. Aerodynamics of wind pollination. *Sci. Amer.* **255** (7): 90–5.

Niklas, K.J. 1992. *Plant Biomechanics*. Chicago: University of Chicago Press.

Nilsson, L.A. 1988. The evolution of flowers with deep corolla tubes. *Nature* **334**: 147–9.

Nilsson, L.A. 1992. Orchid pollination biology. *Trends Ecol. Evol.* **7**: 255–9.

Nilsson, L.A. & Rabakonandrianina, E. 1988. Hawk-moth scale analysis and pollination specialization in the epilithic Malagasy endemic *Aerangis ellisii* (Reichenb. fil.) Schltr. (Orchidaceae). *Bot. J. Linn. Soc.* **97**: 49–61.

Nilsson, L.A., Jonsson, L., Ralison, L. & Randrianjohany, E. 1985. Monophily and pollination mechanisms in *Angraecum arachnites* Schltr. (Orchidaceae) in a guild of long-tongued hawk-moths (Sphingidae) in Madagascar. *Biol. J. Linn. Soc.* **26**: 1–19.

Nilsson, L.A., Jonsson, L., Ralison, L. &

Randrianjohany, E. 1987. Angraecoid orchids and hawkmoths in Central Madagascar: Specialized pollination systems and generalist foragers. *Biotropica* **19**: 310–8.

Nilsson, L.A., Jonsson, L., Rason, L. & Randrianjohany, E. 1986. The pollination of *Cymbidiella flabellata* (Orchidaceae) in Madagascar: A system operated by sphecid wasps. *Nord. J. Bot.* **6**: 411–22.

Nilsson, L.A., Rabakonandrianina, E., Razananaivo, R. & Randriamanindry, J.-J. 1992. Long pollinia on eyes: Hawk-moth pollination of *Cynorkis uniflora* Lindley (Orchidaceae) in Madagascar. *Bot. J. Linn. Soc.* **109**: 145–60.

Noel, A.R.A. 1983. The endothecium – a neglected criterion in taxonomy and phylogeny? *Bothalia* **14**: 833–8.

Nogler, G.A. 1984. Gametophytic apomixis. In Johri, B.M. (ed.), *Embryology of Angiosperms*, 475–518. Berlin: Springer.

Nooteboom, H. 1988. Magnoliaceae. In de Wilde, W.J.J.O. (ed.), *Flora Malesiana*, Ser. I, **10** (3): 561–605. Dordrecht: Kluwer.

Norman, E.M. & Clayton, D. 1986. Reproductive biology of two Florida pawpaws: *Asimina obovata* and *A. pygmaea* (Annonaceae). *Bull. Torrey Bot. Club* **113**: 16–22.

Noronha, H. 1949. Sobre as Aristolochiaceae medicinais. *Rev. Flora Med.* **16** (3): 75–88.

Nur, N. 1976. Studies on pollination in Musaceae. *Ann. Bot.* **40**: 167–77.

Oehler, E. 1927. Entwicklungsgeschichtlich-cytologische Untersuchungen an einigen saprophytischen Gentianaceen. *Planta* **3**: 641–733.

Okada, H. 1990. Reproductive biology of *Polyalthia littoralis* (Annonaceae). *Pl. Syst. Evol.* **170**: 237–45.

Okamoto, M. 1984. Centrifugal ovule inception. I. Sequence of ovule inception in *Silene cucubalus*. *Bot. Mag. (Tokyo)* **97**: 345–53.

Okpon, E.N.U. 1969. Morphological notes on the genus *Cassia* III. Floral ontogeny and aestivation. *Notes Roy. Bot. Gard. Edinburgh* **29**: 339–42.

Olesen, J.M. 1992. Flower mining by moth larvae vs. pollination by beetles and bees in the cauliflorous *Sapranthus palanga* (Annonaceae) in Costa Rica. *Flora* **187**: 9–15.

Olmstead, R.G. 1989. The origin and function of self-incompatibility in flowering plants. *Sex Pl. Reprod.* **2**: 127–36.

Olmstead, R.G., Michaels, H.J., Scott, K.M. & Palmer, J.D. 1992. Monophyly of the Asteridae and identification of their major lineages inferred from DNA sequences of rbcL. *Ann. Missouri Bot. Gard.* **79**: 249–65.

Opler, P.A. 1983. Nectar production in tropical exosystems. In Bentley, B. & Elias, T. (eds), *The Biology of Nectaries*, 30–79. New York: Columbia University Press.

Opler, P.A., Baker, H.G. & Frankie, G.W. 1975. Reproductive biology of some Costa Rican *Cordia* species (Boraginaceae). *Biotropica* **7**: 234–47.

Opler, P.A., Frankie, G.W. & Baker, H.G. 1976. Rain-fall as a factor in the release timing, and synchronization of anthesis by tropical trees and shrubs. *J. Biogeogr.* **3**: 231–6.

Ormond, W.T., Pinheiro, M.C.B. & Cortella de Castells, A.R. 1981. A contribution to the floral biology and reproductive system of *Couroupita guianensis* Aublet (Lecythidaceae). *Ann. Missouri Bot. Gard.* **68**: 514–23.

Ornduff, R. 1978. Reproductive characters and taxonomy. *Syst. Bot.* **3**: 420–7.

Ornduff, R. 1992. Historical perspectives on heterostyly. In Barrett, S.C.H. (ed.), *Evolution and Function of Heterostyly*, 31–9. Berlin: Springer.

Ornduff, R. & Dulberger, R. 1978. Floral enantiomorphy and the reproductive system of *Wachendorfia paniculata* (Haemodoraceae). *New Phytol.* **80**: 427–34.

Osche, G. 1983. Optische Signale in der Coevolution von Pflanze und Tier. *Ber. Deutsch. Bot. Ges.* **96**: 1–27.

Overland, L. 1960. Endogenous rhythm in opening and odor of flowers of *Cestrum nocturnum*. *Am. J. Bot.* **47**: 378–82.

Owen, D.F. 1971. *Tropical Butterflies*. Oxford: Clarendon Press.

Owens, S.J. 1989. Stigma, style, pollen, and the pollen-stigma interaction in Caesalpinioideae. *Monogr. Syst. Bot. Missouri Bot. Gard.* **29**: 113–26.

Owens, S.J. 1990. The morphology of the wet, non-papillate (WN) stigma form in the tribe Caesalpinieae (Caesalpinioideae: Leguminosae). *Bot. J. Linn. Soc.* **104**: 293–302.

Owens, S.J. & Lewis, G.P. 1989. Taxonomic and functional implications of stigma morphology in species, *Chamaecrista* and *Senna* (Leguminosae: Caesalpinioideae). *Pl. Syst. Evol.* **163**: 93–105.

Owens, S.J. & Stirton, C.H. 1989. Pollen, stigma, and style interactions in the Leguminosae. *Monogr. Syst. Bot. Missouri Bot. Gard.* **29**: 105–12.

Pacini, E. 1990. Tapetum and microspore function. In Blackmore, S. & Knox, R.B. (eds), *Microspores: Evolution and Ontogeny*, 213–37. London: Academic Press.

Pacini, E. & Sarfatti, G. 1978. The reproductive calendar of *Lycopersicon peruvianum* Mill. *Bull. Soc. Bot. France, Actual. Bot.*, **125**: 295–9.

Padmanabhan, D., Regupathy, D. & Veni, S.P. 1978. Gynoecial ontogeny in *Enicostemma littorale* Blume. *Proc. Indian Acad. Sci. B* **87**: 83–92.

Paige, K.N. & Whitham, T.G. 1985. Individual and population shifts in flower color by scarlet gilia: A mechanism for pollinator tracking. *Science* **227**: 315–7.

Pandey, K.K. 1960. Evolution of gametophytic and sporophytic systems of self-incompatibility in angiosperms. *Evolution* **14**: 98–115.

Pandey, K.K. 1980. Overcoming incompatibility and promoting genetic recombination in flowering plants. *New Zeal. J. Bot.* **17**: 645–63.

Pascher, A. 1960. Ueber die Wasserkelche von *Datura* und *Anisodus* und über das Vorkommen freien Wassers in den Bälgen von *Paeonia*. *Flora* **148**: 517–28.

Pass, A. 1940. Das Auftreten verholzter Zellen in Blüten und Blütenknospen. *Oesterr. Bot. Z.* **89**: 119–64, 169–210.

Paulus, H.F. 1978. Co-Evolution zwischen Blüten und ihren tierischen Bestäubern. *Sonderbd Naturwiss. Ver. Hamburg* **2**: 51–81.

Paulus, H.F. 1988. Co-Evolution und einseitige Anpassungen in Blüten-Bestäuber-Systemen: Bestäuber als Schrittmacher in der Blütenevolution. *Verh. Deutsch. Zool. Ges.* **81**: 25–46.

Payer, J.-B. 1857. *Traité d'organogénie de la fleur*. Paris: Masson.

Pazy, B. 1984. Insect induced self-pollination. *Pl. Syst. Evol.* **144**: 315–20.

Peakall, R., Handel, S.N. & Beattie, A.J. 1991. The evidence for, and importance of, ant pollination. In Huxley, C.R. & Cutler, D.F. (eds), *Ant-Plant Interactions*, 421–9. Oxford University Press.

Pellmyr, O. 1985. *Cyclocephala*: Visitor and probable pollinator of *Caladium bicolor* (Araceae). *Acta Amaz.* **15**: 269–72.

Pellmyr, O. 1992. Evolution of insect pollination and angiosperm diversification. *Trends Ecol. Evol.* **7**: 46–9.

Pellmyr, O. & Thien, L.B. 1986. Insect reproduction and floral fragrances: Keys to the evolution of the angiosperms. *Taxon* **35**: 76–85.

Pellmyr, O., Thien, L.B., Bergström, G. & Groth, I. 1990. Pollination of New Caledonian Winteraceae: Opportunistic shifts or parallel radiation with their pollinators? *Pl. Syst. Evol.* **173**: 143–57.

Pellmyr, O., Tang, W., Groth, I., Bergström, G. & Thien, L.B. 1991. Cycad cone and angiosperm floral volatiles: Inferences for the evolution of insect pollination. *Biochem. Syst. Ecol.* **19**: 623–7.

Percival, J. 1921. *The Wheat Plant*. London: Duckworth.

Percival, M.S. 1965. *Floral Biology*. Oxford: Pergamon.

Percival, M.S. 1974. Floral ecology of coastal scrub in Southeast Jamaica. *Biotropica* **6**: 104–29.

Pereira-Noronha, M.R., Silberbauer-Gottsberger, I. & Gottsberger, G. 1982. Biologia floral de *Stylosanthes* (Fabaceae) no Cerrados de Botucatu, Estado de São Paulo. Rev. Brasil. Biol. **42**: 595–605.

Periasamy, K. & Sampoornam, C. 1984a. The morphology and anatomy of ovule and fruit development in *Arachis hypogaea* L. *Ann. Bot.* **53**: 399–411.

Periasamy, K. & Sampoornam, C. 1984b. Studies on the hypanthial tube, androecium and pollination in *Arachis hypogaea* L. *Beitr. Biol. Pfl.* **58**: 403–11.

Perry, D.R. 1984. The canopy of the tropical rain forest. *Sci. Am.* **251** (5): 138–47.

Perry, D.R. 1990. Tropical biology. A science on the sidelines. In Head, S. & Heinzman, R. (eds), *Lessons of the Rainforest*, 25–36. San Francisco: Sierra Club Books.

Pettersson, B. 1989. Pollination in the African species of *Nervilia* (Orchidaceae). *Lindleyana* **4**: 33–41.

Pettitt, J.M. 1984. Aspects of flowering and pollination in marine angiosperms. *Oceanogr. Mar. Biol. Ann. Rev.* **22**: 315–42.

Pettitt, J.M. & Jermy, A.C. 1975. Pollen in hydrophilous angiosperms. *Micron* **5**: 377–405.

Pettitt, J.M., Ducker, S.C. & Knox, R.B. 1981. Submarine pollination. *Sci. Am.* **244** (3): 92–101.

Peyritsch, J. 1872. Ueber Pelorienbildungen. *Sitzber. Akad. Wiss. Wien*, Math.-Naturwiss. Cl., Abt. I, **66**: 125–59.

Philbrick, C.T. 1984. Pollen tube growth within vegetative tissues of *Callitriche* (Callitrichaceae). *Am. J. Bot.* **71**: 882–6.

Philbrick, C.T. 1988. Evolution of underwater outcrossing from aerial pollination systems: A hypothesis. *Ann. Missouri Bot. Gard.* **75**: 836–41.

Philipson, W.R. 1974. Ovular morphology and the major classification of the dicotyledons. *Bot. J. Linn. Soc.* **68**: 89–108.

Philipson, W.R. 1977. Ovular morphology and the classification of dicotyledons. *Pl. Syst. Evol.*, Suppl. 1, 123–40.

Philipson, W.R. 1985. Is the grass gynoecium monocarpellary? *Am. J. Bot.* **72**: 1954–61.

Philipson, W.R. 1986. Monimiaceae. In van Steenis, C.G.G.J. (ed.), *Flora Malesiana* I, **10** (2): 255–326. Dordrecht: Nijhoff.

Phillips, R.C., McMillan, C. & Bridges, K.W. 1981. Phenology and reproductive physiology of *Thalassia testudinum* from the western tropical Atlantic. *Aquat. Bot.* **11**: 263–77.

Pickersgill, B. 1983. Dispersal and distribution in crop plants. *Sonderbd. Naturwiss. Ver. Hamburg* **7**: 285–301.

van der Pijl, L. 1930. Uit het leven van enkele gevoelige tropische bloemen, speciall van de 'horlogebloemen'. *Trop. Natuur* **19**: 161–9, 190–6.

van der Pijl, L. 1936. Fledermäuse und Blumen. *Flora* **31**: 1-40.

van der Pijl, L. 1937. Biological and physiological observations on the inflorescence of *Amorphophallus*. *Rec. Trav. Bot. Neerl.* **34**: 157–67.

van der Pijl, L. 1938. Disharmony between asiatic flower-birds and american bird-flowers. *Ann. Jard. Bot. Buitenzorg* **48**: 17–26.

van der Pijl, L. 1939. Over de meeldraden van enkele Melastomataceae. *Trop. Natuur* **28**: 169–72.

van der Pijl, L. 1941a. Houtbij-bloemen bij *Costus*, *Bauhinia*, *Centrosema* en *Thunbergia*. *Trop. Natuur* **30**: 5–14.

van der Pijl, L. 1941b. Flagelliflory and cauliflory as adaptations to bats in *Mucuna* and other plants. *Ann. Bot. Gard. Buitenzorg* **51**: 83–93.

van der Pijl, L. 1953. On the flower biology of some plants from Java – with general remarks on fly-traps (species of *Annona*, *Artocarpus*, *Typhonium*, *Gnetum*, *Arisaema* and *Abroma*). *Ann. Bogor.* **1**: 77–99.

van der Pijl, L. 1954. *Xylocopa* and flowers in the tropics. I-III. *Proc. K. Ned. Akad. Wet.* C **57**: 413–23, 541–62.

van der Pijl, L. 1956. Remarks on pollination by bats in the genera *Freycinetia*, *Duabanga* and *Haplophragma* and on chiropterophily in general. *Acta Bot. Neerl.* **5**: 135–44.

van der Pijl, L. 1960/1961. Ecological aspects of flower evolution, I. Phyletic evolution./ II. Zoophilous flower classes. *Evolution* **14**: 403–16/ **15**: 44–59.

van der Pijl, L. 1969. Evolutionary action of tropical animals on the reproduction of plants. *Biol. J. Linn. Soc.* **1**: 85–96.

van der Pijl, L. 1972. Functional considerations and observations on the flowers of some Labiatae. *Blumea* **20**: 93–103.

van der Pijl, L. & Dodson, C. 1966. *Orchid Flowers. Their Pollination and Evolution*. Coral Gables, Florida: University of Miami Press.

Podoler, H., Galon, I. & Gazit, S. 1984. The role of nitidulid beetles in natural pollination of annona in Israel. *Acta Oecol., Oecol. Appl.* **5**: 369–381.

Pohl, F. 1927. Die anatomischen Grundlagen für die Gleitfallenfunktion von *Stanhopea tigrina* und *S. oculata*. *Jahrb. Wiss. Bot.* **66**: 556–77.

Pohl, F. 1929. Oelüberzüge verschiedener Pflanzenorgane, besonders der Blüten. *Jahrb. Wiss. Bot.* **70**: 565–655.

Pohl, F. 1935. Zwei *Bulbophyllum*-Arten mit besonders bemerkenswert gebauten Gleit- und Klemmfallenblumen. *Beih. Bot. Centralbl. A* **53**: 501–18.

Pohl, F. 1937. Die Pollenerzeugung der Windblütler. Untersuchungen zur Morphologie und Biologie des Pollens VI. *Beih. Bot. Centralbl. A* **56**: 365–470.

Polack, J.M. 1900. Untersuchungen über die Staminodien der Scrophulariaceen. *Oesterr. Bot. Z.* **50**: 33–41, 87–90, 123–32, 164–7.

Polhill, R.M. 1976. Genisteae (Adans.) Benth. and related tribes (Leguminosae). In Heywood, V.H. (ed.), *Botanical Systematics* **1**: 143–368. London: Academic Press.

Polhill, R.M. & Raven, P.H. (eds) 1981. *Advances in Legume Systematics* 1/2. Kew: Royal Botanic Gardens.

Polhill, R.M., Raven, P.H. & Stirton, C.H. 1981. Evolution and systematics of the Leguminosae. In Polhill, R.M. & Raven, P.H.

(eds), *Advances in Legume Systematics* **1**: 1–26. Kew Royal Botanic Gardens.

Polhill, R.M. & Vidal, J.E. 1981. Tribe 1. Caesalpinieae. In Polhill, R.M. & Raven, P.H. (eds), *Advances in Legume Systematics* **1**: 81–95. Kew: Royal Botanic Gardens.

Policansky, D. 1982. Sex change in plants and animals. *Ann. Rev. Ecol. Syst.* **13**: 471–95.

Poncy, O. & Lobreau-Callen, D. 1978. Le genre *Pararistolochia*, Aristolochiaceae d'Afrique tropicale. *Adansonia*, Sér. 2, **17**: 465–94.

Poppendieck, H.-H. 1987. Monoecy and sex changes in *Freycinetia* (Pandanaceae). *Ann. Missouri Bot. Gard.* **74**: 314–20.

Porsch, O. 1908/1909. Die deszendenztheoretische Bedeutung sprunghafter Blütenvariationen und korrelativer Abänderung für die Orchideenflora Südbrasiliens. *Z. Ind. Abst.-Vererb.* **1**: 69–121/ 195–238, 352–76.

Porsch, O. 1924a. Vogelblumenstudien I. *Jahrb. Wiss. Bot.* **63**: 553–706.

Porsch, O. 1924b. Methodik der Blütenbiologie. In Abderhalden, E. (ed.), *Handbuch der biologischen Arbeitsmethoden* XI (1), 395–514. Berlin: Urban & Schwarzenberg.

Porsch, O. 1931. *Crescentia* – eine Fledermausblume. *Oesterr. Bot. Z.* **80**: 31–44.

Porsch, O. 1935. Zur Blütenbiologie des Affenbrotbaumes. *Oesterr. Bot. Z.* **84**: 219–24.

Porsch, O. 1936. Säugetiere als Blumenausbeuter und die Frage der Säugetierblume. III. *Biol. Gen.* **12**: 1–21.

Pötter, U. & Klopfer, K. 1987. Untersuchungen zur Blatt- und Blütenentwicklung bei *Galium aparine* L. (Rubiaceae). *Flora* **179**: 305–14.

Prakash, N. & Alexander, J.H. III. 1984. Self-incompatibility in *Austrobaileya scandens*. In Williams, E.G. & Knox, R.B. (eds), *Pollination '84*, 214–16. Melbourne: School of Botany, University of Melbourne.

Prance, G.T. 1976. The pollination and anthophore structure of some Amazonian Lecythidaceae. *Biotropica* **8**: 235–41.

Prance, G.T. 1980a. A note on the pollination of *Nymphaea amazonum* Mart. & Zucc. (Nymphaeaceae). *Brittonia* **32**: 505–7.

Prance, G.T. 1980b. A note on the probable pollination of *Combretum* by *Cebus* monkeys. *Biotropica* **12**: 239.

Prance, G.T. 1985. The pollination of Amazonian plants. In Prance, G.T. & Lovejoy, T.E. (eds), *Key Environments: Amazonia*, 166–91. New York: Pergamon Press.

Prance, G.T. 1990. Management and conservation of tropical ecosystems requires knowledge of plant/animal interactions: afterword. *Mem. New York Bot. Gard.* **55**: 186–7.

Prance, G.T. 1991. Rates of loss of biological diversity. In Spellerberg, I.F., Goldsmith, F.B. & Morris, M.G. (eds), *The Scientific Management of Temperate Communities for Conservation*, 27–44. Oxford: Blackwell.

Prance, G.T. & Anderson, G.B. 1977. Studies of the floral biology of neotropical Nymphaeaceae. III. *Acta Amaz.* **6**: 163–70.

Prance, G.T. & Arias, R.J. 1975. A study on the floral biology of *Victoria amazonica* (Poepp.) Sowerby (Nymphaeaceae). *Acta Amaz.* **5**: 109–39.

Prance, G.T. & Mori, S.A. 1979. The actinomorphic flowered New World Lecythidaceae. *Flora Neotropica Monograph* **21**: 1–270.

Prance, G.T. & Mori, S.A. 1987. Future Research. In Mori, S.A. (ed.), *The Lecythidaceae of a Lowland Neotropical Forest: La Fumée Mountain, French Guiana. Mem. New York Bot. Gard.* **44**: 156–63.

Prance, G.T. & White, F. 1988. The genera of Chrysobalanaceae – A study in practical and theoretical taxonomy and its relevance to evolutionary biology. *Phil. Trans. Roy. Soc. Lond. B* **320**: 1–184.

Prance, G.T., Idrobo, J.M. & Castano M., O.V. 1983. Mecanismos de polinizacion de *Eschweilera garagarae* Pittier en el Choco, Colombia. *Mutisia* **60**: 1–7.

Pridgeon, A.M. & Stern, W.L. 1983. Ultrastructure of osmophores in *Restrepia* (Orchidaceae). *Am. J. Bot.* **70**: 1233–43.

Primack, R.B. 1985a. Longevity of individual flowers. *Ann. Rev. Ecol. Syst.* **16**: 15–38.

Primack, R.B. 1985b. Patterns of flowering phenology in communities, populations, individuals and single flowers. In White, J. (ed.), *Population Structure of Vegetation*, 571–93. Dordrecht: Junk.

Primack, R.B. 1987. Relationships among flowers, fruits, and seeds. *Ann. Rev. Ecol. Syst.* **18**: 409–30.

Primack, R.B. & Tomlinson, P.B. 1980. Variation in tropical forest breeding systems. *Biotropica* **12**: 229–31.

Primack, R.B., Ducke, N. & Tomlinson, P.B. 1981. Floral morphology in relation to pollination ecology in five Queensland coastal plants. *Austrobaileya* **1**: 346–55.

Proctor, M. & Yeo, P. 1973. *The Pollination of Flowers*. London: Collins.

Proença, C.E.B. 1992. Buzz pollination – older and more widespread than we think? *J. Trop. Ecol.* **8**: 115–20.

Puff, C. (ed.) 1991. The Genus *Paederia* L. (Rubiaceae-Paederieae): A Multidisciplinary Study. *Opera Bot. Belg.* **3**: 1–376.

Puff, C., Robbrecht, E. & Randrianasolo, V. 1984. Observations on the SE African-Madagascan genus *Alberta* and its ally *Nematostylis* (Rubiaceae, Alberteae), with a survey of the species and a discussion of the taxonomic position. *Bull. Jard. Bot. Natl Belg.* **54**: 293–366.

Puri, V. 1948. Studies in floral anatomy. V. On the structure and nature of the corona in certain species of the Passifloraceae. *J. Indian Bot. Soc.* **27**: 130–49.

du Puy, D. & Cribb, P. 1988. *The Genus Cymbidium*. London: Helm.

Qiu, Y.-L., Chase, M.W., Les, D.H. & Parks, C.R. 1993. Molecular phylogenetics of the Magnoliidae: A cladistic analysis of nucleotide sequences of the plastid gene rbcL. *Ann. Missouri Bot. Gard.* **80**: 587–606.

Rachmilevitz, T.& Fahn, A. 1975. The floral nectary of *Tropaeolum majus* L. – The nature of the secretory cells and the manner of nectar secretion. *Ann. Bot.* **39**: 721–8.

Raciborski, M. 1895. Die Schutzvorrichtung der Blütenknospen. *Flora* **81**: 151–94.

Raju, M.V.S. 1954. Pollination mechanism in *Passiflora foetida* L. *Proc. Natl Inst. Sci. India* **20**: 431–6.

Rama Devi, D. 1991. Floral anatomy of *Hypseocharis* (Oxalidaceae) with a discussion on its systematic position. *Pl. Syst. Evol.* **177**: 161–4.

Ramakrishna, T.M. & Arekal, G.D. 1979. Pollination biology of *Calotropis gigantea* (L.) R.Br. *Curr. Sci.* **48**: 212–3.

Raman, S. 1989/1990/1991. The trichomes of the corolla of the Scrophulariaceae. *Beitr. Biol. Pfl.* **64**: 127–40, 141–55, 199–212, 213–9, 357–75, 377–90/ **65**: 223–34, 235–48, 249–62/ **66**: 127–43.

Ramirez, B.W. & Gomez, L.D. 1978. Production of nectar and gums by flowers of *Monstera deliciosa* (Araceae) and of some species of *Clusia* (Guttiferae) collected by New World *Trigona* bees. *Brenesia* **14**: 407–12.

Ramirez, N., Sobrevilla, C., de Enrech, N.X. & Ruiz-Zapata, T. 1984. Floral biology and breeding system of *Bauhinia ungulata* (Leguminosae), a bat-pollination tree in Venezuela 'Llanos'. *Am. J. Bot.* **71**: 273–80.

Ramirez, N., Gil, C., Hokche, O., Seres, A. & Brito, Y. 1990. Biologia floral de una comunidad arbustiva tropical en la Guayana Venezolana. *Ann. Missouri Bot. Gard.* **77**: 383–97.

Ramirez-Domenéch, J.I. & Tucker, S.C. 1990. Comparative ontogeny of the perianth in mimosoid legumes. *Am. J. Bot.* **77**: 624–35.

Ramp, E. 1987. Funktionelle Anatomie des Gynoeciums bei *Staphylea*. *Bot. Helv.* **97**: 89–98.

Ramp, E. 1988. Struktur, Funktion und systematische Bedeutung des Gynoeciums bei Rutaceae und Simaroubaceae. Ph.D. Dissertation, University of Zurich. Zurich: ADAG.

Ramsey, M.W. 1988. Floret opening in *Banksia menziesii* R.Br.; the importance of nectarivorous birds. *Austral. J. Bot.* **36**: 225–32.

Rao, A.N. 1961. Fibrous thickenings in the anther epidermis of *Wormia burbidgei* Hook. *Curr. Sci.* **31**: 426.

Rao, M.M., Rao, K.L. & Bahadur, B. 1986. Significance of corolla handedness and stylar hairs in the pollination biology of some Papilionaceae. In Kapil, R.P. (ed.), *Pollination Biology – An Analysis*, 221–9. New Delhi: Inter-India Publications.

Rao, O.M. & Kumari, O.L. 1979. Germination loci of pollinia and their taxonomic significance. *Geobios* **6**: 163–5.

Rao, V.S. 1963. The epigynous glands of Zingiberaceae. *New Phytol.* **62**: 342–9.

Rao, V.S. 1971. The disk and its vasculature in the flowers of some dicotyledons. *Bot. Not.* **124**: 442–50.

Rao, V.S. & Sirdesmukh, K.B. 1956. The floral anatomy of *Delonix regia*. *J. Univ. Bombay* **25**: 35–44.

Raskin, I., Ehmann, A., Melander, W.R. & Meeuse, B.J.D. 1987. Salicylic acid: A natural inducer of heat production in *Arum* lilies. *Science* **237**: 1601–2.

Rasmussen, F.N. 1982. The gynostemium of the neottioid orchids. *Opera Bot.* **65**: 1–96.

Rasmussen, F.N. 1985a. Orchids. In Dahlgren, R., Clifford, H. & Yeo, P.F. (eds), *The Families of Monocotyledons:*

Structure, Evolution and Taxonomy, 249–74. Berlin: Springer.

Rasmussen, F.N. 1985b. The gynostemium of *Bulbophyllum ecornutum* (J.J. Smith) J.J. Smith (Orchidaceae). *Bot. J. Linn. Soc.* **91**: 447–56.

Rasmussen, F.N. 1986a. On the various contrivances by which pollinia are attached to viscidia. *Lindleyana* **1**: 21–32.

Rasmussen, F.N. 1986b. Ontogeny and phylogeny in Orchidaceae. *Lindleyana* **1**: 114–24.

Rathcke, B. & Lacey, E.B. 1985. Phenological patterns of terrestrial plants. *Ann. Rev. Ecol. Syst.* **16**: 179–214.

Rauh, W. 1979. *Die grossartige Welt der Sukkulenten* (Ed. 2). Berlin: Parey.

Rauh, W. (Ed.) 1984. Anatomisch-biochemische Untersuchungen an Euphorbien, Teil 1. *Trop. Subtrop. Pflanzenwelt* **45**: 1–108. Wiesbaden: Steiner.

Rauh, W. & Reznik, H. 1951. Histogenetische Untersuchungen an Blüten- und Infloreszenzachsen. I.Teil. Die Histogenese becherförmiger Blüten- und Infloreszenzachsen, sowie der Blütenachsen einiger Rosoideen. *Sitzungsber. Heidelberger Akad. Wiss.*, Math.-Naturwiss. Kl., **1951** (3): 1–71.

Raven, P.H. 1973. Why are bird-visited flowers predominantly red? *Evolution* **26**: 674.

Raven, P.H. 1974. *Erythrina* (Fabaceae): Achievements and opportunities. *Lloydia* **37**: 321–31.

Raven, P.H. 1977. *Erythrina* Symposium II. *Erythrina* (Fabaceae: Faboideae): Introduction to Symposium II. *Lloydia* **40**: 401–6.

Raven, P.H. 1979. *Erythrina* Symposium III. *Erythrina* (Fabaceae: Faboideae): Introduction to Symposium III. *Ann. Missouri Bot. Gard.* **66**: 417–21.

Raven, P.H. (ed.), 1980a. *Research Priorities in Tropical Biology*. Washington, D.C.: National Academy of Sciences.

Raven, P.H. 1980b. A survey of reproductive biology in Onagraceae. *New Zeal. J. Bot.* **17**: 575–93.

Raven, P.H. 1982. *Erythrina* (Fabaceae: Faboideae): Introduction to Symposium IV. *Allertonia* **3** (1): 1–6.

Raven, P.H. 1988. Onagraceae as a model of plant evolution. In Gottlieb, L.D. & Jain, S.K. (eds), *Plant Evolutionary Biology*, 85–107. London: Chapman & Hall.

Reese, G. 1973. The structure of the highly specialized carrion-flowers of stapeliads. *Cact. Succ. J.* **45**: 18–29.

Regal, P.J. 1977. Ecology and evolution of flowering plant dominance. *Science* **196**: 622–9.

Regal, P.J. 1982. Pollination by wind and animals: Ecology of geographic patterns. *Ann. Rev. Ecol. Syst.* **13**: 497–524.

Reif, W.-E. 1975. Lenkende und limitierende Faktoren in der Evolution. *Acta Biotheor.* **24**: 136–62.

Reinsch, J. 1926. Ueber die Entstehung der Aestivationsformen von Kelch und Blumenkrone dikotyler Pflanzen und über die Beziehungen der Deckungsweisen zur Gesamtsymmetrie der Blüte. *Flora* **121**: 77–124.

Renner, S.S. 1983. The widespread occurrence of anther destruction by *Trigona* bees in Melastomataceae. *Biotropica* **15**: 251–6.

Renner, S.S. 1984a. Phänologie, Blütenbiologie und Rekombinationssysteme einiger zentralamazonischer Melastomataceen. Doctoral Dissertation, University of Hamburg.

Renner, S.S. 1984b. Reproduktive Isolation sympatrischer *Miconia*-Arten in der Neotropis. In Ehrendorfer, F. (Ed.), *Mitteilungsband Botaniker-Tagung Wien*, 133. Wien: Institut für Botanik, Universität Wien.

Renner, S.S. 1987. Reproductive biology of *Bellucia* (Melastomataceae). *Acta Amaz.* **16/17**: 197–208.

Renner, S.S. 1989a. Floral biological observations on *Heliamphora tatei* (Sarraceniaceae) and other plants from Cerro de la Neblina in Venezuela. *Pl. Syst. Evol.* **163**: 21–9.

Renner, S.S. 1989b. A survey of reproductive biology in neotropical Melastomataceae and Memecylaceae. *Ann. Missouri Bot. Gard.* **76**: 496–518.

Renner, S.S. 1990. Reproduction and evolution in some genera of neotropical Melastomataceae. *Mem. New York Bot. Gard.* **55**: 143–52.

Retallack, G. & Dilcher, D.L. 1981. Arguments for a glossopterid ancestry of angiosperms. *Paleobiology* **7**: 54–67.

Richards, A.J. 1986. *Plant Breeding Systems*. London: Allen & Unwin.

Richards, A.J. 1990. The implications of reproductive versatility for the structure of grass populations. In Chapman, G.P. (ed.), *Reproductive Versatility in the Grasses*, 131–53. Cambridge University Press.

Richards, P.W. 1952. *The Tropical Rain Forest: An Ecological Study*. Cambridge University Press.

Richter, P.H. & Schranner, R. 1978. Leaf arrangement. Geometry, morphogenesis, and classification. *Naturwissenschaften* **65**: 319–27.

Riedl, R. 1977. A systems-analytical approach to macroevolutionary phenomena. *Quart. Rev. Biol.* **52**: 351–70.

Riedl, R. 1978. *Order in Living Organisms*. Chichester: Wiley.

Ritterbusch, A. 1991. Morphologisches Beschreibungsmodell tubiflorer Kronen, ein Beitrag zu Terminologie und Morphologie der Asteriden-Blüte. *Bot. Jahrb. Syst.* **112**: 329–45.

Robacker, D.C., Meeuse, B.J.D. & Erickson, E.H. 1988. Floral aroma. *Bioscience* **38**: 390–8.

Robbrecht, E. 1988. Tropical woody Rubiaceae: characteristic features and progressions: contributions to a new subfamilial classification. *Opera Bot. Belg.* **1**: 1–271.

Robertson, C.R. 1895. The philosophy of flower seasons. *Am. Nat.* **29**: 97–117.

Robinson, H. 1985. Observations on fusion and evolutionary variability in the angiosperm flower. *Syst. Bot.* **10**: 105–9.

Robyns, W. 1931. L'organisation florale des Solanacées zygomorphes. *Mém. Acad. Roy. Belg.*, Cl. Sci., **11** (8): 1–82.

Robyns, W. & Lebrun, J. 1929. Etude critique sur les Labiatacées monadelphes. *Ann. Soc. Sci. Bruxelles*, Sér. B, **49**: 88–106.

Rocha, O.J. & Stephenson, A.G. 1991. Order of fertilization within the ovary in *Phaseolus coccineus* L. (Leguminosae). *Sex Pl. Reprod.* **4**: 126–31.

Rohrhofer, J. 1931. Morphologische Studien an den Staminodien der Bignoniaceae. *Oesterr. Bot. Z.* **80**: 1–30.

Rohweder, O. 1963. Anatomische und histogenetische Untersuchungen an Laubsprossen und Blüten der Commelinaceen. *Bot. Jahrb. Syst.* **82**: 1–99.

Rohweder, O. 1967. Karpellbau und Synkarpie bei Ranunculaceen. *Ber. Schweiz. Bot. Ges.* **77**: 376–432.

Rohweder, O. 1969. Beiträge zur Blütenmorphologie und -anatomie der Commelinaceen mit Anmerkungen zur Begrenzung und Gliederung der Familie. *Ber. Schweiz. Bot. Ges.* **79**: 199–220.

Rohweder, O. 1972. Das Andröcium der Malvales und der 'Konservatismus' des Leitgewebes. *Bot. Jahrb. Syst.* **92**: 155–67.

Rohweder, O. & Endress, P.K. 1983. *Samenpflanzen. Morphologie und Systematik der Angiospermen und Gymnospermen.* Stuttgart: Thieme.

Rohweder, O. & Huber, K. 1974. Centrospermen-Studien. 7. Beobachtungen und Anmerkungen zur Morphologie und Entwicklungsgeschichte einiger Nyctaginaceen. *Bot. Jahrb. Syst.* **94**: 327–59.

Romero, G.A. 1990. Phylogenetic relationships in subtribe Catasetinae (Orchidaceae, Cymbidieae). *Lindleyana* **5**: 160–81.

Romero, G.A. 1992. Non-functional flowers in *Catasetum* orchids (Catasetinae, Orchidaceae). *Bot. J. Linn. Soc.* **109**: 305–13.

Romero, G.A. & Nelson, C.E. 1986. Sexual dimorphism in *Catasetum* orchids: Forcible pollen emplacement and male flower competition. *Science* **232**: 1538–40.

Ronse Decraene, L.-P. 1992. The Androecium of the Magnoliophytina: Characterisation and Systematic Importance. Ph.D. Thesis, Catholic University of Leuven.

Ronse Decraene, L.-P. & Smets, E. 1990a. The systematic relationships between Begoniaceae and Papaveraceae: A comparative study of their floral development. *Bull. Jard. Bot. Natl Belg.* **60**: 229–73.

Ronse Decraene, L.-P. & Smets, E. 1990b. The floral development of *Popowia whitei* (Annonaceae). *Nord. J. Bot.* **10**: 411–20.

Ronse Decraene, L.-P. & Smets, E. 1991a. Morphological studies in Zygophyllaceae. I. The floral development and vascular anatomy of *Nitraria retusa*. *Am. J. Bot.* **78**: 1438–48.

Ronse Decraene, L.-P. & Smets, E. 1991b. The impact of receptacular growth on polyandry in the Myrtales. *Bot. J. Linn. Soc.* **105**: 257–69.

Ronse Decraene, L.-P. & Smets, E. 1992. Complex polyandry in the Magnoliatae: Definition, distribution and systematic value. *Nord. J. Bot.* **12**: 621–49.

Rosen, W.G. 1971. Pistil-pollen interactions in *Lilium*. In Heslop-Harrison, J. (Ed.), *Pollen: Development and Physiology*, 239–54. London: Butterworths.

Ross, R. 1982. Initiation of stamens, carpels and receptacle in the Cactaceae. *Am. J. Bot.* **69**: 369–79.

Roubik, D.W. 1989. *Ecology and Natural History of Tropical Bees*. Cambridge University Press.

Roubik, D.W. & Ackerman, J.D. 1987. Long-term ecology of euglossine orchid-bees

(Apidae: Euglossini) in Panama. *Oecologia* **73**: 321–33.

Rourke, J. & Wiens, D. 1977. Convergent floral evolution in South African and Australian Proteaceae and its possible bearing on pollination by nonflying mammals. *Ann. Missouri Bot. Gard.* **64**: 1–17.

Rowan, M.K. 1974. Bird pollination of *Strelitzia*. *Ostrich* **45**: 40.

Rowley, G.D. 1980. The pollination mechanism of *Adenium* (Apocynaceae). *Natl Cact. Succ. J.* **35**: 2–5.

Rübsamen-Weustenfeld, T. 1991. Morphologische, embryologische und systematische Untersuchungen an Triuridaceae. *Bibl. Bot.* **140**: 1–113.

Rudall, P. 1981. Flower anatomy of the subtribe Hyptidinae (Labiatae). *Bot. J. Linn. Soc.* **83**: 251–62.

Ruggiero, C. & Andrade, V. M. de M. 1989: *Passiflora*. In Halevy, A.H. (ed.), *CRC Handbook of Flowering* VI: 495–506. Boca Raton, Florida: CRC Press.

Rutishauser, A. 1967. *Fortpflanzungsmodus und Meiose apomiktischer Blütenpflanzen*. Wien: Springer.

Rutishauser, A. 1969. Die embryologischen und cytogenetischen Grundlagen der Sameninkompatibilität. *Ber. Schweiz. Bot. Ges.* **79**: 5–48.

Rutishauser, R. 1983. Der Plastochronquotient als Teil einer quantitativen Blattstellungsanalyse bei Samenpflanzen. *Beitr. Biol. Pfl.* **57**: 323–57.

Safwat, F.M. 1962. The floral morphology and the evolution of the pollinating apparatus in Asclepiadaceae. *Ann. Missouri Bot. Gard.* **49**: 95–129.

Sage, T.L., Broyles, S.B. & Wyatt, R. 1990. The relationship between the five stigmatic chambers and two ovaries of milkweed (*Asclepias amplexicaulis* Sm.) flowers: A three-dimensional assessment. *Israel J. Bot.* **39**: 187–96.

Salleh, K.M. 1991. *Rafflesia, magnificent flower of Sabah*. Kota Kinabalu: Borneo Publishing Company.

Salleh, K.M. & Latiff, A. 1989. A new species of *Rafflesia* and notes on other species from Trus Madi Range, Sabah (Borneo). *Blumea* **34**: 111–16.

Sampson, F.B. 1980. Natural hybridism in *Pseudowintera* (Winteraceae). *New Zeal. J. Bot.* **18**: 43–51.

Sampson, F.B. 1983. A new species of *Zygogynum* (Winteraceae). *Blumea* **28**: 353–60.

Sampson, F.B. 1987. Stamen venation in the Winteraceae. *Blumea* **32**: 79–89.

Sampson, F.B. & Kaplan, D.R. 1970. Origin and development of the terminal carpel in *Pseudowintera traversii*. *Am. J. Bot.* **57**: 1185–96.

Sampson, F.B. & Tucker, S.C. 1978. Placentation in *Exospermum stipitatum* (Winteraceae). *Bot. Gaz. (Crawfordsville)* **139**: 215–22.

Sanders, L.C. & Lord, E.M. 1989. Directed movement of latex particles in the gynoecia of three species of flowering plants. *Science* **243**: 1606–8.

Sanders, L.C. & Lord, E.M. 1992. A dynamic role for the stylar matrix in pollen tube extension. *Int. Rev. Cytol.* **140**: 297–318.

Sanderson, M.J. & Donoghue, M.J. 1989. Patterns of variation in levels of homoplasy. *Evolution* **43**: 1781–95.

Sassen, M.M.A. 1974. The stylar transmitting tissue. *Acta Bot. Neerl.* **23**: 99–108.

Satina, S. 1944. Periclinal chimeras in *Datura* in relation to development and structure (A) of the style and stigma (B) of calyx and corolla. *Am. J. Bot.* **31**: 493–502.

Sattler, R. 1973. *Organogenesis of Flowers. A Photographic Text-Atlas*. Toronto: University of Toronto Press.

Sattler, R. 1978. 'Fusion' and 'continuity' in floral morphology. *Notes Roy. Bot. Gard. Edinburgh* **36**: 397–405.

Sattler, R. 1988. Homeosis in plants. *Am. J. Bot.* **75**: 1606-17.

Sattler, R. & Singh, V. 1973. Floral development of *Hydrocleis nymphoides*. *Can. J. Bot.* **51**: 2455–8.

Sattler, R. & Singh, V. 1978. Floral organogenesis of *Echinodorus amazonicus* Rataj and floral construction of the Alismatales. *Bot. J. Linn. Soc.* **77**: 141–56.

Sauer, H. 1933. Blüte und Frucht der Oxalidaceen, Linaceen, Geraniaceen, Tropaeolaceen und Balsaminaceen. *Planta* **19**: 417–81.

Sazima, I. & Sazima, M. 1989. Mamangavas e irapuas (Hymenoptera, Apoidea): Visitas, interações e consequências para polinização do maracuja (Passifloraceae). *Revta Bras. Entomol.* **33**: 109–18.

Sazima, M. & Sazima, I. 1978. Bat pollination of the passion flower, *Passiflora mucronata*, in southeastern Brazil. *Biotropica* **10**: 100–9.

Sazima, M. & Vogel, S. 1989. *Cyphomandra* species (Solanac.) visited by euglossine bees: floral fragrances as a reward for pollen dispersal. *9. Symposium Morphologie, Anatomie und Systematik, Vienna*, Abstr. 52.

Schatz, G. 1987. Pollination systems in neotropical Annonaceae. *XIV International Botanical Congress, Berlin*, Abstr. 5–21–7.

Schatz, G. 1990. Some aspects of pollination biology in Central American tropical forests. In Bawa, K.S. & Hadley, M. (eds), *Reproductive Ecology of Tropical Forest Plants*, 69–84. Paris: UNESCO.

Schemske, D.W. 1976. Pollination specificity in *Lantana camara* and *L. trifolia* (Verbenaceae). *Biotropica* **8**: 260–4.

Schemske, D.W. 1980. The evolutionary significance of extrafloral nectar production by *Costus woodsonii* (Zingiberaceae): An experimental analysis of ant protection. *J. Ecol.* **68**: 959–67.

Schemske, D.W. 1981. Floral convergence and pollinator sharing in two bee-pollinated, tropical herbs. *Ecology* **62**: 946–54.

Schemske, D.W. 1983a. Breeding system and habitat effects on fitness components in three Neotropical *Costus* (Zingiberaceae). *Evolution* **37**: 523–39.

Schemske, D.W. 1983b. Limits to specialization and coevolution in plant-animal mutualisms. In Nitecki, M.H. (ed.), *Coevolution*, 67–109. Chicago: University of Chicago Press.

Schemske, D.W. & Horvitz, C.C. 1984. Variation among floral visitors in pollination ability: A precondition for mutualism specialization. *Science* **225**: 519–21.

Schemske, D.W. & Lande, R. 1987. On the evolution of plant mating systems. *Am. Nat.* **130**: 804–6.

Schick, B. 1980/1982a. Untersuchungen über die Biotechnik der Apocynaceenblüte. I. Morphologie und Funktion des Narbenkopfes/ II. Bau und Funktion des Bestäubungsapparates. *Flora* **170**: 394–432/ **172**: 347–71.

Schick, B. 1982b. Zur Morphologie, Entwicklung, Feinstruktur und Funktion des Translators von *Periploca* L. (Asclepiadaceae). *Trop. Subtrop. Pflanzenwelt* **40**: 511–53. Wiesbaden: Steiner.

Schick, B. 1988/1989. Zur Anatomie und Biotechnik des Bestäubungsapparates der Orchideen. I/II. *Bot. Jahrb. Syst.* **110**: 215–62/ 289–323.

Schick, B. & Remus, S. 1984. Ueber die Integration der Nuptialnectarien in den Bestäubungsapparat der Apocynales. In Ehrendorfer, F. (ed.), *Mitteilungsband Botaniker-Tagung Wien*, 123. Vienna: Institut für Botanik der Universität.

Schill, R. & Pfeiffer, W. 1977. Untersuchungen an Orchideenpollinien unter besonderer Berücksichtigung ihrer Feinskulpturen. *Pollen Spores* **19**: 5–118.

Schill, R. & Jäkel, U. 1978. Beitrag zur Kenntnis der Asclepiadaceen-Pollinarien. *Trop. Subtrop. Pflanzenwelt* **22**: 51–170. Wiesbaden: Steiner.

Schill, R. & Wolter, M. 1986. Ontogeny of elastoviscin in the Orchidaceae. *Nord. J. Bot.* **5**: 575–80.

Schill, R., Baumm, A. & Wolter, M. 1985. Vergleichende Mikromorphologie der Narbenoberflächen bei den Angiospermen; Zusammenhänge mit Pollenoberflächen bei heterostylen Sippen. *Pl. Syst. Evol.* **148**: 185–214.

Schlessman, M.A. 1988. Gender diphasy ('sex choice'). In Lovett Doust, J. & Lovett Doust, L. (eds), *Plant Reproductive Ecology. Patterns and Strategies*, 139–153. Oxford University Press.

Schmid, B. 1992. Phenotypic variation in plants. *Evol. Trends Pl.* **6**: 45–60.

Schmid, R. 1972. Floral bundle fusion and vascular conservatism. *Taxon* **21**: 429–46.

Schmid, R. 1975. Two hundred years of pollination biology: an overview. *Biologist* **57**: 26–35.

Schmid, R. 1976. Filament histology and anther dehiscence. *Bot. J. Linn. Soc.* **73**: 303–15.

Schmid, R. 1985. Functional interpretations of the morphology and anatomy of septal nectaries. *Acta Bot. Neerl.* **34**: 125–8.

Schmid, R. 1988. Reproductive versus extra-reproductive nectaries – historical perspective and terminological recommendations. *Bot. Rev.* **54**: 179–232.

Schmucker, T. 1932. Physiologische und ökologische Untersuchungen an Blüten tropischer *Nymphaea*-Arten. *Planta* **16**: 376–412.

Schnepf, E., Witzig, F. & Schill, R. 1979. Ueber Bildung und Feinstruktur des Translators der Pollinarien von *Asclepias curassavica* und *Gomphocarpus fruticosus* (Asclepiadaceae). *Trop. Subtrop. Pflanzenwelt* **25**: 1–39. Wiesbaden: Steiner.

Schoch-Bodmer, H. 1939. Beiträge zur Kenntnis des Streckungswachstums der Gramineen-Filamente. *Planta* **30**: 168–204.

Schoen, D.J. & Lloyd, D.G. 1984. The selection of cleistogamy and heteromorphic diaspores. *Biol. J. Linn. Soc.* **23**: 303–22.

Schöffel, K. 1932. Untersuchungen über den Blütenbau der Ranunculaceen. *Planta* **17**: 315–71.

Schoute, J.C. 1935. On corolla aestivation and phyllotaxis of floral phyllomes. *Verh. Kon. Akad. Wet. Amsterdam*, Afd. Natuurk., Sect. 2, **34** (4): 1–77.

Schrire, B.D. 1989. A multidisciplinary approach to pollination biology in the Leguminosae. *Monogr. Syst. Bot. Missouri Bot. Gard.* **29**: 183–242.

Schubert, K. 1925. Zur Kenntnis der Blütenblätter-Epidermis. *Bot. Arch.* **12**: 226–89.

Schumann, K. 1890. *Neue Untersuchungen über den Blüthenanschluss*. Leipzig: Engelmann.

Scogin, R. 1983. Visible floral pigments and pollinators. In Jones, C.E. & Little, R.J. (eds), *Handbook of Experimental Pollination Biology*, 160–72. New York: Scientific and Academic Editions.

Scotland, R.W., Endress, P.K. & Lawrence, T.J. 1994. Corolla ontogeny and aestivation in Acanthaceae. *Bot. J. Linn. Soc.* **114**: 49–65.

Scott-Elliot, G.F. 1890. Note on the fertilization of *Musa*, *Strelitzia reginae*, and *Ravenala madagascariensis*. *Ann. Bot.* **4**: 259–63.

Scott-Elliot, G.F. 1891. Notes on the fertilisation of South African and Madagascar flowering plants. *Ann. Bot.* **5**: 333–405.

Scotti, L. 1911. Contribuzioni alla biologia fiorale delle Contortae. *Ann. Bot. (Roma)* **9**: 199–314.

Scribailo, R.W. & Tomlinson, P.B. 1992. Shoot and floral development in *Calla palustris* (Araceae-Calloideae). *Int. J. Pl. Sci.* **153**: 1–13.

Sedgley, M. & Blesing, M.A. 1982. Foreign pollination of the stigma of watermelon (*Citrullus lanatus* (Thunb.) Matsum. and Nakai). *Bot. Gaz. (Crawfordsville)* **143**: 210–15.

Sedgley, M. & Griffin, A.R. 1989. *Sexual Reproduction of Tree Crops*. London: Academic Press.

Seidenfaden, G. 1979. Orchid genera in Thailand VIII. *Bulbophyllum* Thou. *Dansk Bot. Ark.* **33** (3): 1–228.

Seilacher, A. 1974. Fabricational noise in adaptive morphology. *Syst. Zool.* **22**: 451–65.

Sethi, R.S. & Jensen, W.A. 1981. Ultrastructural evidence for secretory phases in the life history of stigmatic hairs of cotton: A dry type stigma. *Am. J. Bot.* **68**: 666–74.

Shaanker, R.U. & Ganeshaiah, K.N. 1990. Pollen grain deposition patterns and stigma strategies in regulating seed number per pod in multi-ovulated species. In Bawa, K.S. & Hadley, M. (eds), *Reproductive Ecology of Tropical Forest Plants*, 165–178. Paris: UNESCO.

Shamrov, I.I. 1990. The ovule of *Gentiana cruciata* (Gentianaceae): Structural-functional aspects of development. *Bot. Zhurn. (Moscow & Leningrad)* **75**: 1363–79.

Shivamurthy, G.R. & Arekal, G.D. 1982. Ontogeny, structure and morphology of the female flower of *Balanophora*. In Periasamy, K. (ed.), *Histochemistry, Developmental and Structural Anatomy of Angiosperms: A Symposium*, 245–51. Tiruchirapalli: P & B Publications.

Shivanna, K.R. 1982. Pollen-pistil interaction and control of fertilization. In Johri, B.M. (ed.), *Experimental Embryology of Vascular Plants*, 131–74. Berlin: Springer.

Shivanna, K.R. & Owens, S.J. 1989. Pollen-pistil interactions (Papilionoideae). *Monogr. Syst. Bot. Missouri Bot. Gard.* **29**: 157–82.

Shrivastava, G.P. & Shrivastava, U. 1991. Coevolution of stamens and carpels in cucurbits and of their insect pollinators. *Acta Horticult.* **288**: 347–53.

Sigmond, H. 1929. Vergleichende Untersuchungen über die Anatomie und Morphologie von Blütenknospenverschlüssen. *Beih. Bot. Centralbl.*, Abt. I, **46**: 1–67.

Silberbauer-Gottsberger, I. 1990. Pollination and evolution in palms. *Phyton (Horn)* **30**: 213–33.

Silberbauer-Gottsberger, I. & Gottsberger, G. 1975. Ueber sphingophile Angiospermen Brasiliens. *Pl. Syst. Evol.* **123**: 157–84.

Silberglied, R.E. 1979. Communication in the ultraviolet. *Ann. Rev. Ecol. Syst.* **10**: 373–98.

Simberloff, D. & Dayan, T. 1991. The guild concept and the structure of ecological communities. *Ann. Rev. Ecol. Syst.* **22**: 115–43.

Simpson, B.B. & Neff, J.L. 1981. Floral rewards: alternatives to pollen and nectar. *Ann. Missouri Bot. Gard.* **68**: 301–22.

Simpson, B.B. & Neff, J.L. 1983. Evolution and diversity of floral rewards. In Jones, C.E. & Little, R.J. (eds), *Handbook of Experimental Pollination Biology*, 142–59. New York: Scientific and Academic Editions.

Simpson, B.B., Neff, J.L. & Dieringer, G. 1990. The production of floral oils by *Monttea* (Scrophulariaceae) and the function of tarsal pads in *Centris* bees. *Pl. Syst. Evol.* **173**: 209–22.

Simpson, M.G. 1990. Phylogeny and classification of the Haemodoraceae. *Ann. Missouri Bot. Gard.* **77**: 722–84.

Sitte, P., Falk, H. & Liedvogel, B. 1980. Chromoplasts. In Czygan, F.-C. (ed.), *Pigments in Plants* (Ed. 2), 117–48. Stuttgart: Fischer.

Skog, L.E. 1976. A study of the tribe Gesnerieae, with a revision of *Gesneria* (Gesneriaceae: Gesnerioideae). *Smithsonian Contrib. Bot.* **29**: 1–182.

Skutch, A.F. 1971. *A Naturalist in Costa Rica*. Gainesville: University of Florida Press.

Skutch, A.F. 1980. *A Naturalist on a Tropical Farm*. Berkeley: University of California Press.

Smets, E. 1986. Localisation and systematic importance of the floral nectaries in the Magnoliatae (dicotyledons). *Bull. Jard. Bot. Natl Belg.* **56**: 51–76.

Smith, A.C. 1981. *Flora Vitiensis Nova*, Vol. 2. Lawai, Kauai, Hawaii: Pacific Tropical Botanical Garden.

Smith, B.W. 1950. *Arachis hypogaea*, aerial flower and subterranean fruit. *Am. J. Bot.* **37**: 802–15.

Smith, C.C., Hamrick, J.L. & Kramer, C.L. 1990. The advantage of mast years for wind pollination. *Am. Nat.* **136**: 154–66.

Smith, G.F., Thron, P.D. & Loots, G.C. 1992. Notes on the microfaunal complement and pollination mechanisms of *Poellnitzia rubriflora* (Asphodelaceae: Alooideae): An example of mite-flower domatia association. *Taxon* **41**: 437–50.

Smith, J.J. 1925. Ephemeral orchids. *Ann. Jard. Bot. Buitenzorg* **35**: 50–70.

Smith, R.M. 1985. A review of Bornean Zingiberaceae: 1 (Alpinieae p.p.). *Notes Roy. Bot. Gard. Edinburgh* **42**: 261–314.

Smith, R.M. 1986. A review of Bornean Zingiberaceae: II (Alpinieae, concluded). *Notes Roy. Bot. Gard. Edinburgh* **43**: 439–66.

Smith, R.M. 1988. A review of Bornean Zingiberaceae: IV (Globbeae). *Notes Roy. Bot. Gard. Edinburgh* **45**: 1–19.

Snow, A.A. & Spira, T.P. 1991. Pollen vigour and the potential for sexual selection in plants. *Nature* **352**: 796–7.

Snow, D.W. 1965. A possible selective factor in the evolution of fruiting seasons in tropical forest. *Oikos* **15**: 274–81.

Snow, D.W. & Snow, B.K. 1980. Relationships between hummingbirds and flowers in the Andes of Colombia. *Bull. Brit. Mus. (Nat. Hist.)*, Zool. Ser., **38**: 105–39.

Sobrevila, C. & Arroyo, M.T. 1982. Breeding systems in a montane tropical cloud forest in Venezuela. *Pl. Syst. Evol.* **140**: 19–37.

Soderstrom, R.T. & Calderon, C.E. 1971. Insect pollination in tropical rain forest grasses. *Biotropica* **3**: 1–16.

Soepadmo, E. & Eow, B.K. 1976. The reproductive biology of *Durio zibethinus* Murr. *Gard. Bull. Singapore* **29**: 25–33.

Soulé, M.E. 1990. The real work of systematics. *Ann. Missouri Bot. Gard.* **77**: 4–12.

Southwick, E.E. 1982. 'Lucky hit' nectar rewards and energetics of plant and pollinators. *Comp. Physiol. Ecol.* **7**: 51–5.

Sporne, K.R. 1977. Girdling vascular bundles in dicotyledonous flowers. *Gard. Bull. Singapore* **29**: 165–73.

Sprengel, C.K. 1793. *Das entdeckte Geheimniss der Natur im Bau und in der Befruchtung der Blumen*. Berlin: Vieweg d. Ae.

Spruce, R. 1908. *Notes of a Botanist on the Amazon and Andes*. London: Macmillan.

Stanton, M.L., Snow, A.A. & Handel, S.N. 1986. Floral evolution: Attractiveness to pollinators increases male fitness. *Science* **232**: 1625–7.

Start, A.N. & Marshall, A.G. 1976. Nectarivorous bats as pollinators of trees in West Malaysia. In Burley, J. & Styles, B.T. (eds), *Tropical Trees. Variation, Breeding and Conservation*, 141–50. London: Academic Press.

Stauffer, H.U. 1961. Beiträge zum Blütendiagramm der Santalales. *Verh. Schweiz. Naturf. Ges.* **141**: 123–5.

Stebbins, G.L. 1950. *Variation and Evolution in Plants*. New York: Columbia University Press.

Stebbins, G.L. 1957. Self-fertilization and variability in the higher plants. *Am. Nat.* **41**: 337–54.

Stebbins, G.L. 1967. Adaptive radiation and trends of evolution in higher plants. *Evol. Biol.* **1**: 101–42.

Stebbins, G.L. 1970. Adaptive radiation in angiosperms. I. Pollination mechanisms. *Ann. Rev. Ecol. Syst.* **1**: 307–26.

Stebbins, G.L. 1974a. *Flowering Plants. Evolution Above the Species Level*. Cambridge, Massachusetts: Harvard University Press.

Stebbins, G.L. 1974b. Adaptive shifts and evolutionary novelty: A compositionist approach. In Ayala, F.J. & Dobzhansky, T. (eds), *Studies in the Philosophy of Biology. Reduction and Related Problems*, 285–306. Berkeley: University of California Press.

Stebbins, G.L. 1976. Seeds, seedlings, and the origin of angiosperms. In Beck, C.B. (ed.), *Origin and Early Evolution of Angiosperms*, 300–11. New York: Columbia University Press.

Stebbins, G.L. 1988. Essays in comparative evolution. The need for evolutionary comparisons. In Gottlieb, L.D. & Jain, S.K. (eds), *Plant Evolutionary Biology*, 3–20. London: Chapman & Hall.

Stebbins, G.L. 1989. Adaptive shifts toward hummingbird pollination. In Bock, J.H. & Linhart, Y.B. (eds), *The Evolutionary Ecology of Plants*, 39–60. Boulder, Colorado: Westview Press.

van Steenis, C.G.G.J. 1927. Malayan Bignoniaceae, their taxonomy and geographic distribution. *Rev. Trav. Bot. Néerl.* **24**: 787–1049.

van Steenis, C.G.G.J. 1942. Gregarious flowering of *Strobilanthes* (Acanthaceae) in Malaysia. *Ann. Roy. Bot. Gard. Calcutta*, 150th Anniv. Vol., 91–7.

van Steenis, C.G.G.J. 1949. Notes on the genus *Wightia* (Scrophulariaceae). *Bull. Bot. Gard. Buitenzorg*, Ser. III, **18**: 213–27.

van Steenis, C.G.G.J. 1977. Bignoniaceae. In van Steenis, C.G.G.J. (ed.), *Flora Malesiana*, Ser. I, **8** (2): 114–86. Alphen: Sijthoff & Noordhoff.

Stein, B.A. 1992. Sicklebill hummingbirds, ants, and flowers. *BioSci.* **42**: 27–33.

Stein, B.A. & Tobe, H. 1989. Floral nectaries in Melastomataceae and their systematic and evolutionary implications. *Ann. Missouri Bot. Gard.* **76**: 519–31.

Steiner, K.E. 1981. Nectarivory and potential pollination by a neotropical marsupial. *Ann. Missouri Bot. Gard.* **68**: 505–13.

Steiner, K.E. 1985. The role of nectar and oil in the pollination of *Drymonia serrulata* (Gesneriaceae) by *Epicharis* bees (Anthophoridae) in Panama. *Biotropica* **17**: 217–29.

Steiner, K.E. 1992. A new *Diascia* species (Scrophulariaceae) from the Richtersveld, South Africa. *South Afr. J. Bot.* **58**: 36–8.

Steiner, K.E. & Whitehead, V.B. 1990. Pollinator adaptation to oil-secreting flowers – *Rediviva* and *Diascia*. *Evolution* **44**: 1701–7.

Stephenson, A.G. 1981. Flower and fruit abortion: proximate causes and ultimate functions. *Ann. Rev. Ecol. Syst.* **12**: 253–79.

Stern, W.L., Curry, K.J. & Pridgeon, A.M. 1987. Osmophores of *Stanhopea* (Orchidaceae). *Am. J. Bot.* **74**: 1323–31.

Stern, W.L., Curry, K.J. & Whitten, W.M. 1986. Staining fragrance glands in orchid flowers. *Bull. Torrey Bot. Club* **113**: 288–97.

Stevens, P.F. 1976. The altitudinal and geographical distributions of flower types in *Rhododendron* section *Vireya*, especially in the Papuasian species, and their significance. *Bot. J. Linn. Soc.* **72**: 1–33.

Stevens, P.F. 1985. Malesian *Vireya* rhododendrons – Towards an understanding of their evolution. *Notes Roy. Bot. Gard. Edinburgh* **43**: 63–80.

Stiles, F.G. 1975. Ecology, flowering phenology, and hummingbird pollination of some Costa Rican *Heliconia* species. *Ecology* **56**: 285–301.

Stiles, F.G. 1977. Coadapted competitors: flowering seasons of hummingbird food plants in a tropical forest. *Science* **198**: 1177–8.

Stiles, F.G. 1978a. Temporal organization of flowering among the hummingbird food-plants of a tropical wet forest. *Biotropica* **10**: 194–210.

Stiles, F.G. 1978b. Ecological and evolutionary implications of bird pollination. *Am. Zool.* **18**: 715–27.

Stiles, F.G. 1979. Notes on the natural history of *Heliconia* (Musaceae) in Costa Rica. *Brenesia* **15**: 151–80.

Stiles, F.G. 1981. Geographical aspects of bird-flower coevolution, with particular reference to Central America. *Ann. Missouri Bot. Gard.* **68**: 323–51.

Stirton, C.H. 1976. *Thuranthos*: Notes on generic status, morphology, phenology and pollination biology. *Bothalia* **12**: 161–5.

Stirton, C.H. 1977a. A note on the flowers of *Halleria lucida* (Scrophulariaceae). *Bothalia* **12**: 223–4.

Stirton, C.H. 1977b. The pollination of *Canavalia virosa* by xylocopid and megachilid bees. *Bothalia* **12**: 225–7.

Stirton, C.H. 1981. Petal sculpturing in papilionoid legumes. In Polhill, R.M. & Raven, P.H. (eds), *Advances in Legume Systematics* **1**: 771–788. Kew: Royal Botanic Gardens.

Stirton, C.H. (Ed.) 1987. *Advances in Legume Systematics* 3. Kew: Royal Botanic Gardens.

Stirton, C.H. & Zarucchi, J.L. (Eds) 1989. Advances in Legume Biology. *Monogr. Syst. Bot. Missouri Bot. Gard.* **29**: 1–842.

Stout, A.B. 1927. The flower behavior of avocados. *Mem. New York Bot. Gard.* **7**: 145–203.

Stout, A.B. 1928. Dichogamy in flowering plants. *Bull. Torrey Bot. Club* **55**: 141–53.

Stratton, D.A. 1989. Longevity of individual flowers in a Costa Rican cloud forest: Ecological correlates and phylogenetic constraints. *Biotropica* **21**: 308–18.

Straw, R.M. 1956. Adaptive morphology of the *Penstemon* flower. *Phytomorphology* **6**: 112–19.

Sun, M. & Ganders, F.R. 1986. Female frequencies in gynodioecious populations correlated with selfing rates in hermaphrodites. *Am. J. Bot.* **73**: 1645–8.

Sussman, R.W. & Raven, P.H. 1978. Pollination by lemurs and marsupials: an archaic co-evolutionary system. *Science* **200**: 731–4.

Sutherland, S. 1986. Patterns of fruit-set: What controls fruit-flower ratios in plants? *Evolution* **40**: 117–28.

Suttle, J.C. & Kende, H. 1978. Ethylene and senescence in petals in *Tradescantia*. *Pl. Physiol.* **62**: 267–71.

Sutton, D.A. 1988. *A Revision of the Tribe Antirrhineae*. London/Oxford: British Museum (Natural History)/Oxford University Press.

Swamy, B.G.L. 1949. Further contributions to the morphology of the Degeneriaceae. *J. Arn. Arb.* **30**: 10–38.

Swamy, B.G.L. & Periasamy, K. 1964. The concept of the conduplicate carpel. *Phytomorphology* **14**: 319–27.

Szalay, F.S. & Bock, W.J. 1991. Evolutionary theory and systematics: Relationships between process and patterns. *Z. Zool. Syst. Evol.-Forsch.* **29**: 1–39.

Takaso, T. & Bouman, F. 1984. Ovule ontogeny and seed development in *Potamogeton natans* L. (Potamogetonaceae), with a note on the campylotropous ovule. *Acta Bot. Neerl.* **33**: 519–33.

Takeoka, Y., Hiroi, K., Kitano, H. & Wada, T. 1991. Pistil hyperplasia in rice spikelets as affected by heat stress. *Sex Pl. Reprod.* **4**: 39–43.

Takhtajan, A.L. 1959. *Die Evolution der Angiospermen*. Jena: Fischer.

Takhtajan, A.L. 1973. *Evolution und Ausbreitung der Blütenpflanzen*. Stuttgart: Fischer.

Takhtajan, A.L. 1976. Neoteny and the origin of flowering plants. In Beck, C.B. (ed.), *Origin and Early Evolution of Angiosperms*, 207–19. New York: Columbia University Press.

Takhtajan, A.L. 1987. *Systema Magnoliophytorum*. Leningrad: Nauka.

Takhtajan, A.L. 1991. *Evolutionary Trends in Flowering Plants*. New York: Columbia University Press.

Tanksley, S.D. & Loaiza-Figueroa, F. 1985. Gametophytic self-incompatibility is controlled by a single major locus on chromosome 1 in *Lycopersicon peruvianum*. *Proc. Natl Acad. Sci. USA* **82**: 5093–6.

Tanner, E.V.J. 1982. Species diversity and reproductive mechanisms in Jamaican trees. *Biol. J. Linn. Soc.* **18**: 263–78.

Taylor, D.W. 1991. Angiosperm ovules and carpels: Their characters and polarities, distribution in basal clades, and structural evolution. *Postilla* **208**: 1–40.

Taylor, D.W. & Hickey, L.J. 1992. Phylogenetic evidence for the herbaceous origin of angiosperms. *Pl. Syst. Evol.* **180**: 137–56.

Taylor, P. 1989. The genus *Utricularia* – A taxonomic monograph. *Kew Bull.*, Add. Ser. XIV: 1–724.

von Teichman, I. & van Wyk, A.E. 1991. Trends in the evolution of dicotyledonous seeds base on character associations, with special reference to pachychalazy and recalcitrance. *Bot. J. Linn. Soc.* **105**: 211–37.

Teppner, H. 1988. *Lathyrus grandiflorus* (Fabaceae-Vicieae): Blüten-Bau, -Funktion und *Xylocopa violacea*. *Phyton (Horn)* **28**: 321–36.

Terborgh, J. 1986. Keystone plant resources in the tropical forest. In Soulé, M.E. (ed.), *Conservation Biology: The Science of Scarcity and Diversity*, 330–44. Sunderland, Massachusetts: Sinauer.

Terborgh, J. 1992. *Diversity and the Tropical Rain Forest*. New York: Scientific American Library.

Terborgh, J. & Stern, M. 1987. The surreptitious life of the saddle-backed tamarin. *Am. Sci.* **75**: 260–9.

Tewari, R.B. & Nair, P.K.K. 1984. Floral morphology of some Indian Papilionaceae – A study of wings and keel. In Nair, P.K.K. (ed.), *Glimpses in Plant Research 6*, 127–57. Delhi: Vikas.

Thiele, E.-M. 1988. Bau und Funktion des Antheren-Griffel-Komplexes der Compositen. *Dissertationes Botanicae* **117**: 1–169. Berlin: Cramer.

Thien, L.B. 1980. Patterns of pollination in the primitive angiosperms. *Biotropica* **12**: 1–13.

Thien, L.B., Pellmyr, O., Yetsu, L.Y., Bergström, G. & McPherson, G. 1990. Polysaccharide food-bodies as pollinator rewards in *Exospermum stipitatum* and other Winteraceae. *Bull. Mus. Natl Hist. Nat., Paris*, Sér. 4, Sect B, *Adansonia*, **12**: 191–7.

Thien, L.B., Bernhardt, P., Gibbs, G.W., Pellmyr, O., Bergström, G., Groth, I. and McPherson, G. 1985. The pollination of *Zygogynum* (Winteraceae) by a moth, Sabatinca (Micropterigidae): An ancient association? *Science* **227**: 540–3.

Thieret, J.W. 1967. Supraspecific classification in the Scrophulariaceae: A review. *Sida* **3**: 87–106.

Le Thomas, A. 1969. Annonacées. In Aubréville, A. (ed.), *Flore du Gabon* **16**: 1–371. Paris: Muséum National d'Histoire Naturelle.

Thompson, J.M. 1927. A study in advancing gigantism with staminal sterility with special reference to the Lecythidaceae. *Publ. Hartley Bot. Lab.* **4**: 3–44.

Thompson, J.M. 1931/1933. Studies in advancing sterility. Part V. The theory of the leguminous strobilus./ Part VI. The theory of scitaminean flowering. *Publ. Hartley Bot. Lab.* **7**: 3–79/ **11**: 3–111.

Thompson, W.R., Meinwald, J., Aneshansley, D. & Eisner, T. 1972. Flavonols: Pigments responsible for ultraviolet absorption in nectar guide of flower. *Science* **177**: 528–30.

Thomson, J.D. 1989. Germination schedules of pollen grains: Implications for pollen selection. *Evolution* **43**: 220–3.

Thorne, R.F. 1992. An updated phylogenetic classification of the flowering plants. *Aliso* **13**: 365–89.

Thorp, R.W. 1979. Structural, behavioral, and physiological adaptations of bees (Apoidea) for collecting pollen. *Ann. Missouri Bot. Gard.* **66**: 788–812.

Thorp, R.W. & Sugden, E.A. 1990. Extrafloral nectaries producing rewards for pollinator attraction in *Acacia longifolia* (Andr.) Willd. *Israel J. Bot.* **39**: 177–86.

Tilton, V.R. 1980a. The nucellar epidermis and micropyle of *Ornithogalum caudatum* (Liliaceae) with a review of these structures in other taxa. *Can. J. Bot.* **58**: 1872–84.

Tilton, V.R. 1980b. Hypostase development in *Ornithogalum caudatum* (Liliaceae) and notes on other types of modifications in the chalaza of angiosperm ovules. *Can. J. Bot.* **58**: 2059–66.

Tilton, V.R. & Horner, H.T. 1980. Stigma, style, and obturator of *Ornithogalum caudatum* (Liliaceae) and their function in the reproductive process. *Am. J. Bot.* **67**: 1113–31.

Tilton, V.R. & Lersten, N.R. 1981. Ovule development in *Ornithogalum caudatum* (Liliaceae) with a review of selected papers on angiosperm reproduction. III. Nucellus and megagametophyte. *New Phytol.* **88**: 477–504.

Todzia, C.A. 1988. Chloranthaceae: *Hedyosmum. Flora Neotropica Monograph* **48**: 1–139.

du Toit, J.T. 1990. Giraffe feeding on *Acacia* flowers: Predation or pollination? *Afr. J. Ecol.* **28**: 63–8.

Toledo, V.M. 1974. Observations on the relationship between hummingbirds and *Erythrina* species. *Lloydia* **37**: 482–7.

Toledo, V.M. & Hernandez, H.M. 1979. *Erythrina oliviae*: A new case of oriole pollination in Mexico. *Ann. Missouri Bot. Gard.* **66**: 503–11.

Tomlinson, P.B. 1962. Phylogeny of the Scitamineae – Morphological and anatomical considerations. *Evolution* **16**: 192–213.

Tomlinson, P.B. 1969. On the morphology and anatomy of turtle grass, *Thalassia testudinum* (Hydrocharitaceae). III. Floral morphology and anatomy. *Bull. Mar. Sci.* **19**: 286–305.

Tomlinson, P.B. 1974. Breeding mechanisms in trees native to tropical Florida – a morphological assessment. *J. Arn. Arb.* **55**: 269–90.

Tomlinson, P.B. 1977. Plant morphology and anatomy in the tropics – the need for integrated approaches. *Ann. Missouri Bot. Gard.* **64**: 685–93.

Tomlinson, P.B. 1980. *The Biology of Trees Native to Tropical Florida*. Allston, Massachusetts: Harvard University Printing Office.

Tomlinson, P.B. 1982. Helobiae (Alismatidae). In Metcalfe, C.R. (ed.), *Anatomy of the Monocotyledons*, Vol. VII. Oxford: Clarendon Press.

Tomlinson, P.B. 1986. *The Botany of Mangroves*. Cambridge University Press.

Tomlinson, P.B. 1990. *The Structural Biology of Palms*. Oxford University Press.

Tomlinson, P.B. 1991. Pollen scavenging. *Natl Geogr. Res. Explor.* **7**: 188–95.

Tomlinson, P.B. 1992. Aspects of cone morphology and development in Podocarpaceae (Coniferales). *Int. J. Pl. Sci.* **153**: 572–88.

Tomlinson, P.B. & Posluszny, U. 1978. Aspects of floral morphology and development in the seagrass *Syringodium filiforme* (Cymodoceaceae). *Bot. Gaz. (Crawfordsville)* **139**: 333–45.

Tomlinson, P.B., Braggins, J.E. & Rattenbury, J.A. 1991. Pollination drop in relation to cone morphology in Podocarpaceae: A novel reproductive mechanism. *Am. J. Bot.* **78**: 1289–1303.

Tomlinson, P.B., Primack, R.B. & Bunt, J.S. 1979. Preliminary observations on floral biology in mangrove Rhizophoraceae. *Biotropica* **11**: 256–77.

Tomlinson, P.B., Sattler, R. & Stevens, P.F. 1984. Homology in modular organisms – Concepts and Consequences. *Syst. Bot.* **9**: 373–409.

Torgaard, S.S. 1924. *Studien über die Morphologie und Baumechanik der Oleaceen-Blüte.* Kalmar: Appeltofft.

Trapp, A. 1954. Staubblattbildung und Bestäubungsmechanismus von *Incarvillea variabilis* Bat. *Oesterr. Bot. Z.* **101**: 208–19.

Trapp, A. 1956. Zur Morphologie und Entwicklungsgeschichte der Staubblätter sympetaler Blüten. *Bot. Studien* **5**: 1–93. Jena: Fischer.

Tremblay, R.L. 1992. Trends in the pollination ecology of the Orchidaceae: Evolution and systematics. *Can. J. Bot.* **70**: 642–50.

Treub, M. 1889. Les bourgeons floraux du *Spathodea campanulata* Beauv. *Ann. Jard. Bot. Buitenzorg* **8**: 38–46.

Treub, M. 1891. Sur les Casuarinées et leur place dans le système naturel. *Ann. Jard. Bot. Buitenzorg* **10**: 145–231.

Troll, W. 1922. Ueber Staubblatt- und Griffelbewegungen und ihre teleologische Deutung. *Flora* **115**: 191–250.

Troll, W. 1928a. *Organisation und Gestalt im Bereich der Blüte.* Berlin: Springer.

Troll, W. 1928b. Zur Auffassung des parakarpen Gynaeceums und des coenokarpen Gynaeceums überhaupt. *Planta* **6**: 255–76.

Troll, W. 1929. *Roscoea purpurea* Sm., eine Zingiberacee mit Hebelmechanismus in den Blüten. Mit Bemerkungen über die Entfaltungsbewegungen der fertilen Staubblätter von *Salvia*. *Planta* **7**: 1–28.

Troll, W. 1951. Botanische Notizen II. *Abh. Akad. Wiss. Lit. Mainz*, Math.-Naturwiss. Kl., **1951**: 25–80.

Tschech, K. 1939. Der Gewebebau grüner Kelchblätter. *Oesterr. Bot. Z.* **88**: 187–99.

Tucker, S.C. 1959. Ontogeny of the inflorescence and flower in *Drimys winteri* var. *chilensis*. *Univ. Calif. Publ. Bot.* **30**: 257–336.

Tucker, S.C. 1972. The role of ontogenetic evidence in floral morphology. *Adv. Pl. Morph.* **1972**: 359–69.

Tucker, S.C. 1987. Floral initiation and development in legumes. In Stirton, C.H. (ed.), *Advances in Legume Systematics* **3**: 183–239. Kew: Royal Botanic Gardens.

Tucker, S.C. 1988a. Loss versus suppression of floral organs. In Leins, P., Tucker, S.C. & Endress, P.K. (eds), *Aspects of Floral Development*, 69–82. Berlin: Cramer.

Tucker, S.C. 1988b. Dioecy in *Bauhinia* resulting from organ suppression. *Am. J. Bot.* **75**: 1584–97.

Tucker, S.C. 1988c. Heteromorphic flower development in *Neptunia pubescens*, a mimosoid legume. *Am. J. Bot.* **75**: 205–24.

Tucker, S.C. 1989. Evolutionary implications of floral ontogeny in legumes. *Monogr. Syst. Bot. Missouri Bot. Gard.* **29**: 59–75.

Tucker, S.C. 1990. Loss of floral organs in *Ateleia* (Leguminosae: Papilionoideae: Sophoreae). *Am. J. Bot.* **77**: 750–61.

Tucker, S.C. 1991. Helical floral organogenesis in *Gleditsia*, a primitive caesalpinioid legume. *Am. J. Bot.* **78**: 1130–49.

Tucker, S.C. 1992a. The developmental basis for sexual expression in *Ceratonia siliqua* (Leguminosae: Caesalpinioideae: Cassieae). *Am. J. Bot.* **79**: 318–27.

Tucker, S.C. 1992b. The role of floral development in studies of legume evolution. *Can. J. Bot.* **70**: 692–700.

Tucker, S.C. 1992c. Research on evolution of flower form among legumes, at Louisiana State University. *Flowering Newsl.* **13**: 21–8.

Tucker, S.C. & Douglas, A. 1994. Ontogenetic evidence and phylogenetic relationships among basal taxa of legumes. In Ferguson, I.K. & Tucker, S.C. (eds), *Advances in Legume Systematics 6. Structural Botany*. Kew: Royal Botanic Gardens.

Tucker, S.C., Sampson, F.B. & Leroy, J.-F. 1979. The gynoecium in winteraceous plants. *Science* **203**: 920–1.

Tucker, S.C., Stein, O.L. & Derstine, K.S. 1985. Floral development in *Caesalpinia* (Leguminosae). *Am. J. Bot.* **72**: 1424–34.

Turner, V. 1982. Marsupials as pollinators in Australia. In Armstrong, J.A., Powell, J.M. & Richards, A.J. (eds), *Pollination and Evolution*, 55–66. Sydney: Royal Botanic Gardens.

Uhl, N.W. 1988. Floral organogenesis in palms. In Leins, P., Tucker, S.C. & Endress, P.K. (eds), *Aspects of Floral Development*, 25–44. Berlin: Cramer.

Uhl, N.W. & Dransfield, J. 1984. Development of the inflorescence, androecium, and gynoecium with reference to palms. In White, R.A. & Dickison, W.C. (eds), *Contemporary Problems in Plant Anatomy*, 397–450. New York: Academic Press.

Uhl, N.W. & Moore, H.E., Jr. 1973. The protection of pollen and ovules in palms. *Principes* **17**: 111–49.

Uhl, N.W. & Moore, H.E., Jr. 1977. Correlations of inflorescence, flower structure, and floral anatomy with pollination in some palms. *Biotropica* **9**: 170–90.

Uhl, N.W. & Moore, H.E., Jr. 1980. Androecial development in six polyandrous genera representing five major groups of palms. *Ann. Bot.* **45**: 57–75.

Ullrich, H. 1935. Ueber die Bewegungen der Staubfäden und des Griffels bei Columneen. *Planta* **24**: 154–9.

Uphof, J.C.T. 1938. Cleistogamic flowers. *Bot. Rev.* **4**: 21–50.

Urbanska, K.M. 1989. Reproductive effort or reproductive offer? – A revised approach to reproductive strategies of flowering plants. *Bot. Helv.* **99**: 49–63.

Valla, J.J. & Cirino, D.R. 1972. Biologia floral del irupé, *Victoria cruziana* D'Orb. (Nymphaeaceae). *Darwiniana* **17**: 477–500.

Vane-Wright, R.I., Humphries, C.J. & Williams, P.H. 1991. What to protect? Systematics and the agony of choice. *Biol. Conserv.* **55**: 235–54.

Vanstone, V.A. & Paton, D.C. 1988. Extrafloral nectaries and pollination of *Acacia pycnantha* Benth. by birds. *Austral. J. Bot.* **36**: 519–32.

Vasil, I. 1974. The histology and physiology of pollen germination and pollen tube growth on the stigma and in the style. In Linskens, H.F. (ed.), *Fertilization in Higher Plants*, 105–18. Amsterdam: North-Holland.

Vasil, I.K. & Johri, M.M. 1964. The style, stigma, and pollen tube. I. *Phytomorphology* **14**: 352–69.

Vavilov, N.I. 1922. The law of homologous series in variation. *J. Genet.* **12**: 47–89.

Venkata Rao, C. 1963. On the morphology of the calyculus. *J. Indian Bot. Soc.* **42**: 618–28.

Venkatesh, C.S. 1956. The form, structure and special modes of dehiscence in anthers of *Cassia*. I/II/III. *Phytomorphology* **6**: 168–76/ 272–7/ **7**: 253–73.

Verbeke, J.A. 1992. Fusion events during floral morphogenesis. *Ann. Rev. Pl. Physiol. Mol. Biol.* **43**: 583–98.

Verdcourt, B. 1979. *A Manual of New Guinea Legumes*. Lae: Office of Forests, Division of Botany (Botany Bull. 11).

Verkerke, W. 1989. Structure and function of the fig. *Experientia* **45**: 612–22.

Vermeulen, P. 1966. The system of the Orchidales. *Acta Bot. Neerl.* **15**: 224–53.

Vink, W. 1970. The Winteraceae of the Old World. I. *Pseudowintera* and *Drimys* – morphology and taxonomy. *Blumea* **18**: 225–354.

Vink, W. 1977. The Winteraceae of the Old World. II. *Zygogynum* - morphology and taxonomy. *Blumea* **23**: 219–50.

Vink, W. 1978. The Winteraceae of the Old world. III. Notes on the ovary of *Takhtajania*. *Blumea* **24**: 521–5.

Vink, W. 1988. Taxonomy in Winteraceae. *Taxon* **37**: 691–8.

Vöchting, H. 1886. Ueber Zygomorphie und deren Ursachen. *Jahrb. Wiss. Bot.* **17**: 297–346.

Vogel, S. 1950. Farbwechsel und Zeichnungsmuster bei Blüten. *Oesterr. Bot. Z.* **97**: 44–100.

Vogel, S. 1954. Blütenbiologische Typen als Elemente der Sippengliederung, dargestellt anhand der Flora Südafrikas. *Bot. Studien* **1**: 1–338. Jena: Fischer.

Vogel, S. 1957. Fledermausblumen in Südamerika. *Oesterr. Bot. Z.* **104**: 491–530.

Vogel, S. 1959. Organographie der Blüten kapländischer Ophrydeen mit Bemerkungen zum Koaptations-Problem I/II. *Abh. Akad. Wiss. Lit. Mainz*., Math.-Naturwiss. Kl. **1959**: 265–532.

Vogel, S. 1960. (Note) In Troll, W., Kommission für Biologische Forschung. *Akad. Wiss. Lit. Mainz, Jahrb.* **1960**: 81–94.

Vogel, S. 1961. Die Bestäubung der Kesselfallen-Blüten von *Ceropegia*. *Beitr. Biol. Pfl.* **36**: 159–237.

Vogel, S. 1963a. Duftdrüsen im Dienste der Bestäubung. Ueber Bau und Funktion der Osmophoren. *Abh. Akad. Wiss. Lit. Mainz*, Math.-Naturwiss. Kl., **1962**: 601–763.

Vogel, S. 1963b. Blütenökotypen und die Gliederung systematischer Einheiten. *Ber. Deutsch. Bot. Ges.* **76**: (98)-(101).

Vogel, S. 1963c. Das sexuelle Anlockungsprinzip der Catasetinen- und Stanhopeen-Blüten und die wahre Funktion ihres sogenannten Futtergewebes. *Oesterr. Bot. Z.* **100**: 308–37.

Vogel, S. 1966a. Parfümsammelnde Bienen als Bestäuber von Orchideen und *Gloxinia*. *Oesterr. Bot. Z.* **113**: 302–61.

Vogel, S. 1966b. Pollination neotropischer Orchideen durch duftstoffhöselnde Prachtbienen-Männchen. *Naturwissenschaften* **53**: 181–2.

Vogel, S. 1968/1969a/b. Chiropterophilie in der neotropischen Flora. Neue Mitteilungen I/II/III. *Flora* **157**: 562–602/ **158**: 185–222/ 289–323.

Vogel, S. 1969c. Flowers offering fatty oil instead of nectar. *XIth Internat. Bot. Congr., Seattle*, Abstr., 229.

Vogel, S. 1969d. Ueber synorganisierte Blütensporne bei einigen Orchideen. *Oesterr. Bot. Z.* **116**: 244–62.

Vogel, S. 1974. Oelblumen und ölsammelnde Bienen. *Trop. Subtrop. Pflanzenwelt* **7**: 283–547. Wiesbaden: Steiner.

Vogel, S. 1975. Mutualismus und Parasitismus in der Nutzung von Pollenträgern. *Verh. Deutsch. Zool. Ges.* **1975**: 102–10.

Vogel, S. 1977. Nektarien und ihre ökologische Bedeutung. *Apidologie* **8**: 321–35.

Vogel, S. 1978a. Evolutionary shifts from reward to deception in pollen flowers. In Richards, A.J. (ed.), *The Pollination of Flowers by Insects*, 89–96. London: Academic Press.

Vogel, S. 1978b. Pilzmückenblumen als Pilzmimeten. *Flora* **167**: 329–98.

Vogel, S. 1980. Florengeschichte im Spiegel blütenökologischer Erkenntnisse. *Rhein.-Westfäl. Akad. Wiss.*, Vortr. N **291**: 7–48.

Vogel, S. 1981a. Bestäubungskonzepte der Monokotylen und ihr Ausdruck im System. *Ber. Deutsch. Bot. Ges.* **94**: 663–75.

Vogel, S. 1981b. Trichomatische Blütennektarien bei Cucurbitaceae. *Beitr. Biol. Pfl.* **55**: 325–53.

Vogel, S. 1981c. Die Klebstoffhaare an den Antheren von *Cyclanthera pedata* (Cucurbitaceae). *Pl. Syst. Evol.* **137**: 291–316.

Vogel, S. 1983. Ecophysiology of zoophilic pollination. In Lange, O.L., Nobel, P.S., Osmond, C.B. & Ziegler, H. (eds), *Encyclopedia of Plant Physiology*, N.S., 12 C, Physiological Plant Ecology III, 559–624. Berlin: Springer.

Vogel, S. 1984. Blütensekrete als akzessorischer Pollenkitt. In Ehrendorfer, F. (Ed.), *Mitteilungsband Botaniker-Tagung Wien*, 123. Vienna: Institut für Botanik, Universität Wien.

Vogel, S. 1986. Oelblumen und ölsammelnde Bienen. Zweite Folge. *Lysimachia* und *Macropis*. *Trop. Subtrop. Pflanzenwelt* **54**: 1–168. Stuttgart: Steiner.

Vogel, S. 1988. Die Oelblumensymbiosen – Parallelismus und andere Aspekte ihrer Entwicklung in Raum und Zeit. *Z. Zool. Syst. Evol.-Forsch.* **26**: 341–62.

Vogel, S. 1989. Fettes Oel als Lockmittel. Erforschung der ölbietenden Blumen und ihrer Bestäuber. In: *Akademie der Wissenschaften und der Literatur Mainz 1949–1989*, 113–30. Stuttgart: Steiner.

Vogel, S. 1990a. Oelblumen und ölsammelnde Bienen. *Folge* **3**: *Momordica*, *Thladiantha* und die Ctenoplectridae. *Trop. Subtrop. Pflanzenwelt* **73**: 1–186. Stuttgart: Steiner.

Vogel, S. 1990b. History of the Malpighiaceae in the light of pollination ecology. *Mem. New York Bot. Gard.* **55**: 130–42.

Vogel, S. 1990c. Radiación adaptiva del síndrome floral en las familias neotropicales. *Bol. Acad. Nac. Ci. Cordoba, Argentina*, **59**: 5–30.

Vogel, S. 1990d. *The Role of Scent Glands in Pollination* (translated by S.S. Renner). Washington, D.C.: Smithsonian Institution.

Vogel, S. & Cocucci, A. 1988. Pollen threads in *Impatiens*: Their nature and function. *Beitr. Biol. Pfl.* **63**: 271–87.

Vogel, S. & Machado, I.C. 1991. Pollination of four sympatric species of *Angelonia* (Scrophulariaceae) by oil-collecting bees in NE. Brazil. *Pl. Syst. Evol.* **178**: 153–78.

Vokou, D., Petanidou, T. & Bellos, D. 1990. Pollination ecology and reproductive potential of *Jankaea heldreichii* (Gesneriaceae); a Tertiary relict on Mt. Olympus, Greece. *Biol. Conserv.* **52**: 125–33.

Volk, O.H. 1950. Zur Kenntnis der Pollinien der Asclepiadaceen. *Ber. Deutsch. Bot. Ges.* **62**: 68–72.

Volk, O.H. 1951. Zur Kenntnis der Stapelieen-Blüten. *Sukkulentenkunde* **4**: 46–59.

Waddle, R.M. & Lersten, N.R. 1974. Morphology of discoid floral nectaries in Leguminosae, especially tribe Phaseoleae

(Papilionoideae). *Phytomorphology* **23**: 152–61.

Wagenitz, G. 1981. Orchideen und Compositen. Vergleich zweier Familien und Evolutionsstrategien. *Ber. Deutsch. Bot. Ges.* **94**: 229–47.

Wagenitz, G. 1992. The Asteridae: Evolution of a concept and its present status. *Ann. Missouri Bot. Gard.* **79**: 209–17.

Wagner, A. 1894. Zur Anatomie und Biologie der Blüte von *Strelitzia reginae*. *Ber. Deutsch. Bot. Ges.* **12**: 53–72.

Wagner, G. 1986. The systems approach: An interface between development and population genetic aspects of evolution. In Raup, D.M. & Jablonski (eds), *Patterns and Processes in the History of Life*, 149–65. Berlin: Springer.

Wagner, G. 1989. The origin of morphological characters and the biological basis of homology. *Evolution* **43**: 1157–71.

Wagner, H.O. 1946. Food and feeding habits of Mexican hummingbirds. *Wilson Bull.* **58**: 69–93.

Waha, M. 1984. Zur Ultrastruktur und Funktion pollenverbindender Fäden bei Ericaceae und andern Angiospermenfamilien. *Pl. Syst. Evol.* **147**: 189–203.

Wainwright, S.A. 1988. *Axis and Circumference. The Cylindrical Shape of Plants and Animals.* Cambridge, Massachusetts: Harvard University Press.

Wake, D.B. 1991. Homoplasy: The result of natural selection, or evidence of design limitations? *Am. Nat.* **138**: 543–67.

Walker, D.B. 1978. Postgenital carpel fusion in *Catharanthus roseus* (Apocynaceae). IV. Significance of the fusion. *Am. J. Bot.* **65**: 119–21.

Walker, J.W. & Walker, A.G. 1985. Ultrastructure of Lower Cretaceous angiosperm pollen and the origin and early evolution of flowering plants. *Ann. Missouri Bot. Gard.* **71**: 464–521.

Walsh, N.E. & Charlesworth, D. 1992. Evolutionary interpretations of differences in pollen tube growth rates. *Quart. Rev. Biol.* **67**: 19–37.

Walter, D.M. 1988. Plant morphology and reproduction. In Lovett Doust, J. & Lovett Doust, L. (eds), *Plant Reproductive Ecology*, 203–27. New York: Oxford University Press.

Walter, K.S. 1983. Orchidaceae (orquideas, orchids). In Janzen, D.H. (ed.), *Costa Rican Natural History*, 282–92. Chicago: University of Chicago Press.

Wannan, B.S. & Quinn, C.J. 1991. Floral structure and evolution in the Anacardiaceae. *Bot. J. Linn. Soc.* **107**: 349–85.

Wanntorp, H.-E. 1974. *Calotropis gigantea* (Asclepiadaceae) and *Xylocopa tenuiscapa* (Hymenoptera, Apidae). Studies in flower morphology and pollination biology. *Svensk Bot. Tidskr.* **68**: 25–32.

Wanntorp, H.-E., Brooks, D.R., Nilsson, T., Nylin, S., Ronquist, F., Stearns, S.C. & Wedell, N. 1990. Phylogenetic approaches in ecology. *Oikos* **57**: 119–32.

Warming, E. 1913. Observations sur la valeur systématique de l'ovule. In *Mindeskrift for Japetus Steenstrup*, XXIV, 1–45. Copenhagen: Lunos.

Waser, N.M. 1986. Flower constancy: Definition, cause, and measurement. *Am. Nat.* **127**: 593–603.

Webb, C.J. 1984. Constraints on the evolution of plant breeding systems and their relevance to systematics. In Grant, W.F. (ed.), *Plant Biosystematics*, 249–70. London: Academic Press.

Webb, C.J. & Lloyd, D.G. 1986. The avoidance of interference between the presentation of pollen and stigmas in angiosperms. II. Herkogamy. *New Zeal. J. Bot.* **24**: 163–78.

Webb, M.C. & Williams, E.G. 1988. The pollen tube pathway in the pistil of *Lycopersicum peruvianum*. *Ann. Bot.* **61**: 415–23.

Weber, A. 1971. Zur Morphologie des Gynoeceums der Gesneriaceae. *Oesterr. Bot. Z.* **119**: 234–305.

Weber, A. 1976a. Morphologie, Anatomie und Ontogenese der Blüte von *Monophyllaea* R.Br. *Bot. Jahrb. Syst.* **95**: 435–54.

Weber, A. 1976b. Wuchsform, Infloreszenz- und Blütenmorphologie von *Epithema* (Gesneriaceae). *Pl. Syst. Evol.* **126**: 287–322.

Weber, A. 1978. Spross-, Infloreszenz- und Blütenbau von *Rhynchoglossum*. *Bot. Jahrb. Syst.* **99**: 1–47.

Weber, A. 1979. *Ornithoboea arachnoidea* – an 'orchid-flowered' gesneriad. *Gloxinian* **29**: 9–12.

Weber, A. 1980. Die Homologie des Perigons der Zingiberaceen. Ein Beitrag zur Morphologie und Phylogenie des Monokotylen-Perigons. *Pl. Syst. Evol.* **133**: 149–79.

Weber, A. 1982. Evolution and radiation of the pair-flowered cyme in Gesneriaceae. *Austral. Syst. Bot. Soc. Newsl.* **30**: 23–41.

Weber, A. 1986. The nocturnal splendour of *Barringtonia*. *Nature Malaysiana* **11** (2): 24–31.

Weber, A. 1989a. *Didymocarpus geitleri*, a remarkable new species of Gesneriaceae with deceptive pollen flowers. *Pl. Syst. Evol.* **165**: 95–100.

Weber, A. 1989b. Family position and conjectural affinities of *Charadrophila capensis* Marloth. *Bot. Jahrb. Syst.* **111**: 87–119.

Weber, A. & Burtt, B.L. 1983. Supplementary notes on '*Didymocarpus corchorifolius* and its allies (Gesneriaceae)'. *Blumea* **31**: 155–9.

Weber, A. & Vogel, S. 1984. The pollination syndrome of *Deplanchea tetraphylla* (Bignoniaceae). *Pl. Syst. Evol.* **154**: 237–50.

Weber, H. 1955. Ueber die Blütenkelche tropischer Rubiaceen. *Abh. Akad. Wiss. Lit. Mainz*, Math.-Naturwiss. Kl. **1955** (11): 449–66.

Weberling, F. 1989. *Morphology of Flowers and Inflorescences*. Cambridge University Press.

Webster, G.L. & Armbruster, W.S. 1991. A synopsis of the neotropical species of *Dalechampia* (Euphorbiaceae). *Bot. J. Linn. Soc.* **105**: 137–77.

Wehner, R. 1976. Polarized light navigation by insects. *Sci. Am.* **235** (1): 106–15.

Weiss, M. 1991. Floral colour changes as cues for pollinators. *Nature* **354**: 227–9.

Wells, M.S. & Lloyd, D.G. 1991. Dichogamy, gender variation and bet-hedging in *Pseudowintera colorata*. *Evol. Ecol.* **5**: 310–26.

Wemple, D.K. & Lersten, N.R. 1966. An interpretation of the flower of *Petalostemon* (Leguminosae). *Brittonia* **18**: 117–26.

Werth, E. 1915. Kurzer Ueberblick über die Gesamtfrage der Ornithophilie. *Bot. Jahrb. Syst.* **53**, Beibl., 314–78.

Werth, E. 1922. Ueber einige bemerkenswerte Formen von Blütennektarien. *Verh. Bot. Ver. Prov. Brandenburg* **64**: 222–9.

Werth, E. 1942. Ueber einige umstrittene und weniger bekannte Bestäubungseinrichtungen tropischer Blumen. *Ber. Deutsch. Bot. Ges.* **60**: 473–94.

Werth, E. 1956a. *Bau und Leben der Blumen. Die blütenbiologischen Bautypen in Entwicklung und Anpassung*. Stuttgart: Enke.

Werth, E. 1956b. Zur Kenntnis des Androeceums der Gattung *Salvia* und seiner stammesgeschichtlichen Wandlung. *Ber. Deutsch. Bot. Ges.* **69**: 381–6.

Westerkamp, C. 1990. Bird-flowers: hovering versus perching exploitation. *Bot. Acta* **103**: 366–71.

Westra, L.Y.T. 1985. Studies in Annonaceae. IV. A taxonomic revision of *Tetrameranthus* R.E. Fries. *Proc. Kon. Ned. Akad. Wet.* C **88**: 449–82.

Whalen, M.D. 1978. Reproductive character displacement and floral diversity in *Solanum* section *Androceras*. *Syst. Bot.* **3**: 77–86.

Wheelwright, N.T. 1983. Fruits and the ecology of the resplendent quetzals. *Auk* **100**: 286–301.

Whitehead, D.R. 1969. Wind pollination in the angiosperms: Evolutionary and environmental considerations. *Evolution* **23**: 28–35.

Whitehead, D.R. 1983. Wind pollination: Some ecological and evolutionary perspectives. In Real, L. (ed.), *Pollination Biology*, 97–108. Orlando, Florida: Academic Press.

Whitehouse, H.L.K. 1950. Multiple-allelomorph incompatibility of pollen and style in the evolution of the angiosperms. *Ann. Bot.* **14**: 199–216.

Whitehouse, H.L.K. 1960. Origin of angiosperms. *Nature* **188**: 957.

Whitmore, T.C. 1990. *An Introduction to Tropical Rain Forests*. Oxford: Clarendon Press.

Whitten, W.M., Young, A.M. & Williams, N.H. 1989. Function of glandular secretions in fragrance collection by male euglossine bees (Apidae: Euglossini). *J. Chem. Ecol.* **15**: 1285–95.

Whitten, W.M., Williams, N.H., Armbruster, W.S., Battiste, M.A., Strekovski,L. & Linquist, N. 1986. Carvone oxide: An example of convergent evolution in euglossine pollinated plants. *Syst. Bot.* **11**: 222–8.

Wiebes, J.T. 1979. Co-evolution of figs and their insect pollinators. *Ann. Rev. Ecol. Syst.* **10**: 1–12.

Wiehler, H. 1978. The genera *Episcia*, *Alsobia*, *Nautilocalyx*, and *Paradrymonia* (Gesneriaceae). *Selbyana* **5**: 11–60.

Wiehler, H. 1983. A synopsis of the neotropical Gesneriaceae. *Selbyana* **6**: 1–219.

Wiens, D. 1984. Ovule survivorship, brood size, life history, breeding systems and reproductive success in plants. *Oecologia* **64**: 47–53.

Wiens, D., Rourke, J.P., Casper, B.B., Rickart, E.A., LaPine, T.R., Peterson, C.J. & Channing, A. 1983. Nonflying mammal pollination of Southern African proteas: A non-coevolved system. *Ann. Missouri Bot. Gard.* **70**: 1–31.

Wiermann, R., Wollenweber, E. & Rehse, C. 1981. 'Yellow flavonols' as components of pollen pigmentation. *Z. Naturforsch. C* **36**: 204–6.

de Wilde, W.J.J.O. 1974. The genera of tribe Passifloreae (Passifloraceae), with special reference to flower morphology. *Blumea* **22**: 37–50.

Willemse, M.T.M. & Franssen-Verheijen, M.A.W. 1986. Stylar development in the open flower of *Gasteria verrucosa* (Mill.) H. Duval. *Acta Bot. Neerl.* **35**: 297–309.

Willemstein, S.C. 1987. An Evolutionary Basis for Pollination Ecology. Ph.D. Thesis, University of Leiden. Leiden: Brill.

Williams, E.G. & Rouse, J.L. 1988. Disparate style lengths contribute to isolation of species in *Rhododendron*. *Austral. J. Bot.* **36**: 183–91.

Williams, E.G. & Webb, M.C. 1987. Enclosure of pollinated flowers weakens the gametophytic self-incompatibility response in *Lycopersicon peruvianum*. *J. Exp. Bot.* **38**: 1756–64.

Williams, L.O. 1970. An overlooked genus of the Scrophulariaceae. *Fieldiana, Bot.*, **32**: 211–14.

Williams, N.H. 1982. The biology of orchids and euglossine bees. In Arditti, J. (ed.), *Orchid Biology. Reviews and Perspectives II*, 119–71. Ithaca: Comstock.

Williams, N.H. 1983. Floral fragrances as cues in animal behavior. In Jones, C.E. & Little, R.J. (eds), *Handbook of Experimental Pollination Biology*, 50–72. New York: Scientific and Academic Editions.

Williams, N.H. & Dodson, C.H. 1972. Selective attraction of male euglossine bees to orchid floral fragrances and its importance in long distance pollen flow. *Evolution* **26**: 84–95.

Williams, N.H. & Whitten, W.M. 1983. Orchid floral fragrances and male euglossine bees: Methods and advances in the last sesquidecade. *Biol. Bull.* **164**: 355–95.

Willson, M.F. 1983. *Plant Reproductive Ecology*. New York: Wiley.

Willson, M.F. & Agren, J. 1989. Differential floral rewards and pollination by deceit in unisexual flowers. *Oikos* **55**: 23–9.

Willson, M.F. & Burley, N. 1983. *Mate Choice in Plants*. Princeton: Princeton University Press.

Willson, M.F. & Melampy, M.N. 1983. *Asclepias curassavica* (bailarina, mata caballo, mal casada, milkweed). In Janzen, D.H. (ed.), *Costa Rican Natural History*, 191–3. Chicago: University of Chicago Press.

Wilms, H.J. 1980. Ultrastructure of the stigma and style of spinach in relation to pollen germination and pollen tube growth. *Acta Bot. Neerl.* **29**: 33–47.

Wilms, H.J. 1981. Pollen tube growth through nucellus into embryo sac of *Spinacia* – An ultrastructural investigation. *Acta Soc. Bot. Polon.* **50**: 191–4.

Wilson, E.O. 1992. *The Diversity of Life*. Cambridge, Massachusetts: Harvard University Press.

Winkler, H. 1906. Beiträge zur Morphologie und Biologie tropischer Blüten und Früchte. *Bot. Jahrb. Syst.* **38**: 233–71.

de Wit, H.C.D. 1956. A revision of the Malaysian Bauhinieae. *Reinwardtia* **3**: 381–541.

Woinarski, J.C.Z., Whitehead, P.J., Bowman, M.J.S. & Russell-Smith, J. 1992. Conservation of mobile species in a variable environment: The problem of reserve design in the Northern Territory, Australia. *Global Ecol. Biogeogr. Lett.* **2**: 1–10.

Wolda, H. & Sabrosky, C.W. 1986. Insect visitors to two forms of *Aristolochia pilosa* in Las Cumbres, Panama. *Biotropica* **18**: 295–9.

Wolfe, L.M. 1987. Inflorescence size and pollinaria removal in *Asclepias curassavica* and *Epidendrum radicans*. *Biotropica* **19**: 86–9.

Woltering, E.J. & van Doorn, W.G. 1988. Role of ethylene in senescence of petals: Morphological and taxonomical relationships. *J. Exp. Bot.* **39**: 1605–16.

Wood, D. 1974. A revision of *Chirita* (Gesneriaceae). *Notes Roy. Bot. Gard. Edinburgh* **33**: 123–205.

Woon, C. & Keng, H. 1979. Observations on stamens of the Dipterocarpaceae. *Gard. Bull. Singapore* **32**: 1–55.

Wootton, J.T. & Sun, I. 1990. Bract liquid as a herbivore defense mechanism for *Heliconia wagneriana* inflorescences. *Biotropica* **22**: 155–9.

Wunderlich, R. 1967. Some remarks on the taxonomic significance of the seed coat. *Phytomorphology* **17**: 301–11.

Wunderlin, R.P. 1983. Revision of the arborescent bauhinias (Fabaceae: Caesalpinioideae: Cercideae) native to Middle America. *Ann. Missouri Bot. Gard.* **70**: 95–127.

Wunderlin, R.P., Larsen, K. & Larsen, S.S. 1981. Tribe 3. Cercideae Bronn (1822). In Polhill, R.M. & Raven, P.H. (eds), *Advances in Legume Systematics* **3**: 107–16. Kew: Royal Botanic Gardens.

Wunderlin, R.P., Larsen, K. & Larsen, S.S. 1987. Reorganization of the Cercideae

(Fabaceae: Caesalpinioideae). *Biol. Skrift. Kong. Dansk. Videnskab. Selsk.* **28**: 1–40.

Wunderlin, U. 1992. Untersuchungen zur vergleichenden Entwicklungsgeschichte von Scrophulariaceen-Blüten. *Diss. Bot.* **188**: 1–313. Berlin: Cramer.

Wyatt, R. 1978. Experimental evidence concerning the role of the corpusculum in *Asclepias* pollination. *Syst. Bot.* **3**: 313–21.

Wyatt, R. 1983. Pollinator-plant interactions and the evolution of breeding systems. In Real, L. (ed.), *Pollination Biology*, 51–95. Orlando, Florida: Academic Press.

Yan, H., Yang, H.-Y. & Jensen, W.A. 1991. Ultrastructure of the micropyle and its relationship to pollen tube growth and synergid degeneration in sunflower. *Sex Pl. Reprod.* **4**: 166–75.

Yeo, P.F. 1972. Miscellaneous notes on pollination and pollinators. *J. Nat. Hist.* **6**: 667–86.

Yeo, P.F. 1975. Some aspects of heterostyly. *New Phytol.* **75**: 147–53.

Yeung, E.C. 1987. Development of pollen and accessory structures in orchids. In Arditti, J. (ed.), *Orchid Biology. Reviews and Perspectives IV*: 193–226. Ithaca: Comstock.

Young, A.M. 1984. Mechanism of pollination by Phoridae (Diptera) in some *Herrania* species (Sterculiaceae) in Costa Rica. *Proc. Entomol. Soc. Washington* **86**: 503–18.

Young, A.M. 1991. *Sarapiqui Chronicle. A Naturalist in Costa Rica.* Washington, D.C.: Smithsonian Institution Press.

Young, A.M., Erickson, B.J., Erickson, E.H., Jr. & Strand, M.A. 1986. Pollination biology of *Theobroma* and *Herrania* (Sterculiaceae). I. Floral biology. *Insect Sci. Appl.* **8**: 151–64.

Young, H.J. 1986. Beetle pollination of *Dieffenbachia longispatha* (Araceae). *Am. J. Bot.* **73**: 931–44.

Young, H.J. 1990. Pollination and reproductive biology of an understory neotropical aroid. In Bawa, K.S. & Hadley, M. (eds), *Reproductive Ecology of Tropical Forest Plants*, 151–64. Paris: UNESCO.

Zandonella, P. 1977. Apports de l'étude comparée des nectaires floraux à la conception phylogénétique de l'ordre des Centrospermales. *Ber. Deutsch. Bot. Ges.* **90**: 105–25.

Zandonella, P. & Piolat, C. 1982. Nectaires et sécrétion du nectar chez *Thunbergia laurifolia* Lindl. (Acanthacées). *Bull. Soc. Bot. France, Actual. Bot.*, **129** (1): 109–17.

Zapata, T.R. & Arroyo, M.T.K. 1978. Plant reproductive ecology of a secondary deciduous tropical forest in Venezuela. *Biotropica* **10**: 221–30.

Zavada, M.S. & Taylor, T.N. 1986. The role of self-incompatibility and sexual selection in the gymnosperm-angiosperm transition: A hypothesis. *Am. Nat.* **128**: 538–50.

Zucchi, R., Sakagami, S.F. & de Camargo, J.M.F. 1969. Biological observations on a Neotropical parasocial bee, *Eulaema nigrita*, with a review on the biology of Euglossinae (Hymenoptera, Apidae). A comparative study. *J. Fac. Sci. Hokkaido Univ.*, Ser. VI, **17**: 271–380.

Zweifel, R. 1939. Cytologisch-embryologische Untersuchungen an *Balanophora abbreviata* Blume und *Balanophora indica* Wall. *Vierteljahrsschr. Naturforsch. Ges. Zürich* **84**: 245–306.

GLOSSARY

(Special terms that are used only for single families are not considered here.)

abaxial: adjective describing a floral organ that is positioned at the opposite side in relation to the axis from which the floral axis has branched.
abscission: the shedding of a plant part by means of a special tissue (abscission layer).
adaxial: adjective describing a floral organ that is positioned at the same side as the axis from which the floral axis has branched.
aestivation: the mutual position of perianth organs in a floral bud (or of vegetative leaves in a vegetative bud).
agamospermy (or apomixis): the production of offspring via seeds without fertilization.
allogamy: pollination and fertilization between different flowers (of the same or of different individuals) (as opposed to autogamy).
anagenesis: evolutionary advancement.
analogy: similarity of parts of organisms of different evolutionary origin that have the same function.
anatropous: adjective describing a curved ovule with the micropyle adjacent to the attachment region and the nucellus not included in the curvature.
androecium: the totality of the stamens of a flower.
androgynophore: a 'stalk' between perianth and androecium bearing androecium and gynoecium.
andromonoecious: adjective describing a species with male and bisexual flowers that occur on the same individuals.
androphore: a 'stalk' between perianth and androecium bearing the androecium in a male flower without a gynoecium.
anemophilous: adapted to pollination by wind.
anther: the pollen-bearing part of a stamen.
anthesis: the open flower phase, or, more precisely, the phase of a flower when pollen is presented and/or the stigma is receptive.
antitropous (new term): adjective describing an ovule that is curved in the opposite direction as compared to the carpel margin from which it originates.
apert: aestivation of perianth organs that are not contiguous.
apocarpous (or choricarpous): adject-

ive describing a gynoecium with the carpels congenitally free.
apomorphic: character state of a clade that is different from the original state in the clade.
apparatus: a functional unit consisting of two or more synorganized organs or different parts of organs.
architecture (construction, gestalt): form (shape) (of a flower) that is dependent on static, constructional constraints.
ascending: a prominent kind of cochlear aestivation, whereby the innermost organ is adaxial.
ascidiate, ascidiform, utriculate: adjective describing a strongly peltate carpel that is congenitally urn-shaped, i.e. from the beginning of development (as opposed to plicate: a carpel that is postgenitally urn-shaped, i.e. by later concrescence of the margins during development).
asymmetric: without a symmetry plane.
athecal: adjective describing an anther that is not differentiated into thecae.
autogamy: pollination and fertilization with pollen of the same flower.
axile placenta: placenta in a syncarpous gynoecium with chambered ovary, in which the ovules originate from the inner angle of each locule.
basifixed: adjective describing an anther with basally attached filament.
bauplan (see organization)
big bang species: species with mass flowering over a short period.
bisexual (perfect, hermaphrodite): adjective describing a flower containing organs of both sexes.
bitegmic: adjective describing an ovule with two integuments.
blade: the upper broadened part of a leaf or a petal.
blossom: a plant part that looks like a flower and functions like a flower. It may be a flower in the strict sense, or an inflorescence, or only part of a flower.
bract: a scale-like leaf, commonly used for the subtending organ of a flower.
bracteole (prophyll): small scale-like leaf, commonly used for the two first organs of a lateral axis (floral axis). In monocotyledons there is commonly only one prophyll.
brush-flower (-blossom): a flower or blossom with numerous long, conspicuous stamen filaments but usually an inconspicuous perianth.
buzz-pollination: a particular kind of pollen collection (and pollination) by means of vibrations by the wing muscles, performed by certain groups of bees.
calyx: the totality of the sepals.
campylotropus: adjective describing a curved ovule, where the nucellus is included in the curvature.
cantharophilous: adapted to pollination by beetles.
carpel: female floral organ.
centrifugal: adjective describing a sequence of origin of floral organs from the centre towards the periphery of the flower.
centripetal: adjective describing a sequence of origin of floral organs from the periphery towards the centre of the flower.
chalaza: the part of the ovule below the insertion of the integuments.
chasmogamy: pollination of an open flower (in contrast to cleistogamy).
chiropterophilous: adapted to pollination by bats.
choricarpous (see apocarpous)

choripetalous: adjective describing a flower with free petals.
clade: a monophyletic group of organisms.
cladogenesis: the evolutionary process of diversification into different groups of organisms.
claw: the narrow basal part of a petal.
cleistogamy: pollination and fertilization within a flower that has not opened.
cochlear, cochleate: adjective describing the kind of imbricate aestivation of a pentamerous corolla, in which one organ is outside, one is inside, and the other three each with one side overlapping and the other overlapped.
collenchyma: a tissue with thick (cellulosic/pectic) walls that are not lignified.
colleter: secretory structure of vegetative and floral buds that secretes various kinds of protective substances for the young organs.
common primordium: a distinct primordium that gives rise to more than one organ.
compitum: the common pollen tube transmitting tract of all carpels of a flower.
conduplicate: adjective describing a carpel that is largely plicate.
congenital fusion: fusion of structural elements from the beginning of development.
connective: sterile tissue between the two thecae of an anther.
construction (see architecture)
contort, twisted: adjective describing the kind of aestivation of sepals or petals in which each organ has one side overlapping and the other overlapped.
convergent evolution: the evolution of similar traits in two or more groups from different origins.
cornucopia species: species with short-lived individual flowers that flower strongly over a period of several weeks.
corolla: the totality of the petals of a flower.
corona, paracorolla: corolla-like structure in addition to the regular corolla, formed between corolla and androecium.
crassinucellar: adjective describing an ovule having a nucellus with more than one cell layer around the meiocyte(s).
cross-zone: the region where the margin of a peltate organ crosses over the median plane of the organ.
cymose: adjective describing a branching system in which each branch has only one or two lateral branches (monochasial or dichasial).
dehiscence: the separation of parts of an organ resulting in an opening of an originally closed chamber (especially in anthers and mature fruits).
descending: adjective describing a prominent kind of a cochlear aestivation in which the outermost organ is adaxial.
dextrorse, contort to the right: with the right sides overlapping, if viewed from outside.
diaspore: the dispersal unit of a plant at seed maturity.
dichasial: adjective describing a branching system in which each branch has two lateral branches.
dichogamy: the phenomenon in bisexual flowers that pollen exposition and stigma receptivity do not occur at the same time.
dioecious: adjective describing a species with unisexual flowers in which an

individual bears only one kind of flowers.

diplostemonous: adjective describing a flower with two whorls of stamens.

disk nectary: a disk-like structure that secretes nectar, mostly around the gynoecium base.

disymmetric: adjective describing a flower with two symmetry planes.

divergence angle: the angle formed by two subsequently initiated floral organs (or leaves).

dorsifixed: adjective describing an anther with dorsally attached filament.

double position (of floral organs): in whorls where instead of an expected single organ two organs appear side by side.

dry stigma: a stigma without secretory activity at anthesis.

elaboration: refinement or evolution of more complexity of parts (organs or organ complexes).

elaiophore: the site where oil is secreted.

embryo: the young sporophyte in a seed.

embryo sac: the female gametophyte, located in the nucellus.

enantiomorphy: the occurrence of two mirror-symmetrical floral morphs in asymmetric flowers (often the style is bent to the side: enantiostyly).

endosperm: the normally triploid nutritive tissue of a seed (in angiosperms) that originates by fusion of two embryo sac nuclei and one of the two male gametes.

endothecium: the tissue in the thecal walls of an anther that serves to open the thecae by means of deformation by desiccation of the partially lignified cell walls.

endothecium-like tissue: tissue of the same differentiation as the endothecium outside of the thecal walls of an anther.

entomophilous: adapted to pollination by insects.

epeltate: adjective describing an organ (usually a carpel) without a congenitally closed base. The base may be open or only postgenitally closed.

epidermis: the outermost cell layer of an organ.

epithelium: an epidermis that is specialized as a secretory surface.

eusyncarpous: adjective describing a kind of a syncarpous gynoecium in which each carpel forms a separate ovary locule.

extrafloral nectary: a nectary that is located outside a flower.

extranuptial nectary: a nectary without a function in pollination biology.

extrorse: adjective describing an anther with the stomia directed to the floral periphery.

fertilization: the fusion of a male and a female gamete.

filament: the often elongated region of a stamen below the anther.

flag: an optically attractive organ that is presented in an effective way (flag or standard used in particular for the adaxial petal in flowers of Fabaceae).

flag-flower: a flower with an optically attractive upper part and the pollination organs enclosed in a lower lip (especially used for flowers of Fabaceae).

floral cup, hypanthium: cup-shaped floral base.

floral nectary: a nectary located on a floral organ.

flower: the reproductive unit that characterizes the angiosperms (and all anthophytes), consisting of stamens and ovule-bearing organs (carpels in angiosperms) and (facultatively) a perianth.

free central placenta: placenta in a syncarpous gynoecium with a single ovary locule, whereby the ovules originate from the base of the locule or from a larger body that originates from the base of the locule.

fruit: the derivative of a flower at seed maturity.

funiculus, funicle: the often narrowed basal region of an ovule, where it is attached to the placenta.

gamete: a haploid cell (or unicellular organism) that fuses with a corresponding one of the opposite sex to form a zygote.

gametophyte: the haploid organism that originates from a spore and produces gametes.

geitonogamy: pollination of a flower with pollen from another flower of the same individual.

gestalt (see architecture)

granulocrine secretion: the secreted substance passes the plasmalemma of the secretory cell in membrane bound vesicles.

guild: a group of (related or unrelated) organisms that occur in the same geographic region and have similar ecological traits (e.g. pollination by the same animals).

gynoecium: the totality of the carpels of a flower.

gynomonoecious: adjective describing a species with female and bisexual flowers occurring on the same individual.

gynophore: a 'stalk' below the gynoecium in a flower.

gynostegium: an organ complex originating by postgenital fusion of androecium and gynoecium (e.g. in Asclepiadaceae).

gynostemium: an organ complex originating by congenital fusion of androecium and gynoecium (e.g. in Aristolochiaceae and Orchidaceae).

haplostemonous: adjective describing a flower with one whorl of stamens.

herkogamy: spatial separation of male and female organs of a flower, impeding self-pollination.

heteranthery: the presence of stamens (anthers) of different shapes within a flower (mainly in buzz-pollinated pollen-flowers).

heterochrony: evolutionary change in the timing of a developmental event relative to others.

heteromorphy: the existence of two distinct morphs of an organ category or of organ complexes in a species.

heterostyly: the presence of floral morphs within a species with two (or more) different style lengths.

histogenesis: the differentiation of a meristematic part into different tissues.

holocrine secretion: secretion results in disintegration of the secretory cells.

homogamy: the phenomenon in bisexual flowers that pollen presentation and stigma receptivity are at the same time (in contrast to dichogamy).

homology: similarity of parts of organisms based on derivation from a condition that originated only once in a common ancestor.

homoplasy: similarity of parts of organisms that has arisen more than once by parallel evolution from a common ancestor.

hydrophilous (hydatophilous): adapted to pollination by water.

inbreeding: the producing of offspring resulting from fertilization by gametes of the same individual.

imbricate: adjective describing aestiv-

ation (of sepals or petals) in which neighbours overlap each other.
indument: covering of hairs.
inferior: adjective describing the position below a certain level of reference, especially used for ovaries where the outer floral organs insert on top of them.
inflorescence: a branching system consisting of flower-bearing axes.
initiation: the process by which an organ is caused to originate.
integument: the covering parts of an ovule.
intercalary elongation: the basal elongation of an organ or of an axis.
introrse: adjective describing an anther with the stomia directed towards the floral centre.
laminar-diffuse placenta: placenta with more than two series of ovules at the flanks of a carpel.
lignified (sclerified, sclerenchymatic): adjective describing a cell or a tissue with cell walls that are reinforced by lignin.
limiting divergence: the most common divergence angle in spiral phyllotaxis (ca. 137°) (representing the limiting value of the 'Fibonacci series').
linear placenta: a placenta with only two rows of ovules.
lip: part of a monosymmetric flower (usually an organ complex composed of more than one petal) that serves as a landing platform for pollinators (lower lip in lip-flowers) or forms a roof over the pollination organs (upper lip in lip-flowers).
lip-flower: a flower with a lower and an upper lip, in which the pollination organs are more or less protected by the upper lip (e.g. in many Scrophulariales).
locule: an internal cavity (in anthers containing pollen, in ovaries containing ovules)
mate choice: selection of gametophytes (gametes).
median: the plane through an axis and the axis from which it originates.
megasporangium: sporangium producing megaspores (i.e. embryo sac producing cells in angiosperms)
meiocyte: a cell that undergoes meiosis.
melittophilous: adapted to pollination by bees.
meristem: an undifferentiated tissue with the capacity of morphogenesis and growth.
mesophyll: the tissues between the upper and the lower epidermis of a leaf or floral organ.
micropyle: orifice of the ovule formed by the apical parts of the integuments, acting as pollen tube entrance.
microsporangium: sporangium producing microspores (pollen grains).
mimicry: the imitation in a part of an organism of something else.
mode (floral, or floral style): totality of special traits of a flower in evolutionary response to its pollinators or other ecological factors.
moniliform hair: a hair consisting of a series of rounded cells
monochasial: adjective describing a branching system in which each axial order has a single lateral branch.
monoecious: adjective describing the presence of unisexual flowers of both sexes on the same individual.
monophilic: adjective describing flowers with a single pollinator species.
monophyletic: adjective describing a group with a common ancestor and containing all descendants of this ancestor.

monosymmetric (zygomorphic): adjective describing a flower with a single symmetry plane.
monotropic: adjective describing an animal species that pollinates only a single species of plants.
morph: one of different distinct forms of flowers in a species.
morphogenesis: the origin of form during development.
my(i)ophilous: adapted to pollination by flies.
nectar: a liquid containing various sugars and (in smaller quantities) also other substances secreted in the flower or on other plant parts.
nectar cover: a protective cover over the site where nectar is located in a flower.
nectar-flower: flower that provides nectar as a reward to pollinators.
nectar guide: an optical mark at the floral entrance guiding pollinators towards the site where nectar is located.
nectar holder: the site in the flower where nectar is located.
nectary: the site where nectar is secreted.
nototribic: adjective describing a flower that is pollinated by the back of the pollinator.
nuptial nectary: a nectary that is involved in pollination biology.
obdiplostemonous: adjective describing a flower with two whorls of stamens in which the stamens of the outer have a larger base than those of the inner; this causes the carpels to alternate with the outer instead of the inner.
oil-flower: a flower that provides oil as a reward to pollinators.
oligophilic: adjective describing flowers with a narrow group of pollinators.
oligotropic: adjective describing animals that pollinate only a small array of plant groups.
ontogenetic spiral: the spiral curve that links all subsequent organ primordia in spiral phyllotaxis.
optical tapetum: a tissue that reflects light and enhances optical brightness of an organ.
organ: a structural and functional element in a flower (e.g. sepal, petal), sometimes also used for a smaller or larger unit of a certain structural and functional individuality.
organization: the complex (totality) of genetically deeply seated features that tend to remain constant and characterize larger taxa.
ornithophilous: adapted to pollination by birds.
orthotropous: adjective describing an ovule that is not curved.
osmophore: a scent producing organ.
outbreeding (outcrossing, xenogamy): the producing of offspring resulting from fertilization of gametes from different individuals.
ovary: the part of the gynoecium containing the ovules.
ovule: the floral part that develops into the seed; it consists of a nucellus, containing the embryo sac, and of one or two integuments.
paedomorphosis: an evolutionary change resulting in retention of a (comparatively) juvenile-like stage up to maturity.
panicle: an inflorescence with several lateral branches of the main axis that may bear several lateral branches again and so on.
papillate cell: an epidermal cell that protrudes from the epidermal surface.
paracarpous: adjective describing a form of a syncarpous gynoecium in

which the ovary contains a single locule with parietal placentation.
paracorolla (see corona)
parallel evolution: evolution in the same direction from a different common ancestral state in closely related groups.
paraphyletic: adjective describing a group of organisms of common origin but not containing all subgroups that are derived from the common ancestor.
parastichies: sets of spiral curves that link adjacent organ primordia in spiral or whorled phyllotaxis (apart from the ontogenetic spiral).
parenchyma: an unspecialized tissue that is no longer meristematic.
parietal placenta: the placentation in a syncarpous ovary with a single locule in which the ovules originate from the congenital fusion region of two adjacent carpels.
pedicel: flower stalk.
peduncle: inflorescence stalk.
peloria: abnormal polysymmetric flower or flower-like structure in plants with normally monosymmetric flowers.
peltate: adjective describing an organ (in flowers mostly a carpel) with a basal cross-zone, i.e. its margin crossing over the median plane.
pentamerous: five-parted.
perfume-flowers: flowers that provide scent substances as a reward to pollinating male euglossine bees.
perianth: the totality of the floral organs outside androecium and gynoecium, floral envelope.
perigon: a perianth that consists of only one kind of organs (tepals).
personate: adjective describing a lip-flower with a closed throat that has to be opened by a pollinator.
petal: organ of the second whorl of the perianth, usually for optical attraction of pollinators.
phalaenophilous: adapted to pollination by moths.
pherophyll (see subtending bract, subtending leaf)
phyllotaxis: arrangement pattern of leaves or floral organs on an axis.
placenta: the site of ovule attachment in the gynoecium.
plastochron: the time lapse between the initiation of two subsequent organ primordia on a floral or vegetative apex.
plesiomorphic: adjective describing a primitive character state.
pleurotribic: adjective describing pollination by lateral body parts of an animal.
plicate: adjective describing the folded (postgenitally fused or unfused) part of a carpel.
pollen: the male gametophyte in the stage where it is dispersed.
pollen-flower: a flower that provides only pollen as a reward to pollinators.
pollenkitt: sticky substances coating the pollen surface, produced by the tapetum.
pollen:ovule ratio: the ratio of pollen number to ovule number as produced in a flower or in a plant.
pollen sac: almost all angiosperm anthers have four pollen sacs, and a pair of pollen sacs forms a theca. In most angiosperms a pollen sac corresponds to a microsporangium containing a sporogenous tissue surrounded by a tapetum. Rarely (in polysporangiate anthers) each pollen sac is subdivided into several sporangia.
pollen tube: the male gametophyte growing from the stigma into an ovule.
pollen tube transmitting tract: the pre-

formed pathway for pollen tubes in a gynoecium.
pollinarium: apparatus for the transport of pollinia (the pollinia included) as occurring in Asclepiadaceae and Orchidaceae.
pollination: the transfer of pollen to a stigma.
pollinium: pollen mass of a pollen sac that is transported as a coherent unit in pollination.
polyad: a compact group of pollen from two or more tetrads (originating in secondarily subdivided pollen sacs).
polyandrous: adjective describing a flower with a large number of stamens.
polyphilic: adjective describing flowers with a large spectrum of pollinators.
polyphyletic: adjective describing a group of organisms that has evolved from different ancestors.
polysporangiate: adjective describing a pollen sac that is secondarily subdivided into more than one sporangium.
polysymmetric (actinomorphic): adjective describing a flower with more than two symmetry planes.
polytropic: adjective describing an animal species that pollinates many different plant groups.
poricidal: opening by a pore.
postgenital fusion: fusion of structural elements that were free from each other at the beginning of development.
primary morphogenesis: the early formation of organs.
primary morphological surface: the surface of organs that is derived from the surface of the original floral or shoot apex.
primordium: the first visible stage of an organ.
procambium strand: early stage of a vascular bundle.
prophyll: the first one or two leaves (bracts) on a lateral shoot.
protandry: the phenomenon in bisexual flowers that pollen is exposed before stigma receptivity.
protogyny: the phenomenon in bisexual flowers that stigmatic receptivity is present before pollen is exposed.
protruding-diffuse placenta: a placenta that is protruding from its surroundings and has more than two series of ovules (either axile or parietal).
pseudanthium: a flower-like structure that is composed of several flowers.
pseudonectary: a glistening body in a flower presumed to mimic a drop of nectar, usually well exposed and located near a more hidden true nectary.
psychophilous: adapted to pollination by butterflies.
quincuncial: adjective describing the kind of imbricate aestivation of five sepals or petals in which two organs are outside, two are inside, and one has one side overlapping and the other overlapped.
raceme: a racemose inflorescence with stalked flowers.
racemose: adjective describing a branching pattern of an inflorescence in which a main axis has a number of lateral flower-bearing axes.
resin-flower: a flower that provides resin as a reward to pollinators.
resupination: a 180° rotation or recurving of a monosymmetric flower during ontogeny.
revolver flower: a floral architecture with several separate entrances towards the reward, so that a pollinator tends to rotate around the floral axis when sitting in the floral centre.

ring primordium: a ring-like meristematic mound giving rise to several organs.
roundabout flower: a floral architecture, where a pollinator moves tangentially around the entire flower to take up the reward.
sapromyophilous: adapted to pollination by flies that are attracted to decaying organic material.
sclerenchymatic, sclerified (see lignified)
secondary morphogenesis: developmental phase during which organs attain their mature shape by differential growth.
secondary morphological surface: an organ surface that is not derived from the surface of the initial floral apex but originates by later separation of parts (e.g. by abscission, dehiscence or rupture).
secondary pollen presentation: presentation of pollen grains to the pollinators not by the anthers but by another floral organ where they have been deposited by the anthers.
seed: the mature organ derived from an ovule containing embryo and endosperm.
self-compatibility: the phenomenon that pollination and fertilization with pollen of the same individual leads to viable seeds.
self-incompatibility: the phenomenon that pollination with pollen of the same individual (or another individual of the same genetic constitution) fails to give rise to viable seeds.
self-pollination, selfing (cf. also inbreeding): pollination and fertilization with pollen of the same individual.
semaphore, semaphyll: an optically attractive organ.
sepal: organ of the outer whorl of the perianth, mostly for protection of young floral organs.
septal nectaries: nectaries that are located in the septa of an ovary.
septum: a wall subdividing a cavity (mainly in anthers and ovaries).
sinistrorse, contort to the left: the left side overlapping, if viewed from outside.
spadix: a spike with a thickened main axis.
spathe: a large bract, usually covering an entire or partial inflorescence in bud.
sphingophilous: adapted to pollination by hawkmoths.
spike: a racemose inflorescence with sessile flowers.
spiral: successively originating organs with the same divergence angle from each other.
sporangium: a spore containing part of an organ.
spore: a haploid cell (organism) originating by meiosis.
sporogenous tissue: the tissue of a sporangium that gives rise to the spores.
sporophyte: the diploid organism derived from a zygote and producing spores.
spur: a hollow outgrowth of an organ.
staggered flowering: the successive flowering of a guild of plant species.
stamen: male floral organ.
staminode: a sterile stamen.
standard (see flag)
stapet: organ complex of a petal and a stamen that are congenitally fused.
steady-state species: species that exhibit constant production of a few flowers each day.
stele: the vascular system in an axis.
sternotribic: adjective describing

pollen transfer by the underside of a pollinator.
stigma: receptive surface of the gynoecium where pollen grains are deposited and germinate.
stoma (water pore): an apparatus consisting of two epidermal cells that leave a pore in the centre functioning as a pathway for liquids or gaseous substances (in flowers especially nectar).
stomium: the dehiscence region of a theca in an anther.
style: often elongated part of the carpel or of the syncarpous gynoecium between stigma and ovary.
style (floral) (see mode)
subtending bract (subtending leaf, pherophyll): a bract (or leaf) that bears a lateral branch in its axil.
superior: the position above a certain level of reference, especially used for ovaries, where the outer floral organs are inserted at the base of them.
sympatric: adjective describing species that occur in the same region.
sympetalous: adjective describing a flower with congenitally united petals.
symplicate: adjective describing the region of a syncarpous gynoecium, where the flanks of neighbouring carpels are congenitally united but the individual carpels are plicate.
synandrous: adjective describing a flower with congenitally united stamens.
synapomorphic: adjective describing a derived character state of a clade that is common to several subclades.
synascidiate: adjective describing the region of a syncarpous gynoecium, where the flanks of neighbouring carpels are congenitally united and the individual carpels are ascidiate.
syncarpous: adjective describing a gynoecium with the carpels congenitally united.
syndrome: a characteristic combination of features correlated with certain ecological factors (often in the context of pollination biology).
synorganization: the intimate (spatial and functional) connection of organs of the same or of different kinds to form a functional apparatus.
synthecal: adjective describing an anther with the two thecae confluent over the apex.
syntropous (new term): adjective describing an ovule that is curved in the same direction as the carpel margin from which it originates.
tapetum: the cell layer surrounding the meiocytes in a microsporangium.
taxon: a group of organisms with a formal name in a system of classification.
tenuinucellar: adjective describing an ovule that has a nucellus with a single cell layer above and besides the meiocyte(s).
tepal: organ of a perianth that is not differentiated into calyx and corolla.
tetrad: the four cells resulting from meiosis of a meiocyte. In some plant groups pollen is transported as compact tetrads.
theca: a lateral pair of pollen sacs of an anther with a common opening slit. Almost all anthers have two thecae.
thermogenic: producing heat.
thyrse: an inflorescence with several lateral branches of the main axes that are branched in a cymose pattern.
translator: the part of a pollinarium apart from the pollinia (mainly used for Asclepiadaceae).
transverse: the plane through the floral axis perpendicular to the median plane.

trap-flower: flower where the pollinators remain trapped for a certain time.
traplining: the behaviour of certain pollinators, which fly over long distances to visit the flowers of scattered individuals of the same species.
trichome: hair.
trimerous: three-parted.
unisexual: adjective describing a flower with only male or female organs.
unitegmic: adjective describing an ovule with a single integument.
unordered (chaotic) phyllotaxis: arrangement of floral organs without a regular pattern.
utriculate (see ascidiate).
valvate: adjective describing (1) aestivation of sepals or petals, whereby the neighbouring organs are laterally contiguous without overlapping each other; (2) opening of a theca by a stomium that bifurcates at both ends or opening of individual pollen sacs by extended, curved dehiscence lines.
valve: the part that is detached by valvate opening.
vascular bundle: tissue differentiated for the transport of substances in plants.
vascular trace: basal vascular bundle(s) of an organ connecting with the stele.
ventrifixed: adjective describing an anther with ventrally attached filament.
viscin threads: thread-like structures of various origin to hold clumps of pollen grains together for pollination.
water calyx: a calyx that is filled with a secretion in bud.
water pore (see stoma).
wet stigma: a stigma that is secretory at anthesis.
whorl: a group of organs arranged more or less at the same level around the axis. They originate more or less simultaneously or in a rapid spiral. Independent of their sequence of origin, the organs of successive whorls alternate in their position, more rarely they are superposed.
xenogamy (see outbreeding).
zygote: the diploid fusion product of two gametes, giving rise to a sporophyte.

APPENDIX

Classes, subclasses, orders and families of the angiosperms (Magnoliophyta) mentioned in the text (after Cronquist 1988, with some modifications)

Class Magnoliopsida (dicotyledons)

Subclass Magnoliidae

Order Magnoliales (Families Winteraceae, Degeneriaceae, Himantandraceae, Eupomatiaceae, Austrobaileyaceae, Magnoliaceae, Annonaceae, Myristicaceae, Canellaceae, Calycanthaceae, Illiciaceae, Schisandraceae)
Order Laurales (Families Chloranthaceae, Trimeniaceae, Monimiaceae, Lauraceae)
Order Piperales (Families Saururaceae, Piperaceae)
Order Aristolochiales (Families Aristolochiaceae, Rafflesiaceae, Hydnoraceae)
Order Ceratophyllales (Family Ceratophyllaceae)
Order Nymphaeales (Families Cabombaceae, Nymphaeaceae)
Order Nelumbonales (Family Nelumbonaceae)
Order Ranunculales (Families Ranunculaceae, Lardizabalaceae, Menispermaceae, Berberidaceae, Papaveraceae)

Subclass Hamamelididae

Order Trochodendrales (Family Trochodendraceae)
Order Hamamelidales (Family Hamamelidaceae)
Order Urticales (Families Moraceae, Urticaceae)
Order Juglandales (Family Juglandaceae)
Order Myricales (Family Myricaceae)
Order Fagales (Families Balanopaceae, Fagaceae)
Order Casuarinales (Family Casuarinaceae)

Subclass Caryophyllidae

Order Caryophyllales (Families Nyctaginaceae, Cactaceae, Chenopodiaceae, Amaranthaceae, Portulacaceae)
Order Polygonales (Family Polygonaceae)
Order Plumbaginales (Family Plumbaginaceae)

Subclass Dilleniidae

Order Dilleniales (Family Dilleniaceae)
Order Theales (Families Ochnaceae, Dipterocarpaceae, Caryocaraceae, Theaceae, Actinidiaceae, Marcgraviaceae, Clusiaceae)
Order Malvales (Families Elaeocarpaceae, Tiliaceae, Sterculiaceae, Bombacaceae, Malvaceae)
Order Lecythidales (Family Lecythidaceae)
Order Nepenthales (Family Nepenthaceae)
Order Violales (Families Flacourtiaceae, Cochlospermaceae, Turneraceae, Passifloraceae, Caricaceae, Cucurbitaceae, Begoniaceae)
Order Capparales (Families Capparaceae, Brassicaceae)
Order Ericales (Families Epacridaceae, Ericaceae, Pyrolaceae, Monotropaceae)
Order Ebenales (Family Styracaceae)
Order Primulales (Families Myrsinaceae, Primulaceae)

Subclass Rosidae

Order Rosales (Families Connaraceae, Cunoniaceae, Pittosporaceae, Escalloniaceae, Bruniaceae, Anisophylleaceae, Crassulaceae, Saxifragaceae, Rosaceae, Chrysobalanaceae, Loasaceae)
Order Fabales, legumes (Families Mimosaceae, Caesalpiniaceae, Fabaceae)
Order Myrtales (Families Lythraceae, Myrtaceae, Onagraceae, Melastomataceae, Combretaceae)
Order Proteales (Family Proteaceae)
Order Rhizophorales (Family Rhizophoraceae)
Order Santalales (Families Olacaceae, Opiliaceae, Santalaceae, Loranthaceae, Viscaceae, Balanophoraceae)
Order Celastrales (Families Celastraceae, Stackhousiaceae, Icacinaceae)
Order Euphorbiales (Families Simmondsiaceae, Euphorbiaceae)
Order Rhamnales (Families Rhamnaceae, Leeaceae, Vitaceae)
Order Linales (Families Erythroxylaceae, Linaceae)
Order Polygalales (Families Malpighiaceae, Polygalaceae, Krameriaceae)
Order Sapindales (Families Staphyleaceae, Melianthaceae, Sapindaceae, Hippocastanaceae, Aceraceae, Anacardiaceae, Simaroubaceae, Meliaceae, Rutaceae, Zygophyllaceae, Coriariaceae)
Order Geraniales (Families Oxalidaceae, Geraniaceae, Tropaeolaceae, Balsaminaceae, Ixonanthaceae)
Order Apiales (Family Araliaceae)

Subclass Asteridae

Order Gentianales (Families Loganiaceae, Gentianaceae, Apocynaceae, Asclepiadaceae)
Order Solanales (Families Solanaceae, Convolvulaceae, Menyanthaceae, Boraginaceae)
Order Lamiales (Families Verbenaceae, Lamiaceae)
Order Scrophulariales (Scrophulariaceae, Globulariaceae, Myoporaceae, Orobanchaceae, Gesneriaceae, Acanthaceae, Pedaliaceae, Martyniaceae, Bignoniaceae, Lentibulariaceae, Callitrichaceae)
Order Campanulales (Families Campanulaceae, Stylidiaceae, Goodeniaceae)
Order Rubiales (Family Rubiaceae)
Order Dipsacales (No families mentioned)
Order Asterales (Family Asteraceae)

Class Liliopsida (monocotyledons)

Subclass Alismatidae

Order Alismatales (Families Butomaceae, Limnocharitaceae, Alismataceae, Hydrocharitaceae)
Order Najadales (Families Potamogetonaceae, Ruppiaceae, Zannichelliaceae, Posidoniaceae, Cymodoceaceae)
Order Triuridales (Family Triuridaceae)

Subclass Arecidae

Order Arecales (Family Arecaceae, palms)
Order Cyclanthales (Family Cyclanthaceae)
Order Pandanales (Family Pandanaceae)
Order Arales (Family Araceae)

Subclass Commelinidae

Order Commelinales (Families Mayacaceae, Commelinaceae)
Order Cyperales (Families Cyperaceae, Typhaceae)
Order Poales (Family Poaceae, grasses)

Subclass Zingiberidae

Order Bromeliales (Family Bromeliaceae)
Order Zingiberales (Strelitziaceae, Heliconiaceae, Musaceae, Lowiaceae, Zingiberaceae, Costaceae, Cannaceae, Marantaceae)

Subclass Liliidae

Order Liliales (Families Philydraceae, Pontederiaceae, Haemodoraceae, Cyanastraceae, Liliaceae, Iridaceae, Velloziaceae, Aloeaceae, Agavaceae, Taccaceae)
Order Orchidales (Families Burmanniaceae, Orchidaceae, orchids)

TAXONOMIC INDEX (PLANTS AND ANIMALS)

Page numbers in italics indicate figures.

GENERAL INDEX

Page numbers in italics indicate figures.

Printed in the United States
By Bookmasters